WATER-QUALITY ENGINEERING
in Natural Systems

WATER-QUALITY ENGINEERING
in Natural Systems

DAVID A. CHIN
University of Miami
Coral Gables, Florida

A JOHN WILEY & SONS, INC., PUBLICATION

Library of Congress Cataloging-in-Publication Data:

Chin, David A.
    Water-quality engineering in natural systems / David A. Chin.
        p.   cm.
    Includes bibliographical references and index.
    ISBN-13: 978-0-471-71830-7 (cloth)
    ISBN-10: 0-471-71830-0 (cloth)
    1. Water quality management.   I. Title.
TD365.C485 2006
628.1'68—dc22   2005023394

Printed in the United States of America

10 9 8 7 6 5 4 3 2 1

To my wife, Linda Sue, for her love and support

CONTENTS

PREFACE

Water-quality engineering is a specialty area in environmental engineering that includes the subspecialties of water treatment, wastewater treatment, and water-quality control in natural systems. This textbook is intended to encompass the latter subspecialty, and the content of this book constitutes baseline knowledge expected of water-quality engineers and managers. The need for competent water-quality engineers and managers is apparent when one realizes that in the United States over 50% of natural surface-water bodies do not meet their designated water uses and statutory water-quality goals. In addition, many shallow aquifers are contaminated by anthropogenic contaminants such as nitrates and organic chemicals, primarily pesticides and solvents. It is clear that water-quality engineering in natural systems will be an important practice area for the foreseeable future.

The practice of water-quality engineering is significantly influenced by laws and regulations, and practitioners must be fully aware of all applicable statutory requirements. The phenomenological foundations of water-quality control in natural systems are the relationships between contaminant concentrations in the aqueous phase and other phases (solid, vapor), the biochemical reactions of the contaminant in the environment, and the flows that transport the contaminant in the environment. The fundamental phenomenological relationships are typically brought together in a single fate and transport equation whose solution is closely tied to the advection–dispersion equation. Although the generic fate and transport equation can be applied in most natural waters, the physical, chemical, and biological differences between various types of water bodies dictate that these water bodies be considered separately to focus more closely on the processes that are important to that particular water body. For example, nutrient enrichment (eutrophication) is a primary concern in lakes and reservoirs, whereas toxic substances released from spills or leaking storage facilities is a primary concern in ground waters. The major categories of natural waters are rivers and streams, lakes and reservoirs, wetlands, ground water, and oceans and estuaries. Aside from assessing the fate and transport of contaminants purposely discharged into natural waters, remediation of contaminated waters also requires an understanding of the relationship between contaminant-generating activities in the surrounding watershed and the contaminant input to the receiving water body. In this regard, terrestrial fate and transport processes and their relationship to various best management practices must be understood and quantified.

The book begins with an introduction to the principles of water-quality control and the laws and regulations relating to water-quality control in the United States. Particular attention is given to use-attainability analyses and the estimation of total maximum daily loads, both of which are essential components of water-quality control in natural systems. Chapter 2 covers the essential components of water-quality standards, including the physical, chemical, and biological measures of water quality. Chapter 3 covers the mathematical formulation of fate and transport processes in aquatic systems, including the derivation of the advection–dispersion equation from first principles and the mathematical solution and properties of this fundamental equation. The advection–dispersion equation is applicable to all natural waters. Chapter 4 covers fate and transport processes in rivers and streams, including lateral and longitudinal mixing from both instantaneous spills and continuous discharges, the fate of volatile organic compounds in streams, and the depletion of dissolved oxygen in streams resulting from the discharge and accumulation of biodegradable organics. Guidelines for the restoration and management of polluted rivers are also provided. Chapter 5 describes water-quality processes in lakes and reservoirs, with particular emphasis on quantitative relationships describing flow and dispersion, sedimentation, eutrophication, nutrient recycling, and thermal stratification. Techniques to control eutrophication, dissolved-oxygen levels, toxic contaminants, acidity, and aquatic plants are all covered. Chapter 6 describes the occurrence, function, and hydrology of wetlands, the delineation of jurisdictional wetlands, and the design, construction, and operation of artificial (constructed) wetlands. Particular attention is given to factors controlling the contaminant-removal efficiencies in constructed wetlands. Chapter 7 covers water-quality-related processes in ground water, including the natural quality of ground water; quantification of sources of ground-water contamination; advection, dispersion, and sorption onto aquifer materials; biochemical decay; and the fate and transport of nonaqueous phase liquids in ground water. Detailed coverage is provided on the application of fate and transport principles to the remediation of contaminated ground water. Chapter 8 covers water-quality processes in oceans and estuaries, with particular emphasis on the design and operation of domestic wastewater outfalls, and water-quality control in estuaries as they relate to the physical, chemical, and biological conditions in the estuary. Chapter 9 covers water-quality-based watershed management where the primary focus is on estimating the contaminant loading on receiving waters from activities within the watershed. Detailed attention is given to sources of pollution and fate and transport processes associated with urban and agricultural watersheds. Atmospheric loading on natural waters due to airshed activities is also covered.

The material covered in this book is most appropriate for seniors and first-year graduate students in environmental and civil engineering programs. Others with backgrounds in environmental science might also find the contents of this book useful.

The practice of water-quality engineering in natural systems as described in this book reflects the reality that the fate and transport of anthropogenic contaminants introduced into natural waters must be understood and manipulated to minimize the impact of contaminant discharges into these waters. By controlling the quality, quality, timing, and distribution of contaminant discharges into the environment, the effects of human activities on natural waters can be controlled. The design of effective remediation measures in contaminated waters is based on these same principles, with additional technological considerations relating to the efficacy of various cleanup systems. The essential background for all these practices is provided.

DAVID A. CHIN

University of Miami

CHAPTER 1

INTRODUCTION

Water is essential for life on Earth, and any changes in the natural quality and distribution of water have ecological impacts that can sometimes be devastating. The sciences of hydrology and ecology are the scientific foundations of water-quality management. *Hydrology* is the science dealing with the occurrence and movement of water, *ecology* is the science dealing with interactions between living things and their nonliving (abiotic) environment or habitat, and the relationship between hydrology and ecology is sometimes called *hydrologic connectivity* (Pringle, 2003).

It is widely recognized that the establishment of new hydrologic connections in the landscape and modification of natural connectivity in highly modified human-dominated landscapes can have significant ecosystem impacts. For example, the modification of free-flowing rivers for energy or water supply and the drainage of wetlands can have a variety of deleterious effects on aquatic ecosystems, including losses in species diversity, floodplain fertility, and biofiltration capability (Gleick, 1993). Specific environmental issues that are of global concern include regional declines in migratory birds and wildlife caused by wetland drainage, bioaccumulation of methylmercury in fish and wildlife in newly created reservoirs, and deterioration of estuarine and coastal ecosystems that receive the discharge of highly regulated silicon-depleted and nutrient-rich rivers.

Water above land surface (in liquid form) is called *surface water*, and water below land surface is called *ground water*. Although surface water and ground water are directly connected, these waters are typically considered as separate systems and managed under different rules and regulations. A key feature of any surface-water body is its *watershed*, which is delineated by topographic high points surrounding the water body, and all surface runoff within the watershed has the potential to flow into the surface-water body. Consequently, surface-water bodies are the potential recipients of all contamination contained in surface runoff from all locations within the watershed. For this reason, the management of water quality in surface waters is best done at the watershed scale rather than at the scale of individual water bodies. This is the *watershed approach* to water-quality management. The main limitations to implementing the watershed approach are rooted in our inability to quantify most of the sub-watershed-scale contaminant-transport processes that are fundamental to implementing watershed models of water quality. Contaminant inputs into surface waters from the atmosphere are also considered in water-quality management plans (Patterson, 2000). In contrast to surface waters, the quality of ground water is influenced primarily by activities on and below the ground surface, and the potential sources of ground-water contamination are determined primarily by overlying land uses and subsurface geology. The concept of a watershed is not applicable to ground water.

Identification of some polluted water bodies are obvious to the casual observer, such as the stream shown in Figure 1.1, but some polluted water bodies are not so obvious, such as an apparently pristine lake that is so contaminated with acid rain that the existence of aquatic life is extremely limited.

To put the water-quality problem in perspective, a comprehensive study of the quality of waters in the United States (USGS, 1999) determined that:

- The highest levels of nitrogen occur in streams and ground waters in agricultural areas. Fifteen percent of samples exceeded the drinking-water standard for nitrate nitrogen, 10 mg/L as N.
- Pesticides, primarily herbicides, are found frequently in agricultural streams and ground water. Pesticides found most frequently include atrazine, metachlor, alochlor, and cyanazine.
- Urban streams have the highest frequencies of occurrence of DDT, chlordane, and dieldrin in fish and sediments. Complex mixtures of pesticides are commonly found in urban streams.

FIGURE 1.1 Polluted stream. (From Wetlands Connection Center, 2005. Photo by Suzanne Dilworth Bates.)

- Concentrations of phosphorus are elevated in urban and agricultural streams. These concentrations commonly exceed 0.1 mg/L.
- Hydrology and land use are the major factors controlling nutrient and pesticide concentrations in major rivers. Concentrations are proportional to the extent of urban and agricultural land use throughout the watershed. Key factors are soils and land slopes.
- Base flow in rivers originating from ground-water sources can be a major source of nutrients and pesticides to streams.

Aside from these general findings, individual states have their own unique problems with diffuse sources of pollution: In Wisconsin, waters are polluted by dairy and cranberry farms; in North Carolina, by hogs; in Maryland, by chickens; in southern Florida, by sugar; in Wyoming, by beef cattle; in Oregon, by clear-cutting; and in California, by irrigation return flows.

1.1 PRINCIPLES OF WATER-QUALITY CONTROL

The water-quality-based approach to pollution control provides a mechanism through which the amount of pollution entering a water body is controlled based on the conditions of the water body and the standards set to protect it. The water-quality-based approach to pollution control, illustrated in Figure 1.2, can be viewed as consisting of the following eight stages:

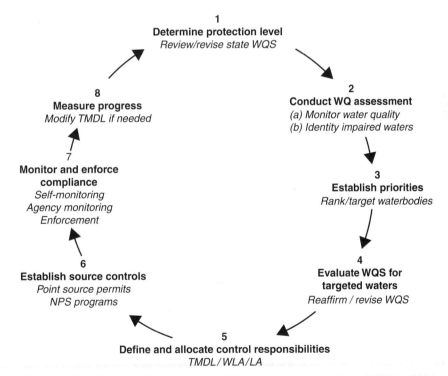

FIGURE 1.2 Water-quality-based approach to pollution control. (From USEPA, 1994.)

1. *Determining the protection level* requires the development of water-quality standards, which are the bases on which the condition of a water body is assessed.

2. *Monitoring and assessing water quality* requires a determination of whether water-quality standards are being met, detects pollution trends, and identifies sources of pollution. According to terminology used in the United States, water bodies not meeting water-quality standards are called *impaired waters*, and a *threatened water body* is one that currently meets water-quality standards but for which existing and readily available data and information on adverse declining trends indicate that water-quality standards will probably not be met at some time in the near future (USEPA, 1991b).

3. *Establishing priorities* involves ranking impaired water bodies based on considerations of risk to human health, aquatic life, and wildlife; degree of public interest and support; the recreational, economic, and aesthetic importance of a particular water body; and other relevant considerations.

4. *Evaluating water-quality standards* involves evaluating the appropriateness of the water quality standards for specific waters. It is possible that generally applied standards may be either over- or underprotective for a specific water body that has not had an in-depth standards analysis.

5. *Defining and allocating control responsibilities* establishes the level of control needed to meet water-quality standards and defines and allocates control responsibilities. The quantity of potential pollutants that can be discharged into any given water body without altering its integrity is called the *waste assimilative capacity*, and the sum of the waste assimilative capacity and the background load is called the *loading capacity* (USEPA, 1991a). Determining the loading capacity of a receiving body is one of the most important steps in any water-quality-based environmental protection or restoration effort. The *total maximum daily load* (TMDL) is defined as the maximum amount of a pollutant that may be discharged into a water body and still meet water-quality standards. Mathematical models and/or monitoring are used to determine TMDLs for impaired water bodies. Pollutant loadings above the TMDL amount will result in waters exceeding the water-quality standards. The TMDLs include *waste-load allocations* (WLAs) for point sources, *load allocations* (LAs) for non-point sources, and a margin of safety (USEPA, 1994). Allocations of pollution loads from point and nonpoint sources are calculated to ensure that water-quality standards are not exceeded, and as such are water-quality-based limits. For each impaired water body, TMDLs are developed by considering all pollution sources within the contributing watershed.

6. *Establishing source control* implements point-source controls through discharge permits, whereas nonpoint-source controls typically are implemented through state and local laws and regulations. In the United States, discharge permits required by the National Pollution Discharge Elimination System (NPDES) place controls on point-source discharges, and these controls may be either technology- or water-quality-based. *Technology-based limits* are determined by the performance of state-of-the-art treatment systems. For example, secondary municipal wastewater treatment plants are capable of producing effluents with biochemical oxygen demand (BOD) and suspended solids values of less than 30 mg/L, and the discharge limits for municipal wastewaters are usually set at these technology limits, even if the receiving water is capable of assimilating higher levels of BOD and/or suspended solids. In cases where the water-quality requirements are more stringent than the technology-based limits, more advanced treatment is required to meet the ambient water-quality criteria. Under these conditions, the allowable contaminant concentration in the discharge is *water-quality-limited*.

In contrast to point-source controls, nonpoint-source controls usually take the form of required best management practices (BMPs) or other management measures. The lands that are most polluting within a watershed are sometimes called *critical lands*, and identification of such lands is one of the most important tasks in planning, watershed management, and establishing and enforcing TMDLs.

7. *Monitoring and enforcing compliance* involves the collection of data for assessing compliance with water-quality-based controls (permits) and for evaluating whether the TMDLs, and control actions based on the TMDLs, are consistent with water-quality standards.

8. *Measuring progress* involves assessing the effectiveness of the contaminant-source controls and determining whether water-quality standards have been attained, water-quality standards need to be revised, or more stringent controls should be applied. If water-quality standards are not met, TMDL allocations must be modified.

In implementing the principles of water-quality control, it is essential to understand the connection between contaminant-source loading and the water quality. An understanding of pollution abatement resulting from physical, chemical, and/or biological controls and an ability to model these processes are essential in controlling the quality of water in natural systems.

1.2 SOURCES OF WATER POLLUTION

Sources of water pollution are broadly grouped into point sources and nonpoint sources. *Point sources* are defined as localized discharges of contaminants and include industrial and municipal wastewater outfalls, septic tank discharges, and hazardous-waste spills. *Nonpoint sources* of pollution include contaminant sources that are distributed over large areas or are a composite of many point sources, including runoff from agricultural operations, the atmosphere, and urban runoff. Surface runoff that collects in storm sewers and discharges via a pipe is still considered nonpoint-source pollution since it originates as diffuse runoff from the land surface. Pollution loads from nonpoint sources are commonly called *diffuse loads*. The most widespread nonpoint-source pollutants in the United States are eroded sediments, fertilizers, and pesticides, associated primarily with agricultural operations (Corwin et al., 1999).

Much of the pollution in waterways is caused by nonpoint-source pollution as opposed to point-source pollution. Cunningham (1988) reported that nonpoint sources were the principal contributors to pollution in 76% of lakes and reservoirs in the United States that failed to meet water-quality standards, and USEPA (1997d) reported that nonpoint sources impaired 65% of streams and 45% of estuaries in the United States that failed to meet water-quality standards. A report on the Danube River basin found that nonpoint sources contributed 60% of the nitrogen and 44% of the phosphorus load to the entire river basin (Commission for European Communities, 1994). In developing countries most pollution generated in large urban centers, from farms, deforestation, and land and wetland conversion could be classified as nonpoint (Novotny, 2003). The Black Sea, Adriatic Sea, Chesapeake Bay, and Gulf of Mexico are all examples of large water bodies affected by excessive inputs of nutrients from farming operations and cities located thousands of kilometers upstream and brought in by large tributaries—the Danube and Volga Rivers for the

Black Sea, the Po River for the Adriatic Sea, the Susquehanna and Potomac Rivers for Chesapeake Bay, and the Mississippi River for the Gulf of Mexico.

Wet-weather discharges refer to discharges that result from precipitation events such as rainfall and snowmelt. Wet-weather discharges include stormwater runoff, combined-sewer overflows (CSOs), and wet-weather sanitary-sewer overflows (SSOs). Stormwater runoff collects pollutants such as oil and grease, chemicals, nutrients, metals, and bacteria as it travels across land. CSOs and wet-weather SSOs contain a mixture of raw sewage, industrial wastewater, and stormwater, and have resulted in beach closings, shellfish bed closings, and aesthetic problems.

1.2.1   Point Sources

The identifying characteristic of point sources is that they discharge pollutants into receiving waters at identifiable single- or multiple-point locations. A typical point source of contamination is shown in Figure 1.3. In most countries, these (point) sources are regulated, their control is mandated, and a permit is required for waste discharge. Point sources of contamination that are of concern in managing surface waters are domestic wastewater discharges, industrial discharges, and spills.

Domestic Wastewater Discharges   Most municipal wastewater treatment plants discharge their effluent into rivers, lakes, or oceans. For river discharges of municipal wastewater, the effect of the effluent on the dissolved oxygen and nutrient levels in the river are usually of most concern. Decreased oxygen levels in rivers can cause harm to the aquatic life, and increased nutrient levels stimulate the growth of algae, which consume

FIGURE 1.3   Point source of pollution. (From *Illinois Leader*, 2004.)

oxygen. For ocean discharges, pathogen and heavy-metal concentrations are usually of most concern. In particular, pathogenic microorganisms discharged into the ocean can infect humans who come in contact with the ocean water in recreational areas such as beaches. Domestic wastewater discharged from septic tanks contains large numbers of pathogenic microorganisms, with viruses of particular concern because of the ability of viruses to move considerable distances in ground water. Approximately 50 million residents of the United States, 29% of the population, dispose of their sewage by individual on-site (septic) systems. Septic tanks represent the highest total volume of wastewater discharged directly to ground water and are the most common source of ground-water contamination (Novotny, 2003).

Properly designed, operated, and maintained sanitary-sewer systems collect and transport all of the sewage that flows into them to publicly owned treatment works (POTWs). However, occasional unintentional discharges of raw sewage from municipal sanitary sewers occur in almost every system. These types of discharges, collectively called *sanitary-sewer overflows* (SSOs), have a variety of causes, including but not limited to extreme weather, improper system operation and maintenance, and vandalism. There are at least 40,000 SSOs each year in the United States. The untreated sewage from SSOs can contaminate receiving waters and cause serious water-quality problems. These SSOs can also back up into basements, causing property damage and threatening public health.

Combined-Sewer Overflows   Combined-sewer systems are designed to collect rainwater runoff, domestic sewage, and industrial wastewater in the same pipe. Most of the time, combined-sewer systems transport all of their wastewater to a sewage treatment plant, where it is treated and discharged to a receiving water body. During periods of heavy rainfall or snowmelt, the wastewater volume in a combined-sewer system can exceed the capacity of the sewer system or treatment plant. For this reason, combined-sewer systems are designed to overflow occasionally and discharge excess wastewater directly to nearby streams, rivers, or other water bodies. These overflows, called *combined-sewer overflows* (CSOs), contain not only stormwater but also untreated human and industrial waste, toxic materials, and debris.

In the United States, combined-sewer systems are remnants of early infrastructure and are typically found in older communities. Combined-sewer systems currently serve approximately 770 communities containing about 40 million people. Most communities with combined-sewer systems (and therefore with CSOs) are located in the northeast and Great Lakes regions and the Pacific northwest. Figure 1.4 provides a rough illustration of the prevalence of combined-sewer systems in the United States.

Stormwater Discharges   Stormwater discharges are generated by runoff from open land and impervious areas such as paved streets, parking lots, and building rooftops during rainfall and snow events. Stormwater runoff often contain pollutants in quantities that could affect water quality adversely. A typical stormwater outlet is shown in Figure 1.5. The primary method to control the quality and quantity of stormwater discharges is through the use of best management practices. Although stormwater runoff is commonly discharged through a single outfall pipe, such discharges are more accurately classified as nonpoint pollutant sources.

Industrial Discharges   There are a wide variety of industrial wastewaters, and elevated levels of nutrients, heavy metals, heat, and toxic organic chemicals are common

FIGURE 1.4 Combined-sewer systems in the United States. (From USEPA, 2005b.)

FIGURE 1.5 Stormwater outlet. (From Rogue Valley Council of Governments, 2000.)

in industrial wastewaters. Some industries provide pretreatment prior to discharging their wastewaters either directly into surface waters or into municipal sewer systems for further treatment in combination with domestic wastewater. In many countries outside the United States, industries are permitted to discharge their wastewater without adequate pretreatment, and the resulting human and environmental impacts are usually noticeable.

FIGURE 1.6   Confined animal feeding operation. (From State of Arkansas, 2005.)

Animal Feeding Operations   Concentrated animal feeding operations (CAFOs) are considered point sources of contamination. To be considered a CAFO, a facility must first be defined as an animal feeding operation (AFO), where animals are kept and raised in confined areas. A typical AFO is shown in Figure 1.6. The following conditions are typical of AFOs: (1) animals have been, are, or will be stabled or confined and fed or maintained for a total of 45 days or more in any 12-month period; and (2) crops, vegetation, forage growth, or postharvest residues are not sustained in the normal growing season over any portion of the lot or facility.

There have been substantial changes in the U.S. animal production industry over the past several decades. Although the total number of AFOs has decreased, overall production has increased. As a result, CAFOs are increasing in size and generating considerably more waste requiring disposal over more limited areas. These operations are expanding to the central and southwestern regions of the United States, where land is less expensive and populations are relatively sparse. CAFO waste releases in the eastern United States have prompted a closer evaluation of their environmental impact on surface waters.

Spills   Spills and accidental or intentional releases can occur in a variety of ways. Transportation accidents on highways and rail freight lines can result in major chemical spills, and accidental releases at petroleum-product storage installations are another common source of accidental spills. Leaks and spills from underground storage tanks and piping into the ground water are of special concern because these releases may remain undetected for long periods of time.

1.2.2   Nonpoint Sources

Nonpoint sources of contamination generally occur over large areas and because of their diffuse nature, are more complex and difficult to control than point sources. Nonpoint-source

TABLE 1.1   Strength of Various Point and Nonpoint Sources[a]

Source	BOD$_5$ (mg/L)	Suspended Solids (mg/L)	Total Nitrogen (mg/L)	Total Phosphorus (mg/L)	Total Coliforms (MPN/100 mL)
Urban stormwater	10–250 (30)	3000–11,000 (650)	3–10	0.2–1.7 (0.6)	10^3–10^8
Construction-site runoff	NA	10,000–40,000	NA	NA	NA
Combined-sewer overflows	60–200	100–1100	3–24	1–11	10^5–10^7
Light industrial area	8–12	45–375	0.2–1.1	NA	10
Roof runoff	3–8	12–216	0.5–4	NA	10^2
Typical untreated sewage	(160)	(225)	(35)	(10)	10^7–10^9
Typical POWT[b] effluent	(20)	(20)	(30)	(10)	10^4–10^6

[a]Numbers in parentheses indicate mean. NA, not available or unreliable.
[b]Publicly owned treatment works with secondary (biological) treatment.

pollution is a direct result of land-use patterns, so many of the solutions to pollution by nonpoint sources lie in finding more effective ways to manage the land and stormwater runoff. Much nonpoint-source pollution occurs during rainstorms and spring snowmelt, resulting in large flow rates that make treatment even more difficult. Federal and state governments do not play a major role in local land-use decisions; therefore, local governments must be responsible for maintaining water quality through nonpoint-source controls. Nonpoint sources of contamination that must generally be considered in managing water bodies are agricultural runoff, livestock operations, urban runoff, landfills, and recreational activities.

Runoff from urban and agricultural areas are the primary sources of surface-water pollution in the United States (USEPA, 2000b). Ground-water contamination originating from septic tanks, leaking underground storage tanks, and waste-injection wells is quite common and are of particular concern when ground water is the source of domestic water supply. The strengths of various sources of water pollution are shown in Table 1.1. It is clearly apparent that pollutants at high concentrations can enter water bodies from a variety of sources, and control of these sources is central to effective water-quality management.

Agricultural Runoff   Application of pesticides, herbicides, and fertilizers are all agricultural activities that influence the quality of both surface and ground waters. In the United States, certain pesticides and herbicides are banned because of their toxicity to humans or their adverse effect on the environment. The application of fertilizers is of major concern because dissolved nutrients in surface runoff accelerate growth of algae and depletion of oxygen in surface waters. Nitrogen, in the form of nitrates, is a contaminant commonly found in ground water underlying agricultural areas and can be harmful to humans. Erosion caused by improper tilling techniques is another agricultural activity that can adversely affect water quality through increased sediment load, color, and turbidity (AWWA, 1990).

FIGURE 1.7   Livestock in a stream. (From State of Arkansas, 2005.)

Livestock   Feedlots have been shown to contribute nitrates to ground water, and pathogenic microorganisms to surface waters. Overgrazing eliminates the vegetative cover that prevents erosion, increasing the sediment loading to surface waters. In some extreme cases, livestock are allowed to wade in and cause direct contamination of streams. Such a circumstance is shown in Figure 1.7. This practice should be avoided as much as possible, since the direct pollution of streams by pathogens such as *Cryptosporidium parvum* is a likely and undesirable consequence.

Urban Runoff   Urban runoff contains contaminants that are washed from pavement surfaces and carried to surface-water bodies. Contaminants contained in urban runoff include petroleum products, metals such as cadmium and lead from automobiles, salt and other deicing compounds, and silt and sediment from land erosion and wear on road and sidewalk surfaces. Bacterial contamination from human and animal sources is also often present. The initial "flushing" of contaminants during storm events typically creates an initial peak in contaminant concentration, with diminishing concentration as pollutants are washed away (AWWA, 1990).

   A major factor associated with impairment of waters in the United States is the amount of impervious area that is directly connected to urban runoff systems (USEPA, 2002b), and such a connection is shown in Figure 1.8. According to Schueler (1995), there is a strong correlation between percent imperviousness of a watershed and stream health. Schueler (1995) further reported that stream degradation can occur when a watershed becomes about 10% impervious, and degradation becomes unavoidable at 30% imperviousness.

Landfills   Leachate from landfills can be a source of contamination, particularly for ground water. Water percolating through a landfill (leachate) contains many toxic constituents and is typically controlled by capping the landfill with a low-permeability cover

FIGURE 1.8   Surface runoff into a drainage system. (From State of Vermont, 2005.)

and installing a leachate-collection system. Many older landfills do not have leachate-collection systems.

Recreational Activities   Recreational activities such as swimming, boating, and camping can have a significant impact on water quality. The impact of human activities has typically been reported in terms of increased levels of pathogenic microorganisms (AWWA, 1990).

1.3   LAWS AND REGULATIONS

Most water-control protocols are motivated directly or indirectly by environmental regulations. Consequently, to gain a proper understanding of water-control practice, one must have a good understanding of environmental regulations.

In the United States, Congress makes *laws*, and to put these laws into effect, Congress authorizes certain government agencies, such as the U.S. Environmental Protection Agency (USEPA), to create and enforce regulations to implement these laws. The creation of a law begins when a *bill* is proposed by a member of Congress. If both houses of Congress approve a bill, it goes to the President, who has the option to either approve (sign) it or veto it. If approved, the new law is called an *act*, and the text of the act is known as a *public statute*. Once an act is passed, the Government Printing Office incorporates the new law into the *United States Code*, which codifies (categorizes and compiles) all federal legislation. Authorized regulatory and enforcement agencies—such as the USEPA—identify the need for regulations to support implementation of the act passed by Congress. Proposed regulations are listed in the *Federal Register* so that the public can provide comments to the agency. After considering all public comments and making appropriate modifications to proposed regulations, a final rule is issued and published in the *Federal Register*. Once a regulation is completed and has been published in the *Federal Register* as a final rule, it is codified by being published in the *Code of Federal Regulations* (CFR). The CFR is the official record of all regulations created by the federal government and is divided into 50 volumes called *titles*, each of which focuses on a particular area. Almost all environmental regulations appear in Title 40, usually abbreviated as simply 40 CFR. Each CFR title is divided into various sections; for example, the baseline regulations dealing with stormwater discharges into waters of the United States are found in 40 CFR 122, and specific permit requirements are found in 40 CFR 122.26. The most important federal laws governing the quality of ground and surface waters in the United States are the Clean Water Act (CWA) and the Safe Drinking Water Act (SDWA).

Programs mandated by both the CWA and SDWA are administered at the federal level by the U.S. Environmental Protection Agency, and these programs are usually implemented through corresponding individual state programs. The CWA allows the delegation of many permitting, administrative, and enforcement aspects of the law to state governments. In states with the authority to implement CWA programs, the USEPA still retains oversight responsibilities. In addition to the implementation of federal laws and regulations, state programs provide protection of public water supplies through water codes, sanitary regulations, regulation of inland wetland areas, and other means of watercourse and ground-water protection (AWWA, 1990). At the local level, municipal and county ordinances can provide significant protection through land-use controls that regulate development activities on key watershed areas.

1.3.1 Clean Water Act

The Clean Water Act, originally enacted in 1972, is intended to restore and maintain the physical, chemical, and biological integrity of waters of the United States. The Clean Water Act applies to surface waters, including rivers, lakes, estuaries, wetlands, and oceans, and focuses primarily on establishing water-quality standards and regulations to control discharges into surface waters. Major requirements of the Clean Water Act are:

- States must set water-quality standards to protect designated uses of all natural water bodies in the state.
- States must prepare and submit (to the USEPA) an annual report on the quality of all waters in the state.
- States must submit (to the USEPA) lists of surface waters that do not meet applicable water-quality standards after implementation of technology-based limitations. Such waters are called *impaired waters*. States must establish total maximum daily loads (TMDLs) for impaired waters on a prioritized schedule that considers the severity of the pollution and the uses to be made of such waters.
- The federal government must set industrywide, technology-based effluent standards for dischargers.
- All dischargers must obtain a permit issued by the federal government or authorized states that specifies discharge limits under the National Pollutant Discharge Elimination System (NPDES). The discharge limits are typically the stricter of the water-quality-based and technology-based limits.

Water-quality standards are the foundation of the water-quality-based pollution control program mandated by the Clean Water Act. Water-quality standards define the goals for a water body by designating its uses, setting criteria to protect those uses, and establishing provisions to protect water bodies from pollutants. Point discharges into water bodies that meet their water-quality standard are typically regulated using technology-based effluent standards, whereas discharges into impaired waters are regulated for each pollutant based on a total maximum daily load (TMDL), which is the amount of a contaminant that a water body can receive and still meet the water-quality standard. For each pollutant, the TMDL includes contributions from both point and nonpoint sources. Dissolved-oxygen concentration and algal biomass (or chlorophyll a) are the most commonly applied assessment endpoints in TMDLs (Tung, 2001).

1.3.2 Safe Drinking Water Act

The Safe Drinking Water Act (SDWA) was enacted in 1974 to protect public health by regulating the nation's public drinking-water supply. The law was amended in 1986 and 1996 and requires many actions to protect drinking water and its sources. The SDWA applies to every public water system in the United States, and there are currently more than 160,000 public water systems. The SDWA does not regulate private wells that serve fewer than 25 persons. Major programs mandated by the SDWA that relate to water-quality control in natural waters are drinking-water standards, source-water protection, and underground injection control.

Drinking-Water Standards   Under the authority of the SDWA, the USEPA has promulgated national primary drinking-water standards for certain radionuclide, microbiological, organic, and inorganic substances. These standards establish maximum contaminant levels (MCLs), which specify the maximum permissible level of a contaminant in water that may be delivered to a user of a public water-supply system. MCLs are established based on consideration of a range of factors, including not only the health effects of the contaminants but also the treatment capability, monitoring availability, and costs. In some cases, MCLs are replaced by treatment techniques (TTs), which are procedures that public water-supply systems must employ to ensure contaminant control. Required treatment techniques for pathogens are contained in the Surface Water Treatment Rule (SWTR) and the coliform standard. The SWTR requires that surface water, and ground water under the direct influence of surface water, undergo some combination of disinfection and filtration to meet the following criteria:

- A minimum of 99.9% of *Giardia lamblia* and 99.99% of viruses are to be killed or inactivated.
- The turbidity must be maintained at less than 5 nephelometric turbidity units (NTU) at all times and less than 1 NTU in 95% of daily samples.
- The total heterotrophic bacterial level, as determined by plate count, must be less than 500 colonies per milliliter.

The coliform standard requires that no more than 5% of water samples test positive for coliforms in a month, and of those positives, that there be no samples that test positive for fecal coliforms.

Secondary drinking water standards are nonenforceable guidelines that are based on potential adverse cosmetic effects, such as skin discoloration and laundry staining, or aesthetic effects such as taste, odor, and color. Drinking-water standards apply to public water systems, which provide water to at least 15 connections or 25 persons at least 60 days out of the year (most cities, towns, schools, businesses, campgrounds, and shopping malls are served by public water systems). The 10% of Americans whose water comes from private wells (individual wells serving fewer than 25 persons) are not protected by these federal standards. People with private wells are responsible for making sure that their own drinking water is safe. Bottled water is regulated by the U.S. Food and Drug Administration as a food product and is required to meet standards equivalent to those that the USEPA sets for drinking water.

Source-Water Protection   Under provisions of the Safe Drinking Water Act, the USEPA also establishes minimum water-quality criteria for water that may be taken into a water-supply system. All states are required to develop source-water assessment plans (SWAPs) to protect public water supplies from contaminant sources within the catchment area of the drinking-water intake. This catchment area defines the source-water protection area (SWPA) of the intake. Key components of SWAPs are as follows:

1. *Delineation of the source-water protection area.* For ground-water systems, available information about ground-water flow and recharge is used to determine the source-water protection area around a well or well field. Such areas are called *wellhead protection areas.* Wellhead protection areas are typically divided into three zones, with an inner zone given the most protection, an intermediate zone where detailed contaminant inventories are conducted and best management practices required, and an outer zone where the potential impact of contaminants on drinking-water intakes is regarded as minimal. The inner zone is usually delineated using a distance criterion, the intermediate zone (also known as the inventory region) delineated by time of travel, and the outer zone delineated by a hydrogeologic boundary. For surface-water intakes drawing water from a stream, river, lake, or reservoir, the land area in the watershed upstream of the intake defines the source-water protection area. Surface-water intakes commonly have two protection zones, an inner zone that is typically delineated by an upstream distance or travel time (based on the average annual high flow) plus a minimum buffer width. Outer zones are typically delineated by watershed boundaries. Water-supply wells classified as ground water under the direct influence of surface water (GWUDI) are treated as surface-water sources for delineation purposes.

2. *Inventory of actual and potential sources of contamination.* Sources of pollutants that could potentially contaminate the water supply are identified. This inventory usually results in a list of facilities and activities within the delineated area that may release contaminants into the underground water supply (for wells) or the watershed of the river or lake (for surface-water sources). Some examples of potential pollutant sources include landfills, underground or aboveground fuel storage tanks, residential or commercial septic systems, urban runoff from streets and lawns, farms and other entities that apply pesticides and fertilizers, and sludge-disposal sites.

3. *Determination of the susceptibility of the water source to contamination.* *Susceptibility* is defined by the USEPA as "the potential for a public water supply to draw water contaminated by inventories sources at concentrations that would pose concern" (USEPA, 1997c). Susceptibility determinations are typically based on the fate and transport of contaminants within the SWPA, and establish the likelihood of contamination of the water supply by potential sources of contamination. This critical step helps local decision makers consider priority approaches for protecting the drinking-water supply from contamination. It is common to prioritize the potential for contamination from the

specific potential contamination sources or from individual chemicals that could pollute the water. In some cases, a susceptibility ranking of high, medium, or low is assigned to each pollutant source.

4. *Releasing the results of the assessments to the public.* After the source-water assessment for a particular water system is completed, the information must be summarized for the public. Such summaries help communities better understand the potential threats to their water supplies and identify priority needs for protecting the water from contamination. Local communities, working in cooperation with local, regional, and state agencies, can use the information gathered through the assessment process to create broader source-water protection programs to address current and future threats to the quality of their drinking-water supplies.

Communities use a wide array of source-water protection methods to prevent contamination of their drinking-water supplies. One management option involves regulations, such as prohibiting or restricting land uses that may release contaminants in critical source-water protection areas. To go along with these regulations, many communities hold local events and distribute information to educate and encourage citizens and businesses to recycle used oil, limit their use of pesticides, participate in watershed cleanup activities, and engage in a variety of other prevention activities. Another approach to source-water protection is to purchase land or create conservation easements to serve as protection zones near drinking-water sources.

The use of prescribed buffer zones and travel times to protect surface-water intakes and the use of travel times to protect ground-water intakes are the most common approaches to controlling the susceptibility of drinking-water intakes to contamination (USEPA, 1997a, b). A typical riparian (river) buffer zone consisting of a forested area is shown in Figure 1.9. Buffer zones can be created through utility ownership or usage restrictions,

FIGURE 1.9   River with an adjacent (riparian) buffer. (From NRCS, 2005g.)

thereby providing for filtration of runoff through natural vegetation. The required buffer areas take into account exposure to contamination and the degree of treatment provided by the buffer. Wetlands are frequently used as buffers for water-supply intakes, and preservation of wetlands maintains the natural filtration and cleansing provided by these critical areas.

Underground Injection Control Program   The underground injection control (UIC) program provides safeguards such that injection wells do not endanger underground sources of drinking water (USDWs). Facilities across the United States discharge into underground formations a variety of hazardous and nonhazardous fluids via more than 800,000 injection wells, and a typical injection well is shown in Figure 1.10. Agricultural operations and the chemical and petroleum industries could not exist without underground injection wells since it is frequently cost-prohibitive to treat and release to surface waters the large amounts of wastes that these industries produce each year. When injection wells are properly sited, constructed, and operated, these wells are an effective and environmentally safe alternative to surface disposal.

An injection well is defined as a bored, drilled, or driven shaft whose depth is greater than the largest surface dimension; a dug hole whose depth is greater than the largest surface dimension; an improved sinkhole; or a subsurface fluid distribution system. This definition covers a wide variety of injection practices, ranging from more than 140,000 technically sophisticated highly monitored wells that pump fluids into isolated formations up to 2 miles below Earth's surface to the far more numerous on-site drainage systems,

FIGURE 1.10   Typical injection well. (From Alberta Energy and Utilities Board, Alberta Geological Survey, 2005.)

such as septic systems, dry wells, and stormwater wells, that discharge fluids a few meters underground.

USEPA groups underground injection wells into five classes for regulatory control purposes. Each class includes wells with similar functions, construction, and operating features, so that technical requirements can be applied consistently to the class:

- *Class I wells:* inject hazardous and nonhazardous fluids (industrial and municipal wastes) into isolated formations beneath the lowermost underground source of drinking water (USDW). Because these wells inject hazardous waste, class I wells are the most strictly regulated and are regulated further under the Resource Conservation and Recovery Act (RCRA).
- *Class II wells:* inject brines and other fluids associated with oil and gas production.
- *Class III wells:* inject fluids associated with solution mining of minerals.
- *Class IV wells:* inject hazardous or radioactive wastes into or above a USDW and are banned unless authorized under other statutes for ground-water remediation.
- *Class V wells:* include all underground injection not included in classes I through IV. Generally, most class V wells inject nonhazardous fluids into or above a USDW and are on-site disposal systems such as floor and sink drains, which discharge to dry wells, septic systems, leach fields, and drainage wells. Injection practices or wells that are not covered by the UIC Program include single-family septic systems and cesspools as well as nonresidential septic systems and cesspools serving fewer than 20 persons that inject only sanitary wastewater.

All injection wells are not waste-disposal wells. Some class V wells, for example, inject surface water to replenish depleted aquifers or to prevent saltwater intrusion. Some class II wells inject fluids for enhanced recovery of oil and natural gas.

Most injection wells have the potential to inject fluids that may cause a public water system to violate drinking water standards. In general, the UIC program prevents contamination of water supplies by setting minimum requirements for state UIC programs. A basic concept of USEPA's UIC program is to prevent contamination by keeping injected fluids within the intended injection zone, or in the case of injection directly or indirectly into a USDW, the fluids must not endanger or have the potential to endanger a current or future public water supply. Most of the minimum requirements that affect the siting of an injection well and the construction, operation, maintenance, monitoring, testing, and finally, the closure of the well, are designed to address these concerns. All injection wells require authorization under general rules or specific permits, and most states have primary enforcement authority (primacy) for the UIC program.

1.4   STRATEGY FOR WATER-QUALITY MANAGEMENT

The strategy for water-quality management in the United States is dictated primarily by the Clean Water Act and has three major components:

1. Designate the use of each natural water body.
2. Establish water-quality criteria to protect the designated uses.
3. Maintain a policy of antidegradation in executing steps 1 and 2.

Regulatory categories of designated uses include public water supply, recreation, and propagation of fish and wildlife; other uses, such as navigation and irrigation, can also be specified. For each designated use, water-quality criteria are established by the state with the approval of the USEPA. Comparing the quality of a water body with the criteria corresponding to its designated use is called *water-body assessment*. If the water-quality criteria are met by the water body, the designated use is attained.

1.4.1 Use-Attainability Analysis

If the water-quality criteria are not met, the designated use has not been attained and a *use-attainability analysis* (UAA) must be done. The primary objectives of a UAA are to determine what maximum pollutant loads [expressed as total maximum daily loads (TMDLs)] are consistent with meeting the water-quality criteria and whether limiting these loads are feasible from a socioeconomic viewpoint. Therefore, a UAA has three components: (1) water-body assessment, (2) determination of total maximum daily load, and (3) socioeconomic analysis. Possible outcomes of a use-attainability analysis are that the designated use (a) is attainable, (b) is not attainable and should be downgraded, or (c) is more than attainable and should be upgraded. Acceptable reasons for changing the designated use and/or water-quality criteria of a water body are:

- Naturally occurring pollutant concentrations prevent attainment of the use.
- Natural, ephemeral, intermittent, or low flow or water levels prevent the attainment of the use unless these conditions are compensated for by the discharge of a sufficient volume of effluent discharge without violating conservation requirements.
- Human-caused conditions or sources of pollution prevent the attainment of the use and cannot be remedied or would cause more environmental damage to correct than to leave in place.
- Dams, diversions, or other types of hydrologic modifications preclude the attainment of the use and its feasibility to restore the water body to its original condition or to operate such modification in a way that would result in the attainment of the use.
- Physical conditions related to the natural features of the water body, such as the lack of proper substrate, cover, flow, depth, pools, riffles, and the like, unrelated to water quality, preclude attainment of aquatic-life protection uses.
- More stringent controls would result in substantial and widespread adverse societal and economic impact.

1.4.2 Total Maximum Daily Load Process

The TMDL process is the preferred planning process in situations where simple control of point sources by permits based on technology-based standards or on waste load allocations alone would fail to accomplish water-quality goals in water-quality-limited receiving waters. The outcome of this process is a watershed management plan that puts controls on sources of pollution located within the watershed. Management options in water-quality-limited (impaired or threatened) waters are as follows (Novotny, 1996, 1999):

1. *Reduction of point-source loads beyond those mandated by technology-based effluent standards.* In this case, the entire burden of abatement is placed on point-source dischargers

by issuing TMDL-based point-discharge (NPDES) permits. These permits are more stringent than those mandated by technology-based effluent standards.

2. *A combination of additional point-source controls and feasible best management practices to reduce nonpoint-source loads.* In this approach, loading capacity from point and nonpoint sources and TMDLs must be determined for both low-flow and wet-weather conditions. This alternative is more efficient and equitable than a sole reliance on point-source controls.

3. *Water-body restoration and waste assimilative capacity enhancement.* For some water-quality-limited watersheds, enough pollutants have accumulated in the sediments that even if all present point and nonpoint discharges are eliminated, the water-quality goals will not be achieved. Assimilative capacity can be increased by means such as low-flow augmentation, contaminated-sediment removal or capping, in-stream aeration, restoration of riparian wetlands, and implementation of riparian buffer strips.

4. *Changing the designated use and associated water-quality standards.* If a use-attainability analysis (UAA) proves that the existing water-quality standards are not attainable, the designated use of the water body, and associated standards, can be adjusted to reflect the attainable optimum use.

An implementation plan must generally be part of a TMDL study, and this plan is subject to approval by a regulatory agency, which in the United States is the U.S. Environmental Protection Agency or a delegated state agency.

SUMMARY

The regulation and management of potential water pollutants are best accomplished by taking into account the relationship between the location and magnitude of these pollutant sources and their water-quality impact. This approach to pollution control, generally referred to as water-quality-based pollution control, requires consideration of the impact on both ground and surface waters. For surface waters, the land area of concern is the watershed, and for ground waters, pollution-handling activities on the overlying land area are of most concern.

Sources of water pollution are grouped into point and nonpoint sources. Point sources discharge pollutants into receiving waters at identifiable locations, typically out of the end of a pipe, and include domestic wastewater discharges, combined sewer overflows, industrial discharges, animal feeding operations, and spills. Nonpoint sources of water pollution, which occur over large areas and tend to be diffuse in nature, include agricultural runoff, livestock, urban runoff, landfills, and recreational activities.

Regulatory control of point and nonpoint sources of water pollution in the United States is based primarily on the authority of the Clean Water Act and the Safe Drinking Water Act. Under the guidelines of the CWA, water-quality-based pollution control requires that each natural water body have a designated use and federally approved water-quality criteria associated with each designated use. Water-quality data collected in monitoring programs are compared with applicable water-quality criteria, and if the criteria are exceeded, the water is called an impaired water and action must be taken either to reduce the pollutant loading to bring the water body into compliance with its designated use, or the designated use must be changed. In these cases, total maximum daily loads (TMDLs) must be calculated that would be consistent with the water body meeting the water-quality criteria associated with its designated use. The TMDL is allocated between point sources, non-point sources, and background loading of contaminants into the polluted water body. Any water body that does not meet its water-quality criteria can technically be referred to as polluted. The Safe Drinking Water Act includes several provisions that complement the Clean Water Act in controlling the quality of natural water bodies, such as the establishment of drinking-water standards that are commonly used as water-quality criteria for ground waters; source-water protection, wellhead protection, and sole-source aquifer programs that require regulation of pollutant sources that could affect drinking water quality; and the underground injection program that protects underground sources of drinking water.

The most fundamental water-quality management issues to be addressed when a water body does not meet its water-quality criteria are related to the following two questions: (1) Is the designated use attainable? and (2) What pollutant loads on the water body would be consistent with attaining the designated use? To answer the first question, a use-attainability analysis must be conducted, and to answer the second question, the TMDL process must be followed.

PROBLEMS

1.1. What is the organizational structure of the U.S. Environmental Protection Agency? Explain how this structure is related to laws governing water quality.

1.2. What is the organizational structure of your state agency responsible for environmental protection? Explain how this structure is related to the federal and state laws governing water quality.

1.3. Briefly summarize the federal regulations authorized by the Clean Water Act.

1.4. The loading capacity of a water body that would be consistent with the water body attaining its designated use is calculated to be 100 kg of phosphorus per year.

(**a**) If a 10% factor of safety is mandated by the appropriate regulatory agency, what is the TMDL for phosphorus?

(**b**) If 40 kg is already allocated to point discharges in the water body and 45 kg is allocated to nonpoint sources, how much loading can be reserved for future growth in the watershed?

CHAPTER 2

WATER-QUALITY STANDARDS

2.1 INTRODUCTION

The acceptable quality of a natural water body depends on its present and future most beneficial use. Inland and marine waters should be aesthetically pleasing and of acceptable quality to support healthy and diverse aquatic ecosystems as well as human recreational uses such as fishing and swimming. Freshwater bodies such as rivers, lakes, and aquifers are typically sources of drinking water, and the quality of these waters should be consistent with the level of water treatment provided prior to distribution in public water-supply systems. Ground-water quality is typically superior to that of surface water in bacteriological content, turbidity, and total organic concentrations. However, the mineral content (hardness, iron, manganese) of ground water may be inferior and require additional treatment prior to public consumption. Ground-water supplies are frequently consumed directly by the public or pumped into water-distribution systems with minimal treatment.

In natural water bodies, water-quality criteria are generally considered from two viewpoints: human health and aquatic life. Human-health-based water-quality criteria are derived from assumptions related to the degree of human contact, quantity of water ingested during human contact, and the amount of aquatic organisms (e.g., fish) consumed that are derived from the water body (USEPA, 1994). Quantitative human-health water-quality criteria are derived from experimental dose–response relationships and acceptable risk (of illness) and are typically stated in terms of concentrations corresponding to an acceptable level of risk. Aquatic-life water-quality criteria are derived from mortality studies of selected organisms exposed to various levels of contamination in the water, and these criteria are stated in terms of acceptable duration and frequency of exposure. In the United States, aquatic-life water-quality criteria are typically stated in the form that aquatic ecosystems will not be impaired significantly if (1) the 1-hour average concentration does not

TABLE 2.1   Factors Used in Water-Body Surveys and Assessments

Physical Measures	Chemical Measures	Biological Measures
In-stream characteristics	Dissolved oxygen	Biological inventory (existing)
Size (width, depth)	Toxicants	Fish
Flow, velocity	Suspended solids	Macroinvertibrates
Annual hydrology	Nutrients	Microinvertibrates
Total volume	Nitrogen	Phytoplankton
Reaeration rates	Phosphorus	Periphyton
Gradient, pools, riffles	Sediment oxygen demand	Macrophytes
Temperature	Salinity	Biological potential analysis
Sedimentation	Hardness	Diversity Indexes
Channel modifications	Alkalinity	HSI models
Channel stability	pH	Tissue analysis
Substrate characteristics	Dissolved solids	Recovery index
Channel debris		Intolerant species analysis
Sludge deposits		Omnivore–carnivore analysis
Riparian characteristics		
Downstream characteristics		

Source: USEPA (1994).

exceed the recommended *acute criterion* more than once every 3 years on average; and (2) the 4-day (96-hour) average concentration of the pollutant does not exceed the recommended *chronic criterion* more than once every 3 years on average (USEPA, 1986). Contaminant limits for acute exposure are higher than contaminant limits for chronic exposure, but there is no general rule as to whether human-health or aquatic-life criteria is more restrictive.

2.2   MEASURES OF WATER QUALITY

Water quality is assessed based on (1) the present and future most beneficial use of the water body, and (2) a set of water-quality criteria that correspond to the intended use of the water body. In addition to specifying a designated use and water-quality criteria to protect the designated use, an antidegradation policy is generally instituted to keep waters that meet water-quality criteria from deteriorating from their current condition, even if the existing water-quality criteria are met in the deteriorated condition. Water-quality criteria are generally formulated to maintain the physical, chemical, and biological integrity of a water body. Measures used to assess the suitability of a water body for various uses are listed in Table 2.1. The interrelationships between the physical, chemical, and biological measures are complex, and alterations in the physical and/or chemical condition generally result in changes in biological condition.

2.2.1   Physical Measures

Physical measures that directly affect the quality of aquatic-life habitat include flow conditions, substrate, in-stream habitat, riparian habitat, and thermal pollution. These measures are described below.

(a)   (b)

FIGURE 2.1 Typical (*a*) pool and (*b*) riffle. (From Organization for the Assabet River, 2005 a, b. Photo by Suzanne Flint.)

Flow Conditions Slope and velocity divide streams into four categories: mountain streams, piedmont streams, valley streams, and plains and coastal streams. *Mountain streams*, sometimes called *trout streams*, have steep gradients and rapid currents; streambeds consisting of rock, boulders, and sometimes sand and gravel; and are well aerated and cool, with temperatures rarely exceeding 20°C. *Piedmont streams* are larger than mountain streams, with depths up to 2 m; have rapid currents with alternating *riffles* (shallow, fast-moving waters) and *pools* (deep slow-moving waters); and streambeds typically consist of gravel. A typical pool and riffle in Elizabeth Brook (Massachusetts) are shown in Figure 2.1 Riffles (rapids) are inhabited by salmonids and pools by rheophilic cyprinids. *Valley streams* have moderate gradient and current with alternating rapids and more extensive quiet waters than in piedmont zones. Trout are still present in rapid stretches. *Plains and coastal streams* include the lower stretches of rivers and canals, have low currents, high temperatures and low dissolved oxygen in the summer, and turbid waters. The main fish species are bream, channel catfish, sunfish, and largemouth bass.

For excellent habitat, flows should exceed $0.05\,m^3/s$ for cold streams and $0.15\,m^3/s$ for warm streams. Poor habitat usually exists for flows less than $0.01\,m^3/s$ in cold streams and $0.03\,m^3/s$ in warm streams. For benthic and fish communities, four general categories of velocity (V) and depth (H) are optimal (Novotny, 2003): slow ($V < 0.3\,m/s$) and shallow ($H < 0.5\,m$); (2) slow ($V < 0.3\,m/s$) and deep ($H > 0.5\,m$); (3) fast ($V > 0.3\,m/s$) and deep ($H > 0.5\,m$); and (4) fast ($V > 0.3\,m/s$) and shallow ($H < 0.5\,m$).

The *pool/riffle ratio* or *bend/run ratio* is calculated by dividing the distance between riffles or bends by the width of the stream, respectively. The pool/riffle ratio is used to classify streams with higher slope (mountain, piedmont, and valley), and the bend/run ratio is used to classify slow-moving lowland streams. An optimum value of these ratios is in the range 5 to 7 (Novotny, 2003); ratios greater than 20 correspond to channels that are essentially straight and are poor habitat for many aquatic species. Disruption of the run–riffle–pool sequence has detrimental consequences on macroinvertebrate and fish populations, while habitat diversity is related directly to the degree of meandering in natural and channelized streams (Zimmer and Bachman, 1976, 1978; Karr and Schlosser, 1977).

(a)   (b)

FIGURE 2.2   (*a*) Sand and (*b*) gravel substrates.

Substrate   *Substrate* is the material that makes up the streambed. Sand and gravel are common substrate materials, and these materials are contrasted in Figure 2.2. The type of substrate is influenced significantly by the velocity of flow in the stream, and a typical relationship between the type of substrate and the velocity in a stream is given in Table 2.2. It is also useful to note that sand settles in streams where velocities are less than 0.25 to 1.2 m/s, gravel settles in streams where velocities are less than 1.2 to 1.7 m/s, and erosion of sand and gravel riverbeds occurs at velocities greater than 1.7 m/s (DeBarry, 2004). Stream velocities below 0.1 m/s are typically categorized as slow, 0.25 to 0.5 m/s as moderate, and greater than 0.5 m/s as swift. In general, clean and shifting sand and silt is the poorest habitat. Bedrock, gravel, and rubble on the one side and clay and mud on the other side, especially when mixed with sand, support increasing biomass. Substrate with more than 50% cobble gravel is regarded as excellent habitat conditions; substrate with less than 10% cobble gravel is regarded as poor habitat. Watercourses with swift velocities (>0.5 m/s) that have cobble and gravel beds have the greatest invertebrate diversity (DeBarry, 2004).

Embeddedness is a measure of how much of the surface area of the larger substrate particles is surrounded by finer sediment. This provides a measure of the degree to which the

TABLE 2.2   Flow Velocity Versus Type of Substrate

Velocity (m/s)	Type of Substrate
<0.2	Silt and bottom muck
0.2–0.4	Silt and sand
0.4–0.5	Sand
>0.5	Gravel and rocks

Source: USEPA (1983b).

FIGURE 2.3   Effect of channel erosion on an in-stream habitat. (From USEPA, 2005d.)

primary substrate (e.g., cobble) is buried in finer sediments. The embeddedness measure allows evaluation of the substrate as a habitat for benthic macroinvertibrates, spawning of fish, and egg incubation. Gravel, cobble, and boulder particles between 0 and 25% fraction surrounded by fine sediments are excellent habitat conditions; gravel, cobble, and boulder particles with greater than 75% fraction surrounded by fine sediments are poor habitat conditions.

In-Stream Habitat   The most common channel-alteration activities are channelization, impounding for navigation and electric energy production, channel straightening, reduction of flow by withdrawals, removal of bank vegetation, and building of vertical embankments and flood walls. The impact of these alterations range from minor to complete destruction of habitat. Channel alteration that causes little or no enlargement of islands or point bars are best for maintaining habitat; channel alterations that cause heavy deposits of fine material, increased bar development, and the filling of most pools with silt have the greatest (negative) impact on habitat. Quantitatively, channel modifications that cause less than 5% of the channel bottom to be affected by scouring and deposition have minimal impact; modifications that cause more than 50% of the channel bottom to be affected and where only large rocks or riffle are exposed have significant impact. Channel alterations that lead to unstable side slopes (>60%) or increased erosion will clearly have negative impacts on in-stream habitat. An example of severe stream-channel erosion is illustrated in Figure 2.3.

Riparian Habitat   In most humid regions of the eastern and southern United States, natural riparian areas consist mostly of forests. In the more arid western parts of the United States, natural riparian areas consist of narrow strips of stream bank vegetation made up of grassland, brush, and other nonforested ecosystems. Modification of these natural riparian ecosystems typically has negative impacts on stream water quality. Forest riparian buffers provide shade that keep stream temperatures low; filter, absorb, and adsorb pollutants; provide an area for sediment deposition; promote microbial decomposition of organic matter and nutrients; minimize or prevent stream bank erosion; provide terrestrial,

FIGURE 2.4 Riparian habitat. (From State of California, 2005b.)

stream bank, and aquatic habitat and species biodiversity; open wildlife cooridors; provide infiltration, which replenishes ground water and cool stream base flow; and provide base-flow attenuation. The preserved riparian area adjacent to the San Joaquin River is shown in Figure 2.4. This extremely valuable riparian habitat passes though miles of farmland in the Central Valley of California.

Reduction or elimination of woods and brush vegetation eliminates wildlife habitat, canopy cover, and shade. Reduction or elimination of shading by stream bank vegetation reduces water quality by increasing sun-energy input, which increases water temperature. Cooler streams contain more oxygen, providing better support for aquatic life. Unshaded streams, partly because of an increase in sunlight and increased stream temperature, promote undesirable filamentous algae, whereas shaded streams support the advantageous diatomatious algae.

Excellent conditions exist when over 80% of the stream bank is covered by vegetation or boulders or cobble; poor habitat conditions exist when less than 25% of the stream bank is covered with vegetation, gravel, or larger material. Shrubs provide excellent stream bank cover. Poor conditions exist when more than 50% of the stream bank has no vegetation and the dominant material is soil or rock.

The reduction or elimination of riparian wetlands reduces habitat for aquatic and terrestrial organisms and deprives the stream of buffering capacity for diffuse pollutant loads from surrounding lands. This can adversely affect the diversity and species composition in streams and other surface waters since riparian wetlands provide cover and shelter for fish and other organisms.

As streams increase in size, the integrated effects of adjacent riparian ecosystems should decrease relative to the overall water quality of the stream. Higher-order streams are more influenced by land use within a watershed than by the riparian buffer conditions. Conversely, first-order streams, or smaller intermittent streams, have little up-gradient contributing drainage area and short contributing flow paths; therefore, the condition of the riparian buffer may have a significant impact on the water quality of the stream.

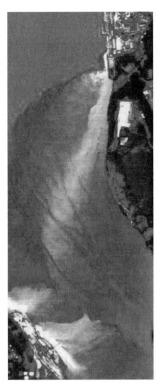

FIGURE 2.5   Thermal plumes from Indian Point and Lovett into the Hudson River. (From Geophysical and Environmental Research Corp., 1988.)

Thermal Pollution   The primary source of thermal pollution is waste heat from nuclear and fossil-fuel electric power plants. Commonly, such discharges are about 10°C above the temperature of the receiving water (Clark, 1997). Approximately half of the fuel energy used by a typical power plant is dissipated as waste heat to waterways, usually to an adjacent water body. Many fish species (e.g., salmon) are extremely sensitive to temperature and cannot adjust readily to warmer waters. Conversely, some fish species thrive in warmer waters near power plants and can be severely harmed by a sudden drop in temperature that usually occurs when a plant shuts down for scheduled maintenance or an unscheduled outage. A hyperspectral image showing thermal plumes from the cities of Indian Point and Lovett (New York) into the Hudson River are shown in Figure 2.5.

In coastal power stations that discharge to temperate seas, heated discharges are generally of little consequence, but in tropical seas, where summer temperatures are already near the thermal death point of many organisms, the increase in temperature can cause substantial loss of life. Most modern power plants are required to install cooling towers that release waste heat to the atmosphere rather than to water bodies.

2.2.2   Chemical Measures

Regulated chemical compounds or combinations of compounds are those that are considered to be toxic to human and aquatic life and that occur or have the potential to occur in the water environment at harmful levels. In the United States, such toxic pollutants are collectively referred to as *priority pollutants*, and the Clean Water Act defines these pollutants as those that after discharge and upon exposure, ingestion, inhalation, or assimilation into any organism, either directly from the environment or indirectly by ingestion through food chains, will cause death, disease, behavioral abnormalities, cancer, genetic mutation, physiological malfunctions (including malfunctions in reproduction), or physical deformations in such organisms or their offspring. The current list of priority pollutants, given in Appendix C.3, contains 129 pollutants, of which 13 are toxic metals and the remaining are mostly organic chemicals. Even though almost all water-quality constituents may become toxic at high enough levels, priority pollutants are either toxic at relatively low levels or at levels that may result from waste discharges. The chemical measures that are included in water-quality standards include priority pollutants plus additional chemical measures that describe the ambient environment, such as dissolved oxygen. Key chemical measures of water quality are described in the following sections.

Dissolved Oxygen Dissolved oxygen (DO) is the amount of molecular oxygen dissolved in water and is one of the most important parameters affecting the health of aquatic ecosystems, fish mortality, odors, and other aesthetic qualities of surface waters. Discharges of oxidizable organic substances into water bodies result in the consumption of oxygen and the depression of dissolved-oxygen levels. If dissolved-oxygen levels fall too low, the effects on fish can range from a reduction in reproductive capacity to suffocation and death. Larvae and juvenile fish are especially sensitive and require higher levels of dissolved oxygen than those required by more mature fish. Oxygen depletion at the lower depths of lakes and reservoirs create reducing conditions in which iron and manganese can be solubilized, and taste and odor problems may also increase because of the release of anoxic and/or anaerobic decay products such as hydrogen sulfide. Nutrient enrichment in surface waters is often signaled by excessive oxygen production, leading to supersaturation in some cases, and by hypoxia or anoxia in deep waters where excessive plant production is consumed.

Saturation levels of DO decrease with increasing temperature, as illustrated in Table 2.3, for a standard atmospheric pressure of 101 kPa. One of the most commonly used empirical equations for estimating the saturation concentration of dissolved oxygen, DO_{sat}, is given by the American Public Health Association (APHA, 1992):

$$\ln DO_{sat} = -139.34411 + \frac{1.575701 \times 10^5}{T_a} - \frac{6.642308 \times 10^7}{T_a^2}$$
$$+ \frac{1.243800 \times 10^{10}}{T_a^3} - \frac{8.621949 \times 10^{11}}{T_a^4} \tag{2.1}$$

where T_a is the absolute temperature (K) of the water. Equation 2.1 is sometimes referred to as the *Benson–Krause equation*. A more compact equation recommended by the U.S. Environmental Protection Agency (USEPA, 1995) is given by

$$\boxed{DO_{sat} = \frac{468}{31.5 + T}} \tag{2.2}$$

where T is the water temperature in °C. Equation 2.2 is accurate to within 0.03 mg/L, as compared with Equation 2.1, on which the values given in Table 2.3 are based (McCutcheon, 1985). The saturation concentration of oxygen in water is affected by the presence of chlorides (salt), which reduce the saturation concentration by about

TABLE 2.3 Saturation of Dissolved Oxygen in Water

Temperature (°C)	Dissolved Oxygen (mg/L)
0	14.6
5	12.8
10	11.3
15	10.1
20	9.1
25	8.2
30	7.5
35	6.9

0.015 mg/L per 100 mg/L chloride at low temperatures (5 to 10°C) and by about 0.008 mg/L per 100 mg/L chloride at higher temperatures (20 to 30°C) (Tebbutt, 1998). The following equation is recommended to account for the effect of salinity on the saturation concentration of dissolved oxygen (APHA, 1992):

$$\ln DO_s = \ln DO_{sat} - S \left(1.764 \times 10^{-2} - \frac{10.754}{T_a} + \frac{2140.7}{T_a^2} \right) \tag{2.3}$$

where DO_s is the saturated dissolved-oxygen concentration at salinity S, where S is in parts per thousand (ppt). For high-elevation streams and lakes, the barometric-pressure effect is important, and the following equation is used to quantify the pressure effect on the saturated dissolved-oxygen concentration:

$$DO_P = DO_{sat} P \frac{[1 - (P_{wv}/P)](1 - \theta P)}{(1 - P_{wv})(1 - \theta)} \tag{2.4}$$

where DO_P is the saturated dissolved-oxygen concentration at pressure P (atm), P_{wv} is the partial pressure of water vapor (atm), which can be estimated using the relation (Lung, 2001)

$$\ln P_{wv} = 11.8671 - \frac{3840.70}{T} - \frac{216,961}{T^2} \tag{2.5}$$

where T is the temperature in °C and θ is given by

$$\theta = 0.000975 - 1.426 \times 10^{-5} T + 6.436 \times 10^{-8} T^2 \tag{2.6}$$

Cool waters typically contain higher levels of dissolved oxygen, and consequently, aquatic life in streams and lakes is usually under more oxygen stress during the warm summer months than during the cool winter months. The minimum dissolved oxygen level needed to support a diverse aquatic ecosystem is typically on the order of 5 mg/L. An illustration of dissolved-oxygen fluctuations in the Neuse River at marker 38 in October 1999 is shown in Figure 2.6.

Example 2.1 (a) Compare the saturation concentration of dissolved oxygen in freshwater at 20°C given by Equation 2.1 to the value given in Table 2.3. (b) How do these values compare

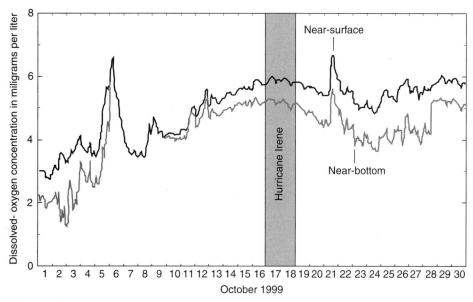

FIGURE 2.6 Dissolved-oxygen concentration in the Neuse River at marker 38, October 1999. (From Bales et al., 2000.)

with the saturation concentration given by Equation 2.2? (c) What would be the effect on the saturation concentration of dissolved oxygen if saltwater intrusion causes the chloride concentration to increase from 0 mg/L to 2500 mg/L? (d) Compare the saturation concentration of dissolved oxygen in freshwater at 20°C in Miami, where atmospheric pressure is 101 kPa, with the saturation concentration in Denver, where atmospheric pressure is 83.4 kPa.

SOLUTION (a) Equation 2.1 gives DO_{sat} in terms of the absolute temperature, T_a, where $T_a = 273.15 + 20 = 293.15$ K. Hence, Equation 2.1 gives

$$\ln DO_{sat} = -139.34411 + \frac{1.575701 \times 10^5}{T_a} - \frac{6.642308 \times 10^7}{T_a^2}$$

$$+ \frac{1.243800 \times 10^{10}}{T_a^3} - \frac{8.621949 \times 10^{11}}{T_a^4}$$

$$= -139.34411 + \frac{1.575701 \times 10^5}{293.15} - \frac{6.642308 \times 10^7}{(293.15)^2}$$

$$+ \frac{1.243800 \times 10^{10}}{(293.15)^3} - \frac{8.621949 \times 10^{11}}{(293.15)^4}$$

$$= 2.207$$

Therefore,

$$DO_{sat} = e^{2.207} = 9.1 \text{ mg/L}$$

This is the same value of DO_{sat} for freshwater given in Table 2.3.

(b) According to Equation 2.2,

$$DO_{sat} = \frac{468}{31.5 + T} = \frac{468}{31.5 + 20} = 9.1 \text{ mg/L}$$

This is the same value (9.1 mg/L) as that given in Table 2.3 and calculated using Equation 2.1. Since Equation 2.2 is supposed to agree with Equation 2.1 within 0.03 mg/L, the calculated result is expected.

(c) The impact of salinity on the saturation concentration of dissolved oxygen is given by Equation 2.3, and the relationship between chloride concentration, c, and salinity, S, in seawater is given by

$$S = 1.80655c \tag{2.7}$$

where S and c are in parts per thousand. In the present case, $c = 2500 \text{ mg/L} = 2.5 \text{ kg/m}^3 = 2.5/1000 = 0.00250 = 2.50 \text{ ppt}$, where the density of water is taken as 1000 kg/m^3. Applying Equation 2.7 to estimate the salinity gives

$$S = 1.80655(2.50) = 4.52 \text{ ppt}$$

and Equation 2.3 gives the corresponding dissolved oxygen as

$$\ln DO_S = \ln DO_{sat} - S\left(1.764 \times 10^{-2} - \frac{10.754}{T_a} + \frac{2140.7}{T_a^2}\right)$$

$$= \ln(9.1) - 4.52\left(1.764 \times 10^{-2} - \frac{10.754}{293.15} + \frac{2140.7}{293.15^2}\right)$$

$$= 2.18$$

which yields

$$DO_S = e^{2.18} = 8.8 \text{ mg/L}$$

Therefore, increasing the chloride concentration from 0 mg/L to 2500 mg/L reduces the saturation concentration of dissolved oxygen from 9.1 mg/L to 8.8 mg/L, a reduction of approximately 3%.

(d) The impact of atmospheric pressure on dissolved-oxygen concentration is given by Equation 2.4. In this case, $DO_{sat} = 9.1 \text{ mg/L}$, $P = 83.4 \text{ kPa} = 83.4/101.325 = 0.823 \text{ atm}$, and P_{wv} is given by Equation 2.5 as

$$\ln P_{wv} = 11.8671 - \frac{3840.70}{T} - \frac{216,961}{T^2}$$

$$= 11.8671 - \frac{3840.70}{20} - \frac{216,961}{20^2}$$

$$= -722.57$$

which yields

$$P_{wv} \approx 0 \, \text{atm}$$

and θ is given by Equation 2.6 as

$$\theta = 0.000975 - 1.426 \times 10^{-5}T + 6.436 \times 10^{-8}T^2$$

$$= 0.000975 - 1.426 \times 10^{-5}(20) + 6.436 \times 10^{-8}(20)^2$$

$$= 0.000716$$

Substituting into Equation 2.4 gives

$$DO_P = DO_{sat} \, P \, \frac{[1 - (P_{wv}/P)](1 - \theta P)}{(1 - P_{wv})(1 - \theta)}$$

$$= 9.1(0.823)\left[\frac{(1 - 0)(1 - 0.000716 \times 0.823)}{(1 - 0)(1 - 0.000716)}\right]$$

$$= 7.5 \, \text{mg/L}$$

This result indicates that the saturation concentration of dissolved oxygen decreases roughly in proportion to the atmospheric pressure. At 20°C, the saturation concentration in Denver (7.5 mg/L) is 18% less than in Miami (9.1 mg/L).

The levels of fish tolerance to low dissolved-oxygen stresses vary: Cold-water fish (e.g., salmon, trout) and biota require higher dissolved-oxygen concentration than do warm-water fish and biota. For example, brook trout may require about 7.5 mg/L of dissolved oxygen, whereas carp can survive at 3 mg/L. As a rule, the more desirable commercial and game fish require higher levels of dissolved oxygen. Federal water-quality criteria recommend 1-day (average) minimum dissolved oxygen concentrations of 8.0 mg/L for early life stages and 4 mg/L for other life stages of cold-water fish; corresponding concentrations are 5.0 and 3.0 mg/L, respectively, for warm-water fish (USEPA, 1986). Dissolved-oxygen criteria for saltwater coastal waters set the 1-day minimum dissolved oxygen for juveniles and adults at 2.3 mg/L (USEPA, 2000c).

Biochemical Oxygen Demand   Bacterial degradation results in the oxidation of organic molecules to stable inorganic compounds. *Biochemical oxygen demand* (BOD) is the amount of oxygen required to biochemically oxidize organic matter present in water. Aerobic bacteria that are responsible for the BOD make use of dissolved oxygen in reactions similar to the following involving glucose ($C_6H_{12}O_6$):

$$C_6H_{12}O_6 + 6O_2 \rightarrow 6H_2O + 6CO_2 \tag{2.8}$$

According to this equation, 6 moles of oxygen would be consumed for every mole of glucose. Waste discharges that contain significant amounts of biodegradable organic matter have high BOD levels and consume significant amounts of oxygen from the receiving waters, thereby reducing the level of dissolved oxygen and producing adverse impacts on aquatic life. If the organic matter is protinaceous, nitrogen and phosphorus are also released

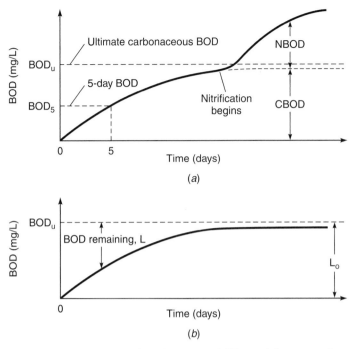

FIGURE 2.7   (*a*) Typical BOD curve; (*b*) carbonaceous BOD remaining versus time. (From Chin, David A., *Water-Resources Engineering.* Copyright © 2000. Reprinted by permission of Pearson Education, Inc., Upper Saddle River, NJ.)

as a result of the decomposition process. Biodegradable organic wastes commonly associated with oxygen consumption in surface waters include human and animal excrement, food wastes, and organic residuals from industrial operations such as paper mills and food-processing plants. The BOD measures the mass of oxygen consumed per unit volume of water and is usually given in mg/L. The wastewater from industrial operations such as pulp mills, sugar refineries, and some food-processing plants may easily have a (5-day) BOD as high as several thousand milligrams per liter. In contrast, raw sewage typically has a (5-day) BOD of about 200 mg/L. A typical BOD curve is illustrated in Figure 2.7(*a*), where the BOD is composed of the carbonaceous BOD (CBOD) and the nitrogenous BOD (NBOD). The CBOD is exerted by heterotrophic organisms that derive their energy for oxidation from an organic-carbon substrate, and NBOD is exerted by nitrifying bacteria that oxidize nitrogenous compounds in the wastewater. The carbonaceous demand is usually exerted first, with a lag in the growth of nitrifying bacteria. Normally, nitrogenous oxidation of raw sewage effluent is only important after 8 to 10 days of oxidation in the presence of excess oxygen; for treated effluents, however, nitrification may be important after a day or two, due to the large number of nitrifying bacteria in the effluent (Tebbutt, 1998).

BOD tests are conducted using 300-mL glass bottles in which a small sample of polluted water is mixed with (clean) oxygen-saturated water containing a phosphate buffer and inorganic nutrients. The mixture is incubated in a stoppered bottle in the dark at 20°C, and the dissolved oxygen in the mixture is measured as a function of time, usually for a minimum of 5 days. Since the sample is incubated in the dark there is no possibility for

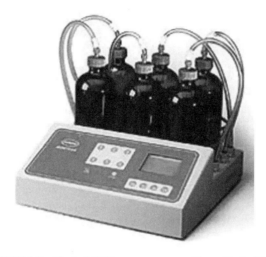

FIGURE 2.8 Typical BOD test apparatus. (From Hach, 2005.)

photosynthesis to occur, so the oxygen concentration must either remain constant or decline. A typical BOD test apparatus is shown in Figure 2.8. A decrease in the measured oxygen concentration may be due to the respiratory activity of organisms and/or the simple chemical oxidation of substances that are unstable in the presence of oxygen, such as ferrous iron or ammonia. Since both biological and chemical processes may cause a decline in oxygen concentration, BOD should be understood to refer to biochemical oxygen demand rather that simply biological oxygen demand. If a problem with nitrification is suspected in the BOD test, a specific nitrification inhibitor can be added to the water sample so that only the carbonaceous BOD is determined.

The cumulative oxygen demand of the polluted water after 5 days is called the 5-*day BOD* and is usually written as BOD_5. The kinetics of carbonaceous BOD (CBOD), illustrated in Figure 2.7(*b*), can be approximated by the following first-order model:

$$\frac{dL}{dt} = -k_1 L \qquad (2.9)$$

where L is the CBOD remaining at time t and k_1 is a rate constant. If L_0 is the CBOD remaining at time $t = 0$, equal to the *ultimate CBOD*, Equation 2.9 can be solved to yield

$$L = L_0 e^{-k_1 t} \qquad (2.10)$$

Since the CBOD at time t is related to L by

$$CBOD = L_0 - L \qquad (2.11)$$

the CBOD as a function of time is given by combining Equations 2.10 and 2.11 to yield

$$\boxed{CBOD = L_0(1 - e^{-k_1 t})} \qquad (2.12)$$

The ultimate CBOD, L_0, can be expressed in terms of the 5-day CBOD, $CBOD_5$, as

$$L_0 = \frac{CBOD_5}{1 - e^{-5k_1}} \qquad (2.13)$$

where both $CBOD_5$ and k_1 are derived from the BOD test data. For secondary-treated municipal wastewaters, k_1 is typically in the range 0.1 to 0.3 day^{-1} at 20°C, which gives a $L_0/CBOD_5$ ratio of approximately 1.6. Schnoor (1996) suggests a value of 1.47 for $L_0/CBOD_5$ in municipal wastewater, and data reported by Lung (2001) indicate that 2.8 may be more typical. On average, biological oxidation is complete in about 60 to 70 days for most domestic wastewaters (Lung, 2001), although little additional oxygen depletion occurs after about 20 days (Vesilind and Morgan, 2004). Municipal-wastewater discharge permits generally allow a maximum $CBOD_5$ of 30 mg/L, and it is recommended that communities discharging treated domestic wastewater into lakes or pristine streams reduce their $CBOD_5$ to below 10 mg/L to protect the indigenous aquatic life (Serrano, 1997).

It is interesting to note that the 5-day BOD was originally chosen as the standard value because the BOD test was devised by sanitary engineers in England, where the River Thames has a travel time to the sea of less than 5 days, so there was no need to consider oxygen demand at longer times.

Example 2.2   The results of a BOD test on secondary-treated sewage give a 5-day BOD of 25 mg/L and a rate constant of 0.2 day^{-1}. (a) Estimate the ultimate carbonaceous BOD and the time required for 90% of the carbonaceous BOD to be exerted. (b) If the ultimate nitrogenous BOD is 20% of the ultimate carbonaceous BOD, estimate the oxygen requirement per cubic meter of wastewater.

SOLUTION   (a) From the data given, $BOD_5 = 25$ mg/L and $k_1 = 0.2$ day^{-1}; hence, the ultimate carbonaceous BOD, L_0, is given by Equation 2.13 as

$$L_0 = \frac{BOD_5}{1 - e^{-5k_1}} = \frac{25}{1 - e^{-5(0.2)}} = 39.5 \text{ mg/L}$$

Letting t^* be the time for the BOD to reach 90% of its ultimate value, Equation 2.12 gives

$$0.9(39.5) = 39.5(1 - e^{-0.2t^*})$$

which gives

$$t^* = 11.6 \text{ days}$$

(b) Since the ultimate nitrogenous BOD is 20% of the ultimate carbonaceous BOD, the ultimate BOD is given by

$$\text{ultimate BOD} = 1.2L_0 = 1.2(39.5) = 47.4 \text{ mg/L}$$

and for $1 \text{ m}^3 = 1000$ L of wastewater, the ultimate mass of oxygen consumed by biochemical reactions is

$$\text{mass of oxygen} = 47.4(1000) = 47{,}400 \text{ mg/m}^3 = 47.4 \text{ g/m}^3$$

If the wastewater is discharged into a surface water, this oxygen will be taken from the ambient water.

If the oxygen concentration falls below about 1.5 mg/L, the rate of aerobic oxidation is reduced (Clark, 1997). In cases where adequate amounts of dissolved oxygen are not available, anaerobic bacteria can oxidize organic molecules without the use of dissolved oxygen, but the end products include compounds such as hydrogen sulfide (H_2S), ammonia (NH_4), and methane (CH_4) which are toxic to many organisms. The process of anaerobic decomposition is much slower than that of aerobic decomposition.

Waste discharges from nonpoint (diffuse) sources rarely cause significant reduction of dissolved oxygen in receiving streams. Exceptions to this include runoff with high concentrations of biodegradable organics from concentrated animal feeding operations (CAFOs) and spring runoff from fields with manure spread on still-frozen soils. River water with BOD_5 less than 2 mg/L can be regarded as unpolluted, and river water with a BOD_5 greater than 10 mg/L is grossly polluted (Hino and Matsuo, 1994; Clark, 1997). Water for salmon or trout should have a BOD_5 value below 3 mg/L, for coarse fish (i.e., other than salmonids) less than 6 mg/L, and drinking water sources may have a value up to 7 mg/L.

Chemical Oxygen Demand   The *chemical oxygen demand* (COD) is the amount of oxygen consumed when the substance in water is oxidized by a strong chemical oxidant. The COD is measured by refluxing a water sample in a mixture of chromic and sulfuric acid for a period of 2 hours. This oxidation procedure almost always results in a larger oxygen consumption than the standard BOD test since many organic substances that are not immediately available as food to aquatic microbes (e.g., cellulose) are readily oxidized by a boiling mixture of chromic and sulfuric acid. Domestic wastewaters typically have a BOD_5/COD ratio of 0.4 to 0.5 (Metcalf & Eddy, 1989). Comparison of BOD and COD results can help identify the occurrence of toxic conditions in a waste stream or indicate the presence of biologically resistant (refractory) wastes. For example, a BOD_5/COD ratio approaching 1 indicates a highly biodegradable waste; a ratio approaching zero suggests a poorly biodegradable material.

Suspended Solids   Suspended solids (SS) is the amount of suspended matter in water. SS is measured in the laboratory by filtering a known volume of water through a 1.2-μm microfiber filter, drying the filter at 105°C, and calculating the SS value by dividing the mass of solids retained on the filter by the volume of water filtered. The concentration of particles in the water that is passed through the 1.2-μm filter is called the *total dissolved solids* (TDS). Particles ranging in size from about 10^{-3} to 1.2 μm are classified as *colloidal solids*. The suspended-solids value is normally expressed in mg/L, and SS concentrations are usually quite high in surface runoff. A high level of SS produces a turbid receiving water, blocks sunlight needed by aquatic vegetation, and clogs the gills of fish. A water plume with high suspended solids is shown in Figure 2.9. The sedimentation of suspended solids in receiving waters can cause a buildup of organic matter in the sediments, leading to an oxygen-demanding sludge deposit. This deposit can also adversely affect fish populations by reducing their growth rate and resistance to disease, preventing the development of eggs and larvae, and reducing the amount of food available on the bottom of the water body. Land erosion from human activities such as mining, construction, logging, and farming is the major cause of suspended sediment in surface runoff.

In the United States, municipal wastewater treatment plants produce effluents with SS values of less than 30 mg/L. It is recommended that communities that discharge treated

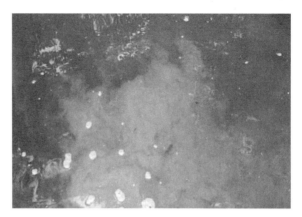

FIGURE 2.9   Water plume with high suspended solids. (From State of North Carolina, 2005b.)

domestic wastewater into lakes or pristine streams reduce their SS values to below 10 mg/L to protect the aquatic life in the receiving water (Serrano, 1997).

Example 2.3   An outfall discharges wastewater into a flood-control lake that is approximately 300 m long, 100 m wide, and 20 m deep. The suspended-solids concentration in the wastewater is 30 mg/L, the wastewater discharge rate is 0.05 m³/s, and the bulk density of the settled solids is 1600 kg/m³. Assuming that all of the suspended solids ultimately settle out in the lake, estimate the time required for 1 cm of sediment to accumulate at the bottom of the lake.

SOLUTION   The SS concentration in the wastewater is $30 \text{ mg/L} = 0.03 \text{ kg/m}^3$, and the discharge flow rate is 0.05 m³/s. Under steady-state conditions, the rate at which suspended solids are discharged into the lake is equal to the rate of sediment accumulation at the bottom of the lake and is given by

$$\text{sediment mass accumulation rate} = 0.03(0.05) = 0.0015 \text{ kg/s}$$

Since the bulk density of the settled solids is 1600 kg/m³, the rate of volume accumulation is

$$\text{sediment volume accumulation rate} = \frac{0.0015}{1600} = 9.38 \times 10^{-7} \text{ m}^3/\text{s}$$

For 1 cm ($= 0.01$ m) to cover the bottom of the 300 m $\times$ 100 m lake, the volume of sediment is $0.01(300)(100) = 300 \text{ m}^3$. Hence, the time required for 300 m³ of sediment to accumulate is given by

$$\text{accumulation time for 1 cm of sediment} = \frac{300}{9.38 \times 10^{-7}}$$

$$= 3.20 \times 10^8 \text{ s} = 3700 \text{ days} = 10.1 \text{ years}$$

It is interesting to note that at this rate of sediment accumulation, the lake will be filled completely in approximately 20,000 years.

Nutrients *Nutrients* are those substances that provide living organisms with the essential elements to sustain growth and life function. Of the approximately 100 elements in the periodic table, about 30 are constituents of living things and can be broadly classified as nutrients. Some of these nutrients are required in relatively large amounts and are termed *macronutrients*, whereas others are needed in only trace quantities and are called *micronutrients*. Despite the fact that some elements are required only in trace quantities, their availability may control the productivity of the entire ecosystem. Included in this group of potentially limiting elements are nitrogen and phosphorus. Both nitrogen and phosphorus are widely used in fertilizers and phosphorus-based household detergents, are commonly found in food-processing wastes and animal and human excrement, and are most responsible for the overenrichment of nutrients in surface waters (Rubin, 2001). Nutrients in fertilizers tend to bind to clay and humus particles in soils and are therefore easily transported to surface waters through erosion and runoff. In addition, fertilizers tend to leach into waterways from the soils of farmlands, lawns, and gardens. Other significant sources of nutrients include malfunctioning septic systems and effluents from sewage treatment plants. Excessive plant growth associated with overenrichment of nutrients causes oxygen depletion, which causes increased stress on aquatic organisms such as fish. An algae bloom fed by nutrient-rich runoff from the coast of Norway is illustrated in Figure 2.10. In most cases, phosphorus is the limiting nutrient in freshwater aquatic systems, such as lakes. In contrast to freshwater systems, nitrogen is mostly the limiting nutrient in the seaward portions of estuarine systems.

Nitrogen Nitrogen is a nutrient that stimulates the growth of algae, and the oxidation of nitrogen species can consume significant amounts of oxygen. There are several forms of nitrogen that can exist in water bodies, including organic nitrogen (e.g., proteins, amino acids,

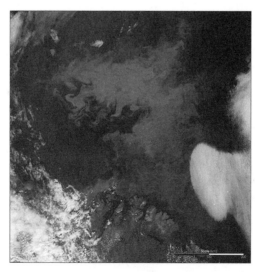

FIGURE 2.10 Algae bloom caused by nutrient-rich runoff from the coast of Norway. (From NASA, 2005c.)

urea), ionized and nonionized ammonia (NH_4^+ and NH_3), nitrite ion (NO_2^-), nitrate ion (NO_3^-), and dissolved nitrogen gas (N_2). For water in contact with the atmosphere, the most fully oxidized state of nitrogen is $+5$ and oxidation of nitrogen compounds proceeds as follows:

$$\text{organic nitrogen} + O_2 \rightarrow NH_3 \text{ nitrogen} + O_2 \rightarrow NO_2 \text{ nitrogen} + O_2 \rightarrow NO_3 \text{ nitrogen}$$

$$(2.14)$$

In aquatic environments, decomposers break down organic nitrogen to release ammonia, a process called *ammonification* or *deamination*, and ammonia (NH_3) is transformed to NO_3 nitrogen in a process called *nitrification*. The nitrification process is a result of the action of the *nitrosomas* and *nitrobacter* bacteria.

Ammonification can occur in sediments, water, and soils. Depending on the pH of the water, dissolved ammonia gas (NH_3) and ammonium ions (NH_4^+) will exist in an equilibrium according to the relation

$$NH_4^+ + OH^- \rightleftharpoons NH_3 + H_2O \qquad (2.15)$$

At pH 7 or below, most of the ammonia nitrogen will be ionized as ammonium, while at pH levels greater than 9 the proportions of nonionized ammonia will increase. The nonionized ammonia is toxic to fish, while the ionized ammonium is a nutrient to algae and aquatic plants and also exerts dissolved-oxygen demand. The nonionized ammonia is a gas that will mostly volatilize from water, and water-quality standards regulate the total ammonia nitrogen ($NH_4^+ + NH_3$). At normal pH values, ammonia-nitrogen occurs in the ammonium form (NH_4^+), and because of the positive charge, it is readily adsorbed by negatively charged (organic and clay) soil particles.

Ammonium ions are converted to nitrite by *Nitrosomonas* bacteria, and nitrite is converted to nitrate by *Nitrobacter* bacteria according to the following reactions:

$$4NH_4^+ + 6O_2 \rightarrow 4NO_2^- + 8H^+ + 4H_2O \qquad (2.16)$$

$$4NO_2^- + 2O_2 \rightarrow 4NO_3^- \qquad (2.17)$$

The combined reaction, called *nitrification*, can be written in the form

$$NH_4^+ + 2O_2 \rightarrow NO_3^- + 2H^+ + H_2O \qquad (2.18)$$

Stoichiometrically, the oxygen requirement for the overall nitrification reaction (Equation 2.18) is 4.56 mg of O_2 per milligram of NH_4^+. However, since the reaction is autotrophic, oxygen is also produced as a result of bacterial growth, and the overall oxygen requirement for nitrification is less than the stoichiometric value. Wezernak and Gannon (1968) measured the O_2 requirement in nitrification as about 4.3 mg of O_2 per milligram of NH_4^+. The following environmental factors affect the nitrification rate (Krenkel and Novotny, 1980; Novotny, 2003):

- The reaction is strictly aerobic. If the oxygen concentration is depleted below 2 mg/L, the reaction rate decreases rapidly.
- The optimum pH range is between 8 and 9. Below a pH of 6, the reaction essentially ceases.

- The nitrifying organisms tend to reside on either sediments or solid surfaces. Hence, the nitrification process tends to occur near the sediment–water interface.
- The growth rate of nitrifying organisms is much less than the growth rate of heterotrophic decomposers. Consequently, if higher concentrations of biodegradable organics are present, the heterotrophs will limit the growth of nitrifiers and nitrification will be suppressed.
- The optimal temperature range is between 20 and 25°C. The rate of the nitrification process decreases if the temperature is less than the optimum.

Plants take up and utilize nitrogen in the form of ammonia or nitrate, which are typically in short supply in agricultural soils, thus leading to requirements for fertilization. Nitrate-nitrogen commonly originates in runoff from agricultural areas with heavy fertilizer usage, whereas organic nitrogen is commonly found in municipal wastewaters. Nitrite-nitrogen does not typically occur in natural waters at significant levels, and its presence would indicate wastewater contamination and/or lack or oxidizing conditions (AWWA, 1990). Excessive nitrate in ground water is of concern if the aquifer is to be used as a drinking-water supply, since nitrate can pose a health threat to infants by interfering with oxygen transfer in the bloodstream.

Under anoxic conditions the nitrate-nitrogen ion becomes the electron acceptor in the organic matter oxidation reaction. This reaction, called *denitrification*, can be represented as (Schindler, 1985; Davis and Masten, 2004)

$$5CH_2O + 4NO_3^- + 4H^+ \rightarrow 5CO_2 + 7H_2O + 2N_2 \tag{2.19}$$

where nCH_2O represents a form of organic carbon, and several forms of organic carbon (e.g., dissolved methane from anaerobic decomposition in sediments) may serve as the source of energy in this reaction. The denitrification process described by Equation 2.19 represents a loss of nitrogen from the water since the nitrogen gas produced volatilizes into the air. A few planktonic algal organisms are capable of fixing N_2-nitrogen in the photosynthetic reaction.

Some environments that contain biodegradable materials lack both oxygen and nitrate to serve as an oxidizing agent, and in such cases, sulfate may serve that role. This occurs in the following reaction where the conversion of one sulfur atom from $+6$ in sulfate (SO_4^{2-}) to -2 in hydrogen sulfide (H_2S) oxidizes two carbon atoms from 0 to $+4$:

$$2CH_2O + 2H^+ + SO_4^{2-} \rightarrow 2CO_2 + 2H_2O + H_2S \tag{2.20}$$

This reaction frequently occurs in stagnant anaerobic marine sediments that are supplied with decaying biomass: for example, from algae or seaweed accumulation. Sulfate is plentiful as a dissolved ion in seawater, and the hydrogen sulfide produced by this reaction may be released as a gas, producing an offensive odor downwind. In the absence of oxygen, nitrate, and sulfate, biomass can still be converted to carbon dioxide, as in the reaction

$$2\,CH_2O \rightarrow CO_2 + CH_4 \tag{2.21}$$

This (anaerobic) biodegradation process, which commonly occurs in anoxic sediments, is generally called *methanogenesis*, since it generates methane (CH_4).

Conditions at the sediment–water interface can have a major impact on the nitrogen balance in aquatic systems. Unless the water at the bottom of a water body is completely anoxic, a thin aerobic sediment layer will form. In the aerobic sediment layer, nitrification should occur. Because the gradient of NO_3^- is downward to the sediment (zero nitrate concentration), the NO_3^- thus formed will diffuse back into the anaerobic sediment, where denitrification is certain (Keeney, 1973). This is referred to as the *nitrification–denitrification process*, by which nitrogen is lost from the system. This process can occur only at the sediment–water interface, and nitrification in free-flowing water is relatively rare. Experiments by Keeney et al. (1975) have shown that about one-third of nitrogen input to a lake may be lost from the water body by the nitrification–denitrification process, and Kadlec (1988) and others have documented similar nitrogen losses for wetlands.

Nitrates tend to travel very easily in ground water and commonly originate from the nitrification of wastewater discharges near the ground surface. This wastewater originates from such sources as septic tanks, livestock, feedlots, waste stabilization ponds or lagoons, waste treatment plant effluents, leaking sewers, and spray irrigation. Nitrogen deposited in the soil from wastewater application occurs largely as ammonium (NH_4^+), which is adsorbed by negatively charged clay particles as it percolates through the ground until negatively charged sites are saturated. In the presence of oxygen, ammonium will be converted to nitrate (NO_3), which will flow freely into the ground water causing a potential hazard. At pH values below 7, nitrates may be adsorbed, whereas at neutral pH, nitrates will not be adsorbed. Once in the saturated zone, nitrates will be reduced to gaseous nitrogen (denitrification) at a minimal rate. Elevated nitrate levels in drinking water cause *methemoglobinemia* or "blue baby" in infants. This sickness is caused by nitrates interacting with the hemoglobin in red blood cells, forming methemoglobin, which unlike hemoglobin, cannot carry sufficient oxygen to the body's cells and tissues. This causes babies to turn blue. Whether nitrates constitute a problem in ground water depends greatly on their source. For example, ammonia-nitrogen seeping through the bottom of a waste stabilization pond would remain unoxidized because oxygen would be at a minimum, whereas nitrates would become abundant in ground water underlying a septic tank tile field (assuming a neutral pH) because the flow is near the surface.

Phosphorus Phosphorus-bearing minerals have low solubility, and thus most surface waters naturally contain very little phosphorus. Phosphorus is normally present in watersheds in extremely small amounts and commonly exists in water as phosphates and organophosphates originating from wastewater discharges, household detergents, and agricultural runoff associated with fertilizer application and concentrated livestock operations. Untreated domestic wastewater contains 5 to 15 mg/L of phosphorus (in both organic and inorganic forms), concentrations more than two orders of magnitude greater than those desired in healthy surface waters (< 0.02 mg/L). Thus, significant phosphorus removal has become required as part of the wastewater-treatment process.

Orthophosphates are salts of phosphoric acid (H_3PO_4) and the only significant form of phosphorus available to plants and algae is the soluble reactive inorganic orthophosphate species ($H_2PO_4^-$, HPO_4^{2-}, PO_4^{3-}) that are incorporated into organic compounds. For water in contact with the atmosphere, the most fully oxidized state of phosphorus is $+5$, and phosphorus in the form of phosphate ($H_2PO_4^-$) from fertilizer, detergents, and organic wastes becomes adsorbed to sediment, which is carried to streams during the erosion—sedimentation process.

Phosphorous is usually the limiting nutrient for the growth of algae in lakes. For lakes in the northern United States to be free of algal nuisances, the generally accepted upper concentration limit is 10 μg/L (0.01 mg/L) of orthophosphate-phosphorus.

Metals Because of the significant (negative) effects that certain toxic metals can have on human health, metal pollution is potentially one of the most serious forms of aquatic pollution. Metals are introduced into aquatic systems by many processes, including the weathering of soils and rocks, atmospheric deposition, volcanic eruptions, and a variety of human activities involving mining, industrial use, and exhaust and tire deposition from automobiles. Metal-rich drainage from mining activities in the Animas River catchment is shown in Figure 2.11. Urban runoff is a major source of zinc (originating from tire wear) in many water bodies, and metals tend to accumulate in bottom sediments. Toxic metals of concern in water bodies include arsenic (Ar), cadmium (Cd), copper (Cu), chromium (Cr), lead (Pb), mercury (Hg), nickel (Ni), and zinc (Zn). At toxic levels, most metals adversely affect the internal organs of the human body. Specific concerns with several metals are described below.

Arsenic Arsenic (Ar) is a naturally occurring element in the environment and its occurrence in ground water is largely the result of minerals dissolving from weathered rocks and soils. High arsenic concentrations in ground water have been documented in parts of Maine, Michigan, South Dakota, Oklahoma, and Wisconsin, where arsenic concentrations exceeding 10 mg/L have been observed. The maximum contaminant limit for arsenic in drinking water is 10 μg/L.

Cadmium Cadmium (Cd) is widely used in metal plating and is an active ingredient of rechargeable batteries. Cadmium causes high blood pressure and kidney damage and is a probable human carcinogen.

FIGURE 2.11 Metal drainage into the Animas River. (From USGS, 2005b.)

Chromium   Chromium (Cr) is a trace constituent in ordinary soils, a natural impurity in coal, and is widely used in the manufacture of stainless steel. Chromium exists in two oxidation states in the environment, $+3$ and $+4$. Cr^{3+} is an essential trace element in human diets, whereas Cr^{4+} causes a variety of adverse health effects, including liver and kidney damage, internal hemorrhage, respiratory disorders, and cancer.

Lead   Lead (Pb) was used extensively in several commercial products before its adverse health effects became well known. It was incorporated in pigments used in house paint and in glazes applied to dishware. Lead was also used in pipes and solder in water-distribution systems and in the gasoline additive tetraethyllead, $(C_2H_5)_4Pb$. Although a substantial decrease in human exposure to lead was achieved by eliminating it from gasoline, there is still a legacy of lead in paint and pipes of old houses and in land near heavily used roadways. A range of adverse health effects result from the accumulation of lead in the bloodstream, including anemia, kidney damage, elevated blood pressure, and central nervous system effects such as mental retardation. Infants and young children are especially susceptible to lead poisoning because they absorb ingested lead more readily than do older humans. Lead is a probable human carcinogen.

Mercury   Mercury (Hg) is a metal of particular concern in surface waters, where the biological magnification of mercury in freshwater food fish is a significant hazard to human health. A significant amount of mercury discharged into the environment is first emitted as an air pollutant, but the most damaging effects typically occur in lakes after the mercury moves through the atmosphere, is deposited into a lake, and then undergoes *methylation*, which is a process in which mercury is bound to a carbon molecule. This process resulted in approximately 1300 fishing advisories in lakes throughout the United States in 1995 (Mihelcic, 1999). Methyl mercury is an especially toxic form of mercury that affects the central nervous system. As far as is now known, human exposure to methyl mercury occurs only through the consumption of contaminated fish and seafood. Mercury is the only contaminant introduced by humans into the sea, apart from pathogens in sewage, that has been responsible for human deaths (Clark, 1997).

   As a general rule, metals of biological concern can be divided into three groups: light metals (e.g., Na, K, Ca) which are normally transported as mobile cations in aqueous solutions, transitional metals (e.g., Fe, Cu, Co, Mn) which are essential at low concentrations but may be toxic at high concentrations, and metalloids (e.g., Hg, Pb, Sn, Se, Ar) which are generally not required for metabolic activity and are toxic at low concentrations.

Synthetic Organic Chemicals   Synthetic organic chemicals (SOCs) include pesticides, PCBs, industrial solvents, petroleum hydrocarbons, surfactants, organometallic compounds, and phenols. Many of these organic substances are hazardous to humans in relatively small concentrations. A list of potentially harmful synthetic organic chemicals can number in the thousands or tens of thousands (Domenico and Schwartz, 1998). More than 1000 new chemicals are synthesized each year and about 60,000 chemicals are in daily use (Ramaswami et al., 2005). Complete toxicity and hazard information is available for only a small percentage of the synthetic chemicals produced by industry and consumed by society.
   Categories of synthetic organic compounds are given in Table 2.4. These pollutants can enter natural water bodies by a variety of mechanisms, such as runoff (pesticides), municipal and industrial discharges (PCBs), chlorination by-products in municipal wastewater

TABLE 2.4   Categories of Organic Priority Pollutants

Category	Description
Pesticides	Generally chlorinated hydrocarbons
Polyclorinated biphenyls (PCBs)	Used in electrical capacitors and transformers, paints, plastics, insecticides, other industrial products
Halogenated aliphatics (HAHs)	Used in fire extinguishers, refrigerants, propellants, pesticides, solvents for oils and greases, and in dry cleaning
Ethers	Used mainly as solvents for polymer plastics
Phthalate esters	Used chiefly in production of polyvinyl chloride and thermoplastics as plasticizers
Monocyclic aromatics (MAHs)	(Excluding phenols, cresols, and pthalates) Used in the manufacture of other chemicals, explosives, dyes, and pigments, and in solvents, fungicides, and herbicides
Phenols	Large-volume industrial compounds used chiefly as chemical intermediates in the production of synthetic polymers, dyestuffs, pigments, herbicides, and pesticides
Polycyclic aromatic hydrocarbons (PAHs)	Used as dyestuffs, chemical intermediates, pesticides, herbicides, motor oils, and fuels
Nitrosamines	Used in the production of organic chemicals and rubber

Source: Council on Environmental Quality (1978).

discharges (HAHs), and spills (PAHs). Pesticides are frequently found in ground and surface waters that receive runoff and infiltration from agricultural areas, and industrial solvents are common contaminants in ground water, resulting from both accidental and illicit spills. The cloud resulting from a chemical (herbicide) spill in the Sacramento River is shown in Figure 2.12. The bulk of SOCs that are of concern in drinking water are pesticides (AWWA, 2003).

When classified according to target species, the most common pesticides may be broadly defined as herbicides, insecticides, or fungicides, depending on whether they are designed to kill plants, insects, or fungi, respectively. These three categories of pesticides account for 57, 14, and 8% by weight, respectively, of conventional pesticide use in the United States (USEPA, 1999b). The two most widely used herbicides in the United States are atrazine and metolachlor, both of which are chlorinated organics. Pesticides such as chlordane and carbofuran are highly persistent in the environment since they do not readily break down in natural ecosystems and thus tend to accumulate in the tissue of organisms near the top of the food chain, such as birds and fish.

An important group of toxic organic compounds are classified as *volatile organic compounds* (VOCs) and include substances such as vinyl chloride, carbon tetrachloride, and trichloroethylene. Chemicals in this class are often used as industrial or household solvents and as ingredients in chemical manufacturing processes. Many of these volatile organic compounds are suspected or known hazards to the health of humans and aquatic ecosystems, and all of these compounds have chemical and physical properties that allow them to move freely between the water and air phases of the environment (Rathbun, 1998). The distinguishing characteristics of VOCs are low molecular weight, high vapor

FIGURE 2.12 Chemical (herbicide) spill in the Sacramento River. (From State of California, 2005a.)

pressure, and low-to-medium water solubility. Because they tend to evaporate easily, the concentration of VOCs in surface waters is typically much lower than that in ground water.

Radionuclides Radionuclides are elements with an unstable atomic nucleus. When radionuclides undergo radioactive decay, energy is released that can damage exposed tissue, and at high doses, radiation exposure can cause acute illness and even death. Most radioactivity in water is associated with natural causes, but there is also a threat of radionuclide contamination from various industrial and medical processes. A radioactive atom is unstable and seeks stability by emitting alpha particles, beta particles, and/or gamma rays. These types of radioactivity are described as follows:

- An alpha (α) particle consists of two protons and two neutrons and is equivalent to the nucleus of a helium ion, designated as ^{4}He. Since the radioactive nucleus loses two positive charges, the atom becomes that of an element two places lower in the periodic table. Alpha particles are relatively slow moving and lose their energy in a short distance; they are stopped by a few centimeters of air or only $40 \, \mu$m of tissue. They are, however, intensely ionizing in the matter through which they pass and can cause more damage to living tissue than particles with a longer path. Nuclei emitting α particles are of biological consequence if they are taken into the body, for example, by ingestion or inhalation. Alpha-particle emission occurs mainly in radioisotopes whose atomic number is greater than 82.

- A beta (β) particle has a mass identical to that of an electron and a charge of $+1$ (positron) or -1 (electron). Beta particles vary widely in their energy but lose most of it within a relatively short distance and can be screened by about 40 mm of tissue. Like α particles, their biological significance is greatest if a β emitter is taken into the body.
- Either alpha or beta particles may be accompanied by gamma (γ) radiation. Gamma rays consist of high-energy electromagnetic radiation and are highly penetrating and strongly ionizing. Living tissues need to be shielded from gamma radiation by a considerable thickness of heavy material such as lead or concrete to absorb the radiation.

The radioactive substances that are of concern as drinking-water contaminants are radium, uranium, radon, and artificial radionuclides (AWWA, 2003). These substances are described briefly as follows.

Radium   Naturally occurring radium is leached into ground water from rock formations, so it is present in water sources in areas where there is radium-bearing rock. It may also be found in surface water as a result of runoff from mining and industrial operations, where radium is present in the soil. The two isotopes of radium of concern in drinking water are radium 226, which emits primarily alpha particles, and radium 228, which emits beta particles and alpha particles from its daughter decay products. Radium is chemically similar to calcium, so about 90% of ingested naturally occurring radium goes to the bones. Consequently, the primary risk from radium ingestion is bone cancer.

Uranium   Naturally occurring uranium is found in some ground-water supplies as a result of leaching from uranium-bearing sandstone, shale, and other rock. Uranium may also occasionally be present in surface water, carried there in runoff from areas with mining operations. Uranium may be present in a variety of complex ionic forms, depending on the pH of the water. The primary adverse effect of uranium is toxicity to human kidneys.

Radon   Radon is a naturally occurring radioactive gas. It cannot be seen, smelled, or tasted. Radon comes from the radioactive decay of uranium and is the direct radioactive decay daughter of radium 226. The highest concentrations of radon are found in soil and rock containing uranium. Significant concentrations, from a health standpoint, may be found in ground water from any type of geologic formation, including unconsolidated formations. The problem with radon from a public water-supply viewpoint is that if radon is present in the water, a significant amount of the gas will be liberated into a building as the water is used. Showers, washing machines, and dishwashers are particularly efficient in transferring radon gas into the air. The radon released from the water adds to the radon that seeps into a building from the soil, adding to the health danger. Inhaled radon is considered to be a cause of lung cancer. The USEPA estimates that between 1 and 5 million homes in the United States may have significantly elevated levels of radon contamination and that between 5000 and 20,000 lung cancer deaths per year may be attributed to all sources of radon (AWWA, 2003).

Artificial Radionuclides   Significant artificial radionuclide levels have been recorded in surface waters as a result of atmospheric fallout following nuclear testing, leaks, and

FIGURE 2.13 Deployment of a radionuclide sampler in the ocean. (From Challenger Oceanic, 2005.)

disasters. Otherwise, surface water typically contains little or no radioactivity. If an accidental discharge of artificial radionuclides occurs from power plants, waste-disposal sites, or medical facilities, the elements most likely to be present are strontium 90 and tritium.

The presence of radionuclides is typically detected by selective deployment of radionuclide samplers, and the deployment of a radionuclide sampler in the ocean is illustrated in Figure 2.13. The effects of excessive levels of radioactivity on the human body include developmental problems, nonhereditary birth defects, genetic defects that might be inherited by future generations, and various types of cancer.

pH The pH of water is defined as the negative log of the hydrogen-ion activity (in mol/L), commonly written as

$$pH = -\log_{10}[H^+] \tag{2.22}$$

Values of pH below 7 are associated with *acidic* waters, while higher values of pH are associated with *basic* or *alkaline* waters. The pH of natural water bodies affect biological and chemical reactions, control the solubility of metal ions, and affect natural aquatic life. For example, the viability of fish species such as brook trout and lake trout is significantly diminished below a pH of 5.5, and most fish species cannot survive in waters with a pH below 5.0. The desirable pH for freshwater aquatic life is in the range 6.5 to 9.0, and 6.5 to 8.5 for marine aquatic life. Most natural waters have a pH near neutral, with 95% of all natural waters having a pH between 6 and 9 (Mihelcic, 1999).

Drainage and runoff from mining operations and discarded mine wastes have long been major sources of acid loadings on streams, lakes, and rivers. Coal mining, for example, releases sulfur-bearing minerals that form dilute sulfuric acid when contacted with process water used in the mining operation, or with rainfall that leaches acidic materials from waste piles and carries them to nearby surface waters (Rubin, 2001).

Acid deposition from the atmosphere directly into surface waters or onto surrounding lands is another important source of acid waters. The northeastern United States and Canada frequently record rainwater pH values between 4 and 5, and this acid rain has caused a complete eradication of some fish species, such as brook trout, in hundreds of lakes in the Adirondaks, the acidification of more than 1350 mid-Atlantic Highland streams, the acidification of approximately 580 streams in the mid-Atlantic Coastal Plain, and the acidification of more than 90% of streams in the New Jersey Pine Barrens (Davis and Masten, 2004). Most rainwater not affected by anthropogenic acid-rain emissions has a pH of approximately 5.6, due to the presence of dissolved carbon dioxide from the atmosphere (Mihelcic, 1999).

2.2.3 Biological Measures

In evaluating what levels of aquatic-life protection are attainable, the biology of a water body should be evaluated. Such evaluations are generally related to maintaining the biological integrity of an ecological system. *Biological integrity* is defined as the ability to support and maintain a balanced, integrated, and adaptive community of organisms having a species composition, diversity, and functional organization comparable to those of natural habitats within a region (Karr and Dudley, 1981). Typical aquatic ecosystems include rivers, lakes, wetlands, and estuaries; and biological monitoring of a stream is illustrated in Figure 2.14.

FIGURE 2.14 Biological monitoring of a stream. (From State of Iowa, 2005.)

The principal biological factors to be considered in evaluating the integrity of an aquatic ecological system are derived from a biological inventory and a biological potential analysis.

Biologists have traditionally placed living things within five *kingdoms*, differentiated primarily by the organization of their nuclear material and by their feeding strategies. *Procaryotic* organisms have their nuclear material distributed throughout the cell, while *eucaryotic* organisms utilize a membrane to segregate the nuclear material. Feeding strategies include absorption (uptake of dissolved nutrients), photosynthesis (fixation of light energy in simple carbohydrates), and ingestion (intake of particulate nutrients). The five kingdoms are as follows:

1. *Monera:* unicellular procaryotes that obtain nutrients strictly by absorption. Bacteria are members of the Monera kingdom.
2. *Protista:* mostly unicellular eucaryotes that obtain food by absorption, photosynthesis, or ingestion. Algae and protozoans are members of the Protista kingdom. Some biologists still debate the classification of algae within the Protista kingdom.
3. *Fungi:* mostly multicellular eucaryotes that obtain food by absorption. Fungi lack *motility*; that is, they lack the ability to move by self-generated propulsion.
4. *Plantae:* multicellular eucaryotes that obtain food by photosynthesis.
5. *Animalia:* multicellular eucaryotes that obtain food by ingestion.

Each kingdom can be further subdivided into phyla, classes, orders, families, genera, and species. A *species* is a group of individuals that possess a common gene pool and that can interbreed successfully. In scientific work, the (Latin) genus–species combination is typically used to identify an organism; however, most organisms have one or more common nonscientific names by which they are frequently referred. For example, *Stizostedion vitreum* is the scientific name for an organism commonly known as the yellow pike (a fish).

In addition to classifying microorganisms by their cellular organization (procaryote or eucaryote) and feeding strategy (absorption or photosynthesis or ingestion), microorganisms are sometimes classified by their principal carbon source, energy source, and their oxygen requirements. These classifications are listed in Table 2.5, where it is shown that microorganisms are either *autotrophic* or *heterotrophic*, depending on whether they use carbon dioxide or organic compounds as their principal carbon source; *phototrophic* or

TABLE 2.5   Classifications of Microorganisms

Characteristic	Label	Attribute
Principal carbon source	Autotrophic	Carbon dioxide
	Heterotrophic	Organic compounds
Energy source	Phototrophic	Sunlight through photosynthesis
	Chemotrophic	Conversion of chemicals
Relationship to oxygen	Obligate aerobe	Requires O_2 to grow
	Obligate anaerobe	Grows in absence of O_2
	Facultative aerobe	Can grow with or without O_2
Cellular organization	Eucaryote	Cell nucleus contains genetic material
	Procaryote	No cell nucleus, genetic material grouped in cell

chemotrophic, depending on whether they use sunlight or chemical energy; and *obligate aerobe*, *obligate anaerobe*, or *facultative aerobe*, depending on whether they require oxygen, can grow in the absence of oxygen, or can grow with or without oxygen, respectively.

Biocriteria and Bioassessment Data The occurrence, condition, and numbers of types of fish, insects, algae, plants, and other organisms are *biological indicators* that provide direct information about the health of specific bodies of water. Using these biological indicators as a way to evaluate the health of a water body is called *biological assessment*. *Biological criteria* or *biocriteria* describe the qualities that must be present to support a desired condition in a water body, and they serve as the benchmarks against which biological-indicator measurements are compared. Biological criteria are narrative or numeric expressions that describe the reference biological integrity (structure and function) of aquatic communities inhabiting waters of a given designated aquatic-life use. Biocriteria are based on the numbers and types of organisms present and are regulatory-based biological measurements. Reference conditions should be the foundation for biocriteria and should represent unimpaired or minimally impaired conditions (e.g., Ohio's "excellent warm-water habitat" or Maine's "class A, as naturally occurs").

Biological indicator species are unique environmental indicators, as they offer a signal of the biological condition in a water body. Biological indicators can serve as early-warning measures of pollution or degradation in an ecosystem. The major groups of biological indicators include (1) fish, (2) invertebrates, (3) periphyton, and (4) macrophytes. Marine environments utilize biological indicators which are different from those used in freshwater bodies.

In marine and estuarine waters benthic macroinvertebrates (e.g., polychaetes) are good indicators of water quality, as their response to pollutants is comparable to those in freshwater systems. Polychaetes (commonly known as worms) are one of the most tolerant marine organisms to stressors (e.g., low oxygen, organic contamination of sediment, and sewage pollution), so they are typically used as biological indicators. In addition, macroinvertebrates also have limited mobility and a long enough life span to both avoid pollutants and assess environmental stressors accurately. Typically, it is much more difficult to assess marine–estuarine conditions, as it is often difficult to evaluate reference conditions in these ecosystems. Typical marine–estuarine indicators include:

- Phytoplankton as indicators of water quality, specifically nutrients (e.g., nitrogen and phosphorus)
- Zooplankton that are sensitive to changes in water quality (e.g., toxic pollution, excess nutrients, and low oxygen) and are useful for future-fisheries health assessment, as they serve as a food source for animals higher up in the food chain
- Benthos that are susceptible to stresses associated with toxic pollution, excess nutrients, and low oxygen
- Submerged aquatic vegetation that serves as a good indicator of water conditions
- Fish

Typical samples of phytoplankton and zooplankton are illustrated in Figure 2.15. It is important to keep the differences between biological water-quality indicators and chemical water-quality indicators in mind when considering the application of biocriteria. Biological water-quality indicators provide direct measures of the cumulative response of

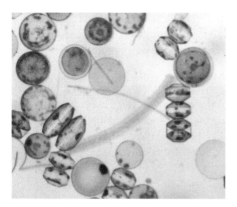

FIGURE 2.15   (*a*) Phytoplankton; (*b*) zooplankton. (From Musèo Nazionale Dell'Antartide, 2005.)

the biological community to all sources of stress. Therefore, biocriteria set the biological quality goal, or target, to which water quality can be managed rather than the maximum allowable level of a pollutant or other water-quality condition in a water body. Physical and chemical water-quality criteria are designed to protect the biological community in a water body from different categories of stress, such as toxic levels of pollutants and unhealthy physical conditions.

One of the most meaningful ways to assess the quality of surface waters is to observe directly the communities of plants and animals that live in them. Because aquatic plants and animals are constantly exposed to the effects of various stressors, these communities reflect not only current conditions, but also stresses and changes in conditions over time and their cumulative impacts. Bioassessment data are invaluable for managing aquatic resources and ecosystems. They can be used to set protection and restoration goals, to decide what to monitor and how to interpret what is found, to identify stresses to the water body and decide how the stresses should be controlled, and to assess and report on the effectiveness of management actions.

Traditional chemical and physical water-quality assessments cannot fully answer questions about the ecological integrity of a water body or determine whether aquatic resources are being protected. Relying on traditional chemistry alone may lead to situations in which meeting chemical standards may not be enough to fully protect the aquatic community, or conversely, to situations in which the community remains in satisfactory condition despite a failure to attain standards. This is illustrated in Figure 2.16 with the six leading causes of aquatic-life use impairment from Ohio rivers and streams. It is apparent that stressors such as poor habitat quality are proving more important than typically regulated pollutants in limiting the attainment of designated aquatic life uses. This and other examples reinforce the need for a comprehensive suite of measures and indicators to characterize the ecological health of a water body. The presence, condition, and numbers of types of fish, invertebrates, amphibians, algae, and plants are data that together provide direct information about the health of specific bodies of water. Bioassessment data can also help distinguish among potential stressors. Establishing credible relationships between stressors and impairments can help identify likely causes of problems. Bioassessment data also serve as a measure to evaluate the effectiveness of

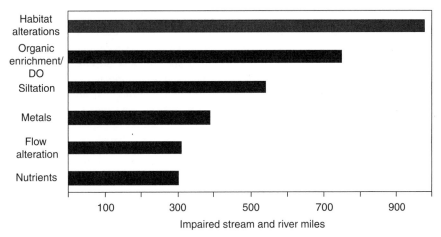

FIGURE 2.16   Leading causes of aquatic-life use impairment in Ohio rivers and streams. (From USEPA, 2005c.)

management actions as reflected in the responses and improved conditions of biological communities.

Biological Inventory   By knowing what species are present in a water body, biologists can analyze the health of a water body. For example, if the fish species present are primarily carnivores, the quality of water is generally higher than in a water body dominated by omnivores (USEPA, 1994). A biological inventory also allows biologists to assess the biological health of a water body by evaluating:

- The species richness or number of species
- The presence of intolerant species
- The proportion of omnivores to carnivores
- The biomass production
- The number of individuals per species

The role of the biologist is critical in evaluating the health of the biota because knowledge of expected richness or expected species comes only from understanding the general biological traits and regimes of the area.

Biological Potential Analysis   A significant step in determining the possible designated uses of a water body requires an evaluation of what communities could potentially exist in a particular water body if pollution were abated or if the physical habitat were modified. A biological potential analysis involves:

- Defining boundaries of fish faunal regions
- Selecting control sampling sites in the reference reaches of each area
- Sampling fish and recording observations at each reference sampling site

- Establishing the community characteristics for the reference reaches of each area
- Comparing the water body in question to the reference reaches

Selection of reference reaches is of critical importance because the characteristics of the aquatic community in reference reaches will be used to establish baseline conditions against which similar reaches (based on physical and hydrological characteristics) are compared.

Human Pathogenic Microorganisms   Pathogenic microorganisms are disease-causing agents that typically originate in the intestines of infected people or animals and are excreted in the feces that enter sewer systems or are deposited onto the ground. Microorganisms are too small to be seen clearly without the aid of a miscoscope, and although there is no precise definition, the size that separates microorganisms from other organisms is approximately 100μm. Pathogenic microorganisms mostly result from urban stormwater runoff, domestic and municipal wastewater discharges, combined sewer overflows, septic systems, and runoff from pastures and animal feedlots. Water uses that are affected by the presence of pathogenic microorganisms include bathing, fishing, and shellfish harvesting. Concern about the presence of pathogenic microorganisms in various water bodies is supported by the following facts (Tebbutt, 1998):

- Each year over 5 million people die from water-related diseases.
- Two million of these annual deaths are children.
- In developing countries, 80% of all illness is water related.
- At any one time, half of the population in developing countries will be suffering from one or more of the main water-related diseases.
- A quarter of the children born in developing countries will have died before the age of 5, the great majority from water-related diseases.

In most cases, simply bathing in a stream, drinking water from a lake, or swimming in an ocean will not cause person to become ill, since one must first come into contact with the pathogens, the pathogens must gain entry to the body, and the dose of pathogens must be sufficient to overcome the human body's natural defenses. Under special circumstances, an infection can develop from a single pathogenic microorganism, but the minimum infective dose for pathogenic bacteria is typically between 100 and 10^6, depending on the species (Majeti and Clark, 1981; National Academy of Sciences, 1996). This indicates that recreational waters need not be absolutely free of pathogens to be relatively safe, but the higher the concentration of pathogens, the greater the probability that health problems will result from contact with the water body.

A list of pathogens that are typically found in natural water bodies, along with their associated diseases and number of associated outbreaks in the United States (1987–1996), are given in Table 2.6. During the period 1987–1996 there were 305 reported outbreaks of waterborne diseases in the United States, and the percentages shown in Table 2.6 are based on this total number. In 28.2% of the outbreaks no specific pathogen was identified, although the disease identified was gastroenteritis (Laws, 2000). Waterborne-disease outbreaks have been averaging about 31 per year in the United States since 1980, and the number of cases associated with these outbreaks have been averaging around 7000 per year. All of the microorganisms shown in Table 2.6 originate primarily from human feces

TABLE 2.6   Pathogenic Microorganisms Commonly Found in Surface Waters

Pathogen Group and Name	Associated Disease	U.S. Outbreaks 1987–1996 (% of Total)
Virus		
Adenoviruses	Upper respiratory and gastrointestinal illness	0.3
Enteroviruses		
Polioviruses	Poliomyelitis	—
Echoviruses	Aseptic meningitis, respiratory infections	—
Coxsackieviruses	Aseptic meningitis, herpangina, myocarditis	—
Other enteroviruses	Encephalitis	—
Hepatitis A virus	Infectious hepatitis	1.6
Reoviruses	Mild upper respiratory and gastrointestinal illness	—
Rotaviruses	Gastroenteritis,[a] diarrhea	—
Norwalk and related gastrointestinal viruses	Gastroenteritis	1.3
Bacterium		
Salmonella typhi	Typhoid fever	—
Salmonella paratyphi	Paratyphoid fever	—
Other salmonella	Salmonellosis, gastroenteritis	0.7
Shigella spp.	Bacterial dysentery[b]	9.2
Vibrio cholerae	Cholera	0.3
Escherichia coli	Gastroenteritis	3.6
Yersinia enterocolitica	Gastroenteritis	—
Campylobacter jejuni	Gastroenteritis	1.0
Cyanobacteria		0.3
Legionella spp.		1.3
Leptospira spp.		0.7
Plesiomonas shigelloides		0.3
Pseudomonas aeruginosa		16.1
Protozoan		
Acanthamoeba castellani	Amebic meningoencephalitis	—
Balantidium coli	Balantidosis (dysentery)	—
Entamoeba histolytica	Amebic dysentery	—
Giardia lamblia	Giardiasis (gastoenteritis), diarrhea	10.8
Cryptosporidium spp.	Cryptosporidiosis, diarrhea	7.5
Naegleria spp.		5.2
Helminth		
Taenia saginata	Tapeworm	—
Ascaris lumbricoides	Ascariasis	—
Schistosoma spp.	Schistosomiasis	1.6

[a]Gastroenteritis is an inflammation of the lining membrane of the stomach and intestines.
[b]Dysentery is an inflamation of the intestine characterized by the frequent passage of feces, usually with blood and mucus.

(AWWA, 1990). Between 1980 and 1996, 401 waterborne disease outbreaks were reported in the United States, with over 750,000 associated cases of disease. (AWWA, 2003). In April 1993, the largest waterborne-disease outbreak in the United States occurred in Milwaukee, Wisconsin, when 400,000 people were exposed to the protozoan parasite *Cryptosporidium parvum.*

Microorganisms of most concern from a water-quality viewpoint are viruses, bacteria, protozoans, helminths, and algae. These microorganisms are described in more detail below.

Viruses   Viruses are complex molecules that typically contain a protein coat surrounding a deoxyribonucleic acid (DNA) or ribonucleic acid (RNA) core of genetic material. Viruses have no independent metabolism and depend on host living cells for reproduction. They range in size from 0.01 to 0.4 μm in diameter. As an example, the structure of the human immunodeficiency virus (HIV) is shown in Figure 2.17. The outer shell of the HIV virus is called the *viral envelope*, and embedded within the viral envelope is a protein complex. Viruses do not exist long outside a human or animal body, but while they exist, they can survive heat, drying, and chemical agents.

More than 130 types of enteric viruses are excreted in human feces and urine. Viral pathogens accounted for about 5% of waterborne disease outbreaks in the United States between 1987 and 1996, and are responsible for diseases such as poliomyelitis, aseptic meningitis, infectious hepatitis, gastroenteritis, upper-respiratory infections, skin rashes, and cardiac pathologies (Mahin and Pancorbo, 1999). In the United States, the viruses most commonly linked with waterborne-disease outbreaks are the infectious hepatitis (hepatitis A) virus and the Norwalk virus. The hepatitis A virus accounts for about 30% of the waterborne-disease outbreaks in the United States linked to viruses, and about 65% of nonbacterial gastroenteritis in the United States is due to the Norwalk and Norwalk-like viruses. In developing countries, rotaviruses cause an estimated 870,000 deaths per year and are detected in 20 to 70% of fecal specimens from children hospitalized with acute diarrhea (Glass et al., 1996).

A practical consideration in the case of viruses is the fact that they are more resistant to standard water-treatment procedures than are the commonly used indicator organisms. Therefore, public water supplies and recreational waters that appear safe based on standard assays may in fact contain dangerous concentrations of viruses (Laws, 2000).

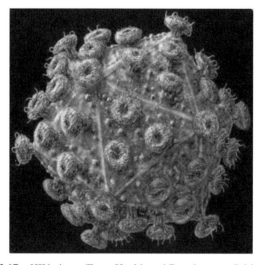

FIGURE 2.17   HIV virus. (From Health and Development Initiative, 2005.)

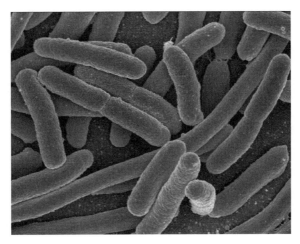

FIGURE 2.18   *Escherichia coli* bacteria. (From State of Michigan, 2005.)

Bacteria   Bacteria are single-celled microorganisms that have a cell wall and a single strand of DNA. Bacteria are members of the Monera kingdom, reproduce by binary fission, are not photosynthetic (like plants), and do not need sunlight to reproduce. They do not have a well-defined nucleus and do not contain chlorophyll. The more common shapes of bacteria are spheres, rod-shaped, spiral and branching threads, and filamentous. Bacteria range in size from 0.1 to $10\,\mu$m in diameter and 2 to $4\,\mu$m in length. Most have flagella, a tail-like structure for movement.

Bacteria are responsible for waterborne diseases such as cholera and typhoid. Pathogenic bacteria of particular concern to humans are *Salmonella, Escherichia coli, Shigella,* and *Legionella. E. coli* bacteria are illustrated in Figure 2.18. Opportunistic bacterial pathogens are not normally a danger to persons in good health, but they can cause sickness or death in those who are in a weakened condition. Particularly at risk are newborns, the elderly, and persons who already have a serious disease. Included among the opportunistic bacteria are *Pseudomonas* spp., *Aeromonas hydrophila, Edwardsiella tarda, Flavobacterium, Klebsiella, Enterobacter, Serratia, Proteus, Providencia, Citrobacter,* and *Acinebacter* (AWWA, 2003). About 16% of the waterborne-disease outbreaks reported between 1987 and 1996 were attributed to the bacterial pathogen *Pseudomonas aeruginosa,* and all of these outbreaks were associated with the use of recreational waters. *P. aeruginosa* is probably best known for its effect on persons suffering from cystic fibrosis. The bacteria that ranked first and third (*P. aeruginosa* ranked second) in terms of the number of waterborne-disease cases from 1987 to 1996 were *Shigella* and *Salmonella,* respectively. These outbreaks involved both drinking water and recreational waters.

Protozoans   Protozoans are single-celled microorganisms with a nucleus but without a cell wall. Protozoans are members of the Protista kingdom, reproduce by fission, and feed on bacteria. Protozoans range in size from $5\,\mu$m to $2\,$cm, and scientists have identified approximately 40,000 species of protozoans. Numerous protozoan species normally inhabit the intestinal tracts of warm-blooded animals, including humans. *Giardia lamblia* and *Cryptosporidium parvum* are the protozoans of most concern in drinking-water supply sources. These protozoans, shown in Figure 2.19, are described briefly below.

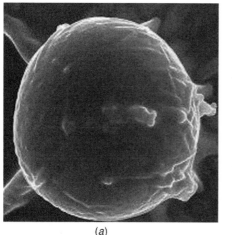
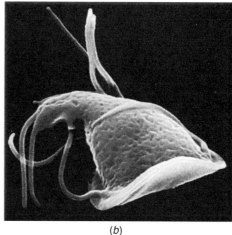

(a)   (b)

FIGURE 2.19   (a) *Cryptosporidium parvum;* (b) *Giardia lamblia.* (From State of Michigan, 2005.)

1. *Giardia lamblia* causes the disease giardiasis (also known as "beaver fever"), which is the most frequently diagnosed waterborne disease in the United States (AWWA, 2003). Symptoms of giardiasis include skin rash, flulike symptoms, diarrhea, fatigue, and severe cramps. The protozoan attaches itself to the upper intestinal tract and produces cysts, which are shed in the feces. One of the major reasons why giardiasis continues to be a problem as a waterborne disease is that the cysts survive well under adverse conditions. *Giardia* cysts are highly resistant to chlorine and can live in cold water for months. Virtually all of the outbreaks of giardiasis in the United States have been reported in mountainous areas, where the frequent use of unfiltered easily contaminated surface waters by hikers and campers has undoubtedly contributed to the prevalence of the disease. Three of the major hosts for *Giardia* are humans, beaver, and muskrat; and beavers are believed to be a particularly significant source of *G. lamblia* in surface waters. Although water is a major means of transmitting giardiasis, the largest percentage of recorded cases is caused by person-to-person contact, primarily in child-care centers. *G. lamblia* has accounted for more outbreaks of gastroenteritis in recent years than any other single pathogen, and it has been estimated that roughly 4% of persons in the United States are carriers, but rates may be as high as 16% in some areas (Hill, 1990).

2. *Cryptosporidium parvum* is a small parasite measuring approximately 3 to 5 μm that has caused several outbreaks of the disease cryptosporidiosis and poses severe health risks. At least 10 species of *Cryptosporidium* are currently recognized, with *C. parvum* the most common form found in humans. In healthy persons, cryptosporidiosis causes 7 to 14 days of diarrhea, with possibly a low-grade fever, nausea, and abdominal cramps. The effects on immunocompromised persons can be life threatening, and no antibiotic treatment currently exists for cryptosporidiosis. Severely immunocompromised persons are encouraged to avoid any contact with water in lakes and streams and should not drink water from these water bodies. *C. parvum* produces eggs called *oocysts* that are not readily killed by chlorination and can be removed effectively only by filtration. A person infected with *C. parvum* may excrete as many as 100 million oocysts per day.

3. *Entamoeba histolytica* is a protozoan that causes a disease of the large intestine called amebiasis. Although of little public health significance in the United States, amebiasis causes millions of cases of diarrhea worldwide and is estimated to account for 100,000 deaths each year (Cohen, 1995).

Helminths A number of parasitic intestinal helminths (worms) are found in domestic sewage. Most helminths enter the human body through the mouth, although a few gain entry through the skin. Contamination of drinking water with these worms can be prevented effectively with modern water-treatment methods that include disinfection; however, swimming or wading is sewage-polluted waters may lead to outbreaks of intestinal-worm infections. Such incidents are of little public health significance in the United States, but are much more common in other parts of the world.

The best-known helminth is the beef tapeworm *Taenia saginata*, which lives in the intestinal tract of humans, and infected persons may discharge as many as 1 million eggs per day in their feces. Symptoms of tapeworm infection include abdominal pain, digestive disturbance, and weight loss. Geldreich (1972) cites a 50% incidence of beef tapeworm infection in East Africa, but the percentage of carriers in the United States is less than 1%. Certain species of the blood fluke *Schistosoma* spend part of their life cycle in humans, where they may cause a debilitating illness known as schistosomiasis. An electron micrograph of a pair of *S. mansoni* is shown in Figure 2.20. Penetration of the skin by larval schistosomae may cause a skin rash known as "swimmer's itch." The five schistosomiasis outbreaks reported in the United States between 1987 and 1996 all involved polluted recreational waters, and in each case the symptom of the infection was dermatitis. The more serious symptoms of the disease, which generally do not develop until 4 to 8 weeks later, include fever, chills, sweating, and headaches.

Algae Algae are a diverse group of simple organisms that are members of the Protista kingdom. Algae are not members of the plant kingdom, but like plants, most algae use the energy of sunlight to make their own food via photosynthesis. Algae lack the roots, leaves,

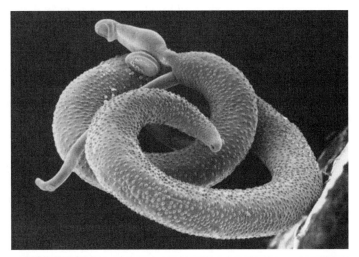

FIGURE 2.20   Pair of *Schistosoma mansoni*. (From Uniformed Services University, 2005.)

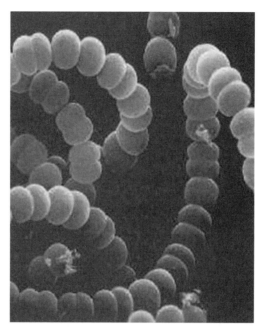

FIGURE 2.21   Algae *Anabaena flos-aquae*. [From Wayne Carmichael (Wright State University), Mark Schneegurt (Wichita State University), and Cyanosite (www.cyanosite.bio.purdue.edu).]

and other structures typical of true plants. Microscopic algae that float or swim, called *phytoplankton*, typically range from 1 to 100 μm in size. Algae found in many natural waters are usually not a health concern; however, certain species may produce endo- or exotoxins which if ingested at high enough concentrations may be harmful. Three species of blue-green algae, *Anabaena flos-aquae*, *Microcystis aeruginosa*, and *Aphanizomenon flos-aquae*, produce exotoxins. The toxin-producing algae *Anabaena flos-aquae* is shown in Figure 2.21. Toxin concentrations during algal blooms have resulted in illness or death in mammals, birds, and fish that have ingested a sufficient dose. Some evidence exists that human exposure in recreational water, and possibly drinking water, has caused contact irritation and possibly gastroenteritis (AWWA, 1990).

Excessive nutrients (nitrogen and phosphorus) in ocean waters often result in the development of *red tides*, which are algae blooms of such intensity (e.g., 50×10^6 cells per liter) that the sea is discolored. The associated coloration is not always red, but may be white, yellow, or brown (Clark, 1997). Many animals, including commercially important fish species, are killed or excluded from the area where red tides occur, either because of clogging of their gills or other structures, or because of the toxic properties of the algae. Blooms of *Phaeocystis* form an unsightly brown foam which when stranded on the shore can be mistaken for sewage pollution. It regularly affects the coastlines of northern France, Belgium, the Netherlands, and Great Britain. A *Phaeocystis* bloom in the upper Adriatic during the summer of 1990 had a devastating economic effect on resorts in the area, which were deserted by the tourists because of the disgusting appearance of the beaches. Marine algae are a major component of phytoplankton, which is at the base of the food chain in the oceans. Along with cyanobacteria, marine algae are consumed by protozoans and microscopic animals, which are, in turn, eaten by fish.

Indicator Organisms   Testing water samples for a wide variety of pathogens is usually not practical, and *indicator organisms* are typically used to provide a measure of fecal contamination from humans or other warm-blooded animals. General criteria for an indicator organism are as follows:

1. The indicator should originate in the digestive tract of humans and warm-blooded animals.
2. The indicator should be present in fecal material in large numbers.
3. The indicator should always be present when the pathogenic organism of concern is present, and should be absent in clean uncontaminated water.
4. The indicator should respond to natural environmental conditions in a manner similar to the pathogens of interest, or survive longer in water that pathogens (AWWA, 2003).
5. The indicator should be easy to isolate, identify, and enumerate.
6. The ratio of indicator to pathogen should be high.
7. The indicator and pathogen should come from the same source.
8. The indicator should not be pathogenic.

A number of microorganisms have been utilized as indicators, including total coliforms, fecal coliforms, *E. coli*, fecal streptococci, *P. aeruginosa*, enterococci, and heterotrophic plate count (HPC).

The coliform group of organisms, collectively referred to as *total coliforms*, is defined as all the anaerobic and facultatively anaerobic, gram-negative, nonspore-forming, rod-shaped bacteria that ferment lactose with gas formation within 48 h at 35°C. One of the USEPA-approved methods for coliform bacteria testing is the membrane filter method, which uses a cellulose membrane material that has pore openings that are approximately 0.47 μm in diameter. Water freely passes through the membrane, but coliform bacteria are retained on the surface. After the water sample is filtered through the membrane, it is incubated for 24 h. at 35°C in a special medium (m-ENDO broth) specific for coliform bacteria. Coliform bacteria produce colonies with a characteristic metallic green sheen; an example of such colonies is shown in Figure 2.22. The definition of coliform bacteria is an operational rather than a taxonomic definition and encompasses a variety of organisms, mostly of intestinal origin. The definition includes *E. coli*, the most numerous facultative bacterium in the feces of warm-blooded animals, plus species belonging to the genera *Enterobacter*, *Klebsiella*, and *Citrobacter*. The latter organisms are present in domestic wastewater but can be derived from plant and animal materials. The coliform group of bacteria meets all criteria for an ideal indicator. These bacteria are generally not pathogenic, yet they are usually present when pathogens are present. Coliform bacteria are more plentiful than pathogens and can often stay alive in the water environment for longer periods of time. Drawbacks to the use of total coliforms as an indicator include their regrowth in water, thus becoming part of the natural aquatic flora. As a rule, water with any detectable coliforms is unsafe to drink (AWWA, 2003).

Fecal coliforms provide stronger evidence of the possible presence of fecal pathogens than do total coliforms. Fecal coliforms are a subgroup of total coliforms, distinguished in the laboratory through elevated temperature tests (43 to 44.5°C, depending on the test). Although the test does determine coliforms of fecal origin, it does not distinguish between

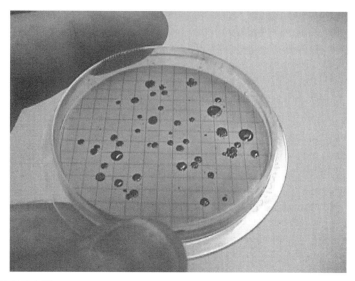

FIGURE 2.22 Membrane filter test for coliforms. (From State of Oregon, 2005.)

human and animal contamination. The membrane filter method can also be used to detect the presence of fecal coliform bacteria. After the water sample is filtered through the filter, it is incubated in a special medium at a temperature of 44.5°C, and fecal coliforms appear as blue colonies. The concentration of fecal coliforms is typically much lower than that of total coliforms. As frames of reference, human feces contain on the order of 13×10^6 fecal coliforms[1] per gram (Droste, 1997) and rivers polluted by combined-sewer outfalls may have fecal coliforms in excess of 1×10^6 cfu per 100 mL (Canale et al, 1993). Unfortunately, the assay for fecal coliforms does not select exclusively for coliforms of human origin but includes thermotolerant members of the Enterobacteriaceae genus *Klebsiella*, which is found infrequently in human feces and when present is usually a minor portion of the coliform population (Cabelli, 1983). Since there are substantial nonfecal sources of *Klebsiella*, the designation of fecal coliform is sometimes regarded as a misnomer (Laws, 2000).

The reliability of total coliform (TC) and fecal coliform (FC) bacteria as indicators of the presence of pathogens in water depends on the persistence of pathogenic microorganisms relative to coliform bacteria. In surface waters, pathogenic bacteria tend to die off faster than coliforms, whereas viruses and protozoans tend to be more persistent. In recreational lakes and streams, FC levels of less than 200 cfu per 100 mL are usually considered acceptable. If no coliforms are detected, the water is presumed to be uncontaminated by sewage.

The concentration of *fecal streptococci* (FS), which includes several species of streptococci that originate from humans, is also a useful indicator of the presence of pathogenic microorganisms. The ratio of FC to FS (FC/FS) has been used as a measure of whether fecal contamination is from human or animal sources. Values of FC/FS greater than 4 indicate human contamination; FC/FS values below 4 indicate contamination primarily from other warm-blooded animals. The ratio FC/FS should be used with caution, since FC and

---

[1] Commonly, stated as colony-forming units (cfu).

FS have different die-off rates, and the FC/FS ratio may give misleading results far down-stream of a contaminant source.

Research on bacteriological indicators has resulted in the USEPA recommending that states use *E. coli* or enterococci as indicators of recreational water quality (USEPA, 1986) instead of fecal coliform because of the better correlation with gastroenteritis in swimmers. However, enterococci are usually found at lower densities than *E. coli* in human feces and sewage effluents, and this fact may compromise their usefulness as indicator organisms (Dufour, 1984).

In ground waters, fecal coliform concentration is not a reliable indicator of contamination by pathogenic microorganisms, since microorganisms larger than viruses seldom travel appreciable distances, and there is no relation between the presence of fecal coliforms and pathogenic microorganisms. The presence of viruses is of particular concern in ground waters, where they originate from septic-tank effluent and because of their small size (typically, 0.01 to 0.03 μm) are able to travel considerable distances in ground water. Since viruses are difficult to detect in ground water and fecal coliforms are not a reliable indicator, protection of drinking-water intakes (wells) from viral contamination is usually achieved by requiring minimum setback distances for viral sources such as septic tanks and minimum disinfection requirements. In the United States, the statute that sets minimum setback distances and minimum disinfection requirements for drinking-water supplies derived from ground water is known as the Ground Water Disinfection Rule.

2.3   U.S. SURFACE-WATER STANDARDS

The U.S. Environmental Protection Agency (USEPA) defines a water-quality standard as the combination of a designated beneficial use of a water body, the water-quality criteria necessary to support that designated use, and protection of the water quality through anti-degradation provisions. When designating water-body uses, states are required to consider extraterritorial effects on their standards. For example, once states revise or adopt standards, upstream jurisdictions will be required when revising their standards and issuing permits, to provide for attainment and maintenance of the downstream standards. Antidegradation provisions prohibit the degradation of water quality even if the degraded water would meet the applicable water-quality criteria.

The terms *water-quality criterion* and *water-quality standard* have different meanings, although the terms are commonly used interchangeably. A *criterion* is a benchmark or reference point on which a judgment is based. Water-quality criteria are expressed in terms of constituent concentrations or narrative statements representing a quality of water that supports a particular designated use. A *standard* applies to a statutory rule, principle, or measure established by an authority. In the United States, the USEPA has responsibility for issuing scientifically developed water-quality criteria and effluent limitations; only states can issue legally binding standards, using the federal water-quality criteria as guidance.

Water-quality criteria and standards are either ambient (receiving water) or effluent (discharge) standards. *Ambient standards* apply to all waters of the United States and depend on their designated use; *effluent standards* apply to discharges into waters of the United States and are based on available and economical technologies for treatment. Ambient standards are formulated to protect human health and the well-being of fish, shellfish, and other aquatic life. Effluent standards typically require mandatory application of certain technologies and hence are *technology-based standards*. Several such

TABLE 2.7 Technology-Based Standards

Source	Technology-Based Standard
Municipal point sources	Maximum effluent BOD_5 and suspended solids limit of 30 mg/L as monthly averages
Combined-sewer overflows	Restriction of the number of overflows or mandatory capture and treatment of a certain portion of wet-weather flow
Urban runoff	Capture and treatment of the first 1.2 to 2.5 cm of runoff; mandatory street sweeping
Unsewered suburban lands	Septic-tank regulations

standards are listed in Table 2.7. Technology-based discharge standards are easy to monitor for compliance, while ambient water-quality standards are difficult to enforce because water quality varies spatially and temporally but can be measured only at discrete locations and time intervals.

2.3.1 Designated Beneficial Uses

The categories of beneficial uses to be considered by states in establishing water-quality standards must include public water supplies, propagation of fish and wildlife, recreation, agricultural use, and navigation (USEPA, 1994).

1. *Public water supplies.* This use includes waters that are a source for drinking water and often includes waters for food processing. Waters for drinking may require treatment prior to distribution in public water systems.

2. *Propagation of fish and wildlife.* This classification is often divided into several more specific subcategories, including cold- and warm-water fish and shellfish. Some coastal states have a use specifically for oyster propagation. This use may also include protection of aquatic flora. Many states differentiate between self-supporting fish populations and stocked fisheries. Wildlife protection should include waterfowl, shorebirds, and other water-oriented wildlife.

3. *Recreation.* Recreational uses have traditionally been divided into primary- and secondary-contact recreation. The primary-contact recreation classification protects people from illness due to activities involving the potential for ingestion of, or immersion in, water. Primary-contact recreation usually includes swimming, water skiing, skin diving, surfing, and other activities likely to result in immersion. The secondary-contact recreation classification is protective when immersion is unlikely. Examples are boating, wading, and rowing. In many northern areas, body-contact recreation is possible only a few months out of the year, and several states have adopted primary-contact recreational uses only for those months when primary-contact recreation actually occurs and have relied on less stringent secondary-contact recreation to protect incidental exposure during the "non-swimming" season.

4. *Agriculture and industry.* The agricultural-use classification defines waters that are suitable for irrigation of crops, consumption by livestock, support of vegetation for range grazing, and other uses in support of farming and ranching.

TABLE 2.8   Classifications of Surface Waters in Florida

Class	Description
I	Potable water supplies
II	Shellfish propagation or harvesting
III[a]	Recreation, propagation, and maintenance of a healthy, well-balanced population of fish and wildlife
IV	Agricultural water supplies
V	Navigation, utility, and industrial use

[a]There are two sets of criteria for class III surface classification. One set of criteria is for predominantly freshwaters (chloride < 1500 mg/L), and the other set of criteria is for predominantly marine waters (chloride > 1500 mg/L).

5. *Navigation.* This use classification is designed to protect ships and their crews and to maintain water quality so as not to restrict or prevent navigation.

States are generally required to classify the uses of all water bodies within their borders and to promulgate water-quality standards that are at least as stringent as the federal criteria. States may adopt uses other than those suggested by the USEPA, such as coral-reef preservation, marinas, ground-water recharge, aquifer protection, and hydroelectric power. Several states have adopted the USEPA surface-water categories described previously, or very similar categories, and associated water-quality criteria. As an example, Florida classifies all surface waters within the state into the five categories shown in Table 2.8, and water-quality criteria have been promulgated for each classification. These classifications, classes I to V, are arranged in order of decreasing degree of protection. As of 2006 there were no class V waters in Florida. Surface waters are subclassified as either freshwaters, open waters, or coastal waters. *Freshwaters* include waters contained in lakes and ponds or in flowing streams above the zone in which tidal actions influence the salinity of the water and where the concentration of chloride ions is less than 1500 mg/L. *Open waters* are typically ocean waters that are seaward of a specified depth contour (5.49 m in Florida), and *coastal waters* are waters that are between freshwaters and open waters. Cases of water bodies in the United States not meeting the water-quality criteria associated with their designated uses are widespread, and Figure 2.23 shows the distribution of rivers, lakes, and estuaries that do not meet their designated uses.

2.3.2 Water-Quality Criteria

National water-quality criteria for surface waters recommended by the USEPA are listed in Appendix C.1. These water-quality criteria cover priority toxic pollutants, nonpriority pollutants, and organoleptic (taste and odor) effects. For waters with multiple-use designations, the criteria must support the most sensitive use. Separate criteria are provided for the protection of human health and aquatic life. Criteria for the protection of human health are needed for water bodies designated for public water supply and when fish ingestion is a concern. For carcinogens, the human-health criteria are usually more stringent than the aquatic-life criteria. In contrast, for noncarcinogens, the aquatic-life criteria are usually more stringent than the human-health criteria. Most state water-quality standards list only the most stringent of the two criteria (human health, aquatic life) for each water-quality parameter.

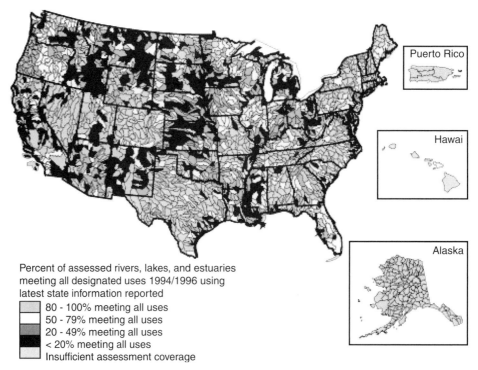

Percent of assessed rivers, lakes, and estuaries
meeting all designated uses 1994/1996 using
latest state information reported

- 80 - 100% meeting all uses
- 50 - 79% meeting all uses
- 20 - 49% meeting all uses
- < 20% meeting all uses
- Insufficient assessment coverage

FIGURE 2.23   Rivers, lakes, and estuaries that meet all designated uses. (From USEPA, 2005f.)

Water-quality regulations in the United States allow individual states to develop numerical water-quality criteria of their own or modify the USEPA's recommended criteria to account for site specificity and other scientifically defensible factors (USEPA 1991a, 1994).

Criteria for Aquatic-Life Protection   The toxicity of various contaminants in aquatic systems are established using *toxicity bioassay tests*. To conduct a toxicity test, the test organisms are placed in containers with various dilutions of toxicants plus one container with toxicant-free water only. The number of organisms surviving and/or unaffected after a specified time interval (e.g., 24, 48, 72, 96) is recorded. A typical aquatic-toxicity test apparatus is shown in Figure 2.24. Ideally, test organisms should include representatives from the following four groups: microorganisms, plants, invertebrates, and fish; and the test organisms should be amenable to captivity, accurately identified, relatively uniform in size, healthy, and acclimated to laboratory conditions.

Test organisms will not respond uniformly to a given dose of toxic compounds because of different sensitivities. Consequently, the most important quantity of interest in the toxicity bioassay test is the dose or concentration at which, after the test period specified, 50% of the test organisms survive or their life functions are not affected by the dose. *Dose* is defined as the amount of toxic compound ingested by the test organism divided by the body mass of the organism, usually expressed in mg/kg. In testing aquatic organisms, the toxicity is related to the concentration in the water, typically expressed in

FIGURE 2.24   Aquatic-toxicity test apparatus. (From USGS, 2005c.)

mg/L or μg/L. USEPA (1991a) recommends at least five dilution concentrations and a control to determine the magnitude of the toxicity. The rationale for using the median (50%) survival rate to measure toxicity is that median tolerance levels are far more reproducible from one subgroup of a species to another than is the limited reproducibility of the tolerance levels of the most sensitive and least sensitive members of a subgroup (Laws, 2000).

The *lethal dose* (LD) or *lethal concentration* (LC) defines the exposure of the test organism that results in death. The 50% survival dose or concentration value, denoted by LD_{50} (dose) or LC_{50} (concentration), is representative of a particular toxic compound or mixture of toxic compounds. The time of exposure is important in toxicity studies and is generally associated with the lethal concentration. For example, the $LC_{50}(48\,h)$ is the concentration of a toxic material at which 50% of the test organisms died after 48 h of continuous constant exposure.

The *effective dose* (ED) or *effective concentration* (EC) are terms used when sublethal effects are considered, such as the impact on reproduction and respiratory stresses. The terms ED_{50} and EC_{50} are used to describe the dose and concentration, respectively, associated with sublethal effects in 50% of the test organisms within the test period.

In establishing chronic water-quality criteria, the *no-observed-effect concentration* (NOEC) and *lowest-observed-effect concentration* (LOEC) are used. The NOEC is the highest concentration of a toxicant that causes no observable effects, and the LOEC is the lowest concentration of a toxicant that causes an observed effect. The NOEC is generally lower than the LOEC, due to gaps in available data.

The aquatic species used to characterize the toxicity of an effluent discharged into a water body will depend on the requirements of regulatory agencies. USEPA (1991a) recommends a minimum of three species be tested, and species commonly used in freshwater toxicity

TABLE 2.9   Species Used for Toxicity Tests in Freshwater

Vertebrates
  Cold water
    Brook trout   *Salvelinus fontinalis*
    Coho salmon   *Oncorhyncus kisutch*
    Rainbow trout   *Salmo gairdneri*
  Warm water
    Bluegill   *Lepomis macrochirus*
    Channel catfish   *Ictalurus punctatus*
    Fathead minnow   *Pimephales promelas*
Invertebrates
  Cold water
    Stoneflies   *Pteronarcys* spp.
    Crayfish   *Pacificastacus leniusculus*
    Mayflies   *Baetis* spp., *Ephemerella* spp.
  Warm water
    Amphipods   *Hyalella* spp., *Gammarus lacustris, G. fasciatus, G. pseudolimnaeus*
    Cladocera   *Daphnia magna, D. pulex, Ceriodaphnia* spp.
    Crayfish   *Orconectes* spp., *Cambarus* spp., *Procambarus* spp.
    Mayflies   *Hexagenia limbata, H. bilineata*
    Midges   *Chironomus* spp.

Source: USEPA (1985b).

tests are shown in Table 2.9. Whenever possible, the test species should be indigenous to the receiving waters. Analysis of species sensitivity in the United States indicates that if toxicity tests are conducted on three particular species (*Daphnia magna*, *Pimephales promelas*, and *Lepomis macrochirus*), the most sensitive of the three will have an LC_{50} within one order of magnitude of the most sensitive of all species. The health of largemouth bass and the green sturgeon, shown in Figure 2.25, have been a concern in several locations around the United States.

(a)   (b)

FIGURE 2.25 (*a*) Largemouth bass; (*b*) green sturgeon. (From Klamath Resource Information System, 2005. Photo by Pat Higgins.)

The development of national water-quality criteria for the protection of aquatic organisms is a process that uses information from many areas of aquatic toxicology. Water-quality criteria for aquatic life contain two measures: a *criterion maximum concentration* (CMC) to protect against acute (short-term) effects such as death; and a *criterion continuous concentration* (CCC) to protect against chronic (long-term) effects such as carcinogenesis, teratogenesis, reproductive toxicity, and mutagenesis. USEPA derives acute criteria from 48- to 96-h tests of lethality or immobilization, and chronic criteria are derived from longer-term (often greater than 28-day) tests that measure survival, growth, or reproduction. For acute criteria, the USEPA recommends an averaging period of 1 h. That is, to protect against acute effects, the 1-h average exposure should not exceed the CMC. For chronic criteria, the USEPA recommends an averaging period of 4 days. That is, the 4-day average exposure should not exceed the CCC. To assess the attainment of aquatic-life criteria, it is necessary to specify the allowable frequency for exceeding the criteria, since it is statistically impossible to project that criteria will never be exceeded. Based on these considerations, the USEPA recommends an average frequency for excursions above acute and chronic criteria not to exceed once in 3 years. The recommended exceedance frequency of 3 years is the USEPA's best scientific judgment of the average amount of time it would take an unstressed system to recover from a pollution event in which exposure to a toxicant exceeded the criterion. Since the aquatic-life criteria recommended by the USEPA are for national guidance, they are intended to be protective of the vast majority of aquatic communities in the United States.

Derivation of the Criterion Maximum Concentration The toxicity database for determining the CMC differs for freshwater and marine organisms (USEPA, 1985b). The criterion in freshwater must be based on acute tests with freshwater animals in at least eight different families, including all of the following categories: (1) salmonid fish, (2) nonsalmonid fish, (3) a third vertebrate family, (4) planktonic crustaceans, (5) benthic crustaceans, (6) insects, (7) a family not included among vertebrates or insects, and (8) a family in any order of insect or any phylum not already represented. For marine organisms, the criterion should be based on acute tests with saltwater animals in at least eight different families subject to the following five constraints: (1) at least two different vertebrate families are included, (2) at least one species is from a family not included among the vertebrates and insects, (3) either the Mysidae or Penaeidae family or both are included, (4) there are representatives from at least three other families not included among the vertebrates, and (5) at least one other family is represented. Using acute toxicity data of these organisms, the following calculations are used to determine the CMC:

1. Calculate the geometric mean of all LC_{50} (EC_{50}) values for each species, yielding the *species mean acute value* (SMAV).
2. Determine the geometric mean of all SMAV values within a genus, called the *genus mean acute value* (GMAV).
3. The GMAV values are ranked from lowest to highest, and the genus ranks are converted to cumulative probabilities by dividing each rank by $N + 1$, where N is the number of genera in the list. Plot the logarithm of the GMAVs for the four genera

with cumulative probabilities closest to 0.05 against the square root of the cumulative probability. Fit a least-squares regression line to the data. The concentration of the toxicant corresponding to a cumulative probability of 0.05 on the regression line is called the *final acute value* (FAV).

4. The criterion maximum concentration is derived from the final acute value using the relation

$$\text{CMC} = a \times \text{FAV} \tag{2.23}$$

where a corrects the FAV values derived from the 50% lethality value LC_{50} rather than from a threshold lethal (zero mortality) effective concentration (USEPA, 1991a). A value of 0.5 is recommended for a (Connolly and Thomann, 1991).

Although most water-quality engineers are not directly involved in the derivation of a CMC, it is important that all understand how this acute criterion is obtained so that applications involving the CMC can be fully understood.

Example 2.4   Tests were sponsored by USEPA to assess the aquatic toxicity of the pesticide dieldrin in salt water. The results of these tests are given in in Table 2.10 in terms of the SMAV and GMAV for several saltwater aquatic organisms. Use these results to estimate the FAV and the CMC for dieldrin.

TABLE 2.10   Mean Acute Values of Dieldrin in Seawater

Species	SMAV (ppb)	GMAV (ppb)
Sphaeroides maculatus, northern puffer	34.0	34.0
Crassostrea virginica, eastern oyster	31.2	31.2
Mugil cephalus, striped mullet	23.0	23.0
Palaemonetes vulgaris, grass shrimp	50.0 ⎫	20.7
Palaemonetes vulgaris, grass shrimp	8.6 ⎭	
Morone saxatilis, striped bass	19.7	19.7
Pagurus longicarpus, hermit crab	18.0	18.0
Gassterosteus aculatus, threespine stickleback	14.2	14.2
Palaemon macrodactylus, Korean shrimp	10.8	10.8
Cyprinodon variegatus, sheepshead minnow	10.0	10.0
Crangon septemspinosa, sand shrimp	7.0	7.0
Fundulus heteroclitus, mummichog	8.9 ⎫	6.7
F. majalis, striped killifish	5.0 ⎭	
Thalassoma bifasciatum, bluehead	6.0	6.0
Menidia menidia, Atlantic silverside	5.0	5.0
Mysidopsis bahia, mysid shrimp	4.5	4.5
Micrometrus minimus, dwarf perch	3.5	3.5
Cyamatogaster aggregata, shiner perch	2.3	2.3
Oncorhynchus tshawytscha, chinook salmon	1.5	1.5
Anguilla rostrata, American eel	0.9	0.9
Panaeus duorarum, pink shrimp	0.7	0.7

Source: USEPA (1980).

TABLE 2.11 Results for Example 2.4

GMAV (ppb)	Rank	Cumulative Probability, P	$\sqrt{P}$	GMAV (ppb)	Rank	Cumulative Probability, P	$\sqrt{P}$
0.7	1	0.05	0.22	10.0	11	0.55	0.74
0.9	2	0.10	0.32	10.8	12	0.60	0.77
1.5	3	0.15	0.39	14.2	13	0.65	0.81
2.3	4	0.20	0.45	18.0	14	0.70	0.84
3.5	5	0.25	0.50	19.7	15	0.75	0.87
4.5	6	0.30	0.55	20.7	16	0.80	0.89
5.0	7	0.35	0.59	23.0	17	0.85	0.92
6.0	8	0.40	0.63	31.2	18	0.90	0.95
6.7	9	0.45	0.67	34.0	19	0.95	0.97
7.0	10	0.50	0.71				

SOLUTION   Table 2.11 shows the GMAVs ranked from lowest to highest, where denoting each rank by m, the genus ranks are converted to cumulative probabilities, P, using the relation

$$P = \frac{m}{N+1}$$

where N is the total number of data points ($N = 19$) and for plotting purposes, $\sqrt{P}$ is also calculated.

The cumulative probabilities associated with the four most sensitive genera are 0.05, 0.10, 0.15, and 0.20; and the GMAVs and the least-squares regression line fitted to these four points are shown in Figure 2.26. It is apparent that the regression line actually gives an excellent fit to the entire data set, although it was fit only to the four lowest GMAV values. According to the regression line, the dieldrin concentration corresponding to a cumulative probability of 5% is 0.63 ppb. Therefore, the FAV for

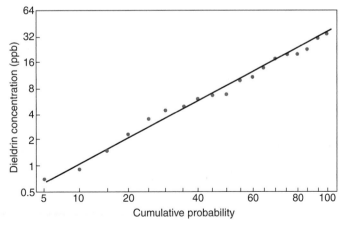

FIGURE 2.26   Cumulative probability distribution of dieldrin GMAV.

dieldrin in salt water is 0.63 ppb. The CMC is derived from the FAV using Equation 2.23, which gives

$$\text{CMC} = a \times \text{FAV} = 0.5 \times 0.63 \text{ ppb} = 0.32 \text{ ppb}$$

where a has been taken as 0.5. Therefore, the toxicity data for dieldrin indicate that a CMC value of 0.32 ppb would be appropriate.

Derivation of the Criterion Continuous Concentration To determine the CCC, the following two values are determined from toxicity tests on individual aquatic species: (1) the no-observed-effect concentration (NOEC); and (2) the lowest-observed-effect concentration (LOEC). These measures are related to the long-term viability of the species and include sublethal effects such as the loss of reproductive capability, reduction in mobility, and change in feeding and reduction of metabolic rates. To establish the CCC, the following steps are followed (Novotny, 2003),

1. For each species for which toxicity data are available, compute the geometric mean of the NOEC and the LOEC. This is called the *chronic value* of the species.
2. If sufficient chronic values are available (at least three species), they are analyzed in the same way as outlined in the procedure described previously for CMC. In this case, the 5% cumulative probability chronic value is called the *final chronic value* (FCV), and the CCC is given by

$$\text{CCC} = \text{FCV} \qquad (2.24)$$

3. In the absence of a sufficient chronic toxicity database, the ratios of LC_{50} or EC_{50} to the chronic value, called the *acute/chronic toxicity ratio* (ACR), is determined for each species. ACR values are needed from at least three families, including one fish, one invertebrate, and one acutely sensitive species. The final ACR (FACR) is calculated in one of four ways:
    a. If no major trend is apparent among the ACRs for the different species and the species mean ACRs lie within a factor of 10 for a number of species, the FACR is the geometric mean of all the species ACRs available for both fresh- and saltwater species.
    b. If the species mean ACRs seem to be correlated with the species mean acute values (SMAVs), the FACR should be taken to be the ACR of species whose acute values lie close to the FAV.
    c. In the case of acute toxicity tests conducted on metals and possibly other substances with embryos and larvae of barnacles, bivalve mollusks, sea urchins, lobsters, crabs, shrimp, and abalones, the FACR is assumed to be 2.
    d. If the most appropriate species mean ACRs are less than 2, the FACR is assumed to be 2.

    After determining the the final acute/chronic toxicity ratio (FACR) using one of the foregoing methods, the FCV is determined using the relation

$$\text{FCV} = \frac{\text{FAV}}{\text{FACR}} \qquad (2.25)$$

and the CCC is taken as equal to FCV, as indicated by Equation 2.24.

TABLE 2.12   Data for Example 2.5

Species	Acute Value (ppb)	Chronic Value (ppb)	Acute/Chronic Ratio
Salmo gairdneri, rainbow trout	2.5	0.22	11
Poecilia reticulata, guppy	4.1	0.45	9.1
Mysidopsis bahia, mysid shrimp	4.5	0.73	6.2

Example 2.5   The acute/chronic ratios (ACRs) for dieldrin toxicity to three species of aquatic animals are shown in Table 2.12, where the SMAV and the chronic value for each species are used to calculate the ACR. If the final acute values for dieldrin in fresh and salt water are 0.48 and 0.71 ppb, respectively, determine the final chronic values for dieldrin in fresh and salt water, and estimate the appropriate CCC.

SOLUTION   Since the ACRs for the three species differ by less than a factor of 2, the FACR is taken as the geometric mean of the three species ACRs; hence,

$$FACR = [11(9.1)(6.2)]^{1/3} = 8.5$$

Since the FAVs for fresh and salt water are 0.48 and 0.71 ppb, the corresponding FCVs are given by

$$FCV = \frac{FAV}{FACR}$$

$$\Rightarrow \frac{0.48}{8.5} = 0.056 \text{ ppb (freshwater)}, \quad \frac{0.71}{8.5} = 0.084 \text{ ppb (salt water)}$$

Since the CCC is taken as equal to the FCV, the given data lead to a dieldrin CCC of 0.056 ppb in freshwater and 0.084 ppb in salt water.

Toxicity to Plants   The USEPA guidelines described previously for calculating acute and chronic concentrations of a toxicant pertain only to aquatic animals, not to plants. However, because plants are the primary producers of organic matter in most aquatic food chains, it is critical that the effects of toxicants on aquatic plants be assessed. The typical aquatic plant *Vallisneria gigantea*, also known as giant eel grass, is shown in Figure 2.27. According to the USEPA guidelines (USEPA, 1986), a substance is toxic to a plant at a given concentration if the growth of the plant is decreased in a 96-h or longer experiment. Growth may be measured in a variety of ways, including photosynthetic rate, change in dry weight, or change in chlorophyll concentration. The *final plant value* (FPV) is the lowest concentration of the toxicant that reduces the plant growth in an appropriate experiment. In cases where the final plant value is available, the CCC is taken as the smaller of the final plant value and the final chronic value.

Example 2.6   Concentrations of the pesticide dieldrin that reduce growth in aquatic plants are shown in Table 2.13. *S. quadricaudata* is a freshwater species and the others are salt-water species. (a) Use this limited data set to estimate the FCV of dieldrin in freshwater and

FIGURE 2.27   *Vallisneria gigantea* (aquatic). (From San Marcos Growers, 2005.)

TABLE 2.13   Data for Example 2.6

Species	Effects	Concentration (ppb)
Scenedesmus quadricaudata	22% reduction in biomass in 10 days	100
Navicula seminulum	50% reduction in growth in 5 days	12,800
Wolffla papulifera	Reduced population growth in 12 days	10,000
Agmenellum quadruplication	Reduced growth rate	950

seawater. (b) If the FCV of dieldrin in freshwater is 0.056 ppb, estimate the appropriate CCC for dieldrin in freshwater.

SOLUTION   (a) Since the FPV is defined as the lowest concentration that produces an adverse effect, the FPV of dieldrin in freshwater is 100 ppb and the FPV in salt water is 950 ppb. (b) The appropriate CCC in freshwater is the lesser of the FCV (0.056 ppb) and the FPV (100 ppb) and is therefore equal to 0.056 ppb.

The FCV is not always less than the FPV.

Criteria for Human-Health Protection   A substance is a carcinogen if an abnormal incidence of cancer occurs in experimental animals exposed to the substance compared to a control group that in not exposed to the substance. The excess incidence of cancer can be plotted as a function of the dose and an analytical function fit to the data. The analytical function typically used in practice has the form (Crump, 1984)

$$A(d) = 1 - \exp(-q_1 d - q_2 d^2 - \cdots - q_k d^k) \qquad (2.26)$$

where $A(d)$ is the extra risk over background associated with a dose d and q_i are constants determined by a least-squares curve-fitting procedure. The model given by Equation 2.26

is based on the assumption that tumors are formed as a result of a sequence of biological events, and this model is called the *multistage model*. An important characteristic of Equation 2.26 is the fact that at low doses it reduces to the form

$$A(d) = q_1 d \qquad (2.27)$$

and hence conforms to the assumption that there is no threshold dose below which the chemical exerts no carcinogenic effects. The levels of exposure for most humans are likely to fall in the range where $A(d)$ can be assumed to be directly proportional to d, as indicated by Equation 2.27. For carcinogenic risk greater than 0.01, the following equation is preferable:

$$A(d) = 1 - \exp(-q_1 d) \qquad (2.28)$$

The parameter q_1, called the *slope factor*, is equal to the incremental risk per unit dose, typically expressed in per mg/kg·day. Since animal studies are only capable of detecting risks on the order of 1%, estimating cancer risks at low dose levels where Equation 2.27 is applicable represents considerable extrapolation of the experimental data.

Human-health criteria in the federal water-quality criteria (Appendix C.1) for priority and nonpriority pollutants are based on considerations of cancer potency or systemic toxicity, exposure, and risk characterization. The USEPA typically considers only exposures to a pollutant that occur through ingestion of water and contaminated fish and shellfish. Beaches are of particular concern since the opportunity for ingestion of contaminated water and the consumption of fish exposed to these waters can be significant. For this reason, the water quality at beaches are measured frequently, and beaches are closed when the concentration of pathogenic-organism indicators exceeds acceptable levels. Beach closings are a fairly regular occurrence in some locations, and Figure 2.28 shows a typical sign that is posted when a beach is closed.

FIGURE 2.28   Beach closed due to human-health risk. (From USEPA, 2005g.)

The consumption of contaminated fish tissue is of serious concern because the presence of even extremely low ambient concentrations of bioaccumulative pollutants (sublethal to aquatic life) in surface waters can result in residue concentrations in fish tissue that can pose a human-health risk. Chemicals that are hydrophobic tend to be lipophilic and therefore tend to partition into the fat tissue; this process is called *bioaccumulation*. The danger of bioaccumulation can be illustrated by the classic case of DDT (Clark, 1997). The pesticide DDT (dichlorodiphenyltrichloroethane) has about the same toxicity as aspirin; a lethal dose of aspirin is about 100 tablets and the same quantity of DDT is also lethal. But it is possible to take 0.5 to 1.0 g of aspirin a day indefinitely without ill effects because it is excreted. DDT is not excreted (it bioaccumulates), so a lethal dose can be acquired after repeated exposure.

The USEPA's human-health criteria assume a human body weight of 70 kg and a consumption of 6.5 g of fish and shellfish per day. The ratio of the contaminant concentrations in fish tissue to that in water is measured by the *bioconcentration factor* (BCF), and the units of BCF are $(\mu g/g)/(\mu g/L) = L/g$ expressed for body weight. Since many toxic chemicals tend to accumulate primarily in the lipid (fat) tissue of organisms, the BCF may be normalized by the weight fraction of the lipid. Bioconcentration is the direct absorption of a chemical into an individual organism by nondietary routes (e.g., from water into a fish by the gills). *Bioaccumulation* refers to the accumulation of chemical species both by exposure to contaminated water (bioconcentration) and ingestion of contaminated food (USEPA, 1994). Bioaccumulation is measured by the *bioaccumulation factor* (BAF), which is defined as the ratio of the contaminant concentration in the fish tissue (as a result of bioaccumulation) to the concentration of the contaminant in the surrounding water. BCF and BAF have the same units and assume that equilibrium conditions have been attained between the organism (fish) and the surrounding water. There is a concentrating effect in each successive layer up the food chain, and the BAF/BCF ratio at each level is called the *food-chain multiplier* (FM). Values of FM range from 1 to 100, with the highest ratios applying to organisms in higher trophic levels and to chemicals with high sorption coefficients. The bioaccumulation and bioconcentration factors for a substance are related as follows:

$$BAF = FM \times BCF \qquad (2.29)$$

Values of FM as a function of sorption coefficient and trophic level can be found in Thomann (1987, 1989).

The *Integrated Risk Information System* (IRIS) is an electronic database of the USEPA that provides chemical-specific information on the relationship between chemical exposure and estimated human health effects. IRIS contains two types of quantitative risk values: the *oral reference dose* (RfD) and the *carcinogenic potency estimate* or *slope factor*. The RfD is a threshold below which systemic toxic effects are unlikely to occur, and the slope factor is defined as a plausible upper-bound estimate of the probability of a response-per-unit intake of a chemical over a lifetime. The slope factor, which was defined as q_1 in Equation 2.28, is used to estimate an upper-bound probability of a person developing cancer as a result of a lifetime of exposure to a particular level of a potential carcinogen.

Human-Health Criteria Based on Noncarcinogenic Effects   The RfD is the primary human-health measure of noncarcinogenic effects. The RfD is expressed in units of milligrams of toxicant per kilogram of human body weight per day. RfDs are derived from the *no-observed-adverse-effect level* (NOAEL) or the lowest-observed-adverse-effect level (LOAEL) identified from chronic or subchronic human-epidemiology studies or animal-exposure studies. Uncertainty factors are applied to the NOAEL or LOAEL to account for uncertainties in the data associated with transfer from animals to humans, variability among individuals, and other uncertainties in developing the data.

The equation for deriving human-health criteria based on noncarcinogenic effects is given by (USEPA, 1994)

$$c = \frac{(RfD \times WT) - (DT + IN) \times WT}{WI + (FC \times L \times FM \times BCF)} \tag{2.30}$$

where c is the human-health water-quality criterion (mg/L), RfD is the oral reference dose (mg toxicant/kg human body weight per day), WT is the weight of an average human adult (typically, 70 kg), DT is the dietary exposure other than fish (mg toxicant/kg human body weight per day), IN is the inhalation exposure (mg toxicant/kg human body weight per day), WI is the average human adult water intake (typically, 2 L/day), FC is the daily fish consumption (kg fish/day), L is the ratio of lipid fraction to fish tissue consumed (typically, 3%), FM is the food-chain multiplier, and BCF is the bioconcentration factor (mg toxicant/kg fish divided by mg toxicant/L water) for fish with a 3% lipid content. If the designated use of a water body is anything other than as a drinking-water source, factor WI in Equation 2.30 can be deleted. The product of the reference dose (RfD) and weight of an average human adult (WT) is commonly called the *acceptable daily intake* (ADI). The RfD values associated with oral ingestion of several noncarcinogenic substances are listed in Table 2.14.

Example 2.7   Available data on the noncarcinogenic effects of cadmium indicate a reference dose of 0.69 μ/kg·day and a fish-consumption rate of 6.5 g/day in areas where cadmium contamination could be a problem. If the FM is taken as 10 and the BCF for fish with a 3% lipid content is 122 L/g, estimate the human-health criterion for cadmium in aquatic systems.

SOLUTION   From the data given, RfD = 0.69 μg/kg·day, FC = 6.5 g/day = 0.0065 kg/day, FM = 10, and BCF = 122 L/g. Taking WT = 70 kg, WI = 2 L/day, and L = 3% = 0.03, and assuming that there is no exposure to cadmium other than drinking water and eating fish, then DT = 0, IN = 0, and Equation 2.30 gives the indicated human-health criterion for cadmium as

$$c = \frac{(RfD \times WT) - (DT + IN) \times WT}{WI + (FC \times L \times FM \times BCF)} = \frac{(0.69 \times 70) - (0 + 0)(70)}{2 + [0.0065(0.03)(10)(122)]} = 21 \, \mu g/L$$

These results indicate that a maximum concentration of 21 µg/L in aquatic waters would be protective of human health. It is interesting to note that the actual water-quality criterion for cadmium is 5 μg/L.

The USEPA uses the *hazard quotient* (HQ) to measure the level of concern for potential noncancer effects. This parameter is defined by the relation

$$HQ = \frac{intake}{RfD} \tag{2.31}$$

TABLE 2.14   RfD Values for Noncarcinogenic Effects

Substance	Reference Dose, RfD (mg / kg·day)
Acetone	0.1
Arsenic	0.0003
Barium	0.05
Cadmium	0.0005
Chlordane	0.0005
Chloroform	0.01
Chromium(VI)	0.003
Cyanide	0.02
1,1-Dichloroethlyene	0.009
Hydrogen cyanide	0.02
Methylene chloride	0.06
Methyl mercury	0.0001
Napthalene	0.02
Pentachlorophenol	0.03
Phenol	0.6
Silver	0.003
Tetrachloroethylene	0.01
1,2,4-Trichlorobenzene	0.02
Trichloroethylene	0.006
Toluene	0.2
Xylenes	2.0

Source: USEPA IRIS database.

As a rule, the greater the value of HQ above unity, the greater the level of concern. A cumulative *hazard index* (HI) for a contaminated site is calculated by adding the HQs for all chemicals of concern over all exposure pathways.

Human-Health Criteria Based on Carcinogenic Effects   The equation for deriving human-health criteria based on carcinogenic effects is given by

$$c = \frac{RL \times WT}{q_1 \, [WI + FC \times L \times (FM \times BCF)]} \tag{2.32}$$

where c is the water-quality criterion (mg/L), RL is the acceptable-risk level (10^{-x}) where x is usually in the range 4 to 6, WT is the weight of an average human adult (typically, 70 kg), q_1 is the slope factor or carcinogenic potency factor (kg·day/mg), WI is the average human adult water intake (typically, 2 L/day), FC is the daily fish consumption (kg fish/day), L is the ratio of lipid fraction to fish tissue consumed (typically, 3%), FM is the food-chain multiplier, and BCF is the bioconcentration factor (mg toxicant/kg fish divided by mg toxicant/L water) for fish with a 3% lipid content. A carcinogenicity risk of 10^{-6} is used in deriving human-health water-quality criteria; however, alternative risk levels may be obtained by moving the decimal point; for example, for a risk level of 10^{-5}, move the decimal point in the recommended criterion one place to the right (USEPA, 2002a).

TABLE 2.15 Slope Factors for Potential Carcinogens

Substance	Slope Factor (per mg/kg·day)
Arsenic	1.5
Benzene	0.029
Benzo[*a*]pyrene	7.3
Carbon tetrachloride	0.13
Chlordane	0.35
Chloroform	0.0061
DDT	0.34
1,1-Dichloroethylene	0.6
Dieldrin	16.0
Heptachlor	4.5
Hexachloroethane	0.014
Methylene chloride	0.0075
Polychlorinated biphenyls	1.0
2,3,7,8-TCDD (dioxin)	1.5×10^5
Tetrachloroethylene	0.052
Trichloroethylene	0.011
Vinyl chloride	1.9

Source: USEPA IRIS database.

Despite the uncertainties in human- and animal-health studies, consensus values of cancer slope factors are available. These values are based on the deliberations of expert panels reviewing the available studies and the overall weight of evidence. Upper 95% confidence level estimates are often used for slope factors, to provide a degree of conservatism in the evaluation. The slope factors associated with oral ingestion of several carcinogenic substances are listed in Table 2.15.

Example 2.8 Data available on the carcinogenic effects of Cr(IV) indicate a potency factor of 41 per mg / kg·day and a fish-consumption rate of 7 g/day in areas where Cr(IV) contamination could be a problem. If the FM is taken as 10 and the BCF for fish with a 3% lipid content is 100 L/g, estimate the human-health criterion for Cr(IV) in aquatic systems.

SOLUTION   From the data given, $q_1 = 41$ per mg/kg·day, FC $= 7$ g/day $= 0.007$ kg / day, FM $= 10$, and BCF $= 100$ L/g. Taking WT $= 70$ kg, WI $= 2$ L/day, RL $= 10^{-6}$, and $L = 3\% = 0.03$, Equation 2.32 gives the indicated human-health criterion for Cr(IV) as

$$c = \frac{\text{RL} \times \text{WT}}{q_1(\text{WI} + \text{FC} \times L \times \text{FM} \times \text{BCF})} = \frac{10^{-6} \times 70}{41(2 + (0.007)(0.03)(10)(100))}$$

$$= 7.72 \times 10^{-7} \, \text{mg/L} \approx 0.001 \, \mu\text{g/L}$$

These results indicate that a maximum concentration of $0.001 \, \mu$g/L in aquatic waters would be protective of human health. This indicates that small amounts of Cr(IV) in aquatic waters would pose a significant health risk to humans. This result is not reflected in the U.S. federal water-quality criteria in that there is no human-health aquatic criterion for Cr(IV), the effect of Cr(IV) on aquatic life is the focus of the federal water-quality criteria.

Potential carcinogenic health effects are estimated by calculating the individual excess lifetime cancer risk (IELCR) from the carcinogenic intake rate, I_c, and the slope factor, q_1, according to the relation

$$IELCR = I_c q_1 \tag{2.33}$$

The total lifetime cancer risk is estimated by summing the IELCRs for all chemicals over all potential exposure pathways. The USEPA considers IELCRs in the range 10^{-6} to 10^{-4} to be acceptable for regulatory purposes. It is important to keep in mind that these numbers represent excess cancer risks, that is, the incremental risk of developing cancer due to contaminant exposure that is above general background.

2.3.3 Antidegradation Policy

All water-quality standards must include an antidegradation policy. The United States uses a three-tiered antidegradation approach. Tier 1 maintains and protects existing uses and water-quality conditions necessary to support such uses. Tier 1 requirements are applicable to all surface waters. Tier 2 maintains and protects "high-quality" waters—water bodies where existing conditions are better than necessary to support "fishable/swimmable" uses. Water quality can be lowered in such waters; however, state and tribal programs identify procedures that must be followed and questions that must be answered before a reduction in water quality can be allowed. In no case may water quality be lowered to a level that would interfere with existing or designated uses. Tier 3 maintains and protects water quality in outstanding national resource waters (ONRWs). Except for certain temporary changes, water quality cannot be lowered in such waters. ONRWs generally include the highest-quality waters of the United States and waters of exceptional ecological significance. Decisions regarding which water bodies qualify to be ONRWs are made by states and authorized Indian tribes.

Antidegradation implementation procedures identify the steps and questions that must be addressed when regulated activities are proposed that may affect water quality. The specific steps to be followed depend on which tier of antidegradation applies.

2.3.4 General Water-Quality Management Practices

Mixing Zones In many cases, discharges into surface waters do not meet the water-quality criteria of the receiving water body. Whenever such discharges are deemed to be in the public interest, a *mixing zone* is delineated surrounding the discharge location to allow a sufficient amount of dilution such that the water-quality criteria are met outside the mixing zone. A typical wastewater discharge into the ocean is shown in Figure 2.29, where the increasing dilution of the effluent with distance from the discharge point is clearly evident.

The USEPA defines a mixing zone as a limited area or volume of water where initial dilution of a discharge takes place and where numeric water-quality criteria can be exceeded but acutely toxic conditions are prevented. In establishing mixing zones, it is generally required that these zones (1) do not impair the integrity of the water body as a whole; (2) there is no lethality to organisms passing through the mixing zone; and (3) there

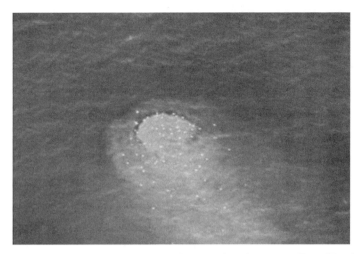

FIGURE 2.29   Spreading of a wastewater discharge into the ocean. (From Wood, 2005a.)

are no significant health risks considering likely pathways of exposure, and they do not endanger critical areas such as drinking-water supplies, recreational areas, breeding grounds, and areas with sensitive biota (USEPA, 1994). Typically, mixing zones are not allowed for certain persistent, carcinogenic, mutagenic, teratogenic, or other highly toxic chemicals that bioaccumulate in the food chain, such as mercury, dioxins, polychlorinated biphenyls, and various pesticides. Eliminating mixing zones for bioaccumulative pollutants may be appropriate when they encroach on areas used for fish harvesting, particularly of stationary species such as shellfish.

Most states allow regulatory mixing zones as a matter of policy. The regulatory mixing zone allows for the initial dilution of a discharge rather than imposing strict end-of-pipe concentration requirements for NPDES water-quality permits for conventional and toxic discharges. In theory, the regulatory mixing zone allows for efficient natural pollutant assimilation and in practice can be used as long as the integrity of the water body as a whole is not impaired. Individual (state) water-quality standards generally describe a methodology for determining the location, size, shape, outfall design, and in-zone quality of mixing zones. Appropriate considerations for delineating mixing zones are as follows (USEPA, 1994):

1. *Location.* Water-quality standards must be met at the edge of the regulatory mixing zone during design-flow conditions, and a continuous zone of passage must be provided that meets water-quality criteria for free-swimming and drifting organisms. In river systems, reservoirs, lakes, estuaries, and coastal waters, zones of passage are defined as continuous water routes of such volume, area, and quality as to allow passage of free-swimming and drifting organisms so that no significant effects are produced on their populations.

2. *Size.* The area or volume of a mixing zone must be limited to an area or volume as small as practicable that will not interfere with the designated use or with the established community of aquatic life in the segment for which the use is designated.

3. *Shape.* The shape of a mixing zone should be a simple configuration that is easy to locate in a body of water and avoids impingement on biologically important areas. In lakes, a circle with a specified radius is generally preferable, but other shapes may be specified in the case of unusual site requirements. Most states provide spatial dimensions to limit the areal extent of the mixing zones.

4. *In-zone quality.* In the general case where a state has both acute and chronic aquatic-life criteria, as well as human-health criteria, independently established mixing-zone specifications may apply to each of the three types of criteria. Typically, chronic aquatic-life criteria are not met within the mixing zone, and if acute aquatic-life criteria are met within the mixing zone, no lethality should result from temporary passage of aquatic life through the mixing zone. Additional protections required within the mixing zones include freedom from contaminants that settle to form objectionable deposits; freedom from floating debris, oil, scum, and other material in concentrations that form nuisances; freedom from substances in concentrations that produce objectionable color, odor, taste, or turbidity; and freedom from contaminants that produce undesirable aquatic life or result in a dominance of nuisance species.

The state of Florida defines a mixing zone as "a volume of surface water containing the point or area of discharge and within which an opportunity for the mixture of wastes with receiving surface waters has been afforded" (Florida DEP, 1996). Mixing zones in freshwater streams in Florida are limited to 10% of the length of the stream or 800 m, whichever is less (Florida DEP, 1995). In lakes, estuaries, bays, lagoons, bayous, sounds, and coastal waters, the mixing zones are limited to 10% of the total area of the water body or $125,600 \, \text{m}^2$ (= area enclosed by a circle of radius 200 m), whichever is less, and in open ocean waters the area of the mixing zone must be less than or equal to $502,655 \, \text{m}^2$ (= area enclosed by a circle of radius 400 m).

Low Flows   Water-quality standards should protect water quality for designated uses in critical low-flow conditions. States may designate a critical low flow below which numerical water-quality criteria do not apply (USEPA, 1994), and the USEPA recommends that numeric water-quality criteria apply to all flows that are greater than or equal to the flows specified in Table 2.16, where the following definitions are used: 1Q10 is the lowest 1-day flow, with an average exceedance frequency of once in 10 years; 1B3 is biologically based and indicates an allowable exceedance for 1 day every 3 years; 7Q10 is the lowest average 7-consecutive-day low flow, with an average exceedance frequency of once in 10 years;

TABLE 2.16   USEPA Recommendations for Low Flows

Aquatic Life	
Acute criteria (CMC)	1Q10 or 1B3
Chronic criteria (CCC)	7Q10 or 4B3
Human Health	
Noncarcinogens	30Q5
Carcinogens	Harmonic mean flow

Source: USEPA (1994).

(a)   (b)

FIGURE 2.30   (*a*) Normal and (*b*) low-flow conditions in the Flint River. (From USGS, 2005d.)

4B3 is biologically based and indicates an allowable exceedance for 4 consecutive days once every 3 years; 30Q5 is the lowest average 30-consecutive-day low flow with an average exceedance frequency of once in 5 years; and the *harmonic mean flow* is a long-term mean flow value calculated by dividing the number of daily flows analyzed by the sum of the reciprocals of those flows. Flow conditions in the Flint River near Culloden, Georgia, at normal flow and low flow (during the June 2000 drought) are shown in Figure 2.30. The lower flow velocities and lower water levels are clearly evident during low-flow conditions.

Monitoring of Water Quality   Monitoring is done to assess the physical, chemical, and biological integrity of water bodies. Physical monitoring may reveal a part of the problem, chemical monitoring reveals only the water-quality situation at the time samples are taken, and biological monitoring best reveals the water-quality problem but may not indicate the cause (Novotny, 2003). Water-quality sampling is typically done by experienced field personnel; Figure 2.31 shows a typical sample-collection effort.

Since aquatic-life water-quality standards are stated in terms of the average duration and frequency of exceedance, any monitoring protocol must collect sufficient data to assess whether such probabilistic criteria are being met. Ideally, required water-quality monitoring data would consist of continuous concentration measurements over periods of many years. In reality, water-quality data are collected over short intervals such as once per week, once per month, or once per year, and appropriate statistical techniques must be used to analyze such data to verify that statistical water-quality standards are being met.

In the case of verifying compliance with the CMC criterion, the monitoring objective typically requires validation that the 1-day averaged concentration with a recurrence interval 3 years (i.e., the 1B3 concentration) is less than or equal to the CMC. For a set of 1-day averaged measurements, the allowable exceedance probability, P_{CMC}, of the CMC is given by

$$P_{\mathrm{CMC}} = \frac{1}{3 \text{ years} \times 365 \text{ days}} = 0.001 \qquad (2.34)$$

To assess whether the measured (daily-averaged) concentration data indicate an exceedance probability greater than 0.001, a theoretical probability distribution is fitted to

FIGURE 2.31 Water-quality sampling. (From USGS, 2005e.)

the measured data, and the exceedance probability of the CMC is determined from this fitted theoretical distribution. The conventional procedure for fitting a theoretical probability distribution to measured data is widely used in hydrology and consists of the following steps (Chin, 2006): (1) rank the measured data; (2) assign an exceedance probability, $P(c)$ to the measurement, c, of rank m by

$$P(c) = \frac{m}{N+1} \qquad (2.35)$$

where N is the total number of measurements; and (3) fit a theoretical probability distribution to $P(c)$. The most common fitted probability distribution is the lognormal distribution. The approach outlined here for compliance with the CMC can also be used for assessing compliance with the CCC, but in the latter case the measured concentration data must be continuous 4-day averages, and the allowable exceedance probability, P_{CCC}, of the CCC is given by

$$P_{CCC} = \frac{1}{3 \text{ years} \times (365/4) \text{ days}} = 0.004 \qquad (2.36)$$

In some cases, the CMC and CCC are given in terms of allowable excursions below a specified threshold level, such as in the case of dissolved oxygen. In these cases, the fitted theoretical distribution is used to assess whether the measured data indicate excursions below the threshold with allowable frequency.

Verifying compliance with human-health criteria is simpler than for CMC and CCC, since the human-health criteria simply require that all observations be less than the stated criterion.

Ocean Discharges Ocean discharges are point discharges into the waters of the United States and, as such, require NPDES permits. Ocean discharges of treated domestic wastewater are typically accomplished using multiport diffusers, and an example of such a discharge is shown in Figure 2.32. The individual port discharges are shown within the boxed area, where each individual port discharge produces a "boil" at the ocean surface. Since ports in multiport diffusers are typically equally spaced, the conditions shown in Figure 2.32 indicate that one of the ports may be clogged.

The technology-based discharge requirement that is typically part of the NPDES permit for ocean discharges is for at least secondary treatment of domestic wastewater by publicly owned treatment works (POTWs). Some municipalities with POTWs have argued that the secondary-treatment requirement for marine disposal of domestic wastewater might be unnecessary on the grounds that marine POTWs usually discharge into deeper waters with large tides and substantial currents, which allow for greater dilution and dispersion than do their freshwater counterparts. The Clean Water Act allows for a case-by-case review of treatment requirements for marine dischargers. Eligible POTW applicants that meet a set of environmentally stringent criteria receive a modified National Pollutant Discharge Elimination System (NPDES) permit waiving the secondary-treatment requirements for BOD, SS, and pH. Even with a secondary-treatment waiver, dischargers must meet water-quality standards on the boundary of the mixing zone, establish a monitoring program to assess impacts, and provide a minimum of primary or equivalent treatment.

FIGURE 2.32 Ocean discharge from a multiport diffuser. (From Wood, 2005b.)

TABLE 2.17 USEPA Classifications of Ground Waters

Class	Description
I	Groundwaters of unusually high value that are in need of special protective measures because they have a relatively high potential for contaminants to enter and/or be transported within the groundwater flow system. The goal of the USEPA protection strategy for class I aquifers is to keep pollutant concentrations below the drinking-water MCLs.
II	Groundwaters that are currently used or potentially available for drinking water and other beneficial use. The class II designation allows a limited siting of hazardous-waste facilities and while ensuring quality suitable for drinking, permits exemptions to requirements under certain circumstances to allow less stringent standards when the protection of human health and the environment can be demonstrated.
III	Groundwaters that are not potential sources of drinking water and are of limited beneficial use either because they are saline or because they are otherwise contaminated beyond levels that allow drinking or other beneficial use. Regulations for the protection of class III groundwaters are more lenient and allow siting of hazardous waste facilities, but they include technical requirements and monitoring rules to protect public health.

Source: USEPA (1984b).

2.4 U.S. GROUND-WATER STANDARDS

Ground waters are generally classified according to their present and future most beneficial uses, and the USEPA recommends the three-part classification system shown in Table 2.17 for ground waters in the United States.

In states that rely heavily on ground water for their water supply, narrower classifications are usually adopted. As an example, the five classes of ground waters in Florida are shown in Table 2.18 arranged in order of degree of protection required, with class G-I having the most stringent water-quality criteria and class G-IV the least. Since many consumers utilize untreated ground water that is pumped directly from a well, water-quality criteria for ground waters classified for potable water use are generally the same as *primary* and *secondary* drinking-water quality standards that apply to public water supplies

TABLE 2.18 Classifications of Ground Waters in Florida

Class	Description
G-I	Potable water use, ground water in single-source aquifers that have a total dissolved solids content of less than 3000 mg/L
G-II	Potable water use, ground water in aquifers that have a total dissolved solids content of less than 10,000 mg/L
G-III	Nonpotable water use, ground water in unconfined aquifers that have a total dissolved solids content of 10,000 mg/L or greater.
G-IV	Nonpotable water use, ground water in confined aquifers that have a total dissolved solids content of 10,000 mg/L or greater.

at the tap. Common exceptions to the primary drinking-water standards as applied to ground water are the total coliform and asbestos requirement. In Florida, ground-water standards require a total coliform count of less than 4 per 100 mL (compared to zero for drinking water) and the asbestos drinking-water regulation does not apply. Primary drinking-water standards are health-based requirements, and secondary standards are aesthetic-based requirements. The (federal) drinking-water criteria in the United States can be found on the Web at http://www.epa.gov/safewater/mcl.html, and the current drinking-water criteria are listed in Appendix C.2. In cases where the quality of native ground water is poorer than the drinking-water standards, the quality of the (uncontaminated) native ground water serves as the water-quality standard.

2.5 BACKGROUND WATER QUALITY

Water-quality standards cannot be enforced if their violation is a result of natural causes. Background water quality is obtained from water-quality and biological measurements of reference water bodies located in the same ecoregion. *Ecoregions* represent regions with common physical geography, geology, soils, vegetation, land use, wildlife, and climate (Omernik, 1987; Gallant et al., 1989). Ecoregions are intended to provide a spatial framework for ecosystem assessment, research, inventory, monitoring, and management. These regions delimit large areas within which local ecosystems recur more or less throughout the region in a predictable pattern. By observing the behavior of the various kinds of systems within a region, it is possible to predict the behavior of an unvisited system. This provides an extrapolation mechanism for identifying areas from which site-specific knowledge on ecosystem behavior can be applied, and they also suggest areas within which similar responses and management strategies are applicable. Ecoregion classifications have been developed by the U.S. Forest Service (Baily, 1995) and the U.S. Environmental Protection Agency (Omernik, 1987, 1995). In the context of water quality, the USEPA system is preferred. The distribution of major ecoregions in the state of Texas according to the USEPA classification system is shown in Figure 2.33 Natural water quality within an ecoregion will vary less than between ecoregions.

The U.S. Geological Survey (USGS) has established a benchmark network of monitoring stations in watersheds that are among the least affected by human activities. These stations are found across the United States and are located primarily in national parks, wilderness areas, state parks, forests, and similar areas protected from development. Water quality measured at these stations provide the best approximation of background levels of contaminants in surface waters located in various ecoregions across the United States. There are four general types of native undisturbed lands (excluding mountains): arid land, prairie, wetland, and woodland. Although the background water quality in these lands varies, there are some common characteristics:

- Streams in arid lands are often ephemeral and typically have very high sediment content during intense but infrequent storm events, the salinity of these streams may be elevated, and nutrient content is low.
- Prairie streams have elevated solids content during wet-weather flow, and the nutrient content is usually low.

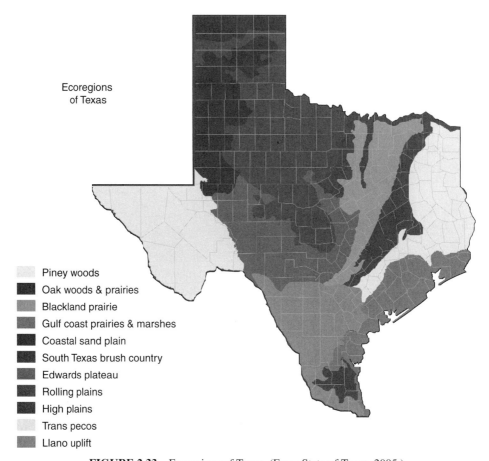

FIGURE 2.33 Ecoregions of Texas. (From State of Texas, 2005.)

- Forested land, including mountain forests, typically exhibit the best water quality, have low mineral content, and have almost no organic content. Lowland forest streams have higher organic content; often, these streams originate in wooded wetlands and the water may contain residues of organic decomposition occurring in soils and wetlands, with measurable BOD and COD.
- Streams draining wetlands and wetland water bodies have higher organic content and may also have low dissolved oxygen caused by the decomposition of organic matter. Humic substances increase the color and turbidity of these water bodies.

There are no background levels for organic chemicals such as pesticides, PAHs, halogenated hydrocarbons and volatile organic chemicals since these substances are all human-made and are not naturally occurring. The apparent background levels of these substances originate primarily from atmospheric fallout.

In summary, background water quality is controlled primarily by the morphological, geological, and geographical characteristics, land cover, soil type, and other ecological factors associated with the surrounding area. Background water quality should be measured rather than estimated.

2.6   COMPUTER CODES

In engineering practice, the use of computer models to apply the fundamental principles and protocols covered in this chapter is sometimes essential. There are usually a variety of models to choose from for a particular application, but in doing work to be reviewed by regulatory agencies, models developed and maintained by agencies of the U.S. government have the greatest credibility and, perhaps more important, are almost universally acceptable in supporting permit applications and defending design protocols. A secondary guideline in choosing a model is that the simplest model that will accomplish the design objectives should be given the most serious consideration. A widely used model developed and endorsed by an agency of the U.S. government is described briefly here.

AQUATOX is a freshwater ecosystem simulation model that was developed and is maintained by the USEPA (2004a,b). AQUATOX predicts the fate of various pollutants, such as nutrients and organic toxicants, and their effects on the ecosystem, including fish, invertebrates, and aquatic plants. AQUATOX is a valuable tool for ecologists, water-quality modelers, and anyone involved in performing ecological risk assessments for aquatic ecosystems. AQUATOX simulates the transfer of biomass, energy, and chemicals from one compartment of the ecosystem to another. It does this by computing each of the most important chemical or biological processes simultaneously for each day of the simulation period; therefore, it is known as a process-based or mechanistic model. AQUATOX can predict the environmental fate of chemicals in aquatic ecosystems and their direct and indirect effects on the resident organisms. AQUATOX has the potential to establish causal links between chemical water quality and biological response and aquatic life uses.

SUMMARY

Water quality can be measured in a variety of ways, and these measures can be classified as either physical, chemical, or biological. Physical measures that are related to water quality in streams include flow conditions, substrate characteristics, stream habitat, riparian habitat, and thermal pollution. Changes in any of these physical measures have the potential to affect significantly the quality of water in a stream. Chemical measures related to water quality include dissolved oxygen, biochemical oxygen demand, chemical oxygen demand, suspended solids, nutrient levels, toxic metals, synthetic organic chemicals, radionuclides, and pH. Water-quality criteria exist for all of these chemical measures. Biological measures of water-quality include number of species, presence of intolerant species, proportion of omnivores to carnivores, biomass production, and number of individuals per species. Biological measures, also called bioindicators, are compared with biological criteria in performing bioassessments. Although physical, chemical, and biological measures all give measures of water quality, biological measures give the best indication of the overall condition of a water body, and physical and chemical measures give indications of possible causes of biological impacts.

In the United States, water-quality standards have been developed for all natural water bodies, including ground and surface waters. Each natural water body is classified according to its designated use, and water-quality criteria are associated with each designated use. Surface-water standards are binary standards that consider contaminant levels that are protective of aquatic life and contaminant levels that are protective of human health. Aquatic-life standards are stated in terms of (1) the criterion maximum concentration, which is the

1-h average concentration with an allowable exceedance probability of once in 3 years; and (2) criterion continuous concentration, which is the 4-day average concentration with an allowable exceedance probability of once in 3 years. Water-quality criteria for human-health protection consider both the degree of ingestion of surface water and the consumption of fish derived from the surface water. In ground-water standards, drinking-water criteria are usually applicable, since many ground waters serve as direct sources of drinking water. Both ground-water and surface-water monitoring programs are necessary to measure background water quality, since water-quality standards cannot be set higher than the background water quality.

PROBLEMS

2.1. Saltwater intrusion into a river has increased the average chloride concentration to 3000 mg/L. If the summer water temperature is around 25°C and the winter temperature is around 15°C, compare the saturated dissolved oxygen level in the summer with the level in the winter.

2.2. At what temperature does the saturation concentration of dissolved oxygen in water fall below the minimum desirable level of 5 mg/L?

2.3. A BOD test on an industrial wastewater indicates that the 5-day BOD is 49 mg/L and that the ultimate carbonaceous BOD is 75 mg/L. Estimate the decay factor.

2.4. Analyses of an industrial wastewater indicate that nitrogenous BOD (NBOD) begins after 10 days of incubation and can be described by the same exponential function as the carbonaceous BOD (CBOD). If the rate constant is 0.1 day^{-1} and the 5-day BOD is 20 mg/L, estimate the total BOD after 20 days and the ultimate BOD.

2.5. What are the water-quality criteria for your state?

2.6. Use IRIS to obtain the oral reference dose (RfD) and the carcinogenic potency estimate or slope factor for five regulated contaminants.

CHAPTER 3

FATE AND TRANSPORT
IN AQUATIC SYSTEMS

3.1 MIXING OF DISSOLVED CONSTITUENTS

Diffusion and dispersion are the processes by which a tracer spreads within a fluid. *Diffusion* is the random advection of tracer molecules on scales smaller than some defined length scale. At small (microscopic) length scales, tracers diffuse primarily through Brownian motion of the tracer molecules, whereas at larger scales, tracers are diffused by random macroscopic variations in the fluid velocity. In cases where the random macroscopic variations in velocity are caused by turbulence, the diffusion process is called *turbulent diffusion*. Where spatial variations in the macroscopic velocity are responsible for the mixing of a tracer, the process is called *dispersion*. It is common practice to use the terms *diffusion* and *dispersion* interchangeably to describe the larger-scale mixing of contaminants in natural water bodies. In open waters, spatial variations in the macroscopic velocity are usually associated with shear flow and shoreline geometry, whereas in ground waters, macroscopic (seepage) velocity variations are caused by the complex pore geometry and spatial variations in hydraulic conductivity. The case of turbulent diffusion in an aqueous environment with a steady uniform mean flow is illustrated in Figure 3.1. In this case, the ambient velocity field consists of a large-scale mean flow with small-scale fluctuations about the mean. The fluctuations occur over length scales smaller than the width of the plume. Regardless of the mechanism responsible for the spatial variations in velocity, whenever these velocity variations are either truly random or spatially uncorrelated over a defined *mixing scale*, the mixing process is described by Fick's law (Fick, 1855), which can be stated in the generalized form

$$q_i^d = -D_{ij}\frac{\partial c}{\partial x_j} \tag{3.1}$$

Water-Quality Engineering in Natural Systems, by David A. Chin
Copyright © 2006 John Wiley & Sons, Inc.

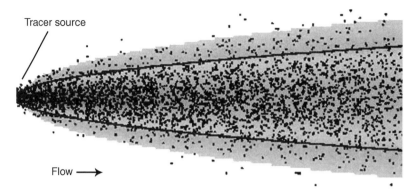

Flow ⟶

FIGURE 3.1 Turbulent diffusion of tracer particles in uniform flow. (From NOAA, 2005a.)

where q_i^d is the dispersive mass flux $[M/L^2T]$[1] in the x_i-direction, D_{ij} is the dispersion coefficient tensor, and c is the tracer concentration. In cases where the dispersion coefficient varies with direction, the dispersion process is called *anisotropic*; in cases where the dispersion coefficient is independent of direction, the dispersion process is called *isotropic*. Hence, for isotropic dispersion, Fick's law is given by

$$q_i^d = -D\frac{\partial c}{\partial x_i} \tag{3.2}$$

Whereas the Fickian relation given by Equation 3.2 parameterizes the mixing effect of velocity variations with correlation length scales smaller than some defined mixing scale or support scale, tracer molecules are also advected by larger-scale fluid motions. The mass flux associated with the larger-scale (advective) fluid motions is given by

$$q_i^a = V_i c \tag{3.3}$$

where q_i^a is the advective tracer mass flux $[M/L^2T]$ in the x_i-direction and V_i is the large-scale fluid velocity in the x_i-direction. Since tracers are transported simultaneously by both advection and dispersion, the total flux of a tracer within a fluid is the sum of the advective and diffusive fluxes given by

$$q_i = q_i^a + q_i^d = V_i c - D\frac{\partial c}{\partial x_i} \tag{3.4}$$

where q_i is the tracer flux in the x_i-direction. Equation 3.4 can also be written in vector form as

$$\mathbf{q} = \mathbf{V}c - D\nabla c \tag{3.5}$$

where $\mathbf{q}$ is the flux vector and $\mathbf{V}$ is the large-scale fluid velocity. The expression of the tracer flux in terms of an advective and diffusive component must generally be associated

---

[1]Dimensional units are shown in brackets throughout.

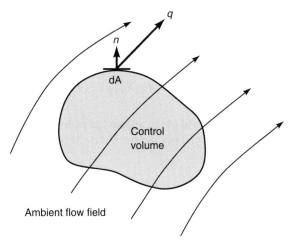

FIGURE 3.2 Control volume in a fluid transporting a tracer. (From Chin, David A., *Water-Resources Engineering*. Copyright © 2000. Reprinted by permission of Pearson Education, Inc., Upper Saddle River, NJ.)

with a length scale, L, that is a measure of the averaging volume used to estimate the advective velocity, $\mathbf{V}$, and the diffusion coefficient, D. The main distinction between advection and diffusion or dispersion is that advection is associated with a net movement of the center of mass of a tracer, whereas diffusion and dispersion are associated with spreading about the center of mass.

Consider the finite control volume shown in Figure 3.2, where this control volume is contained within the fluid transporting the tracer. In accordance with the law of conservation of mass, the net flux [M/T] of tracer mass into the control volume is equal to the rate of change of tracer mass [M/T] within the control volume. The law of conservation of mass can be put in the form

$$\frac{\partial}{\partial t}\int_V c\,dV + \int_S \mathbf{q}\cdot\mathbf{n}\,dA = \int_V S_m\,dV \tag{3.6}$$

where V is the volume of the control volume, c is the tracer concentration, S is the surface area of the control volume, $\mathbf{q}$ is the flux vector (Equation 3.5), $\mathbf{n}$ is the unit outward normal to the control volume, and S_m is the mass flux per unit volume originating within the control volume. Equation 3.6 can be simplified using the divergence theorem, which relates a surface integral to a volume integral by the relation

$$\int_S \mathbf{q}\cdot\mathbf{n}\,dA = \int_V \nabla\cdot\mathbf{q}\,dV \tag{3.7}$$

Combining Equations 3.6 and 3.7 leads to the result

$$\frac{\partial}{\partial t}\int_V c\,dV + \int_V \nabla\cdot\mathbf{q}\,dV = \int_V S_m\,dV \tag{3.8}$$

Since the control volume is fixed in space and time, the derivative of the volume integral with respect to time is equal to the volume integral of the derivative with respect to time, and Equation 3.8 can be written in the form

$$\int_V \left(\frac{\partial c}{\partial t} + \nabla \cdot \mathbf{q} - S_m \right) dV = 0 \tag{3.9}$$

This equation requires that the integral of the quantity in parentheses must be zero for any arbitrary control volume, and this can only be true if the integrand itself is zero. Following this logic, Equation 3.9 requires that

$$\frac{\partial c}{\partial t} + \nabla \cdot \mathbf{q} - S_m = 0 \tag{3.10}$$

This equation can be combined with the expression for the mass flux given by Equation 3.5 and written in the expanded form

$$\frac{\partial c}{\partial t} + \nabla \cdot (\mathbf{V}c - D\nabla c) = S_m \tag{3.11}$$

which simplifies to

$$\frac{\partial c}{\partial t} + \mathbf{V} \cdot \nabla c + c(\nabla \cdot \mathbf{V}) = D\nabla^2 c + S_m \tag{3.12}$$

This equation applies to all tracers in all fluids. In the case of incompressible fluids, which is typical of the water environment, conservation of fluid mass requires that

$$\nabla \cdot \mathbf{V} = 0 \tag{3.13}$$

and combining Equations 3.12 and 3.13 yields the following diffusion equation for incompressible fluids with isotropic dispersion:

$$\boxed{\frac{\partial c}{\partial t} + \mathbf{V} \cdot \nabla c = D\nabla^2 c + S_m} \tag{3.14}$$

In cases where there are no external sources or sinks of tracer mass (a conservative tracer), S_m is zero and Equation 3.14 becomes

$$\boxed{\frac{\partial c}{\partial t} + \mathbf{V} \cdot \nabla c = D\nabla^2 c} \tag{3.15}$$

If the dispersion coefficient, D, is anisotropic, the principal components of the dispersion coefficient can be written as D_i, and the diffusion equation becomes

$$\boxed{\frac{\partial c}{\partial t} + \sum_{i=1}^{3} V_i \frac{\partial c}{\partial x_i} = \sum_{i=1}^{3} D_i \frac{\partial^2 c}{\partial x_i^2} + S_m} \tag{3.16}$$

where x_i are the principal directions of the dispersion coefficient tensor. Equation 3.16 is the most commonly used relationship describing the mixing of contaminants in aquatic environments, and it is known as either the *advection–dispersion* or the *advection–diffusion equation*, with the former term being more appropriate in most cases.

3.2 PROPERTIES OF THE DIFFUSION EQUATION

In cases where the tracer mass is neither created nor destroyed, such as occurs in chemical and biochemical reactions, the contaminant is called *conservative*, and $S_m = 0$ in Equation 3.16. In this case, the transport of the contaminant is described by

$$\frac{\partial c}{\partial t} + \sum_{i=1}^{3} V_i \frac{\partial c}{\partial x_i} = \sum_{i=1}^{3} D_i \frac{\partial^2 c}{\partial x_i^2} \tag{3.17}$$

Consider now a change of variables from x_i and t to x_i' and t, where the new variables x_i' are defined by

$$x_i' = x_i - V_i t \tag{3.18}$$

where V_i is a constant mean velocity in the x_i-direction, and the x_i'-coordinate measures locations relative to the mean position of the tracer particles, given by $V_i t$. The derivatives in the (x_i, t) space are related to the derivatives in the (x_i', t) space by the following relations derived from the chain rule,

$$\frac{\partial(\cdot)}{\partial x_i} = \sum_{j=1}^{3} \frac{\partial(\cdot)}{\partial x_j'} \frac{\partial x_j'}{\partial x_i} \tag{3.19}$$

and

$$\frac{\partial(\cdot)}{\partial t} = \sum_{j=1}^{3} \frac{\partial(\cdot)}{\partial x_j'} \frac{\partial x_j'}{\partial t} + \frac{\partial(\cdot)}{\partial t} \tag{3.20}$$

where $(\cdot)$ represents any scalar function of x and t. Combining Equations 3.18 to 3.20 yields

$$\frac{\partial(\cdot)}{\partial x_i} = \frac{\partial(\cdot)}{\partial x_i'} \tag{3.21}$$

and

$$\frac{\partial(\cdot)}{\partial t} = -\sum_{j=1}^{3} V_j \frac{\partial(\cdot)}{\partial x_j'} + \frac{\partial(\cdot)}{\partial t} \tag{3.22}$$

Substituting Equations 3.21 and 3.22 into the advection–diffusion equation, Equation 3.17, yields the transformed equation in (x_i', t)-space,

$$\frac{\partial c}{\partial t} = \sum_{i=1}^{3} D_i \frac{\partial^2 c}{\partial x_i'^2} \tag{3.23}$$

which is more commonly written in the Cartesian form

$$\frac{\partial c}{\partial t} = D_x \frac{\partial^2 c}{\partial x'^2} + D_y \frac{\partial^2 c}{\partial y'^2} + D_z \frac{\partial^2 c}{\partial z'^2} \qquad (3.24)$$

Equation 3.24 occurs widely in engineering applications and is generally referred to as the *diffusion equation*. This equation has been studied in detail throughout the mathematical literature, particularly in the context of heat conduction, and analytical solutions for a wide variety of initial and boundary conditions are available in many textbooks. Using these solutions together with the transformation given by Equation 3.18 provides many useful results to describe the mixing process when the mean flow is steady and spatially uniform.

3.2.1 Fundamental Solution in One Dimension

Consider the case where the tracer is distributed uniformly in the y and z directions and diffusion occurs only in the x-direction. The diffusion equation is then given by

$$\frac{\partial c}{\partial t} = D_x \frac{\partial^2 c}{\partial x^2} \qquad (3.25)$$

If a tracer of mass M is introduced instantaneously at $x = 0$ at time $t = 0$ (well mixed over y and z), and tracer concentrations are always equal to zero at $x = \pm\infty$, the initial and boundary conditions are given by

$$c(x, 0) = \frac{M}{A} \delta(x) \qquad (3.26)$$

and

$$c(\pm\infty, t) = 0 \qquad (3.27)$$

where A is the area in the yz plane over which the contaminant is well mixed, and $\delta(x)$ is the Dirac delta function, which is defined by

$$\delta(x) = \begin{cases} \infty, & x = 0 \\ 0, & x \neq 0 \end{cases} \quad \text{and} \quad \int_{-\infty}^{+\infty} \delta(x)\, dx = 1 \qquad (3.28)$$

A graph of the Dirac delta function, centered at x_0 (where $x_0 = 0$ in Equation 3.28), is illustrated in Figure 3.3. The solution to Equation 3.25 subject to initial and boundary conditions given by Equations 3.26 and 3.27 can be obtained using the Fourier transform, $f(k, t)$, defined as

$$f(k, t) = \frac{1}{\sqrt{2\pi}} \int_{-\infty}^{\infty} c(x, t) e^{ikx}\, dx \qquad (3.29)$$

and the inverse Fourier transform defined as

$$c(x, t) = \frac{1}{\sqrt{2\pi}} \int_{-\infty}^{\infty} f(k, t) e^{-ikx}\, dk \qquad (3.30)$$

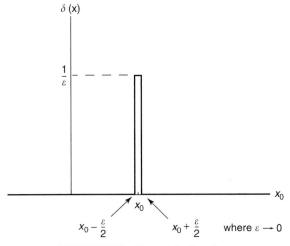

FIGURE 3.3   Dirac delta function.

In applying the Fourier transform to solve Equation 3.25, it is useful to note that integration by parts yields the identity

$$\frac{1}{\sqrt{2\pi}} \int_{-\infty}^{\infty} \frac{\partial^2 c}{\partial x^2} e^{ikx}\, dx = -\frac{k^2}{\sqrt{2\pi}} \int_{-\infty}^{\infty} c(x, t) e^{ikx}\, dx = -k^2 f(k, t) \qquad (3.31)$$

Multiplying Equation 3.25 by

$$\frac{e^{ikx}}{\sqrt{2\pi}}$$

and integrating yields the homogeneous ordinary differential equation

$$\frac{df}{dt} + k^2 D_x f = 0 \qquad (3.32)$$

and the initial condition derived from Equation 3.26 is

$$f(k, 0) = \frac{M}{A\sqrt{2\pi}} \qquad (3.33)$$

The solution to Equation 3.32 subject to the initial condition given by Equation 3.33 can easily be shown to be

$$f(k, t) = \frac{M}{A\sqrt{2\pi}} e^{-k^2 D_x t} \qquad (3.34)$$

This is the solution of the diffusion equation, Equation 3.25, in the Fourier-transformed (k, t)-space, and therefore the solution in the (x, t)-space, $c(x, t)$, can be obtained by

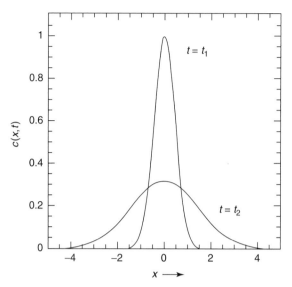

FIGURE 3.4   One-dimensional diffusion.

substituting Equation 3.34 into the inverse Fourier transform relation, Equation 3.30, to yield

$$c(x,\, t) = \frac{M}{A\sqrt{4\pi D_x t}} \exp\left(-\frac{x^2}{4D_x t}\right) \tag{3.35}$$

This result indicates that the concentration distribution resulting from the instantaneous introduction of a mass M is in the form of a Gaussian distribution with a variance growing with time, as illustrated by a plot of Equation 3.35 given in Figure 3.4. To verify that the concentration distribution given by Equation 3.35 is Gaussian, consider the equation for a Gaussian distribution given by

$$f(x) = \frac{M_0}{\sigma\sqrt{2\pi}} \exp\left[-\frac{1}{2}\left(\frac{x-\mu}{\sigma}\right)^2\right] \tag{3.36}$$

where μ is the mean of the distribution, σ is the standard deviation, and M_0 is the total area under the curve. Note that a *normal* distribution is the same as a Gaussian distribution, except that the area under the curve, M_0, is equal to unity. Comparing the fundamental solution of the diffusion equation, Equation 3.35, to the Gaussian distribution, Equation 3.36, it is clear that the fundamental solution is Gaussian with the mean and standard deviation given by

$$\mu = 0 \tag{3.37}$$

$$\sigma = \sqrt{2D_x t} \tag{3.38}$$

This result demonstrates that a mass of contaminant released instantaneously into a stagnant fluid will attain a concentration distribution that is Gaussian, with a maximum concentration remaining at the location the mass was released, and the standard deviation of the distribution grows in proportion to the square root of the elapsed time since the release.

The one-dimensional advection–diffusion equation in a moving fluid is given by

$$\frac{\partial c}{\partial t} + V\frac{\partial c}{\partial x} = D_x\frac{\partial^2 c}{\partial x^2} \tag{3.39}$$

where V is the fluid velocity in the x-direction. Equation 3.39 transforms to

$$\frac{\partial c}{\partial t} = D_x\frac{\partial^2 c}{\partial x'^2} \tag{3.40}$$

in the x'–t domain, where $x' = x - Vt$. The initial and boundary conditions corresponding to an instantaneous release at $x = 0$ and $t = 0$ with boundaries infinitely far away from the release location are

$$c(x', 0) = \frac{M}{A}\delta(x') \tag{3.41}$$

and

$$c(\pm\infty, t) = 0 \tag{3.42}$$

where A is the area in the yz plane over which the contaminant is well mixed. The solution to Equation 3.40 is the same as the fundamental solution for a stationary fluid and is therefore given by

$$c(x', t) = \frac{M}{A\sqrt{4\pi D_x t}}\,\exp\left(-\frac{x'^2}{4D_x t}\right) \tag{3.43}$$

which in the x–t domain is given by

$$\boxed{c(x, t) = \frac{M}{A\sqrt{4\pi D_x t}}\,\exp\left[-\frac{(x - Vt)^2}{4D_x t}\right]} \tag{3.44}$$

The concentration distribution described by Equation 3.44 and illustrated in Figure 3.5 describes the mixing of a tracer released instantaneously into a flowing fluid, where the tracer undergoes one-dimensional diffusion. If the fluid is stagnant, $V = 0$ and the resulting concentration distribution is symmetrical around $x = 0$ and is described by Equation 3.35.

Example 3.1 One hundred kilograms of a contaminant is spilled into a small river and instantaneously mixes across the entire cross section of the river. The cross section of the river is approximately trapezoidal in shape, with a bottom width of 5 m, side slopes of 2:1

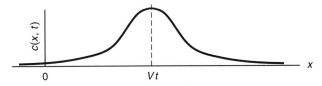

FIGURE 3.5 Solution to a one-dimensional advection–diffusion equation. (From Chin, David A., *Water-Resources Engineering.* Copyright © 2000. Reprinted by permission of Pearson Education, Inc., Upper Saddle River, NJ.)

(H:V), and a depth of flow of 3 m. The discharge in the river is estimated as $30\,m^3/s$, and the dispersion coefficient for mixing along the river is estimated as $10\,m^2/s$. Estimate (a) when the maximum contaminant concentration will be observed at a park recreation area 10 km downstream of the spill, and (b) the maximum concentration expected at the park. (c) If a safe level of this contaminant in recreational waters is $10\,\mu g/L$, how long after the spill can the park expect to resume normal operations?

SOLUTION   (a) From the data given, $M = 100\,kg$, $D_x = 10\,m^2/s$, and the flow rate, Q, in the river is $30\,m^3/s$. The cross-sectional area, A, of the river is given by

$$A = by + my^2$$

where $b = 5\,m$, $y = 3\,m$, and $m = 2$; hence,

$$A = 5(3) + 2(3)^2 = 33\,m^2$$

and

$$V = \frac{Q}{A} = \frac{30}{33} = 0.909\ \text{m/s}$$

The distance, x_m, of the maximum concentration from the spill location at any time, t, is given by

$$x_m = Vt$$

Therefore, for $x_m = 10\,km = 10{,}000\,m$,

$$t = \frac{x_m}{V} = \frac{10{,}000}{0.909} = 11{,}000\ \text{s} = 3.06\,\text{h}$$

Hence, the park can expect to see the peak contaminant concentration 3.06 h after the spill occurs.

(b) The maximum contaminant concentration at any time (t) and location (x) is given by Equation 3.44 for $x = Vt$ as

$$c(x, t) = \frac{M}{A\sqrt{4\pi D_x t}} = \frac{100}{33\sqrt{4\pi(10)(11{,}000)}} = 2.58 \times 10^{-3}\ \text{kg/m}^3 = 2.58\ \text{mg/L}$$

Hence, the maximum contaminant concentration observed at the recreation area is expected to be 2.58 mg/L.

(c) When the concentration at the recreation area is $10\,\mu g/L = 10^{-5}\,kg/m^3$, Equation 3.44 requires that

$$c(x, t) = \frac{M}{A\sqrt{4\pi D_x t}}\exp\left[-\frac{(x - Vt)^2}{4D_x t}\right]$$

$$10^{-5} = \frac{100}{33\sqrt{4\pi(10)t}}\exp\left[-\frac{(10{,}000 - 0.909t)^2}{4(10)t}\right]$$

which yields $t = 9400\,$s and $12{,}850\,$s. Clearly, the concentration is above $10\,\mu g/L$ from $t = 9400$ to $12{,}850\,$s, and the park water is expected to be safe when $t > 12{,}850\,s = 3.57\,h$ after the spill.

In general, water-quality measurements would be required to ensure the safety of the water at the recreational area prior to resuming normal operations.

3.2.2 Principle of Superposition

The principle of superposition states that if a homogeneous differential equation is linear and there are several solutions that satisfy the equation, the sum of the solutions also satisfies the equation, and this sum is called the *total solution* of the linear differential equation. Additionally, if the boundary and initial conditions of the individual solutions are also linear, the boundary and initial conditions of the total solution are equal to the sum of the boundary and initial conditions of the individual solutions. To illustrate the principle of superposition, consider that c_1 and c_2 are separate solutions to the one-dimensional diffusion equation; therefore,

$$\frac{\partial c_1}{\partial t} = D_x \frac{\partial^2 c_1}{\partial x^2} \tag{3.45}$$

and

$$\frac{\partial c_2}{\partial t} = D_x \frac{\partial^2 c_2}{\partial x^2} \tag{3.46}$$

Adding these equations yields

$$\frac{\partial c_1}{\partial t} + \frac{\partial c_2}{\partial t} = D_x \frac{\partial^2 c_1}{\partial x^2} + D_x \frac{\partial^2 c_2}{\partial x^2} \tag{3.47}$$

and invoking the linearity property leads to

$$\frac{\partial (c_1 + c_2)}{\partial t} = D_x \frac{\partial^2 (c_1 + c_2)}{\partial x^2} \tag{3.48}$$

which demonstrates that the sum of the solutions to the diffusion equation (i.e., $c_1 + c_2$) is also a solution to the diffusion equation. To demonstrate the effect of linearity on the boundary conditions of the total solution, suppose that the boundary conditions on c_1 and c_2 are

$$c_1(\infty, t) = 0 \tag{3.49}$$

$$c_2(\infty, t) = 0 \tag{3.50}$$

Then clearly the boundary condition of $c_1 + c_2$ is

$$(c_1 + c_2)(\infty, t) = 0 \tag{3.51}$$

Similarly, if the initial conditions corresponding to instantaneous releases of masses M_1 and M_2 at $x = x_1$ and $x = x_2$ are

$$c_1(x, 0) = \frac{M_1}{A}\delta(x - x_1) \qquad (3.52)$$

$$c_2(x, 0) = \frac{M_2}{A}\delta(x - x_2) \qquad (3.53)$$

the initial condition of $c_1 + c_2$ is

$$(c_1 + c_2)(x, 0) = \frac{M_1}{A}\delta(x - x_1) + \frac{M_2}{A}\delta(x - x_2) \qquad (3.54)$$

This result demonstrates that the principle of superposition can be used to determine the contaminant distribution resulting from two simultaneous mass inputs at two different locations, based on the contaminant distribution resulting from mass releases at single locations. Several additional applications of the principle of superposition are given in the following sections.

Distributed Source   Consider the initial spatial distribution of a contaminant given in Figure 3.6, where the initial one-dimensional concentration distribution is given by

$$c(x, 0) = f(x) \qquad (3.55)$$

This initial concentration distribution is equivalent to an infinite number of adjacent instantaneous sources of mass $f(x)A\,dx$ located along the x-axis between x_L and x_R, where A is the cross-sectional area over which the contaminant is well mixed. For each of these

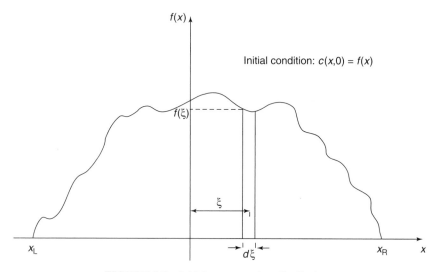

FIGURE 3.6   Initial concentration distribution.

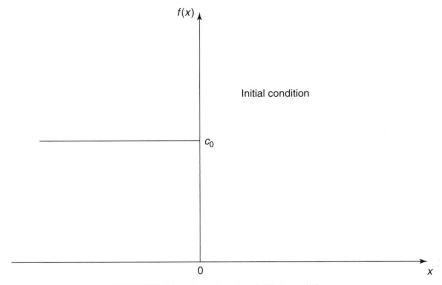

FIGURE 3.7   Step-function initial condition.

incremental sources located a distance ξ away from the origin, the fundamental solution of the one-dimensional diffusion equation is applicable, and the resulting concentration distribution is given by

$$c(x, t) = \frac{f(\xi)\, d\xi}{\sqrt{4\pi D_x t}} \exp\left[-\frac{(x - \xi)^2}{4D_x t} \right] \tag{3.56}$$

Using the principle of superposition to sum the solutions for all the incremental sources, results in the total solution

$$c(x, t) = \int_{x_L}^{x_R} \frac{f(\xi)\, d\xi}{\sqrt{4\pi D_x t}} \exp\left[-\frac{(x - \xi)^2}{4D_x t} \right] \tag{3.57}$$

A frequently encountered initial concentration distribution is the step function shown in Figure 3.7, which is described by

$$c(x, 0) = f(x) = \begin{cases} c_0 & x \leq 0 \\ 0 & x > 0 \end{cases} \tag{3.58}$$

and substituting this initial condition into Equation 3.57 yields

$$c(x, t) = \int_{-\infty}^{0} \frac{c_0\, d\xi}{\sqrt{4\pi D_x t}} \exp\left[-\frac{(x - \xi)^2}{4D_x t} \right] \tag{3.59}$$

Changing variables from x to u, where

$$u = \frac{x - \xi}{\sqrt{4D_x t}} \tag{3.60}$$

Equation 3.59 becomes

$$c(x, t) = \frac{c_0}{\sqrt{\pi}} \int_{x/\sqrt{4D_x t}}^{\infty} e^{-u^2} \, du \qquad \bullet \qquad (3.61)$$

This integral cannot be evaluated analytically, but is similar to a special function in mathematics called the *error function*, erf(*z*), which is defined as

$$\text{erf}(z) = \frac{2}{\sqrt{\pi}} \int_0^z e^{-\xi^2} \, d\xi \qquad (3.62)$$

and values of this function are tabulated in Appendix E.1. It is useful to note the property

$$\text{erf}(-z) = -\text{erf}(z) \qquad (3.63)$$

and the limits

$$\text{erf}(0) = 0 \quad \text{and} \quad \text{erf}(\infty) = 1 \qquad (3.64)$$

Comparing the solution of the diffusion equation, Equation 3.61, with the definition of the error function, Equation 3.62, leads to

$$c(x, t) = \frac{c_0}{\sqrt{\pi}} \left[\int_0^{\infty} e^{-u^2} \, du - \int_0^{x/\sqrt{4D_x t}} e^{-u^2} \, du \right]$$

$$= \frac{c_0}{\sqrt{\pi}} \left[\frac{\sqrt{\pi}}{2} - \frac{\sqrt{\pi}}{2} \text{erf}\left(\frac{x}{\sqrt{4D_x t}} \right) \right] \qquad (3.65)$$

$$= \frac{c_0}{2} \left[1 - \text{erf}\left(\frac{x}{\sqrt{4D_x t}} \right) \right]$$

A further simplification of Equation 3.65 comes from the definition of the *complementary error function*, erfc(*z*), where

$$\text{erfc}(z) = 1 - \text{erf}(z) \qquad (3.66)$$

The solution given by Equation 3.65 can therefore be written in the form

$$\boxed{c(x, t) = \frac{c_0}{2} \text{erfc}\left(\frac{x}{\sqrt{4D_x t}} \right)} \qquad (3.67)$$

and the concentration distribution, $c(x, t)$, is illustrated in Figure 3.8. The corresponding solution for a fluid moving with a velocity V is given by

$$\boxed{c(x, t) = \frac{c_0}{2} \text{erfc}\left(\frac{x - Vt}{\sqrt{4D_x t}} \right)} \qquad (3.68)$$

This concentration distribution is identical to that illustrated in Figure 3.8, with the exception that the origin moves with the fluid at a velocity V.

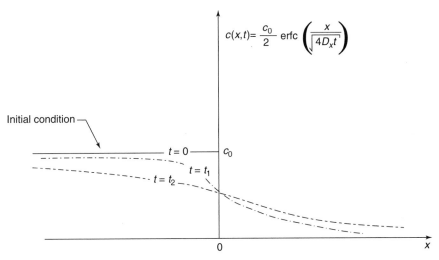

FIGURE 3.8   Diffusion from a step-function initial condition.

Example 3.2   A long drainage canal is gated at the downstream end and is designed to retain the runoff from an agricultural area. The runoff into the canal is expected to infiltrate into the ground water. After a severe storm, the concentration of a toxic pesticide in the channel rises to 5 mg/L and is distributed uniformly throughout the channel. Because of the threat of flooding, the gate at the downstream end of the canal is opened and water in the canal flows downstream at a velocity of 20 cm/s. (a) If the longitudinal dispersion coefficient in the canal is 5 m²/s, give an expression for the concentration as a function of time at a location 400 m downstream of the gate. (b) How long after the gate is opened will the concentration at the downstream location be equal to 1 mg/L?

SOLUTION   (a) From the data given, $c_0 = 5$ mg/L $= 0.005$ kg/m³, $V = 20$ cm/s $= 0.20$ m/s, and $D_x = 5$ m²/s. At $x = 400$ m, Equation 3.68 gives the concentration as a function of time as

$$c(x, t) = \frac{c_0}{2} \, \text{erfc} \left(\frac{x - Vt}{\sqrt{4D_x t}} \right)$$

$$c(400, t) = \frac{0.005}{2} \, \text{erfc} \left(\frac{400 - 0.20t}{\sqrt{4(5)t}} \right) = 0.0025 \, \text{erfc} \left(\frac{8.94 - 0.0447t}{\sqrt{t}} \right)$$

   (b) When the concentration 400 m downstream of the gate is equal to 1 mg/L $= 0.001$ kg/m³,

$$0.001 = 0.0025 \, \text{erfc} \left(\frac{8.94 - 0.0447t}{\sqrt{t}} \right)$$

which leads to

$$\text{erfc} \left(\frac{8.94 - 0.0447t}{\sqrt{t}} \right) = 0.4$$

or

$$\text{erf}\left(\frac{8.94 - 0.0447t}{\sqrt{t}}\right) = 0.6$$

Using the error function tabulated in Appendix E.1,

$$\frac{8.94 - 0.0447t}{\sqrt{t}} = 0.595$$

which leads to

$$t = 80.5\,\text{s}$$

Therefore, the concentration 400 m downstream of the gate will reach 1 mg/L approximately 81 s after the gate is opened.

Transient Source Suppose that a contaminant source is located along the x-axis, and the source injects the contaminant with a time-varying mass flux, $\dot{m}(t)$, uniformly over a cross-sectional area A in a stagnant fluid. This scenario is equivalent to a mass of $\dot{m}(t)\,dt$ being released during every consecutive time interval dt. The concentration distribution, $dc(x, t)$, resulting from an instantaneous mass of $\dot{m}(\tau)\,d\tau$ released from $x = 0$ at time τ is given by

$$dc(x, t) = \frac{\dot{m}(\tau)\,d\tau}{A\sqrt{4\pi D_x(t-\tau)}}\exp\left[-\frac{x^2}{4D_x(t-\tau)}\right] \tag{3.69}$$

and superimposing all of the resulting concentration distributions yields

$$c(x, t) = \int_0^t \frac{\dot{m}(\tau)\,d\tau}{A\sqrt{4\pi D_x(t-\tau)}}\exp\left[-\frac{x^2}{4D_x(t-\tau)}\right] \tag{3.70}$$

This equation describes the concentration distribution resulting from a transient source at $x = 0$. In the case of a spatially distributed transient source with mass flux $\dot{m}(x, t)$, the principle of superposition indicates that the resulting concentration distribution is given by

$$\boxed{c(x, t) = \int_0^t \int_{x_L}^{x_R} \frac{\dot{m}(\xi, \tau)\,d\xi\,d\tau}{A\sqrt{4\pi D_x(t-\tau)}}\exp\left[-\frac{(x-\xi)^2}{4D_x(t-\tau)}\right]} \tag{3.71}$$

where x_L and x_R are the upper and lower bounds of the tracer source location.

Impermeable Boundaries In the superposition examples cited so far, the boundary conditions have required that the tracer concentration approaches zero as x approaches infinity. Therefore, the superimposed concentration distributions all have boundary conditions in which the tracer concentration approaches zero as x approaches infinity. In cases where impermeable boundaries exist in relatively close proximity to the tracer source, the diffusion equation must satisfy boundary conditions that require zero mass flux across the

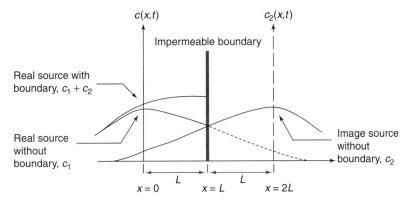

FIGURE 3.9   Impermeable boundary condition.

impermeable boundary. This scenario is illustrated in Figure 3.9. Since mass flux, $\dot{M}_x$, is governed by the Fickian-type diffusion equation

$$\dot{M}_x = -D_x \frac{\partial c}{\partial x} \tag{3.72}$$

an impermeable boundary requires that the concentration gradient, $\partial c/\partial x$, be equal to zero at the boundary. Referring to Figure 3.9, if the tracer source is located at $x = 0$ and the impermeable boundary is located at $x = L$, then if an identical source is superimposed at $x = 2L$, the resulting (superimposed) concentration distribution has the following properties within the domain $x \in (-\infty, L]$: (1) satisfies the diffusion equation (via linearity); (2) satisfies the initial condition of an instantaneous mass release at $x = 0$; and (3) $\partial c/\partial x = 0$ at $x = L$, due to the symmetry of the superimposed solutions around $x = L$. These results demonstrate that the symmetrical placement of an *image source* produces a solution that satisfies the diffusion equation as well as the required initial and boundary conditions. Applying this result to the case of an instantaneous source of mass, M, located at a distance L from an impermeable boundary, indicates that the resulting concentration distribution is given by

$$c(x, t) = \frac{M}{A\sqrt{4\pi D_x t}} \left[\exp\left(-\frac{x^2}{4D_x t} \right) + \exp\left(-\frac{(x - 2L)^2}{4D_x t} \right) \right] \tag{3.73}$$

Example 3.3   One kilogram of a contaminant is spilled into an open channel at a location 50 m from the end of the channel. The channel has a rectangular cross section 10 m wide and 2 m deep, and the dispersion coefficient along the channel is estimated as 10 m²/s. (a) Assuming that the contaminant is initially well mixed across the channel, express the concentration as a function of time at the end of the channel. (b) How long will it take for the contaminant concentrations 25 m upstream (in the direction of the channel end) to be 10% higher than the concentration 25 m downstream of the spill (in the direction away from the channel end)?

SOLUTION   (a) The concentration distribution is given by

$$c(x, t) = \frac{M}{A\sqrt{4\pi D_x t}} \left[\exp\left(-\frac{x^2}{4D_x t} \right) + \exp\left(-\frac{(x - 2L)^2}{4D_x t} \right) \right]$$

where the x-coordinate is measured from the spill location in the direction of the end of the channel. From the data given, $M = 1$ kg, $A = 10$ m $\times 2$ m $= 20$ m^2, $D_x = 10$ m^2/s, $L = 50$ m, and $x = 50$ m (at end of channel). The concentration as a function of time at the end of the channel is therefore given by

$$c(50, t) = \frac{1}{20\sqrt{4\pi(10)t}}\left[\exp\left(-\frac{50^2}{4(10)t}\right) + \exp\left(-\frac{[50 - 2(50)]^2}{4(10)t}\right)\right]$$

which reduces to

$$c(50, t) = \frac{0.00892}{\sqrt{t}}\exp\left(-\frac{62.5}{t}\right)$$

(b) The concentration 25 m upstream from the spill ($x = 25$ m) is given by

$$c(25, t) = \frac{1}{20\sqrt{4\pi(10)t}}\left[\exp\left(-\frac{25^2}{4(10)t}\right) + \exp\left(-\frac{[25 - 2(50)]^2}{4(10)t}\right)\right]$$

$$= \frac{0.00446}{\sqrt{t}}\left[\exp\left(-\frac{15.6}{t}\right) + \exp\left(-\frac{141}{t}\right)\right]$$

and the concentration 25 m downstream from the spill ($x = -25$ m) is given by

$$c(-25, t) = \frac{1}{20\sqrt{4\pi(10)t}}\left[\exp\left(-\frac{25^2}{4(10)t}\right) + \exp\left(-\frac{[-25 - 2(50)]^2}{4(10)t}\right)\right]$$

$$= \frac{0.00446}{\sqrt{t}}\left[\exp\left(-\frac{15.6}{t}\right) + \exp\left(-\frac{391}{t}\right)\right]$$

When $c(25, t)$ is 10% higher than $c(-25, t)$, then

$$\frac{c(25, t)}{c(-25, t)} = 1.1$$

or

$$\frac{\exp(-15.6/t) + \exp(-141/t)}{\exp(-15.6/t) + \exp(-391/t)} = 1.1$$

Solving this equation for t yields $t = 55$ s. Therefore, after 55 s the concentrations 25 m upstream and downstream of the spill differ by 10%.

3.2.3 Solutions in Higher Dimensions

Solutions to the one-dimensional diffusion equation shown so far have been intended to demonstrate some useful analytical procedures for obtaining solutions to relatively complicated problems. Many of these solutions are obtained by superimposing the fundamental solution of the one-dimensional diffusion equation. These analytical procedures are also applicable in higher dimensions, as illustrated in the following sections.

Two-Dimensional Diffusion The fundamental diffusion problem in two dimensions is given by

$$\frac{\partial c}{\partial t} = D_x \frac{\partial^2 c}{\partial x^2} + D_y \frac{\partial^2 c}{\partial y^2} \tag{3.74}$$

with initial and boundary conditions

$$c(x, y, 0) = \frac{M}{L} \delta(x, y) \tag{3.75}$$

$$c(\pm\infty, \pm\infty, t) = 0 \tag{3.76}$$

where M is the mass of contaminant injected, L is the length over which the mass is uniformly distributed in the z-direction, and $\delta(x, y)$ is the two-dimensional Dirac delta function defined by

$$\delta(x, y) = \begin{cases} \infty, & x = 0, y = 0 \\ 0, & \text{otherwise} \end{cases} \quad \text{and} \quad \int_{-\infty}^{+\infty}\int_{-\infty}^{+\infty} \delta(x)\, dx\, dy = 1 \tag{3.77}$$

The solution to the fundamental two-dimensional diffusion problem is given by (Carslaw and Jaeger, 1959)

$$c(x, y, t) = \frac{M}{4\pi t L \sqrt{D_x D_y}} \exp\left(-\frac{x^2}{4D_x t} - \frac{y^2}{4D_y t}\right) \tag{3.78}$$

The principle of superposition can be applied to the fundamental two-dimensional solution of the diffusion equation to yield the concentration distribution, $c(x, y, t)$, resulting from an initial mass distribution, $g(x, y)$, as

$$c(x, y, t) = \int_{x_1}^{x_2}\int_{y_1}^{y_2} \frac{g(\xi, \eta)\, d\xi\, d\eta}{4\pi t L \sqrt{D_x D_y}} \exp\left[-\frac{(x-\xi)^2}{4D_x t} - \frac{(y-\eta)^2}{4D_y t}\right] \tag{3.79}$$

where the contaminant source is located in the region $x \in [x_1, x_2]$, $y \in [y_1, y_2]$. An example of diffusion from a two-dimensional rectangular source is shown in Figure 3.10, where taking $D_x = D_y$ leads to a symmetrical diffusion pattern. Superposition in time can also be applied to yield the concentration distribution, $c(x, y, t)$, resulting from a continuous mass input $\dot{m}(t)$ as

$$c(x, y, t) = \int_0^t \frac{\dot{m}(\tau)\, d\tau}{4\pi(t-\tau)L\sqrt{D_x D_y}} \exp\left[-\frac{x^2}{4D_x(t-\tau)} - \frac{y^2}{4D_y(t-\tau)}\right] \tag{3.80}$$

where the transient source is located at $x = 0$, $y = 0$. In the case of a distributed transient source, $\dot{m}(x, y, t)$, the resulting concentration distribution, $c(x, y, t)$, is given by

$$c(x, y, t) = \int_0^t\int_{x_1}^{x_2}\int_{y_1}^{y_2} \frac{\dot{m}(\xi, \eta, t)\, d\xi\, d\eta\, dt}{4\pi(t-\tau)L\sqrt{D_x D_y}} \exp\left[-\frac{(x-\xi)^2}{4D_x(t-\tau)} - \frac{(y-\eta)^2}{4D_y(t-\tau)}\right] \tag{3.81}$$

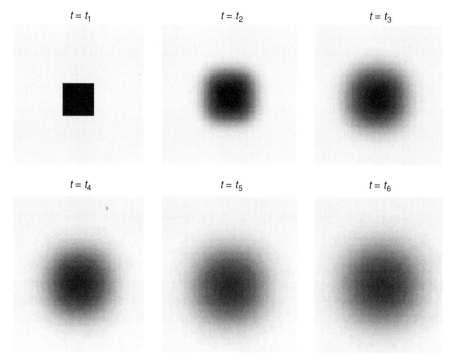

FIGURE 3.10   Two-dimensional diffusion from a finite source.

Example 3.4   One kilogram of a contaminant is spilled at a point in a 4-m-deep reservoir and is instantaneously mixed over the entire depth. (a) If the diffusion coefficients in the N-S and E-W directions are 5 and 10 m²/s, respectively, calculate the concentration as a function of time at locations 100 m north and 100 m east of the spill. (b) What is the concentration at the spill location after 5 min?

SOLUTION   (a) The concentration distribution is given by

$$c(x, y, t) = \frac{M}{4\pi t L \sqrt{D_x D_y}} \exp\left(-\frac{x^2}{4D_x t} - \frac{y^2}{4D_y t}\right)$$

From the data given, $M = 1$ kg, $L = 4$ m, $D_x = 5$ m²/s (N-S), and $D_y = 10$ m²/s (E-W). At 100 m north of the spill, $x = 0$ m, $y = 100$ m, and the concentration as a function of time is given by

$$c(0, 100, t) = \frac{1}{4\pi t(4)\sqrt{5(10)}} \exp\left(-\frac{100^2}{4(5)t}\right) = \frac{0.00281}{t} \exp\left(-\frac{500}{t}\right) \quad \text{kg/m}^3$$

At 100 m east of the spill, $x = 100$ m, $y = 0$ m, and the concentration as a function of time is given by

$$c(100, 0, t) = \frac{1}{4\pi t(4)\sqrt{5(10)}} \exp\left(-\frac{100^2}{4(10)t}\right) = \frac{0.00281}{t} \exp\left(-\frac{250}{t}\right) \quad \text{kg/m}^3$$

(b) At the spill location, $x = 0$ m and $y = 0$ m, and the concentration as a function of time is given by

$$c(0, 0, t) = \frac{1}{4\pi t(4)\sqrt{5}(10)} = \frac{0.00281}{t}$$

At $t = 5$ min $= 300$ s,

$$c(0, 0, 300) = \frac{0.00281}{300} = 9.37 \times 10^{-6} \text{ kg/m}^3 = 9.37 \ \mu\text{g/L}$$

Therefore, the concentration at the spill location after 5 min is $9.37 \ \mu\text{g/L}$.

Example 3.5   A vertical diffuser discharges industrial wastewater at a rate of 5 m³/s uniformly over a 5-m-deep reservoir. If the wastewater contains 50 mg/L of a toxic contaminant for 24 h, calculate the concentration of the contaminant as a function of time during this 24-h period at a distance of 100 m from the outfall. Assume that the diffusion coefficient in the reservoir is equal to 10 m²/s.

SOLUTION   The concentration as a function of time is given by

$$c(x, y, t) = \int_0^t \frac{\dot{m}(\tau)\, d\tau}{4\pi(t-\tau)L\sqrt{D_x D_y}} \exp\left[-\frac{x^2}{4D_x(t-\tau)} - \frac{y^2}{4D_y(t-\tau)}\right]$$

In this case, the two-dimensional contaminant mass flux, $\dot{m}$, is

$$\dot{m} = Qc = 5(50 \times 10^{-3}) = 0.25 \text{ kg/s}$$

Since $L = 5$ m, $D_x = D_y = 10$ m²/s, and at 100 m from the outfall $x^2 + y^2 = 100^2$ m², the concentration 100 m from the outfall, $c_{100}(t)$, is given by

$$c_{100}(t) = \int_0^t \frac{0.25}{4\pi(t-\tau)(5)(10)} \exp\left[-\frac{100^2}{4(10)(t-\tau)}\right] d\tau$$

$$= 0.000398 \int_0^t \frac{1}{t-\tau} \exp\left(-\frac{250}{t-\tau}\right) d\tau \qquad (3.82)$$

This equation can be integrated numerically to yield the magnitude of the concentration 100 m from the outfall as a function of time. It is interesting to note that Equation 3.82 can also be expressed in terms of the well function, $W(u)$, commonly used in ground-water applications, and tabulated values or approximations to $W(u)$ can be used to aid integration.

Three-Dimensional Diffusion   The fundamental diffusion problem in three dimensions is given by

$$\boxed{\frac{\partial c}{\partial t} = D_x \frac{\partial^2 c}{\partial x^2} + D_y \frac{\partial^2 c}{\partial y^2} + D_z \frac{\partial^2 c}{\partial z^2}} \qquad (3.83)$$

with initial and boundary conditions

$$c(x, y, z, 0) = M\delta(x, y, z) \tag{3.84}$$

$$c(\pm\infty, \pm\infty, \pm\infty, t) = 0 \tag{3.85}$$

where $\delta(x, y, z)$ is the three-dimensional Dirac delta function defined by

$$\delta(x, y, z) = \begin{cases} \infty, & x = 0, y = 0, z = 0 \\ 0, & \text{otherwise} \end{cases} \quad \text{and} \quad \int_{-\infty}^{+\infty}\int_{-\infty}^{+\infty}\int_{-\infty}^{+\infty} \delta(x)\, dx\, dy\, dz = 1 \tag{3.86}$$

The solution to the fundamental three-dimensional diffusion problem is given by (Carslaw and Jaeger, 1959)

$$c(x, y, z, t) = \frac{M}{(4\pi t)^{3/2}\sqrt{D_x D_y D_z}} \exp\left(-\frac{x^2}{4D_x t} - \frac{y^2}{4D_y t} - \frac{z^2}{4D_z t}\right) \tag{3.87}$$

The principle of superposition can be applied to the fundamental three-dimensional solution of the diffusion equation to yield the concentration distribution, $c(x, y, z, t)$, resulting from an initial mass distribution (per unit volume), $g(x, y, z)$, as

$$c(x, y, z, t) = \int_{x_1}^{x_2}\int_{y_1}^{y_2}\int_{z_1}^{z_2} \frac{g(\xi, \eta, \zeta)\, d\xi\, d\eta\, d\zeta}{(4\pi t)^{3/2}\sqrt{D_x D_y D_z}} \exp\left[-\frac{(x-\xi)^2}{4D_x t} - \frac{(y-\eta)^2}{4D_y t} - \frac{(z-\zeta)^2}{4D_z t}\right]$$

$$\tag{3.88}$$

where the contaminant source is located in the region $x \in [x_1, x_2]$, $y \in [y_1, y_2]$, $z \in [z_1, z_2]$. Superposition in time can also be applied to yield the concentration distribution, $c(x, y, z, t)$, resulting from a continuous mass input $\dot{m}(t)$ as

$$c(x, y, z, t) = \int_0^t \frac{\dot{m}(\tau)\, d\tau}{[4\pi(t-\tau)]^{3/2}\sqrt{D_x D_y D_z}} \exp\left[-\frac{x^2}{4D_x(t-\tau)} - \frac{y^2}{4D_y(t-\tau)} - \frac{z^2}{4D_z(t-\tau)}\right]$$

$$\tag{3.89}$$

where the transient source is located at $x = 0$, $y = 0$, $z = 0$. In the case of a distributed transient source, $\dot{m}(x, y, z, t)$, the resulting concentration distribution, $c(x, y, z, t)$, is given by

$$c(x, y, z, t) = \int_0^t\int_{x_1}^{x_2}\int_{y_1}^{y_2}\int_{z_1}^{z_2} \frac{\dot{m}(\xi, \eta, \zeta, \tau)\, d\xi\, d\eta\, d\zeta\, d\tau}{[4\pi(t-\tau)]^{3/2}\sqrt{D_x D_y D_z}}$$

$$\times \exp\left[-\frac{(x-\xi)^2}{4D_x(t-\tau)} - \frac{(y-\eta)^2}{4D_y(t-\tau)} - \frac{(z-\zeta)^2}{4D_z(t-\tau)}\right] \tag{3.90}$$

Example 3.6 One kilogram of a toxic contaminant is released deep into the ocean and spreads in all three coordinate directions. The N-S, E-W, and vertical diffusion coefficients are 10, 15, and $0.1 \text{ m}^2/\text{s}$, respectively. (a) Find the concentration at a point 100 m north,

100 m east, and 10 m above the release point as a function of time. (b) What is the concentration at the release point after 24 h?

SOLUTION   (a) The concentration as a function of time is given by

$$c(x, y, z, t) = \frac{M}{(4\pi t)^{3/2}\sqrt{D_x D_y D_z}} \exp\left(-\frac{x^2}{4D_x t} - \frac{y^2}{4D_y t} - \frac{z^2}{4D_z t}\right)$$

In this case, $M = 1\,\mathrm{kg}$, $D_x = 10\,\mathrm{m^2/s}$, $D_y = 15\,\mathrm{m^2/s}$, $D_z = 0.1\,\mathrm{m^2/s}$, and therefore the concentration as a function of time at $x = 100\,\mathrm{m}$, $y = 100\,\mathrm{m}$, $z = 10\,\mathrm{m}$, is given by

$$c(100, 100, 10, t) = \frac{1}{(4\pi t)^{3/2}\sqrt{10(15)(0.1)}} \exp\left[-\frac{100^2}{4(10)t} - \frac{100^2}{4(15)t} - \frac{10^2}{4(0.1)t}\right]$$

$$= \frac{0.00580}{t^{3/2}} \exp\left(-\frac{667}{t}\right) \quad \mathrm{kg/m^3}$$

(b) The concentration, c_0, at the release point, $x = 0$, $y = 0$, $z = 0$, is given by

$$c_0 = \frac{1}{(4\pi t)^{3/2}\sqrt{10(15)(0.1)}} = \frac{0.00580}{t^{3/2}} \quad \mathrm{kg/m^3}$$

and at $t = 24\,\mathrm{h} = 86{,}400\,\mathrm{s}$ the concentration at the release point, c_0, is given by

$$c_0 = \frac{0.00580}{86{,}400^{3/2}} = 2.29 \times 10^{-10}\,\mathrm{kg/m^3} = 2.29 \times 10^{-4}\,\mu\mathrm{g/L}$$

Concentrations at this level would not be detectable, and the contaminant has dissipated, for all practical purposes.

Example 3.7   Ten kilograms of a contaminant in the form of a $1\,\mathrm{m} \times 2\,\mathrm{m} \times 2\,\mathrm{m}$ parallepiped is released into the deep ocean. The N-S, E-W, and vertical diffusion coefficients are 10, 5, and $0.05\,\mathrm{m^2/s}$, respectively. Find the concentration as a function of time at a location 50 m north, 50 m east, and 10 m above the centroid of the initial mass release.

SOLUTION   The concentration distribution is given by

$$c(x, y, z, t) = \int_{x_1}^{x_2}\int_{y_1}^{y_2}\int_{z_1}^{z_2} \frac{g(\xi, \eta, \zeta)\,d\xi\,d\eta\,d\zeta}{(4\pi t)^{3/2}\sqrt{D_x D_y D_z}} \exp\left[-\frac{(x-\xi)^2}{4D_x t} - \frac{(y-\eta)^2}{4D_y t} - \frac{(z-\zeta)^2}{4D_z t}\right]$$

where the initial concentration distribution, $g(\xi, \eta, \zeta)$, is given by

$$g(\xi, \eta, \zeta) = \begin{cases} \dfrac{10\,\mathrm{kg}}{(1)(2)(2)\,\mathrm{m^3}} = 2.5\,\mathrm{kg/m^3}, & -0.5 \le \xi \le 0.5,\ -1.0 \le \eta \le 1.0,\ -1.0 \le \zeta \le 1.0 \\ 0, & \text{otherwise} \end{cases}$$

In this case, $D_x = 10 \, \text{m}^2/\text{s}$, $D_y = 5 \, \text{m}^2/\text{s}$, and $D_z = 0.05 \, \text{m}^2/\text{s}$, and at $x = 50 \, \text{m}$, $y = 50 \, \text{m}$, $z = 10 \, \text{m}$, the concentration as a function of time is given by

$$c(50, 50, 10, t) = \frac{2.5}{(4\pi t)^{3/2} \sqrt{10(5)(0.05)}}$$

$$\times \int_{-0.5}^{0.5} \int_{-1}^{1} \int_{-1}^{1} \exp\left[-\frac{(50-\xi)^2}{4(10)t} - \frac{(50-\eta)^2}{4(5)t} - \frac{(10-\zeta)^2}{4(0.05)t}\right] d\xi \, d\eta \, d\zeta$$

$$= \frac{0.0355}{t^{3/2}} \int_{-0.5}^{0.5} \int_{-1}^{1} \int_{-1}^{1} \exp\left[-\frac{(50-\xi)^2}{40t} - \frac{(50-\eta)^2}{20t} - \frac{(10-\zeta)^2}{0.2t}\right] d\xi \, d\eta \, d\zeta$$

This integral can be evaluated numerically to determine the values of $c(50, 50, 10, t)$ for specified values of t.

3.2.4 Moment Property of the Diffusion Equation

Consider the Cartesian form of the diffusion equation given by Equation 3.24. Multiplying this equation by x'^2 and integrating over x' between $\pm\infty$ yields

$$\int_{-\infty}^{\infty} \frac{\partial c}{\partial t} x'^2 \, dx' = D_x \int_{-\infty}^{\infty} x'^2 \frac{\partial^2 c}{\partial x'^2} \, dx' + D_y \int_{-\infty}^{\infty} x'^2 \frac{\partial^2 c}{\partial y'^2} \, dx' + D_z \int_{-\infty}^{\infty} x'^2 \frac{\partial^2 c}{\partial z'^2} \, dx' \quad (3.91)$$

To evaluate these integrals, assume that the tracer concentrations are equal to zero at $x' = \pm\infty$, which means that

$$\frac{\partial c}{\partial x'_i} = 0 \qquad \text{at } x'_i = \pm\infty \qquad (3.92)$$

and

$$c = 0 \qquad \text{at } x'_i = \pm\infty \qquad (3.93)$$

Applying these conditions to Equation 3.91 and integrating by parts yields

$$\frac{\partial}{\partial t} \int_{-\infty}^{\infty} x'^2 c \, dx' = 2D_x \int_{-\infty}^{\infty} c \, dx' + D_y \frac{\partial^2}{\partial y'^2} \int_{-\infty}^{\infty} x'^2 c \, dx' + D_z \frac{\partial^2}{\partial z'^2} \int_{-\infty}^{\infty} x'^2 c \, dx' \quad (3.94)$$

Integrating Equation 3.94 with respect to y' from $-\infty$ to $+\infty$, applying Equations 3.92 and 3.93, and simplifying yields

$$\frac{\partial}{\partial t} \int_{-\infty}^{\infty} \int_{-\infty}^{\infty} x'^2 c \, dx' dy' = 2D_x \int_{-\infty}^{\infty} \int_{-\infty}^{\infty} c \, dx' dy' + D_z \frac{\partial^2}{\partial z'^2} \int_{-\infty}^{\infty} \int_{-\infty}^{\infty} x'^2 c \, dx' dy' \quad (3.95)$$

Integrating Equation 3.95 with respect to z' from $-\infty$ to $+\infty$, applying Equations 3.92 and 3.93, and simplifying yields

$$\frac{\partial}{\partial t} \int_{-\infty}^{\infty} \int_{-\infty}^{\infty} \int_{-\infty}^{\infty} x'^2 c \, dx' dy' dz' = 2D_x \int_{-\infty}^{\infty} \int_{-\infty}^{\infty} \int_{-\infty}^{\infty} c \, dx' dy' dz' \quad (3.96)$$

The integral term on the right-hand side of Equation 3.96 is equal to the total mass, M, of the tracer, where

$$M = \int_{-\infty}^{\infty}\int_{-\infty}^{\infty}\int_{-\infty}^{\infty} c\, dx'\, dy'\, dz' \tag{3.97}$$

and for conservative tracers M is a constant for the duration of the diffusion process. Equation 3.96 can therefore be written in the form

$$\frac{d}{dt}\left(\frac{1}{M}\int_{-\infty}^{\infty}\int_{-\infty}^{\infty}\int_{-\infty}^{\infty} x'^2 c\, dx'\, dy'\, dz'\right) = 2D_x \tag{3.98}$$

where the partial derivative with respect to time has been replaced by the total derivative with respect to time, since the quantity being differentiated depends only on time.[2] The variance of the concentration distribution along the x'-axis, $\sigma_{x'}^2$, is given by

$$\sigma_{x'}^2 = \frac{1}{M}\int_{-\infty}^{\infty}\int_{-\infty}^{\infty}\int_{-\infty}^{\infty} x'^2 c\, dx'\, dy'\, dz' \tag{3.99}$$

and the integrated diffusion equation, Equation 3.98, can therefore be written in the form

$$\frac{d\sigma_{x'}^2}{dt} = 2D_x \tag{3.100}$$

or

$$\boxed{D_x = \frac{1}{2}\frac{d\sigma_{x'}^2}{dt}} \tag{3.101}$$

This rather remarkable result indicates that in a uniform flow field the diffusion coefficient, D_x, is equal to one-half the rate of growth of variance, $\sigma_{x'}^2$, regardless of the initial conditions. Similar results are obtained for D_y and D_z by multiplying the original diffusion equation by y'^2 and z'^2 prior to integration, yielding

$$\boxed{D_y = \frac{1}{2}\frac{d\sigma_{y'}^2}{dt}} \tag{3.102}$$

and

$$\boxed{D_z = \frac{1}{2}\frac{d\sigma_{z'}^2}{dt}} \tag{3.103}$$

The practical utility of these equations, which relate the diffusion coefficients to the variances of the tracer-concentration distributions, is that the variances can be measured using tracer studies, and then the diffusion coefficients are equal to one-half of the rate of growth of the respective variances. These measured (and validated) diffusion coefficients can then be used in the analysis and design of systems to control contaminant transport.

---

[2]The spatial dimensions have been removed by integration.

TABLE 3.1   Data for Example 3.8

Time, t (h)	$\sigma_{x'}^2$ (cm^2)	$\sigma_{y'}^2$ (cm^2)
0	10^4	10^4
3	3.0×10^7	2.7×10^7
6	1.4×10^8	1.3×10^8
9	3.7×10^8	3.3×10^8
12	7.2×10^8	6.5×10^8

TABLE 3.2   Results for Example 3.8

Time, t (h)	$D_{x'}$ (cm^2/s)	$D_{y'}$ (cm^2/s)
1.5	1.4×10^3	1.2×10^3
4.5	5.1×10^3	4.8×10^3
7.5	1.1×10^4	9.3×10^4
10.5	1.6×10^4	1.5×10^4

Example 3.8   A 10-kg slug of Rhodamine WT dye is released into the ocean, and the concentration distribution of the dye is measured every 3 h for the 12-h duration of daylight when the dye can be seen. The horizontal variances of the dye cloud as a function of time is given in Table 3.1. Estimate the horizontal diffusion coefficients.

SOLUTION   In accordance with Equations 3.101 and 3.102, the diffusion coefficients can be approximated by

$$D_{x'} \approx \frac{1}{2} \frac{\Delta \sigma_{x'}^2}{\Delta t} \quad \text{and} \quad D_{y'} \approx \frac{1}{2} \frac{\Delta \sigma_{y'}^2}{\Delta t}$$

Therefore, between $t = 0$ and $t = 3$ h,

$$D_{x'} \approx \frac{1}{2} \frac{3 \times 10^7 - 10^4}{(3-0)3600} = 1.4 \times 10^3 \text{ cm}^2/\text{s}$$

and

$$D_{y'} \approx \frac{1}{2} \frac{2.7 \times 10^7 - 10^4}{(3-0)3600} = 1.2 \times 10^3 \text{ cm}^2/\text{s}$$

These diffusion coefficients can be taken as the approximate values at $t = (0 + 3)/2 = 1.5$ h. Repeating this analysis for subsequent time intervals, the diffusion coefficients as a function of time are given in Table 3.2. Note that the diffusion coefficients are clearly not steady and are increasing with time. As the plume expands, it experiences a wider variation of ocean currents; hence, mixing occurs at a more rapid rate with increasing time.

3.2.5   Nondimensional Form

Consider the general advection–diffusion equation given by

$$\frac{\partial c}{\partial t} + \sum_{i=1}^{3} V_i \frac{\partial c}{\partial x_i} = \sum_{i=1}^{3} D_i \frac{\partial^2 c}{\partial x_i^2} + S_m \tag{3.104}$$

where V_i are the components of the ambient velocity and S_m is the source flux of tracer per unit volume $[M/L^3 T]$. A reference concentration, C, such as the background concentration of the contaminant, can usually be defined, along with a reference velocity, V, and a reference length, L, which characterizes the dimension of the space in which the contaminant is moving. The concentration, c, coordinates, x_i, time, t, and velocity components, V_i, can be normalized relative to these reference variables to yield the following nondimensional variables:

$$c^* = \frac{c}{C} \tag{3.105}$$

$$x_i^* = \frac{x_i}{L} \tag{3.106}$$

$$t^* = \frac{t}{L/V} \tag{3.107}$$

$$V_i^* = \frac{V_i}{V} \tag{3.108}$$

Substituting Equations 3.105 to 3.108 into Equation 3.104, taking $S_m = 0$ (for a conservative substance), and simplifying yields

$$\frac{\partial c^*}{\partial t^*} + \sum_{i=1}^{3} V_i^* \frac{\partial c^*}{\partial x_i^*} = \sum_{i=1}^{3} \frac{D_i}{VL} \frac{\partial^2 c^*}{\partial x_i^{*2}} \tag{3.109}$$

The utility of this nondimensional representation is that all the terms involving nondimensional variables are on the order of one, since each of the nondimensional variables have been normalized by a reference quantity that is characteristic of the ambient environment. Consequently, the only terms whose magnitudes are not fixed are the diffusion terms, whose magnitudes are on the order of D_i/VL. This nondimensional quantity represents the ratio of diffusive transport to advective transport and is called the *Peclet number*, denoted by Pe. Therefore, defining the Peclet number by

$$\text{Pe}_i = \frac{D_i}{VL} \tag{3.110}$$

Equation 3.109 can be written as

$$\boxed{\frac{\partial c^*}{\partial t^*} + \sum_{i=1}^{3} V_i^* \frac{\partial c^*}{\partial x_i^*} = \sum_{i=1}^{3} \text{Pe}_i \frac{\partial^2 c^*}{\partial x_i^{*2}}} \tag{3.111}$$

According to Equation 3.111, whenever the component Peclet numbers, Pe_i, are small, the diffusion process can be neglected, and the contaminant transport is dominated by advection. Conversely, for large Peclet numbers, diffusion is the dominant process. The important result here is that the Peclet number is particularly useful as an indicator of the dominant transport process, which must be represented accurately in order to properly describe the transport of a contaminant in a particular environment.

In cases where the contaminant of interest is not conservative, the source term, S_m, in the advection–diffusion equation (Equation 3.104) is nonzero. In the special case where the nonconservative contaminant exhibits first-order decay, the source term, S_m, can be expressed as

$$S_m = -kc \tag{3.112}$$

and the nondimensional advection–diffusion equation can be written in the form

$$\frac{\partial c^*}{\partial t^*} + \sum_{i=1}^{3} V_i^* \frac{\partial c^*}{\partial x_i^*} = \sum_{i=1}^{3} \mathrm{Pe}_i \frac{\partial^2 c^*}{\partial x_i^{*2}} + k^* c^* \tag{3.113}$$

where

$$k^* = \frac{kL}{V} \tag{3.114}$$

The variable k^* is the ratio of the relative magnitudes of the reaction rate and the advection rate and is sometimes called the *Damkohler number* (Ramaswami et al., 2005) and denoted by "Da," where

$$\mathrm{Da} = \frac{kL}{V} \tag{3.115}$$

The Damkohler number is sometimes used to contrast the relative magnitudes of the reaction rate and diffusion rate, in which case

$$\mathrm{Da}_i = \frac{kL^2}{D_i} \tag{3.116}$$

where D_i is the i-component of the diffusion coefficient. For either definition of the Damkohler number, if $\mathrm{Da} \ll 1$, chemical reactions can be neglected.

3.3 TRANSPORT OF SUSPENDED PARTICLES

The advection–dispersion equation, Equation 3.16, is appropriate for describing the fate and transport of dissolved contaminants that are advected with the same velocity as the ambient water. In the case of suspended particles, the settling of the particles is influenced by the size, shape, and density of the particles in addition to the ambient flow velocity. The process by which suspended particles settle to the bottom of water bodies is called *sedimentation*, and the settling velocity, v_s, of suspended particles with diameters less than or equal to 0.1 mm ($=100\,\mu$m) can be estimated by the *Stokes equation*, which is given by (Yang, 1996)

$$v_s = \alpha \frac{(\rho_s/\rho_w - 1)g\phi^2}{18\,v_w} \tag{3.117}$$

where α is a dimensionless form factor that measures the effect of particle shape ($\alpha = 1$ for spherical particles), ρ_s is the density of the suspended particle, ρ_w is the density of the

TABLE 3.3   Typical Settling Velocities in Natural Waters

Particle Type	Diameter (μm)	Settling Velocity (m/day)
Phytoplankton		
Cyclotella meneghiniana	2	0.08 (0.24)[a]
Thalassiosira nana	4.3–5.2	0.1–0.28
Scenedesmus quadricauda	8.4	0.27 (0.89)
Asterionella formosa	25	0.2 (1.48)
Thalassiosira rotula	19–34	0.39–0.21
Coscinodiscus lineatus	50	1.9 (6.8)
Melosira agassizii	54.8	0.67 (1.87)
Rhizosolenia robusta	84	1.1 (4.7)
Particulate organic carbon	1–10	0.2
	10–64	1.5
	>64	2.3
Clay	2–4	0.3–1
Silt	10–20	3–30

Source:

[a]Numbers in parentheses are for the stationary phase of microbial growth.

ambient water, g is gravity, ϕ is the particle diameter, and v_w is the kinematic viscosity of the ambient water. The form factor, α, sometimes called the *sphericity*, is defined as the ratio of the surface area of a sphere having the same volume as the particle to the surface area of the particle. Particles in natural waters have complex shapes, typically $\alpha < 1$. The settling velocity given by Stokes' equation (Equation 3.117) is called *Stokes' velocity*. Settling velocities that are of interest in natural waters are given in Table 3.3 (Wetzel, 1975; Burns and Rosa, 1980; Chapra, 1997). As a general rule, settling is unimportant for particles smaller than about 1 μm in diameter. Particles in water must be larger than about 10 μm in diameter to settle through distances of several centimeters in time scales of an hour or less, and particles in this class are sometimes called *settleable solids* (Nazaroff and Alvarez-Cohen, 2001). In cases where the ambient water moves with a horizontal velocity, V, suspended particles tend to move horizontally at the velocity, V, and vertically at the settling velocity.

Suspended solids in natural waters have two primary sources: surface runoff from drainage basins and as products of photosynthesis. The suspended-solids concentration in natural waters typically range from below 1 mg/L in clear waters to over 100 mg/L in highly turbid waters. An example of a highly turbid stream resulting from surface runoff is shown in Figure 3.11. Suspended solids derived from photosynthetic processes tend to be higher in organic matter and less dense than suspended solids derived from surface runoff. Caution should be used in applying the Stokes equation to calculate the settling velocity of living particles, since some phytoplankton, such as blue-green algae, can become buoyant due to the development of internal gas vacuoles.

Many contaminants sorb strongly onto suspended particles, so that prediction of the fate and transport of suspended sediments is essential for describing the fate and transport of these contaminants in natural waters. Heavy metals and hydrophobic organic compounds, such as PCBs, are two classes of contaminants that sorb strongly onto suspended sediments.

FIGURE 3.11   Turbid runoff in a stream. (From Town of Chapel Hill, 2005.)

Example 3.9   Analysis of water from a lake indicates a suspended-solids concentration of 50 mg/L. The suspended particles are estimated to have an approximately spherical shape with an average diameter of 4 μm and a density of 2650 kg/m^3. (a) If the water temperature is 20°C, estimate the settling velocity of the suspended particles. (b) If the suspended particles are mostly clay, compare your estimate of the settling velocity with the data in Table 3.3. (c) If there is 1 g of heavy-metal ion per kilogram of suspended particles, determine the rate at which heavy metals are being removed from the lake by sedimentation.

SOLUTION   (a) From the data given, $\alpha = 1$ (spherical particles), $\rho_s = 2650$ kg/m^3, $\rho_w = 998$ kg/m^3 at 20°C, $\phi = 4\,\mu$m $= 4 \times 10^{-6}$ m, and $v_w = 1.00 \times 10^{-6}$ m^2/s. Substituting into the Stokes equation (Equation 3.117) gives

$$v_s = \alpha \frac{(\rho_s/\rho_w - 1)g\phi^2}{18 v_w} = (1)\,\frac{(2650/998 - 1)(9.81)(4 \times 10^{-6})^2}{18(1.00 \times 10^{-6})}$$

$$= 1.44 \times 10^{-5}\,\text{m/s} = 1.25\,\text{m/day}$$

(b) This result is consistent with the settling velocities for clay-sized particles shown in Table 3.3, which indicates that a 4-μm clay particle will have a settling velocity on the order of 1 m/day.

(c) Since the concentration, c, of suspended particles is 50 mg/L $= 0.05$ kg/m^3, the rate at which sediment is accumulating on the bottom of the lake is given by

$$\text{removal rate of suspended particles} = v_s c = 1.25(0.05) = 0.0625\,\text{kg/day} \cdot \text{m}^2$$

Since heavy metals are attached to the sediment at the rate of 1 g/kg, the removal rate of heavy metals is given by

$$\text{removal rate of heavy metals} = 1(0.0625) = 0.0625 \, \text{g/day} \cdot \text{m}^2$$

SUMMARY

The transport of dissolved contaminants in water bodies is caused by spatial and temporal variations in the the velocity field. Mixing is associated with small-scale variations and is characterized by the Fickian diffusion equation, while advection is associated with larg-scale velocity variations. The length scale that delineates mixing from advection is the size of the contaminant cloud. The collective effects of mixing and advection are represented by the advection–diffusion equation, which is also known as the advection–dispersion equation (ADE). The latter term is usually more appropriate. Fate effects such as sorption onto solid media and biochemical decay are represented as a source term in the ADE. The fundamental solution to the ADE is for the case of an instantaneous release of a mass of conservative substance into an infinitely large medium. Many other solutions to the ADE can be expressed in terms of the fundamental solution in either one, two, or three dimensions by invoking the principle of superposition. A useful property of the ADE is the moment property, which guarantees that for conservative tracers, each component of the dispersion coefficient is equal to one-half of the rate of change of the component variance of the concentration distribution. This property is particularly useful in estimating the dispersion coefficient using conservative tracers. Expressing the ADE in terms of nondimensional variables indicates that the Peclet number is a useful measure of the magnitude of the dispersive flux relative to the advective flux. Dispersion can be neglected in cases where the Peclet number is small. The ADE is applicable only to dissolved contaminants, and the transport of suspended solids is typically predicted using Stokes law.

PROBLEMS

3.1. Show that $\nabla \cdot \mathbf{V} = 0$ for incompressible fluids, where $\mathbf{V}$ is the velocity field in the fluid.

3.2. Fifty kilograms of toxic material is spilled uniformly across a canal. The canal is trapezoidal with a bottom width of 3 m, side slopes of 2.5:1 (H:V), and a depth of flow of 1.7 m. The discharge in the canal is 16 m³/s, and the longitudinal dispersion coefficient is 7 m²/s.

   (a) How soon after the spill will the peak concentration be observed 12 km downstream of the spill, and what is the maximum concentration expected at that location?

   (b) If a safe level of this contaminant in recreational waters is 15 μg/L, for approximately how long will the water at the downstream location be unsafe?

   (c) What length of canal is contaminated 2 h after the spill?

3.3. Water stored in a reservoir behind a gate contains a toxic contaminant at a concentration of 1 mg/L. If the gate is opened and water flows downstream at a velocity of 30 cm/s and with a longitudinal dispersion coefficient of $10\,m^2/s$, how long after the gate is opened will the concentration of the contaminant 1 km downstream be equal to $10\,\mu g/L$?

3.4. Two kilograms of a contaminant is spilled into an open channel at a location 20 m from the end of the channel. The channel has a rectangular cross section 8 m wide and 2 m deep, and the diffusion coefficient along the channel is estimated as $5\,m^2/s$.

   (a) Assuming that the contaminant is initially well mixed across the channel, express the concentration as a function of time at the end of the channel.

   (b) How long will it take for the contaminant concentrations 10 m upstream (in the direction of the channel end) to be 20% higher than the concentration 10 m downstream of the spill (in the direction away from the channel end)?

3.5. Two kilograms of a contaminant is spilled at a point in a 3-m-deep reservoir. The contaminant is spilled at $x = 0$, $y = 0$, and is instantaneously mixed over the entire depth (in the z-direction).

   (a) If the diffusion coefficients in the x (E-W) and y (N-S) directions are 10 and $20\,m^2/s$, respectively, calculate the concentration as a function of time at a point 50 m north and 50 m east of the spill.

   (b) What is the concentration at the spill location after 1 min?

3.6. A vertical diffuser discharges wastewater at a rate of $3\,m^3/s$ uniformly over a 4-m-deep reservoir. If the wastewater contains 100 mg/L of a conservative tracer for 20 h, calculate the concentration of the contaminant as a function of time during this 20-h period at a distance of 150 m from the outfall. Assume that the diffusion coefficient in the reservoir is equal to $15\,m^2/s$.

3.7. A bridge crosses a river 4.5 km upstream of a water-supply intake, and in recent years, there have been several contaminant spills at the river crossing, primarily on the side of the river opposite the water-supply intake. The river is 30 m wide, 3 m deep, has an average flow of $13.5\,m^3/s$, the longitudinal dispersion coefficient in the river is estimated to be $1.27\,m^2/s$, and the transverse coefficient is $0.0127\,m^2/s$. What mass of contaminant spilled at the bridge (on the opposite side to the intake) would lead to a contaminant concentration at the intake equal to 1 mg/L?

3.8. Five kilograms of a toxic contaminant is released deep into the ocean and spreads in all three coordinate directions.

   (a) If the N-S, E-W, and vertical diffusion coefficients are 15, 20, and $0.5\,m^2/s$, respectively, find the concentration at a point 50 m north, 50 m east, and 5 m above the release point as a function of time.

   (b) What is the concentration at the release point after 12 h? Assume no reactions and no advection.

3.9. Eight kilograms of a contaminant in the form of a $1\,m \times 1\,m \times 1\,m$ parallelepiped is released into the deep ocean. If the N-S, E-W, and vertical diffusion coefficients are 15, 10, and $0.1\,m^2/s$, respectively, find the concentration as a function of time at a location 25 m north, 25 m east, and 5 m above the centroid of the initial mass release.

3.10. A submarine releases 100 kg of waste at a location 25 m below the surface of the ocean. If the ambient current is 30 cm/s to the north and the components of the diffusion coefficient are 12, 5, and $1\,m^2/s$ in the N-S, E-W, and vertical directions, determine the maximum concentration as a function of time and the concentration at the release location after 1 h. Utilize the principle of superposition to account for the presence of the ocean surface. Explain how you would account for the solubility of the waste in your calculations.

3.11. The suspended solids in a 200 m × 200 m lake is measured to be 45 mg/L, and the average settling velocity is estimated as 0.1 m/day.

  (**a**) Estimate the rate at which sediment mass is accumulating on the bottom of the lake.

  (**b**) If the suspended-solids concentration remains fairly steady and the water leaving the lake does not have a significant suspended sediment content, at what rate is sediment mass entering the lake?

3.12. The phytoplankton *Coscinodiscus lineatus* has a typical diameter of 50 μm and an estimated density of 1600 kg/m^3. Assuming that the phytoplankton is approximately spherical and the water temperature is 20°C, estimate the settling velocity using the Stokes equation. Compare your result with the settling velocity given in Table 3.3, and provide possible reasons for any discrepancy.

3.13. A stormwater outfall discharges runoff into a pristine river (with negligible dissolved solids) such that the suspended-solids concentration of the combined water just downstream of the outfall is 100 mg/L. The settling velocity of the sediment is estimated to be 2 m/day, the flow velocity in the river is 0.4 m/s, and the river is 10 m wide and 2 m deep.

  (**a**) How far downstream from the outfall will it be before the suspended sediment all settles out?

  (**b**) Estimate the rate at which sediment is accumulating downstream of the outfall.

3.14. Consider the (common) case in which an outfall discharges treated domestic wastewater at a rate of 80 L/s with a suspended-solids concentration of 30 mg/L into a river. The suspended solids are composed of predominantly silt particles, and the river has a trapezoidal shape with a flow depth of 4 m, a bottom width of 6 m, and side slopes of 2:1 (H:V). The mean velocity in the river is 3 cm/s.

  (**a**) Estimate the distance from the outfall within which most of the suspended particles are deposited on the bottom.

  (**b**) Estimate the rate at which sediment is accumulating on the bottom within 500 m of the outfall.

CHAPTER 4

RIVERS AND STREAMS

4.1 INTRODUCTION

Streams are natural drainage channels that collect surface-water runoff and ground-water seepage from the surrounding area. Hydrologists classify the hierarchy of streams according to the number of tributaries upstream using a classification system originally proposed by Horton (1945) and later refined by Strahler (1957). First-order streams are the initial, smallest tributaries; the stream located just below where two first-order streams combine is a second-order stream; the stream located below the confluence of two second-order streams is a third-order stream; and so on. In most cases, streams are called *rivers* when they become seventh-order streams or higher (DeBarry, 2004). Alternative definitions of a river vary from a fifth- to a ninth-order stream (Ramaswami et al., 2005). The Amazon River, which is the largest river in the world, is a twelfth-order stream.

Rivers and streams have long been the primary sources of drinking water to support human populations, and the water quality of rivers and streams has been studied more extensively and longer than any other bodies of water (Shifrin, 2005). Threats to the water quality in streams and rivers commonly originate from end-of-pipe discharges, such as illustrated in Figure 4.1. The four most prevalent water-quality problems affecting rivers and streams in the United States are siltation, nutrients, pathogens, and oxygen-depleting substances (USEPA, 1996c). *Siltation* refers to the accumulation of small soil particles (silt) on the bottom of the river, causing suffocation of fish eggs and destruction of aquatic insect habitats, and damaging the food web that supports fish and other wildlife. Siltation can occur as a result of agriculture, urban runoff, construction, and forest operations. Nutrient pollution generally refers to elevated quantities of nitrogen and phosphorus in the water. Excessive nutrients cause increased plant and algae growth, resulting in reduced oxygen levels and depleted populations of fish and other desirable aquatic species. Municipal and

Water-Quality Engineering in Natural Systems, by David A. Chin
Copyright © 2006 John Wiley & Sons, Inc.

FIGURE 4.1 Contaminant discharge into the Alpenrhein River, Germany. (From Socolofsky and Jirka, 2005.)

industrial wastewater discharges and runoff from agricultural lands, forestry operations, and urban areas are major nutrient sources. Pathogen contamination of surface waters can cause human-health problems ranging from simple skin rash to acute gastroenteritis. Common sources of pathogens include inadequately treated municipal wastewater, agricultural and urban runoff, and wildlife fecal material. The discharge of oxygen-demanding material into rivers and streams causes a depletion of dissolved oxygen downstream of the discharge location. Oxygen depletion can cause problems for aquatic life and have a severe impact on the natural biota of a river or stream. Sources of oxygen-demanding substances include municipal and industrial wastewater as well as agricultural and urban runoff.

Treated sewage effluent, industrial wastewaters, and stormwater runoff are discharged routinely into inland streams, and in most cases the wastewater discharges do not meet the ambient water-quality standards of the stream. In such cases, *regulatory mixing zones* are usually permitted in the vicinity of the discharge location. Within these mixing zones, dilution processes reduce the contaminant concentrations to levels that meet the ambient water-quality criteria. The spatial extent of the mixing zone is usually restricted in size. In Florida, for example, mixing zones in streams are restricted to 800 m in length or 10% of the total length of the stream, whichever is less (Florida DEP, 1995). Wastes can be discharged into streams either through multiport diffusers, which distribute the effluent over a finite portion of the stream width, or through single-port outfalls that discharge the effluent at a "point" in the stream.

Consider the case of a single-port discharge into a stream, such as from an open-ended pipe. The initial mixing of the pollutant is determined by the momentum and buoyancy of the discharge. As the discharge is diluted, the momentum and buoyancy of the discharge are dissipated, and further mixing of the discharge plume is dominated by ambient velocity variations in the stream. As the pollutant cloud extends over the depth and width of the stream, parts of the cloud expand into areas with significantly different longitudinal mean velocities; the pollutant cloud is "stretched" apart in the longitudinal direction in addition to diffusing in all three coordinate directions. This stretching of the cloud is due primarily to vertical and transverse variations in the longitudinal mean velocity. The process of stretching is commonly referred to as *shear dispersion* or simply, *dispersion*. Contaminant discharges from point sources spread in the vertical and transverse directions by turbulent diffusion until the pollutant is well mixed across the stream cross section, at which time almost all of the mixing will be caused by longitudinal shear dispersion. Field results have shown that in most cases transverse variations in the mean velocity (across the channel) have a much greater effect on shear dispersion than vertical variations in the mean velocity. Typical flow velocities in rivers and streams range from 0.1 to 1.5 m/s, corresponding to channel slopes of 0.02 to 1%.

4.2   TRANSPORT PROCESSES

In most cases, contaminants are introduced into rivers over a particular subarea of the river cross section, For example, pipe discharges are typically over a small area along the side of the river, whereas submerged multiport diffuser discharges are over a larger portion of the river cross section. In these cases, two distinct zones are identified: the initial zone where the contaminant mixes across the cross section of the channel, and the well-mixed zone, where the contaminant is well-mixed across the cross section and further mixing is associated with longitudinal dispersion in the flow direction.

4.2.1   Initial Mixing

The turbulent velocity fluctuations in the vertical and transverse directions in rivers are on the same order of magnitude as the *shear velocity*, u_*, which is defined by

$$u_* = \sqrt{\frac{\tau_0}{\rho}} \tag{4.1}$$

where τ_0 is the mean shear stress on the (wetted) perimeter of the stream and ρ is the density of the fluid. The boundary shear stress, τ_0, can be expressed in terms of the Darcy–Weisbach friction factor, f, and the mean (longitudinal) flow velocity, V, by (Chin, 2006)

$$\tau_0 = \frac{f}{8}\rho V^2 \tag{4.2}$$

Combining Equations 4.1 and 4.2 leads to the following expression for the shear velocity:

$$\boxed{u_* = \sqrt{\frac{f}{8}}V} \tag{4.3}$$

The friction factor, f, can generally be estimated from the channel roughness, hydraulic radius, and Reynolds number using the Colebrook equation, which for open channels can be approximated by (Chin, 2006)

$$\frac{1}{\sqrt{f}} = -2\log_{10}\left(\frac{k_s}{12R} + \frac{2.5}{\mathrm{Re}\sqrt{f}}\right) \tag{4.4}$$

where k_s is the roughness height in the channel, R is the hydraulic radius, defined as the flow area divided by the wetted perimeter, and Re is the Reynolds number, defined by

$$\mathrm{Re} = \frac{4VR}{v} \tag{4.5}$$

where v is the kinematic viscosity of the fluid in the channel, usually water. Equation 4.4 should be used with caution in estimating the friction factor, f, in mountain streams, where the presence of transient storage zones (pools), stagnant zones due to bed irregularity, and the existence of hyporheic zones in the gravel bed all require special consideration (Meier and Reichert, 2005). The hyporheic zone is the area below the streambed between the rocks and cobbles. For mountain streams, the friction factor can be estimated using a method suggested by Bathurst (1985).

Based on a theoretical analysis of turbulent mixing, Elder (1959) showed that the average value of the vertical turbulent diffusion coefficient, ε_v, in a wide open channel can be estimated by the relation

$$\boxed{\varepsilon_v = 0.067 d u_*} \tag{4.6}$$

where d is the depth of flow in the channel. The theoretical expression for the vertical turbulent diffusion coefficient, ε_v, given by Equation 4.6 has been confirmed experimentally in a laboratory flume by Jobson and Sayre (1970). The coefficient in Equation 4.6 ($= 0.067$) is sometimes given as 0.1 (Martin and McCutcheon, 1998). Vertical mixing can be enhanced locally by secondary currents (notably at sharp bends) and by obstacles in the flow (such as bridge piers), although neither of these mechanisms has been quantified adequately (Rutherford, 1994).

Experimental results in straight rectangular channels indicate that the transverse turbulent diffusion coefficient, ε_t, can be estimated by the relation

$$\boxed{\varepsilon_t = 0.15 d u_* \qquad \text{(straight uniform channels)}} \tag{4.7}$$

where the coefficient of 0.15 can be taken to have an error bound of $\pm 50\%$ (Fischer et al., 1979). Experimental results yielded $\varepsilon_t = 0.08 d u_*$ to $0.24 d u_*$ (Lau and Krishnappan, 1977). Bends, sidewall irregularities such as the deflection bars shown in Figure 4.2, and variations in channel shape found in natural streams all serve to increase the transverse turbulent diffusion coefficient over that given by Equation 4.7. A more typical estimate of the transverse diffusion coefficient in natural streams is

$$\boxed{\varepsilon_t = 0.6 d u_* \qquad \text{(natural streams)}} \tag{4.8}$$

FIGURE 4.2   Deflection bars on the Little Blue River, Kansas. (From NRCS, 2005b.)

where the coefficient of 0.6 can be taken to have an error bound of $\pm 50\%$ and is sometimes approximated as 1 (Martin and McCutcheon, 1998). Experimental studies continue to provide data on which more precise methods of estimating transverse turbulent diffusion coefficients are being developed (e.g., Boxall and Guymer, 2003). Bends in natural streams cause secondary (transverse) currents that can increase the transverse turbulent diffusion coefficient considerably beyond that given by Equation 4.8. For example, Yotsukura and Sayre (1976) estimated ε_t as $3.4du_*$ in a stretch of the Missouri River containing 90° and 180° bends. If a reliable estimate of ε_t is needed in any particular case, it is recommended that field experiments be conducted to measure it directly (Roberts and Webster, 2002).

Many mixing-zone analyses assume instantaneous cross-sectional mixing and then calculate longitudinal variations in cross-sectionally averaged concentrations downstream from the discharge location. Considering a stream of characteristic depth d and width w, the time scale, T_d, for mixing over the depth of the channel can be estimated by

$$T_d = \frac{d^2}{\varepsilon_v} \tag{4.9}$$

and the distance, L_d, downstream from the discharge point to where complete mixing over the depth occurs is given by

$$L_d = VT_d = \frac{Vd^2}{\varepsilon_v} \tag{4.10}$$

Similarly, the time scale, T_w, for mixing over the width, w, can be estimated by

$$T_w = \frac{w^2}{\varepsilon_t} \tag{4.11}$$

and the corresponding downstream length scale, L_w, to where the tracer is well mixed over the width can be estimated by

$$L_w = VT_w = \frac{Vw^2}{\varepsilon_t} \tag{4.12}$$

Combining Equations 4.10 and 4.12 leads to

$$\frac{L_w}{L_d} = \left(\frac{w}{d}\right)^2 \frac{\varepsilon_v}{\varepsilon_t} \tag{4.13}$$

According to Equations 4.6 and 4.8, it can reasonably be expected that in natural streams $\varepsilon_t \approx 10\varepsilon_v$, in which case Equation 4.13 can be approximated by

$$\boxed{\frac{L_w}{L_d} = 0.1\left(\frac{w}{d}\right)^2} \tag{4.14}$$

Since channel widths in natural streams are usually much greater than channel depths, typically width/depth ≥ 20 (Koussis and Rodríguez-Mirasol, 1998), for single-port discharges the downstream distance to where the tracer becomes well mixed across the stream can be expected to be at least an order of magnitude greater than the distance to where the tracer becomes well mixed over the depth. Therefore, mixing over the depth occurs well in advance of mixing over the width. An aerial view of mixing across a river is given in Figure 4.3. The length scale to where the tracer can be expected to be well mixed over the width is given by Equation 4.12. Fischer and colleagues (1979) used field measurements to estimate the actual distance, L'_w, for a single-port discharge located on the side of a channel to mix completely across a stream as

$$\boxed{L'_w = 0.4\frac{Vw^2}{\varepsilon_t}} \tag{4.15}$$

FIGURE 4.3 Mixing across a river. (From Universität Karlsruhe, 2005.)

In estimating the coefficient in Equation 4.15, Fischer and colleagues defined *complete mixing* as the condition in which the tracer concentration is within 5% of its mean everywhere in the cross section. Equation 4.15 can be applied at any discharge location in a stream, where w is taken as the width over which the contaminant is to be mixed to achieve complete cross-sectional mixing. For example, if a multiport outfall of length L is placed in the center of a stream of width W, full cross-sectional mixing occurs when the contaminant mixes over a width, $w = (W - L)/2$, and the downstream distance, L'_w, to complete cross-sectional mixing is given by

$$L'_w = 0.1 \frac{V(W - L)^2}{\varepsilon_t} \tag{4.16}$$

Clearly, cross-sectional mixing can be accelerated by using multiport diffusers rather than single-port outlets.

Example 4.1 A municipality discharges wastewater from the side of a stream that is 10 m wide and 2 m deep. The average flow velocity in the stream is 1.5 m/s, and the friction factor is estimated to be 0.03 (calculated using the Colebrook equation). (a) Estimate the time for the wastewater to become well mixed over the channel cross section. (b) How far downstream from the discharge location can the effluent be considered well mixed across the stream?

SOLUTION (a) From the data given, $f = 0.03$ and $V = 1.5$ m/s. Therefore, the shear velocity, u_*, is given by Equation 4.3 as

$$u_* = \sqrt{\frac{f}{8}} V = \sqrt{\frac{0.03}{8}} (1.5) = 0.092 \text{ m/s}$$

Since $d = 2$ m, the vertical and transverse diffusion coefficients are

$$\varepsilon_v = 0.067 d u_* = 0.067(2)(0.092) = 0.012 \text{ m}^2/\text{s}$$
$$\varepsilon_t = 0.6 d u_* = 0.6(2)(0.092) = 0.11 \text{ m}^2/\text{s}$$

The time scale for vertical mixing, T_d, is given by

$$T_d = \frac{d^2}{\varepsilon_v} = \frac{2^2}{0.012} = 333 \text{ s} = 5.6 \text{ min}$$

and the time scale for transverse mixing, T_w, is given by

$$T_w = \frac{w^2}{\varepsilon_t} = \frac{10^2}{0.11} = 909 \text{ s} = 15 \text{ min}$$

The discharge is well mixed over the channel cross section when it is well mixed over both the depth and the width, which in this case occurs after about 15 min.

(b) In a time interval of 15 min (= 909 s), the discharged effluent travels a distance, VT_w, given by

$$VT_w = 1.5(909) = 1364 \text{ m}$$

The Fischer et al. (1979) relation given by Equation 4.15 indicates that the actual downstream distance required for complete cross-sectional mixing is $0.4VT_w = 0.4(1364) = 546$ m.

Example 4.2   Estimate the distance downstream to where the wastewater described in Example 4.1 is well mixed across the stream if (a) the wastewater is discharged from the center of the stream, and (b) the wastewater is discharged through a 5-m-long multiport diffuser placed in the middle of the stream.

SOLUTION   From the previous analysis: $\varepsilon_v = 0.012$ m^2/s, $\varepsilon_t = 0.11$ m^2/s, and the time scale, T_s, for transverse mixing over a distance s is given by

$$T_s = \frac{s^2}{\varepsilon_t}$$

In Example 4.1 the wastewater was discharged from the side of the channel, so the mixing width, s, was the width of the channel, w.

(a) If the wastewater is discharged from the center of the stream, the mixing width, s, for the wastewater to become well mixed over the channel cross section is given by

$$s = \frac{w}{2} = \frac{10}{2} = 5\,\text{m}$$

and the corresponding time scale, T_s, is given by

$$T_s = \frac{5^2}{0.11} = 227\,\text{s} = 3.8\,\text{min}$$

Since the flow velocity, V, is 1.5 m/s, the downstream distance, L, for the wastewater to become well mixed is

$$L = 0.4VT_s = 0.4(1.5)(227) = 136\,\text{m}$$

(b) If the wastewater is discharged from a 5-m-long diffuser centered in the stream, the mixing width, s, for the wastewater to become well mixed over the channel cross section is given by

$$s = \frac{w - 5}{2} = \frac{10 - 5}{2} = 2.5\,\text{m}$$

and the corresponding time scale, T_s, is given by

$$T_s = \frac{2.5^2}{0.11} = 56.8\,\text{s} = 0.95\,\text{min}$$

Since the flow velocity, V, is 1.5 m/s, the downstream distance, L, for the wastewater to become well mixed is

$$L = 0.4VT_s = 0.4(1.5)(56.8) = 34\,\text{m}$$

The results of this example illustrate that a 5-m-long outfall diffuser located at the center of the stream will cause the wastewater to mix much more rapidly over the stream cross section than a point discharge at the center of the stream.

The previous examples have illustrated why discharges from multiport outfalls generally achieve complete cross-sectional mixing more rapidly than single-port discharges. Also, at any given distance from the outfall, discharges from multiport outfalls are mixed over a larger portion of the channel width than discharges from single-port outfalls, resulting in multiport outfalls achieving greater dilutions. Consider the case where the mass flux of contaminant from an outfall (single port or multiport) is given by $\dot{M}$ and the contaminant is mixed over a width, w, in a river; then conservation of mass requires that

$$\dot{M} = c_m V w d \tag{4.17}$$

where c_m is the mean concentration of the mixture, V is the mean velocity in the mixed portion of the river, and d is the mean depth in the mixed portion of the river. Rearranging Equation 4.17 gives the mean concentration, c_m, of the mixture by

$$\boxed{c_m = \frac{\dot{M}}{A_m V}} \tag{4.18}$$

where A_m is the area over which the contaminant is mixed ($= wd$).

Example 4.3 An industrial wastewater outfall discharges effluent at a rate of $3\,\text{m}^3/\text{s}$ into a river. The chromium concentration in the wastewater is $10\,\text{mg/L}$, the average velocity in the river is $0.5\,\text{m/s}$, and the average depth of the river is $3\,\text{m}$. Tracer tests in the river indicate that when the wastewater is discharged through a single port in the middle of the river, $100\,\text{m}$ downstream of the outfall the plume will be mixed over a width of $4\,\text{m}$; if a 4-m-long multiport outfall is used, the plume will be well mixed over a width of $8\,\text{m}$. Compare the dilution achieved by the single-port and multiport outfall at a location $100\,\text{m}$ downstream of the discharge.

SOLUTION From the data given, $Q_o = 3\,\text{m}^3/\text{s}$, $c_0 = 10\,\text{mg/L} = 0.01\,\text{kg/m}^3$, and the mass flux, $\dot{M}$, of chromium released at the outfall is given by

$$\dot{M} = Q_o c_0 = 3(0.01) = 0.03\,\text{kg/s}$$

For the single-port discharge, the plume is mixed over an area, A_m, of $4\,\text{m} \times 3\,\text{m} = 12\,\text{m}^2$. Since $V = 0.5\,\text{m/s}$, the average concentration of the mixed river water $100\,\text{m}$ downstream of the outfall is given by Equation 4.18 as

$$c_m = \frac{\dot{M}}{A_m V} = \frac{0.03}{12(0.5)} = 0.005\,\text{kg/m}^3 = 5\,\text{mg/L}$$

For the multiport discharge, the plume is mixed over an area, A_m, of $8\,\text{m} \times 3\,\text{m} = 24\,\text{m}^2$, and the average concentration of the mixed river water is given by

$$c_m = \frac{\dot{M}}{A_m V} = \frac{0.03}{24(0.5)} = 0.0025\,\text{kg/m}^3 = 2.5\,\text{mg/L}$$

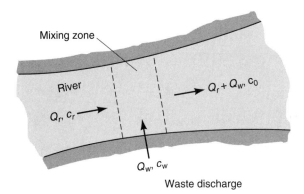

Waste discharge

FIGURE 4.4 Initial mixing of stream discharges. (From Chin, David A., *Water-Resources Engineering.* Copyright © 2000. Reprinted by permission of Pearson Education, Inc., Upper Saddle River, NJ.)

Therefore, at a location 100 m downstream of the discharge, the multiport outfall gives a dilution of 10/2.5 = 4, compared with a dilution of 10/5 = 2 for the single-port outfall.

One-dimensional stream models frequently assume that the discharged contaminant is uniformly mixed over the stream cross section in the vicinity of the discharge location. Analyses described previously demonstrate how this assertion can be assessed quantitatively. The assumption of complete cross-sectional mixing in the vicinity of a discharge location is justified in cases where outfall diffusers span the width of the stream and/or the discharge rate of contaminated water into the stream is comparable to the stream discharge. For single-port discharges on the sides of wide streams, where the discharge rate of contaminated water is small compared with the river discharge, considerable distances may be necessary for complete mixing across a river.

Consider the idealized waste discharge shown in Figure 4.4, where the river flow rate upstream of the wastewater outfall is Q_r, with a contaminant concentration, c_r, and the wastewater discharge rate is Q_w, with a contaminant concentration c_w. Assuming that the river flow and wastewater discharge are completely mixed in the *mixing zone* shown in Figure 4.4, conservation of contaminant mass requires that

$$Q_r c_r + Q_w c_w = (Q_r + Q_w)c_0 \tag{4.19}$$

where c_0 is the concentration of the wastewater/river mixture downstream of the mixing zone. Rearranging Equation 4.19 gives the following expression for the concentration of the diluted wastewater after it is completely mixed with the ambient river water:

$$\boxed{c_0 = \frac{Q_r c_r + Q_w c_w}{Q_r + Q_w}} \tag{4.20}$$

In addition to estimating the concentrations of pollutants in mixed waters, Equation 4.20 can also be used to estimate contaminant concentrations downstream of the confluence of two or more streams, an example of which is shown in Figure 4.5. This spectacular picture shows the joining of three different rivers. The Danube, with very high particulate concentration, on the left joins with the two rivers on the right. The larger of the two rivers

FIGURE 4.5   Confluence of three rivers. (From Universität Karlsruhe, 2005.)

carries a higher particulate load; thus, the darkest (cleanest) smallest river is also visible. Notice how sharp the boundaries are separating the various river flows.

In the case of mixing waters having different temperatures, a heat balance rather than a mass balance must be done. The heat content (also called *enthalpy*), *H*, of a mass of a substance is given by

$$H = mc_p T \qquad (4.21)$$

where m is the mass of substance (kg), c_p is the specific heat at constant pressure (J/kg · K), and T is the temperature (K). The specific heat, c_p, of natural waters varies only slightly with temperature, and for natural waters, a value of 4190 J/kg · K is a satisfactory approximation (Davis and Masten, 2004). Equating the heat content of the mixed water to the sum of the heat contents of the river and waste discharge, and assuming that the density and specific heat remains constant, gives

$$Q_r \rho_r c_p T_r + Q_w \rho_w c_p T_w = (Q_w + Q_r)\rho_m c_p T_m \qquad (4.22)$$

where ρ_r, ρ_w, and ρ_m are the densities of the river, waste discharge, and mixed waters, respectively (assumed to all be equal), and T_r, T_w, and T_m are the temperatures of the river, waste discharge, and mixed waters, respectively. Simplifying Equation 4.22 yields

$$T_0 = \frac{Q_r T_r + Q_w T_w}{Q_r + Q_w} \qquad (4.23)$$

This equation is commonly used to determine the temperature of water mixtures where the densities and specific heats of the individual components of the mixture are approximately equal to the density and specific heat of the mixture.

In models that use the one-dimensional (along-stream) advection–dispersion equation to describe the mixing of contaminants in rivers, an initial concentration c_0 is usually assumed to occur at the wastewater discharge location. It is clear from Equation 4.20 that for a given wastewater discharge, lower river flows will result in higher concentrations of the diluted wastewater. In analyzing the fate and transport of municipal and industrial waste discharges into rivers, the 7-day average low flow with a return period of 10 years is usually used as the design flow in the river (Thomann and Mueller, 1987). Such flows are typically written using the notation aQb, where a is the number of days used in the average and b is the return period in years of the minimum a-day average low flow. Therefore, $7Q10$ is the 7-day average low flow with a return period of 10 years.

Example 4.4 The dissolved-oxygen concentration in a river upstream of a municipal wastewater outfall is 10 mg/L. (a) If the $7Q10$ river flow upstream of the outfall is $50\,m^3/s$, and the outfall discharges $2\,m^3/s$ of wastewater with a dissolved-oxygen concentration of 1 mg/L, estimate the dissolved-oxygen concentration in the river after complete mixing. (b) If the upstream river temperature is 10°C and the wastewater temperature is 20°C, estimate the temperature of the mixed river water.

SOLUTION (a) From the data given, $Q_r = 50\,m^3/s$, $c_r = 10\,mg/L$, $Q_w = 2\,m^3/s$, $c_w = 1\,mg/L$, and the concentration of dissolved oxygen in the mixed river water is given by Equation 4.20 as

$$c_0 = \frac{Q_r c_r + Q_w c_w}{Q_r + Q_w} = \frac{50(10) + 2(1)}{50 + 2} = 9.7 \text{ mg/L}$$

(b) Since $T_r = 10°C$ and $T_w = 20°C$, the temperature, T_0, of the mixture is given by Equation 4.23 as

$$T_0 = \frac{Q_r T_r + Q_w T_w}{Q_r + Q_w} = \frac{50(10) + 2(20)}{50 + 2} = 10.4°C$$

Therefore, in this case, the wastewater discharge has a relatively small effect on the dissolved oxygen and temperature of the river.

4.2.2 Longitudinal Dispersion

Longitudinal mixing in streams is caused primarily by shear dispersion, which results from the "stretching effect" of both vertical and transverse variations in the longitudinal component of the stream velocity. The *longitudinal dispersion coefficient* is used to parameterize the longitudinal mixing of a tracer that is well mixed across a stream, in which case the advection and dispersion of the tracer is described by the one-dimensional advection–dispersion equation

$$\frac{\partial c}{\partial t} + V \frac{\partial c}{\partial x} = \frac{\partial}{\partial x}\left(K_L \frac{\partial c}{\partial x}\right) + S_m \tag{4.24}$$

where c is the cross-sectionally averaged tracer concentration, V is the mean velocity in the stream, K_L is the longitudinal dispersion coefficient, x is the coordinate measured along the stream, and S_m is the net influx of tracer mass per unit volume of water per unit time. If the tracer is conservative, $S_m = 0$. If the tracer undergoes first-order decay,

$$S_m = -kc \qquad (4.25)$$

where k is a *first-order rate constant* or *decay factor*. Combining Equations 4.24 and 4.25 gives the following equation for one-dimensional (longitudinal) advection and dispersion of a substance that undergoes first-order decay:

$$\boxed{\frac{\partial c}{\partial t} + V\frac{\partial c}{\partial x} = \frac{\partial}{\partial x}\left(K_L\frac{\partial c}{\partial x}\right) - kc} \qquad (4.26)$$

Fischer and colleagues (1979) have demonstrated that the longitudinal dispersion coefficient is proportional to the square of the distance over which a velocity shear flow profile extends. Since natural streams typically have widths that are at least 10 times the depth, the longitudinal dispersion coefficient associated with transverse variations in the mean velocity can be expected to be on the order of 100 times larger than the longitudinal dispersion coefficient associated with vertical variations in the mean velocity. Consequently, vertical variations in the mean velocity are usually neglected in deriving expressions relating the longitudinal dispersion coefficient to the velocity distribution in natural streams. Several empirical and semiempirical formulas have been developed to estimate the longitudinal dispersion coefficient, K_L, in open-channel flow. The expressions proposed by Fischer and colleagues (1979), Liu (1977), Koussis and Rodríguez-Mirasol (1998), Iwasa and Aya (1991), Seo and Cheong (1998), and Deng et al. (2001) are listed in Table 4.1, where $\bar{d}$ is the mean depth of the stream, u_* is the shear velocity given by Equation 4.3, and w is the top width of the stream. In applying the equations in Table 4.1 it is important to keep in mind that these equations apply to wide streams ($w \gg \bar{d}$), where longitudinal dispersion is dominated by transverse variations in the mean velocity and the dispersion caused by vertical variations in mean velocity is relatively small. If only vertical variations in the mean velocity are considered and the velocity distribution is assumed to be logarithmic, the longitudinal dispersion coefficient is given by (Elder, 1959)

$$\frac{K_L}{\bar{d}u_*} = 5.93 \qquad (4.27)$$

Therefore, for the equations given in Table 4.1 to be applicable, the calculated values of $K_L/\bar{d}u_*$ must be greater than 5.93; otherwise, vertical variations in velocity dominate the dispersion process and Equation 4.27 should be used to estimate the K_L. Fischer et al. (1979) reported that values of $K_L/\bar{d}u_*$ typically found in rivers are on the order of 20. It is important to note that the formulas for longitudinal dispersion coefficients given in Table 4.1 do not include parameters that measure the sinuosity, sudden contractions, expansions, and dead zones of water that are characteristic of natural streams, and these characteristics tend to increase the dispersion coefficient relative to straight open channels. Since data from both natural streams and straight open channels were used in deriving the formulas in Table 4.1, these formulas yield a range of values, with lower limits

TABLE 4.1   Estimates of the Longitudinal Dispersion Coefficient in Rivers

Formula	Reference
$\dfrac{K_L}{d\,u_*} = 0.011\left(\dfrac{w}{\overline{d}}\right)^2\left(\dfrac{V}{u_*}\right)^2$	Fischer et al. (1979)
$\dfrac{K_L}{d\,u_*} = 0.18\left(\dfrac{w}{\overline{d}}\right)^2\left(\dfrac{V}{u_*}\right)^{0.5}$	Liu (1977)
$\dfrac{K_L}{d\,u_*} = 0.6\left(\dfrac{w}{\overline{d}}\right)^2$	Koussis and Rodríguez-Mirasol (1998)
$\dfrac{K_L}{d\,u_*} = 2.0\left(\dfrac{w}{\overline{d}}\right)^{1.5}$	Iwasa and Aya (1991)
$\dfrac{K_L}{d\,u_*} = 5.915\left(\dfrac{w}{\overline{d}}\right)^{0.620}\left(\dfrac{V}{u_*}\right)^{1.428}$	Seo and Cheong (1998)
$\dfrac{K_L}{d\,u_*} = 0.01875\left[0.145 + \dfrac{1}{3520}\dfrac{V}{u_*}\left(\dfrac{w}{\overline{d}}\right)^{1.38}\right]^{-1}\left(\dfrac{w}{\overline{d}}\right)^{5/3}\left(\dfrac{V}{u_*}\right)^2$	Deng et al. (2001)

characteristic of straight open channels and higher values characteristic of sinuous natural streams.

In a sensitivity analysis performed by Deng et al. (2001), it was shown that the estimated dispersion coefficient is most sensitive to the mean velocity, V, with the top width, w, average depth, $\overline{d}$, and shear velocity, u_*, in decreasing order of sensitivity. The relative sensitivity of V is about twice that of w, which is roughly twice that of $\overline{d}$. Typical values of K_L are on the order of 0.05 to 0.3 m²/s for small streams (Genereux, 1991) to greater than 1000 m²/s for large rivers (Wanner et al., 1989).

Example 4.5   Stream depths and vertically-averaged velocities have been measured at 1-m intervals across a 10-m-wide stream; the results are tabulated in Table 4.2. If the friction factor of the flow is 0.03, estimate the longitudinal dispersion coefficient using the expressions in Table 4.1.

TABLE 4.2   Data for Example 4.5

Distance from Side, y (m)	Depth, d (m)	Velocity, v (m/s)	Distance from Side, y (m)	Depth, d (m)	Velocity, v (m/s)
0	0.0	0.0	6	2.4	1.6
1	0.20	0.30	7	1.5	1.0
2	0.90	0.60	8	0.75	0.50
3	1.2	0.80	9	0.45	0.30
4	2.1	1.4	10	0.0	0.0
5	3.0	2.0			

SOLUTION The flow area, A, can be estimated by summing the trapezoidal areas between the measurement locations, which yields

$$A = (0 + 0.2 + 0.9 + 1.2 + 2.1 + 3.0 + 2.4 + 1.5 + 0.75 + 0.45 + 0)(1) = 12.5\,\text{m}^2$$

The average velocity, V, can be estimated by

$$V = \frac{1}{A}\sum_{i=1}^{10} v_i A_i$$

where v_i and A_i are the velocity and area increments measured across the channel. Therefore,

$$V = \frac{1}{12.5}[(0.3)(0.2 \times 1) + (0.6)(0.9 \times 1) + (0.8)(1.2 \times 1) + (1.4)(2.1 \times 1)$$

$$+ (2.0)(3.0 \times 1) + (1.6)(2.4 \times 1) + (1.0)(1.5 \times 1) + (0.5)(0.75 \times 1)$$

$$+ (0.3)(0.45 \times 1)]$$

$$= 1.3\,\text{m/s}$$

Since $f = 0.03$ and $V = 1.3$ m/s, the shear velocity, u_*, is given by Equation 4.3 as

$$u_* = \sqrt{\frac{f}{8}} V = \sqrt{\frac{0.03}{8}}\,(1.3) = 0.080\,\text{m/s}$$

and the average depth, $\bar{d}$, is given by

$$\bar{d} = \frac{A}{w} = \frac{12.5}{10} = 1.25\,\text{m}$$

Substituting $\bar{d} = 1.25$ m, $u_* = 0.080$ m/s, $w = 10$ m, and $V = 1.3$ m/s into the formulas in Table 4.1 yields the results shown in Table 4.3. These results indicate that if the channel reach is relatively straight and uniform, the longitudinal dispersion coefficient is expected to be on the order of $10\,\text{m}^2/\text{s}$, while if the channel reach is sinuous with contractions, expansions, and dead zones, the dispersion coefficient is expected to be on the order of $100\,\text{m}^2/\text{s}$.

TABLE 4.3 Results for Example 4.5

Method	K_L (m^2/s)
Fischer et al. (1979)	19
Liu (1977)	5
Koussis and Rodríguez-Mirasol (1998)	4
Iwasa and Aya (1991)	5
Seo and Cheong (1998)	115
Deng et al. (2001)	63

4.3 SPILLS

Spills of contaminants in rivers are typically associated with major accidents on transportation routes across or adjacent to rivers, although other mechanisms, such as illicit dumping and spikes in continuous wastewater discharges, are also possible. Spills are characterized by the introduction of a large mass of contaminant in a very short period of time.

4.3.1 Governing Equation

The governing equation for the longitudinal dispersion of contaminants that are well mixed over the cross sections of rivers and streams and undergo first-order decay is given by Equation 4.26. The solution of Equation 4.26 for the case in which a mass, M, of contaminant is instantaneously mixed over the cross section of the stream at time $t = 0$ is given by

$$c(x, t) = \frac{Me^{-kt}}{A\sqrt{4\pi K_L t}} \exp\left[-\frac{(x - Vt)^2}{4K_L t}\right] \tag{4.28}$$

where c is the contaminant concentration in the stream, x is the distance downstream of the spill, t is the time since the spill, A is the cross-sectional area of the stream, V is the average velocity in the stream, and k is the first-order decay factor. The concentration distribution described by Equation 4.28 is illustrated in Figure 4.6.

In many cases we are concerned with the maximum concentration at a distance x downstream of an instantaneous spill. This maximum concentration can be derived by taking the logarithm of both sides of Equation 4.28 and setting the partial derivative of the result with respect to t equal to zero. In the case of a conservative contaminant ($k = 0$), the time, t_0, of peak concentration at location x is given by (Hunt, 1999)

$$\frac{Vt_0}{x} = -\frac{K_L}{Vx} + \sqrt{\left(\frac{K_L}{Vx}\right)^2 + 1} \tag{4.29}$$

In most cases, $x \gg K_L/V$ and Equation 4.29 gives

$$t_0 = \frac{x}{V} \tag{4.30}$$

Example 4.6 Ten kilograms of a conservative contaminant ($k = 0$) are spilled in a stream that is 15 m wide, 3 m deep (on average), and has an average velocity of 35 cm/s. If the

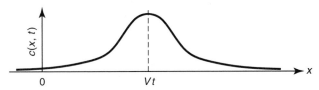

FIGURE 4.6 Concentration distribution resulting from an instantaneous spill. (From Chin, David A., *Water-Resources Engineering.* Copyright © 2000. Reprinted by permission of Pearson Education, Inc., Upper Saddle River, NJ.)

contaminant is rapidly mixed over the cross section of the stream. (a) Derive an expression for the contaminant concentration as a function of time 500 m downstream of the spill. (b) If a peak concentration of 4 mg/L is observed 500 m downstream of the spill, estimate the longitudinal dispersion coefficient in the stream. (c) Using the result in part (b), what would be the maximum contaminant concentration 1 km downstream of the spill? (d) If the detection limit of the contaminant is 1 μg/L, how long after release will the contaminant be detected 1 km downstream of the release point?

SOLUTION (a) Since the width of the stream, w, is 15 m and the average depth, $\bar{d}$, is 3 m, the cross-sectional area, A, of the stream is given by

$$A = w\bar{d} = 15(3) = 45 \, \text{m}^2$$

The concentration distribution resulting from an instantaneous spill of a conservative contaminant ($k = 0$) is given by Equation 4.28 as

$$c(x, t) = \frac{M}{A\sqrt{4\pi K_L t}} \exp\left[-\frac{(x-Vt)^2}{4K_L t} \right]$$

In this case, $V = 35 \, \text{cm/s} = 0.35 \, \text{m/s}$, $M = 10 \, \text{kg}$, and $x = 500 \, \text{m}$; hence, the concentration as a function of time 500 m downstream of the spill site is given by

$$c(500, t) = \frac{10}{45\sqrt{4\pi K_L t}} \exp\left[-\frac{(500 - 0.35t)^2}{4K_L t} \right]$$

$$= \frac{0.0627}{\sqrt{K_L t}} \exp\left[-\frac{(500 - 0.35t)^2}{4K_L t} \right] \quad \text{kg/m}^3$$

(b) The maximum concentration at any x occurs (approximately) at a time t_0 given by

$$t_0 = \frac{x}{V}$$

In this case $x = 500 \, \text{m}$ and $V = 0.35 \, \text{m/s}$, which gives

$$t_0 = \frac{500}{0.35} = 1430 \, \text{s}$$

The (maximum) concentration, c_0, at time t_0 is given by

$$c_0 = \frac{M}{A\sqrt{4\pi K_L t_0}} = \frac{0.0627}{\sqrt{K_L t_0}} \quad \text{kg/m}^3$$

When $c_0 = 4 \, \text{mg/L} = 0.004 \, \text{kg/m}^3$ and $t_0 = 1430 \, \text{s}$, then

$$0.004 = \frac{0.0627}{\sqrt{K_L(1430)}}$$

which gives

$$K_L = 0.172 \, \text{m}^2/\text{s}$$

This gives $K_L/V = 0.172/0.35 = 0.49$ m, and therefore all locations of interest in this problem ($x = 500$ m, 1000 m) satisfy the relation $x \gg K_L/V$, and therefore the assumption that the maximum concentration occurs at $t_0 = x/V$ is valid.

(c) The maximum concentration 1 km downstream from the release point occurs at time t_1, where

$$t_1 = \frac{x_1}{V}$$

In this case, $x_1 = 1000$ m and $V = 0.35$ m/s and therefore

$$t_1 = \frac{1000}{0.35} = 2860 \text{ s}$$

The maximum concentration, c_1, occurring 1 km downstream from the release point is given by

$$c_1 = \frac{0.0627}{\sqrt{K_L t_1}} = \frac{0.0627}{\sqrt{0.172(2860)}} = 0.00283 \text{ kg/m}^3 = 2.83 \text{ mg/L}$$

(d) The concentration 1 km downstream of the release point as a function of time is given by

$$c(1000, t) = \frac{0.0627}{\sqrt{K_L t}} \exp\left[-\frac{(1000 - 0.35t)^2}{4K_L t} \right] \qquad \text{kg/m}^3$$

When $c(1000, t) = 1 \ \mu\text{g/L} = 10^{-6} \text{ kg/m}^3$ and $K_L = 0.172 \text{ m}^2/\text{s}$,

$$10^{-6} = \frac{0.0627}{\sqrt{0.172t}} \exp\left[-\frac{(1000 - 0.35t)^2}{4(0.172)t} \right]$$

which gives $t = 2520$ s. This time is relatively close to the time that the maximum concentration occurs (2860 s, for a difference of 340 s = 5.7 min). It should be noted that there are two times when the concentration at $x = 1000$ m is equal to 1 μg/L: once before and once after the arrival of the peak concentration.

The dispersion problem consists primarily of predicting downstream contaminant concentrations resulting from a known or assumed spill. In some cases, the (inverse) problem might consist of determining the spill characteristics that resulted in measured downstream concentrations. This practical problem is mathematically ill-posed and does not have a unique solution. However, techniques are currently being developed to address this problem using geostatistics (Boano et al., 2005).

Field Measurement of K_L   Field studies are the most accurate method for determining the longitudinal dispersion coefficients in rivers. These studies typically involve releasing a conservative tracer, such as rhodamine WT dye or lithium, into a river and measuring the concentration distribution as a function of time at several downstream locations along the river. A typical rhodamine dye plume is shown as the dark area in the middle of the stream in Figure 4.7. According to the moment property of the advection–diffusion equation presented in Section 3.2.4, the dispersion coefficient is related to the variance of the

FIGURE 4.7   Rhodamine dye in a river. (From Fondriest Environmental, 2005.)

concentration distribution by Equations 3.101 to 3.103. In the context of the one-dimensional dispersion equation that is applicable to rivers, the moment property is given by Equation 3.101, which can be written as

$$K_L = \frac{1}{2}\frac{d\sigma_x^2}{dt} \tag{4.31}$$

where K_L is the longitudinal dispersion coefficient and σ_x^2 is the variance of the concentration distribution in the streamwise, x, direction. Equation 4.31 is useful in determining K_L from field tests in which a slug of tracer is introduced instantaneously into a stream and the concentration as a function of time is measured at two downstream locations, x_1 and x_2. If V is the mean flow velocity in the channel, then at any location, x, in the channel, the travel time, t, of the tracer can be estimated by the relation

$$t = \frac{x}{V} \tag{4.32}$$

where Equation 4.32 assumes that transport due to dispersion is negligible compared to advective transport, which corresponds to flows that have high Peclet numbers. Equation 4.32 requires that the variance of the temporal distribution of concentration, σ_t^2, be related to the spatial variance of the concentration distribution, σ_x^2, by

$$\sigma_t^2 = \frac{1}{V^2}\sigma_x^2 \tag{4.33}$$

Combining the finite-difference form of Equation 4.31 with Equations 4.32 and 4.33 gives

$$K_L = \frac{1}{2}\frac{\sigma_x^2(t_2) - \sigma_x^2(t_1)}{t_2 - t_1} = \frac{1}{2}\frac{V^2\sigma_{t_2}^2 - V^2\sigma_{t_1}^2}{x_2/V - x_1/V} \tag{4.34}$$

which simplifies to the following useful expression for K_L:

$$K_L = \frac{V^3}{2} \frac{\sigma_{t2}^2 - \sigma_{t1}^2}{x_2 - x_1}$$

(4.35)

where σ_{t1}^2 and σ_{t2}^2 are the temporal variances of the concentration distributions measured at distances x_1 and x_2 downstream of the spill, respectively. Equation 4.35 is useful in estimating the longitudinal dispersion coefficient in open channels based on measurements of concentration versus time at two locations along the channel.

Example 4.7 A dye study is conducted to estimate the longitudinal dispersion coefficient in a river. The dye is released instantaneously from a bridge across the entire width of the river, and the dye concentrations as a function of time at locations 400 and 700 m downstream of the bridge are shown in Table 4.4. Estimate the longitudinal dispersion coefficient in the river and verify that mass is conserved between the upstream and downstream locations.

SOLUTION   The longitudinal dispersion coefficient is given by Equation 4.35 as

$$K_L = \frac{V^3}{2} \frac{\sigma_{t2}^2 - \sigma_{t1}^2}{x_2 - x_1}$$

where $x_1 = 400$ m and $x_2 = 700$ m. The mean velocity, V, and the temporal variances at x_1 and x_2, σ_{t2}^2 and σ_{t1}^2, must be estimated from the measured data. The mean velocity, V, can be estimated by

$$V = \frac{x_2 - x_1}{\bar{t}_2 - \bar{t}_1}$$

TABLE 4.4   Data for Example 4.7

Time (min)	Concentration at $x = 400$ m (mg/L)	Concentration at $x = 700$ m (mg/L)	Time (min)	Concentration at $x = 400$ m (mg/L)	Concentration at $x = 700$ m (mg/L)
0	0	0	14	0.11	2.3
1	0	0	15	0.03	3.3
2	0	0	16	0	5.9
3	0.10	0	17	0	8.4
4	0.17	0	18	0	6.1
5	0.39	0	19	0	3.5
6	1.4	0	20	0	2.1
7	2.9	0	21	0	1.3
8	7.1	0	22	0	0.65
9	10.5	0.01	23	0	0.21
10	7.3	0.06	24	0	0.09
11	3.6	0.14	25	0	0.04
12	1.8	0.69	26	0	0
13	0.53	1.1	27	0	0

where $\bar{t}_1$ and $\bar{t}_2$ are the mean travel times to x_1 and x_2, respectively. Hence,

$$\bar{t}_1 = \frac{1}{\sum_{i=1}^{27} c_{1i}} \sum_{i=1}^{27} t_i c_{1i}$$

where t_i and c_{1i} are the measurement times and concentrations at $x = 400\,\text{m}$. Since

$$\sum_{i=1}^{27} c_{1i} = 35.9\,\text{mg/L}$$

and

$$\sum_{i=1}^{27} t_i c_{1i} = 326\,\text{mg} \cdot \text{min/L}$$

therefore

$$\bar{t}_1 = \frac{1}{35.9}(326) = 9.1\,\text{min}$$

At $x = 700\,\text{m}$,

$$\bar{t}_2 = \frac{1}{\sum_{i=1}^{27} c_{2i}} \sum_{i=1}^{27} t_i c_{2i}$$

where t_i and c_{2i} are the measurement times and concentrations at $x = 700\,\text{m}$. Since

$$\sum_{i=1}^{27} c_{2i} = 35.9\,\text{mg/L}$$

and

$$\sum_{i=1}^{27} t_i c_{2i} = 612\,\text{mg} \cdot \text{min/L}$$

therefore

$$\bar{t}_2 = \frac{1}{35.9}(612) = 17.0\,\text{min}$$

The mean velocity in the river is therefore given by

$$V = \frac{x_2 - x_1}{\bar{t}_2 - \bar{t}_1} = \frac{700 - 400}{17.0 - 9.1} = 38.0\,\text{m/min} = 0.63\,\text{m/s}$$

The variance of the concentration distribution at $x = 400\,\text{m}$, σ_{t1}^2, is given by

$$\sigma_{t1}^2 = \frac{1}{\sum_{i=1}^{27} c_{1i}} \sum_{i=1}^{27} (t_i - \bar{t}_1)^2 c_{1i}$$

where

$$\sum_{i=1}^{27} (t_i - \bar{t}_1)^2 c_{1i} = 95.4\,\text{mg} \cdot \text{min}^2/\text{L}$$

Hence,

$$\sigma_{t1}^2 = \frac{1}{35.9}(95.4) = 2.66\,\text{min}^2 = 9580\,\text{s}^2$$

The variance of the concentration distribution at $x = 700\,$m, σ_{t2}^2, is given by

$$\sigma_{t2}^2 = \frac{1}{\sum_{i=1}^{27} c_{2i}} \sum_{i=1}^{27} (t_i - \bar{t}_2)^2 c_{2i}$$

where

$$\sum_{i=1}^{27} (t_i - \bar{t}_2)^2 c_{2i} = 175\,\text{mg} \cdot \text{min}^2/\text{L}$$

Hence,

$$\sigma_{t2}^2 = \frac{1}{35.9}(175) = 4.87\,\text{min}^2 = 17{,}500\,\text{s}^2$$

Substituting into Equation 4.35 to determine the longitudinal dispersion coefficient, K_L, gives

$$K_L = \frac{V^3}{2} \frac{\sigma_{t2}^2 - \sigma_{t1}^2}{x_2 - x_1} = \frac{(0.63)^3}{2} \frac{17{,}500 - 9580}{700 - 400} = 3.30\,\text{m}^2/\text{s}$$

Conservation of mass requires that the areas under the concentration versus time curves at the upstream and downstream measurement locations are equal. The foregoing calculations show that the areas under the concentration versus time curves at the upstream and downstream locations ($\Delta t \sum_{i=1}^{27} c_{1i}$ and $\Delta t \sum_{i=1}^{27} c_{2i}$, respectively) are both equal to $35.9\,\text{mg} \cdot \text{min/L}$ (where $\Delta t = 1\,$min); hence, mass is conserved.

4.3.2 Fate of Volatile Organic Compounds in Streams

Contaminant spills in rivers and streams frequently consist of volatile organic compounds (VOCs). These spills can be either directly into the river, or in a drainage area that is subsequently washed off into the river. An example of an oil spill is a river, contained by a boom, is shown in Figure 4.8. The fate and transport of VOCs in rivers and streams are affected by several processes, including advection, dispersion, volatization, microbial degradation, sorption, hydrolysis, aquatic photolysis, chemical reaction, and bioconcentration. In most cases, volatization and dispersion are the dominant processes affecting the concentrations of VOCs in streams (Rathbun, 1998).

Dispersion is parameterized by the longitudinal dispersion coefficient, K_L, which can be calculated using any of the formulas listed in Table 4.1. *Volatization* is the movement of a substance from the bulk water phase across the water–air interface into the air, and can be described by the first-order relation

$$\frac{dc}{dt} = k_v(c_e - c) \tag{4.36}$$

FIGURE 4.8   Spill of diesel fuel. (From U.S. Materials Management Service, 2005.)

where c is the concentration of the VOC in water [M/L^3], t is time [T], k_v is the volatization coefficient [T^{-1}], and c_e is the concentration of the VOC in the water [M/L^3] if the water were in equilibrium with the partial pressure of the VOC in the air. In most cases, VOC concentrations in the air above streams are negligible and therefore c_e is also negligible. Under these circumstances, Equation 4.36 becomes

$$\frac{dc}{dt} = -k_v c \tag{4.37}$$

which indicates that volatization of VOCs in streams can be accounted for by a first-order decay process with a decay factor equal to k_v. Therefore, downstream concentrations resulting from the spill of a mass M of VOC in a river can be estimated using Equation 4.28, which can be written as

$$c(x,t) = \frac{Me^{-k_v t}}{A\sqrt{4\pi K_L t}} \exp\left[-\frac{(x-Vt)^2}{4K_L t}\right] \tag{4.38}$$

where A is the cross-sectional area of the river, K_L is the longitudinal dispersion coefficient, x is the distance downstream from the spill, and V is the mean velocity in the river. It is important to keep in mind that Equation 4.38 assumes that the spilled mass, M, is initially well mixed across the river. Using the *two-film model* proposed by Lewis and Whitman (1924) and further explored by Rathbun (1998), the volatization coefficient, k_v, can be estimated using the semiempirical relation

$$k_v = \frac{1}{\bar{d}}\left(\frac{1}{\Phi k_2 \bar{d}} + \frac{RT}{H\Psi k_3}\right)^{-1} \tag{4.39}$$

where k_v is in day^{-1}, $\bar{d}$ is the mean depth of the stream (m), k_2 is the reaeration coefficient for oxygen in water (day^{-1}), Φ and Ψ are constants that depend on the particular VOC (dimensionless), R is the ideal gas constant (J/K · mol), T is the temperature (K), H is the Henry's law constant of the VOC (Pa · m³/mol), and k_3 is the mass-transfer coefficient for evaporation of water from a stream (m/day). The reaeration coefficient for oxygen in water, k_2, can be estimated using empirical equations such as those listed in Table 4.8; Φ, Ψ, and H for several VOCs are listed in Table 4.5. The ideal gas constant, R, is 8.31 J/K · mol, and the mass-transfer coefficient, k_3, can be estimated using the empirical equation (Rathbun and Tai, 1983)

$$k_3 = (416 + 156V_w) \exp[0.00934(T - 26.1)] \qquad (4.40)$$

where k_3 is in m/day, V_w is the wind speed over the stream (m/s), and T is the temperature (°C). It should be noted that Henry's law is given by

$$p_e = Hc_e \qquad (4.41)$$

where p_e is the equilibrium partial pressure of the substance in the gas phase (Pa) and c_e is the equilibrium concentration of the substance in the water (mol/m³). The values of

TABLE 4.5 Parameters Used to Estimate Volatization Coefficient, k_v

Compound	Φ	Ψ	Henry's Law Constant, H^a (Pa · m³/mol)	A^b	B^b
Benzene	0.638	0.590	507	17.06	3194
Chlorobenzene	0.601	0.499	311	15.00	2689
Chloroethane	0.694	0.645	1030	15.80	2580
Trichloromethane (chloroform)	0.645	0.485	310	22.94	5030
1,1-Dichloroethane	0.643	0.529	465	17.01	3137
1,2-Dichloroethane	0.643	0.529	112	10.16	1522
Ethylbenzene	0.569	0.512	559	23.45	4994
Methyl *tert*-butyl ether (MTBE)	0.583	0.558	64.3[c]	30.06[d]	7721[d]
Dichloromethane (methylene chloride)	0.697	0.568	229	20.01	4268
Naphthalene	0.560	0.470	56.0[c]	—	—
Tetrachloroethene	0.585	0.417	1390	22.18	4368
Methylbenzene (toluene)	0.599	0.547	529	16.66	3024
1,1,1-Trichloroethane	0.605	0.461	1380	18.88	3399
Trichloroethene (TCE)	0.617	0.464	818	19.38	3702
Chloroethene (vinyl chloride)	0.709	0.510	2200	17.67	2931
1,2-Dimethylbenzene (*o*-xylene)	0.569	0.512	409	17.07	3220
1,4-Dimethylbenzene (*p*-xylene)	0.569	0.512	555	15.00	2689

Source: Rathbun (1998).

[a]All values are at 20°C, except where indicated.

[b]All values are for temperatures in the range 10 to 30°C and were derived from Ashworth et al. (1988), except where indicated.

[c]At 25°C.

[d]For temperatures in the range 20 to 50°C, derived from Robbins et al. (1993).

the Henry's law constant, H, given in Table 4.5 are mostly at a temperature of 20°C, and the variation of H (Pa · m³/mol) with temperature can be described by the empirical equation

$$\ln H = A - \frac{B}{T} \qquad (4.42)$$

where H is in Pa · m³/mol, T is the temperature in K, and values of A and B for several VOCs are listed in Table 4.5. Additional values of A and B for many other VOCs may be found in Rathbun (1998).

Example 4.8   A stream has a mean velocity of 20 cm/s, an average width of 10 m, and an average depth of 1 m. The water and air temperatures are both 24°C, and the average wind speed is 3 m/s. (a) If the longitudinal dispersion coefficient in the river is 1.4 m²/s, estimate the maximum concentration 5 km downstream of a location where 10 kg of TCE has been spilled into the stream. (b) Would your result change significantly if volatization is neglected?

SOLUTION   (a) First, calculate the volatization coefficient, k_v, of TCE using Equation 4.39, where

$$k_v = \frac{1}{\bar{d}} \left(\frac{1}{\Phi k_2 \bar{d}} + \frac{RT}{H\Psi k_3} \right)^{-1}$$

From the data given, $\bar{d} = 1$ m, $T = 24°C = 297.15$ K, and it is known that $R = 8.31$ J/K · mol. From Table 4.5, $\Phi = 0.617$, $\Psi = 0.464$, $A = 19.38$, and $B = 3702$. Using Equation 4.42, the Henry's law constant, H, at 297.15 K is given by

$$\ln H = A - \frac{B}{T} = 19.38 - \frac{3702}{297.15} = 6.922$$

which gives $H = 1014$ Pa · m³/mol. From the data given, $V = 0.2$ m/s and $\bar{d} = 1$ m; therefore, both the O'Connor and Dobbins (1958) and Owens et al. (1964) formulas in Table 4.8 are appropriate for calculating k_2 at 20°C. According to the O'Connor and Dobbins formula,

$$k_2 = 3.93 \frac{V^{0.5}}{\bar{d}^{1.5}} = 3.93 \left(\frac{0.2^{0.5}}{1^{1.5}} \right) = 1.76 \text{ day}^{-1}$$

and according to the Owens et al. formula,

$$k_2 = 5.32 \frac{V^{0.67}}{\bar{d}^{1.85}} = 5.32 \left(\frac{0.2^{0.67}}{1^{1.85}} \right) = 1.81 \text{ day}^{-1}$$

Taking the smaller value of k_2 to be conservative gives $k_2 = 1.76 \text{ day}^{-1}$ at 20°C. At 24°C, k_2 is estimated (using Equation 4.46) as

$$k_{2_T} = k_{2_{20}} (1.024^{T-20}) = 1.76(1.024)^{24-20} = 1.94 \text{ day}^{-1}$$

The mass transfer coefficient, k_3, at $T = 24°C$ and for a wind speed, V_w, of 3 m/s is given by Equation 4.40 as

$$k_3 = (416 + 156V_w)\exp[0.00934(T - 26.1)]$$
$$= (416 + 156 \times 3)\exp[0.00934(24 - 26.1)] = 867 \text{ m/day}$$

Substituting the values of $\bar{d}$, Φ, k_2, R, T, H, Ψ, and k_3 into Equation 4.39 gives

$$k_v = \frac{1}{\bar{d}}\left(\frac{1}{\Phi k_2 \bar{d}} + \frac{RT}{H\Psi k_3}\right)^{-1}$$
$$= \frac{1}{1}\left[\frac{1}{0.617(1.94)(1)} + \frac{8.31(297.15)}{1014(0.464)(867)}\right]^{-1} = 1.19 \text{ day}^{-1}$$

The maximum concentration at a distance x downstream of the spill location is derived from Equation 4.38 by taking $t = x/V$ which yields

$$c_{max}(x) = \frac{Me^{-k_v x/V}}{A\sqrt{4\pi K_L x/V}}$$

From the data given, $M = 10$ kg, $x = 5$ km $= 5000$ m, $V = 0.2$ m/s $= 17{,}280$ m/day, $A = w\bar{d} = 10(1) = 10$ m^2, and $K_L = 1.4$ m^2/s $= 120{,}960$ m^2/day. Hence, at a location 5 km downstream of the spill, the maximum concentration is given by

$$c_{max}(5000) = \frac{10e^{-1.19(5000/17{,}280)}}{10\sqrt{4\pi(120{,}960)(5000)/17{,}280}} = 0.00107 \text{ kg/m}^3 = 1.07 \text{ mg/L}$$

(b) If volatization is neglected, $k_v = 0$ and

$$c_{max}(5000) = \frac{10}{10\sqrt{4\pi(120{,}960)(5000)/17{,}280}} = 0.00151 \text{ kg/m}^3 = 1.51 \text{ mg/L}$$

Therefore, volatization decreases the maximum concentration by 0.44 mg/L or 29%. Hence volatization is a significant process.

It should be noted that the solubility of TCE is in the range 1000 to 1100 mg/L (Appendix B). In parts of the stream where the TCE concentrations equal or exceed the solubility, TCE is likely to exist in the nonaqueous (pure) phase, and probably below the water surface since the density of the pure substance (1460 kg/m^3) is significantly greater than that of water (998 kg/m^3).

4.4 CONTINUOUS DISCHARGES

Continuous discharges of contaminant-laden wastewater into rivers typically occur from domestic wastewater treatment plants and industrial plants. This is illustrated in Figure 4.9, which shows the West Linn sewage treatment plant discharging effluent into the Willamette River (Oregon) via a multiport diffuser. The continuous discharge into rivers and streams of wastewaters with high biochemical oxygen demand (BOD) depletes the dissolved

FIGURE 4.9   Discharge of treated domestic wastewater into a river. (From Citizens for Safe Water, 2005.)

oxygen in the ambient water and can sometimes cause severe stress on aquatic life. The oxygen demand exerted by the wastewater is partially met by oxygen transfer from the atmosphere at the surface of the stream, a process that is commonly referred to as *aeration*. The concentration of dissolved oxygen in natural waters is a primary indicator of overall water quality and the viability of aquatic habitat.

4.4.1   Oxygen Demand of Wastewater

The oxygen demand of wastewaters is typically measured by the BOD, and the associated rate of (de)oxygenation, S_1 [M/L^{3}T], is commonly described by a first-order reaction of the form

$$S_1 = -k_1 L \tag{4.43}$$

where k_1 is a reaction-rate constant [T^{-1}], and L is the BOD remaining [M/L^3]. The reaction-rate constant, k_1, depends primarily on the nature of the waste, the ability of the indigenous organisms to use the waste, and the temperature; typical values of k_1 at 20°C are shown in Table 4.6. For temperatures other than 20°C, the values of k_1 given in Table 4.6 must be adjusted, and the following adjustment is generally used:

$$\boxed{k_{1_T} = k_{1_{20}} \theta^{T-20}} \tag{4.44}$$

where T is the temperature of the stream, k_{1_T} and $k_{1_{20}}$ are the values of k_1 at temperatures T and 20°C, respectively, and θ is a dimensionless temperature coefficient. There are variations in the value for θ used in practice, with Thomann and Mueller (1987) recommending $\theta = 1.04$, Tchobanoglous and Schroeder (1985) and Novotny (2003) recommending 1.047, and Schroepfer et al. (1964) recommending $\theta = 1.135$ for typical domestic wastewater at temperatures between 4 and 20°C, and $\theta = 1.056$ for temperatures between 20 and 30°C. The latter values are widely accepted in practice (Mihelcic, 1999), and the fact that $\theta > 1$ in Equation 4.44 means that BOD reactions occur more rapidly at higher temperatures.

TABLE 4.6   Typical Deoxygenation Rate Coefficients

Type of Water	Ranges of k_1 at 20°C (day^{-1})
Untreated wastewater	0.35–0.7
Treated wastewater	0.10–0.35
Polluted river	0.10–0.25
Unpolluted river	<0.05

Source: Data from Thomann and Mueller, (1987), Kiely (1997), Davis and Masten (2004).

Temperature conditions selected for waste assimilative capacity evaluation should correspond to the average temperature of the warmest month of the year (Novotny, 2003). The reaction-rate constant, k_1, sometimes called the *in-stream deoxygenation rate*, is inversely proportional to the level of treatment provided prior to effluent release into the river or stream, as indicated in Table 4.6. The lower rate constants for treated sewage compared with raw sewage result from the fact that easily degradable organics are more completely removed than less readily degradable organics during wastewater treatment. Lung (2001) reported that when the Metropolitan Wastewater Treatment Plant in St. Paul, Minnesota, discharged primary effluent into the Upper Mississippi River, the in-stream value of k_1 was 0.35 day^{-1}, when the plant was upgraded to secondary treatment, k_1 decreased to 0.25 day^{-1}, and further upgrading all the way up to installing a nitrification process dropped the value of k_1 to 0.073 day^{-1}.

4.4.2   Reaeration

The rate at which oxygen is transferred from the atmosphere into a stream, defined as the *reaeration rate*, S_2 [M/L^{3}T], is commonly described by an equation of the form

$$S_2 = k_2 (c_s - c) \qquad (4.45)$$

where k_2 is the *reaeration constant* [T^{-1}], c_s is the dissolved-oxygen saturation concentration [M/L^3], and c is the actual concentration of dissolved oxygen in the stream. Typical values of k_2 at 20°C are given in Table 4.7, which indicates that reaeration coefficients

TABLE 4.7   Typical Reaeration Constants

Water Body	Ranges of k_2 at 20°C (day^{-1})
Small ponds and backwaters	0.10–0.23
Sluggish streams and large lakes	0.23–0.35
Large streams of low velocity	0.35–0.46
Large streams of normal velocity	0.46–0.69
Swift streams	0.69–1.15
Rapids and waterfalls	>1.15

Source: Tchobanoglous and Schroeder (1985).

TABLE 4.8   Empirical Formulas for Estimating Reaeration Constant, k_2, at 20°C

Formula	Field Conditions	Reference
$k_2 = 3.93 \dfrac{V^{0.5}}{\bar{d}^{1.5}}$	$0.3\,\text{m} < \bar{d} < 9\,\text{m}$, $0.15\,\text{m/s} < V < 0.50\,\text{m/s}$	O'Connor and Dobbins (1958)
$k_2 = 5.23 \dfrac{V}{\bar{d}^{1.67}}$	$0.6\,\text{m} < \bar{d} < 3\,\text{m}$, $0.55\,\text{m/s} < V < 1.50\,\text{m/s}$	Churchill et al. (1962)
$k_2 = 5.32 \dfrac{V^{0.67}}{\bar{d}^{1.85}}$	$0.1\,\text{m} < \bar{d} < 3\,\text{m}$, $0.03\,\text{m/s} < V < 1.50\,\text{m/s}$	Owens et al. (1964)
$k_2 = 3.1 \times 10^4 V S_0$	$0.3\,\text{m} < \bar{d} < 0.9\,\text{m}$, $0.03\,\text{m}^3/\text{s} < Q < 0.3\,\text{m}^3/\text{s}$	Tsivoglou and Wallace (1972)
$k_2 = 517(VS)^{0.524}\, Q^{-0.242}$	Pool and riffle streams, $Q < 0.556\,\text{m}^3/\text{s}$	Melching and Flores (1999)
$k_2 = 596(VS)^{0.528}\, Q^{-0.136}$	Pool and riffle streams, $Q > 0.556\,\text{m}^3/\text{s}$	Melching and Flores (1999)
$k_2 = 88(VS)^{0.313}\, D^{-0.353}$	Channel-control streams, $Q < 0.556\,\text{m}^3/\text{s}$	Melching and Flores (1999)
$k_2 = 142(VS)^{0.333}\, D^{-0.66}$ $\times W^{-0.243}$	Channel-control streams, $Q > 0.556\,\text{m}^3/\text{s}$	Melching and Flores (1999)

typically vary from 0.1 day^{-1} for small ponds to greater than 1.15 for rapids and water-falls. In small rivers, rapids play a major role in maintaining high dissolved-oxygen levels and eliminating rapids by dredging or damming a river can have a severe effect on dissolved oxygen. Several of the most popular empirical formulas for estimating k_2 at 20°C are given in Table 4.8, where the units of k_2 are day^{-1}, the average stream velocity, V, is in m/s, the average stream depth, $\bar{d}$, is in meters, and the average channel slope, S_0, is dimensionless. For temperatures other than 20°C, use

$$k_{2_T} = k_{2_{20}} \theta^{T-20} \tag{4.46}$$

where T is the temperature of the stream, k_{2_T} and $k_{2_{20}}$ are the values of k_2 at temperatures T and 20°C respectively, and θ is the temperature coefficient, which is commonly taken to be in the range 1.024 to 1.025. Other, less popular empirical formulas for estimating k_2 may be found in Flores (1998). In practice, the empirical formula proposed by O'Connor and Dobbins (1958) has the widest applicability and provides reasonable estimates of k_2 at 20°C in most cases. Churchill and colleagues' (1962) formula applies to depths similar to those of the O'Connor and Dobbins model, but for faster streams. Owens and colleagues' (1964) formula is used for shallower streams; in small streams, the formula proposed by Tsivoglou and Wallace (1972) compares best with observed values (Thomann and Mueller,

1987). The last four empirical formulas in Table 4.8 are the most recent and are based on the most comprehensive data set. According to their authors they are the most accurate, with standard errors in the range 44 to 61%, compared to standard errors of 65 to 115% in the other methods (Bennett and Rathbun, 1972). The formulas listed in Table 4.8 all give values of k_2 that approach zero as the depth of the stream increases, implying that reaeration becomes negligible for deep bodies of water. This is certainly not the case, since when water motion is less significant, wind becomes the dominating factor in reaeration. The reaeration constant typically has a minimum value in the range

$$k_{2_{\min}} = \frac{0.6}{d} \quad \text{to} \quad \frac{1.0}{d} \qquad (4.47)$$

If the calculated value of k_2 falls below the range of minimum values given in Equation 4.47, $k_2 = 0.6/\bar{d}$ should be used.

Example 4.9 A river with riffles and pools has a width of 20 m, a mean depth of 5 m, a slope of 0.00003, and an estimated flow rate of 47 m³/s. (a) Estimate the reaeration constant using the applicable equation(s) in Table 4.8. (b) If the temperature of the river is 20°C and the dissolved-oxygen concentration is 5 mg/L, estimate the reaeration rate. (c) Determine the mass of oxygen added per day per meter along the river.

SOLUTION (a) From the data given, $\bar{d} = 5$ m, $w = 20$ m, $S = 0.00003$, $Q = 47$ m³/s, and the average velocity, V, is given by

$$V = \frac{Q}{w\bar{d}} = \frac{47}{20(5)} = 0.47 \text{ m/s}$$

The O'Connor and Dobbins (1958) formula and the Melching and Flores (1999) empirical equation are the only applicable relations in Table 4.8. According to the O'Connor and Dobbins (1958) model,

$$k_2 = 3.93 \frac{V^{0.5}}{\bar{d}^{1.5}} = 3.93 \left[\frac{(0.47)^{0.5}}{(5)^{1.5}} \right] = 0.24 \text{ day}^{-1}$$

According to the Melching and Flores (1999) empirical equation,

$$k_2 = 596(VS)^{0.528} \, Q^{-0.136} = 596(0.47 \times 0.00003)^{0.528} \, (47)^{-0.136} = 0.97 \text{ day}^{-1}$$

According to Table 4.7, the reaeration rate calculated using the O'Connor and Dobbins (1958) formula is typical of sluggish streams and large lakes, whereas the reaeration rate calculated using the Melching and Flores (1999) empirical equation is typical of swift streams. Given that the stream velocity is fairly swift (47 cm/s), the value of $k_2 = 0.97$ day^{-1} given by the Melching and Flores (1999) empirical equation is more characteristic of the actual conditions. The calculated values of k_2 (0.24 day^{-1}, 0.97 day^{-1}) exceed the range of minimum values of k_2 given by Equation 4.47 as 0.6/5 − 1.0/5 or 0.12 − 0.2 day^{-1}.

(b) The reaeration rate, S_2, is given by Equation 4.45 as

$$S_2 = k_2 \, (c_s - c)$$

where $c = 5$ mg/L, and the saturation concentration, c_s, at 20°C is given in Table 2.3 as 9.1 mg/L. Therefore, the reaeration rate, S_2, is given by

$$S_2 = 0.97(9.1 - 5) = 3.98 \text{ mg/L} \cdot \text{day} = 3980 \text{ mg/m}^3 \cdot \text{day}$$

(c) The volume of river water per meter is given by

$$\text{volume of river water} = w\overline{d}(1) = 20(5)(1) = 100 \text{ m}^3$$

Hence, the mass of oxygen added per day per meter along the river is $3980(100) = 398{,}000$ mg/day · m $= 398$ g/day · m.

Great care should be taken in estimating k_2 since this is typically the dominant parameter affecting the reliability of simulated dissolved oxygen concentrations in streams (Brown and Barnwell, 1987; Melching and Yoon, 1997). The most accurate estimates of k_2 are obtained from field measurements using the gas tracer method, and such measurements are strongly recommended in cases where allowable-discharge estimates have significant economic and environmental consequences.

4.4.3 Streeter–Phelps Model

The total flux of oxygen into river water, S_m [M/L^{3}T], can be estimated by adding the (de)oxygenation rate due to biodegradation, S_1, to the oxygen flux due to reaeration, S_2, to yield

$$S_m = -k_1 L + k_2(c_s - c) \tag{4.48}$$

The concentration distribution of oxygen in rivers can be described by combining the advection–dispersion equation (Equation 4.24) with the source flux given by Equation 4.48 to yield

$$\frac{\partial c}{\partial t} + V\frac{\partial c}{\partial x} = \frac{\partial}{\partial x}\left(K_L \frac{\partial c}{\partial x}\right) - k_1 L + k_2(c_s - c) \tag{4.49}$$

where the flow area, A, in the river is assumed to be constant. Assuming steady-state conditions ($\partial c/\partial t = 0$) and that the dispersive flux of oxygen is much less than oxygen fluxes due to reaeration and deoxygenation, Equation 4.49 becomes

$$V\frac{\partial c}{\partial x} = -k_1 L + k_2(c_s - c) \tag{4.50}$$

Instead of dealing with the oxygen concentration, c, it is convenient to deal with the *oxygen deficit*, D, defined by

$$D = c_s - c \tag{4.51}$$

Combining Equations 4.50 and 4.51 and replacing the partial derivative by the total derivative (since c is only a function of x) yields the following differential equation that describes the oxygen deficit in the river:

$$\frac{dD}{dx} = \frac{k_1}{V}L - \frac{k_2}{V}D \tag{4.52}$$

The BOD remaining at any time, t, since release will follow the first-order reaction

$$\frac{dL}{dt} = -k_1 L \tag{4.53}$$

which can be solved independently of Equation 4.52 to yield

$$L = L_0 \exp(-k_1 t) \tag{4.54}$$

where L_0 is the BOD remaining at time $t = 0$. The time since release, t, is related to the distance traveled by

$$t = \frac{x}{V} \tag{4.55}$$

and hence the remaining BOD, L, at a distance x downstream of the wastewater discharge is obtained by combining Equations 4.54 and 4.55 to give

$$L = L_0 \exp\left(-k_1 \frac{x}{V}\right) \tag{4.56}$$

The differential equation describing the oxygen deficit in a river is therefore given by the combination of Equations 4.52 and 4.56, and the simultaneous solution of these equations with the boundary condition that $D = D_0$ at $x = 0$ is given by

$$D(x) = \frac{k_1 L_0}{k_2 - k_1} \left[\exp\left(-\frac{k_1 x}{V}\right) - \exp\left(-\frac{k_2 x}{V}\right) \right] + D_0 \exp\left(-\frac{k_2 x}{V}\right) \tag{4.57}$$

This equation was originally derived by Streeter and Phelps (1925) and later summarized by Phelps (1944) for studies of pollution in the Ohio River. Equation 4.57 is commonly referred to as the *Streeter–Phelps equation*, and a plot of the Streeter–Phelps equation is commonly referred to as the *Streeter–Phelps oxygen-sag curve*. The reason for using the term *sag curve* is apparent from a plot of the oxygen deficit, $D(x)$, as a function of distance, x, from the source as illustrated in Figure 4.10. According to the Streeter–Phelps equation (Equation 4.57), oxygen consumption for biodegradation begins immediately after the waste is discharged, at $x = 0$, with the oxygen deficit in the stream increasing from its initial value of D_0. Since reaeration is proportional to the oxygen deficit, the reaeration rate increases as the oxygen deficit increases, and at some point the reaeration rate becomes

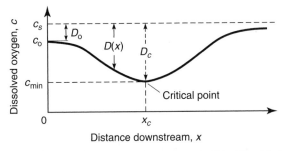

FIGURE 4.10 Streeter–Phelps oxygen-sag curve. (From Chin, David A., *Water-Resources Engineering*. Copyright © 2000. Reprinted by permission of Pearson Education, Inc., Upper Saddle River, NJ.)

equal to the rate of oxygen consumption. This point is called the *critical point*, x_c, and beyond the critical point the reaeration rate exceeds the rate of oxygen consumption, resulting in a gradual decline in the oxygen deficit. The critical point, x_c, can be derived from Equation 4.57 by taking $dD/dx = 0$, which leads to

$$x_c = \frac{V}{k_2 - k_1} \ln \left[\frac{k_2}{k_1} \left(1 - \frac{D_0(k_2 - k_1)}{k_1 L_0} \right) \right] \tag{4.58}$$

and the corresponding critical oxygen deficit, D_c, is given by

$$D_c = \frac{k_1}{k_2} L_0 \exp \left(-\frac{k_1 x_c}{V} \right) \tag{4.59}$$

The location of the critical point and the oxygen concentration at that point are of greatest interest because this is where dissolved-oxygen conditions are the worst. If the value of x_c calculated using Equation 4.58 is less than or equal to zero, the minimum oxygen deficit occurs at the discharge location. This is a typical occurrence when well-treated domestic sewage is discharged into small rivers (Lung and Sobeck, 1999). It is important to keep in mind that the calculated oxygen deficit, D, using Equation 4.57 can (mathematically) exceed the saturation concentration of dissolved oxygen, even though this is physically impossible. If the oxygen deficit calculated from Equation 4.57 is greater than the saturation concentration of dissolved oxygen, this means that all the oxygen was depleted at some earlier time and the actual dissolved oxygen at the selected location is zero.

Example 4.10 A multiport outfall discharges wastewater into a slow-moving river that has a mean velocity of 3 cm/s. After initial mixing, the dissolved-oxygen concentration in the river is 9.5 mg/L and the temperature is 15°C. If the ultimate BOD of the mixed river water is 30 mg/L, the rate constant for BOD at 20°C is 0.6 day^{-1}, and the reaeration rate constant at 20°C is 0.8 day^{-1}, estimate the minimum dissolved oxygen and the critical location in the river.

SOLUTION   At $T = 15°C$, the saturation concentration of oxygen is 10.1 mg/L (Table 2.3), and hence the initial oxygen deficit, D_0, is $10.1 - 9.5 = 0.6$ mg/L. The BOD rate constant at 15°C, $k_{1_{15}}$, is given by

$$k_{1_{15}} = k_{1_{20}}(1.04^{T-20}) = 0.6(1.04)^{15-20} = 0.48 \text{ day}^{-1}$$

The reaeration rate constant at 15°C, $k_{2_{15}}$, is given by

$$k_{2_{15}} = k_{2_{20}}(1.024^{T-20}) = 0.8(1.024)^{15-20} = 0.72 \text{ day}^{-1}$$

Since $L_0 = 30$ mg/L and $V = 3$ cm/s $= 2592$ m/day, Equation 4.58 gives the location, x_c, of the critical oxygen deficit as

$$
\begin{aligned}
x_c &= \frac{V}{k_2 - k_1} \ln \left[\frac{k_2}{k_1} \left(1 - \frac{D_0(k_2 - k_1)}{k_1 L_0} \right) \right] \\
&= \frac{2592}{0.72 - 0.48} \ln \left[\frac{0.72}{0.48} \left(1 - \frac{0.6(0.72 - 0.48)}{0.48(30)} \right) \right] \\
&= 4270 \text{ m}
\end{aligned}
$$

and Equation 4.59 gives the critical oxygen deficit, D_c, as

$$D_c = \frac{k_1}{k_2} L_0 \exp\left(-\frac{k_1 x_c}{V}\right)$$

$$= \frac{0.48}{0.72}(30)\exp\left(-\frac{0.48(4270)}{2592}\right) = 9.0 \text{ mg/L}$$

Hence the minimum dissolved oxygen level in the stream is $10.1 - 9.0 = 1.1$ mg/L. This level of dissolved oxygen will be devastating to the aquatic ecosystem.

When using the Streeter–Phelps dissolved-oxygen sag curve to determine the adequacy of wastewater treatment prior to discharge, it is important to use the river conditions that will result in the least dissolved oxygen concentration. Usually, these conditions occur in the late summer when river flows are low and temperatures are high. A frequently used criterion is the 10-year 7-day low flow, which is equal to the conditions in which the 7-day average low flow has a return period of 10 years. This is typically called the *7Q10 flow*.

The Streeter–Phelps model, Equation 4.57, assumes that the river is mixed completely and uniformly in the transverse directions (i.e., over the width and the depth), and the longitudinal dispersive flux (of dissolved oxygen) is negligible compared with the advective flux. The relative importance of advection to dispersion is measured by the Peclet number, Pe, which can be defined as

$$\text{Pe} = \frac{VL}{K_L} \tag{4.60}$$

where V is the stream velocity, L is the characteristic length scale, and K_L is the longitudinal dispersion coefficient. For a nonconservative substance with a rate constant k, the appropriate length scale is $L = V/k$ and the Peclet number given by Equation 4.60 can be expressed as

$$\text{Pe} = \frac{V^2}{kK_L} \tag{4.61}$$

The assumption that longitudinal dispersion can be neglected is justified when $\text{Pe} > 10$, and dispersion cannot be neglected when $\text{Pe} < 1$. For intermediate values of Pe, both advection and dispersion must be taken into account. Most riverine flows are advection dominated (Rubin and Atkinson, 2001).

4.4.4 Other Considerations

Aside from assuming complete transverse mixing and neglecting longitudinal dispersion, several additional assumptions are implicit in the Streeter–Phelps model, the most important of which are (1) that mixing occurs rapidly over the cross section of the channel, and (2) that biochemical oxygen demand and reaeration are the only significant oxygen sources and sinks. If mixing does not occur rapidly, as measured by the length scale for cross-sectional mixing (Equation 4.15), significant oxygen depletion may occur prior to cross-sectional mixing, and Equation 4.20 will not give an accurate measure of the initial concentration for the one-dimensional oxygen-sag model. Under these circumstances, a more detailed two- or three-dimensional numerical model of cross-sectional mixing is necessary.

Nitrification   In streams and rivers with travel times greater than about 5 days, the ultimate BOD, L_0, in the Streeter–Phelps equation (Equation 4.57) must include both carbonaceous and nitrogenous demand. In these cases it is sometimes assumed that L_0 can be calculated as (Vesilind and Morgan, 2004)

$$L_0 = a(\text{CBOD}_5) + b(\text{TKN}) \tag{4.62}$$

where CBOD_5 is the 5-day carbonaceous BOD, TKN is the total Kjeldahl nitrogen (= organic nitrogen + ammonia) in mg/L, and a and b are constants. Typical values for a and b are 1.2 and 4.0, respectively. In some cases, the contribution of nitrogenous BOD to the ultimate BOD is taken as 4.33 TKN rather than 4.0 TKN (Krenkel and Novotny, 1980; Novotny, 2003).

Instead of taking L_0 given by Equation 4.62 and using it in the Streeter–Phelps equation to calculate the oxygen deficit, it is sometimes recognized that the decay coefficient of nitrogenous BOD is different from the decay coefficient of carbonaceous BOD, k_1, in which case the dissolved oxygen deficit can be expressed in the form

$$D(x) = D_{\text{SF}}(x) + \Delta D_N(x) \tag{4.63}$$

where $D_{\text{SF}}(x)$ is the oxygen deficit predicted by the Streeter–Phelps equation (Equation 4.57), and $\Delta D_N(x)$ is the additional oxygen deficit caused by nitrification and is given by (Novotny, 2003)

$$\Delta D_N(x) = \frac{k_N L_{0N}}{k_2 - k_N} \left[\exp\left(-\frac{k_N x}{V}\right) - \exp\left(-\frac{k_2 x}{V}\right) \right] \tag{4.64}$$

where k_N is the first-order nitrification rate (day^{-1}), and L_{0N} is the initial nitrogenous BOD, which can be taken as

$$L_{0N} = 4.33 \times \text{TKN} \tag{4.65}$$

The magnitude of the nitrification reaction rate constant, k_N, has been reported to range from 0.1 to 15.8 day^{-1} (Ruane and Krenkel, 1975). Temperature affects nitrification differently than other processes, and the thermal factor for nitrification, θ, is around 1.1.

Example 4.11   Treated domestic wastewater is discharged into a river with an average stream velocity of 30 cm/s and an estimated reaeration coefficient of 0.5 day^{-1}. The TKN of a wastewater discharge is 0.8 mg/L and the nitrification rate constant is estimated to be 0.6 day^{-1}. (a) Estimate the oxygen deficit due to nitrification 15 km downstream of the wastewater discharge. (b) If the theoretical oxygen deficit neglecting nitrification is 6.2 mg/L, what is the percent error incurred by neglecting nitrification?

SOLUTION   (a) From the data given, V = 30 cm/s = 0.30 m/s = 25,920 m/day, $k_2 = 0.5$ day^{-1}, TKN = 0.8 mg/L, and $k_N = 0.6$ day^{-1}. The ultimate nitrogenous BOD, L_{0N}, is given by Equation 4.65 as

$$L_{0N} = 4.33 \times \text{TKN} = 4.33 \times 0.8 = 3.5 \text{ mg/L}$$

Substituting into Equation 4.64 yields the oxygen deficit due to nitrification at $x = 15\,\text{km} = 15{,}000\,\text{m}$ as

$$\Delta D_N = \frac{k_N L_{0N}}{k_2 - k_N}\left[\exp\left(-\frac{k_N x}{V}\right) - \exp\left(-\frac{k_2 x}{V}\right)\right]$$

$$= \frac{0.6(3.5)}{0.5 - 0.6}\left[\exp\left(-\frac{0.6(15{,}000)}{25{,}920}\right) - \exp\left(-\frac{0.5(15{,}000)}{25{,}920}\right)\right]$$

$$= 0.9\,\text{mg/L}$$

(b) The incremental oxygen deficit due to nitrification is 0.9 mg/L. Since the theoretical oxygen deficit obtained by neglecting nitrification is 6.2 mg/L, the actual oxygen deficit is 6.2 mg/L + 0.9 mg/L = 7.1 mg/L, and the error incurred by neglecting nitrification is given by

$$\text{error} = \frac{0.9}{7.1}\times 100 = 13\%$$

An error of 13% incurred by neglecting nitrification is quite significant, and this result indicates that nitrification is a significant process and should not be neglected.

A significantly increased oxygen demand associated with nitrification emphasizes the need for wastewater treatment plants to stimulate complete nitrification (conversion of nitrogen to NO_3) or denitrification (conversion to N_2 gas) prior to wastewater discharge.

Photosynthesis, Respiration, and Benthic Oxygen Demand   The Streeter–Phelps analysis assumes that the oxygen demand on a stream is caused by the BOD of the wastewater discharge and that this oxygen demand is supplied by the dissolved oxygen in the stream that is in turn resupplied by atmospheric aeration. Besides biochemical oxygen demand and reaeration, other sources and sinks of oxygen that are distributed along rivers include photosynthesis, respiration of photosynthetic organisms, benthic oxygen demand, and BOD from distributed (diffuse) sources along the river.

Photosynthesis by phytoplankton, particularly algae, contributes oxygen to the water during (bright) daylight hours, and respiration removes oxygen at night and on very cloudy days. The process of photosynthesis can be represented by the reaction

$$CO_2 + H_2O + \Delta \rightarrow C(H_2O) + O_2 \tag{4.66}$$

where Δ is the energy derived from the sun, and $C(H_2O)$ is a general representation of organic carbon; for example, glucose is $C_6H_{12}O_6$ or $6C(H_2O)$. The process of respiration can be represented by the (reverse) reaction

$$C(H_2O) + O_2 \rightarrow CO_2 + H_2O + \Delta \tag{4.67}$$

Microbial ecologists refer to the respiration described in Equation 4.67 as *aerobic respiration* because oxygen is utilized as the electron acceptor. When oxygen is absent, *anaerobic respiration* takes place, utilizing a variety of other compounds as electron acceptors. Since the amount of photosynthetic and respiration activity depends on the amount and intensity of sunlight, there is a diurnal (daily) and seasonal variation in dissolved oxygen amounts contributed by photosynthesis and respiration. In fact, the diurnal variation can sometimes be so extreme that the stream is supersaturated with oxygen during the afternoon and severely depleted of oxygen just before dawn. Oxygen produced by photosynthesis above saturation during bright sunshine is released to the atmosphere and is lost from the water. Photosynthesis and respiration are a major source and sink of oxygen (respectively), particularly in slow-moving streams and lakes, and can be expected to be significant for algal concentrations in excess of $10\,g/m^3$ (dry mass).

Quantification of oxygen fluxes associated with photosynthesis and respiration is difficult, and is generally dependent on such variables as temperature, nutrient concentration, sunlight, turbidity, and whether the plants are floating (phytoplankton) or on the bottom (macrophytes, periphyton). Reported photosynthetic oxygen production rates, S_p^* (averaged over 24 h) range from 0.3 to $3\,g/m^2\cdot day$ for moderately productive surface waters up to $10\,g/m^2\cdot day$ for surface waters that have a significant biomass of aquatic plants (Thomann and Mueller, 1987). Reported respiration rates, S_r^*, have approximately the same range as photosynthetic oxygen production rates. Empirical equations that have been proposed for estimating photosynthetic oxygen production and respiration rates are (Di Toro, 1975)

$$S_p = 0.25\text{Chl}a \tag{4.68}$$

and

$$S_r = 0.025\text{Chl}a \tag{4.69}$$

where S_p and S_r are the average daily oxygen production and respiration rates, respectively, in $mg/L\cdot day$, and Chl$a$ is the chlorophyll a concentration in $\mu g/L$. As an approximation, the ratio of algal biomass (dry weight) to algal chlorophyll a is 100:1. Aside from using Equations 4.68 and 4.69 to estimate the photosynthetic production and respiration rates from chlorophyll a concentrations, production and respiration rates can be measured directly using the light- and dark-bottle method described in *Standard Methods* (APHA, 1992) or the delta method described in Di Toro (1975).

Benthic oxygen demand or *sediment oxygen demand* (SOD) results primarily from the deposition of suspended organics and native benthic organisms in the vicinity of wastewater discharges and can be a major sink of dissolved oxygen in heavily polluted rivers and streams. Most benthic sludge undergoes anaerobic decomposition, which is a relatively slow process; however, aerobic decomposition can occur at the interface between the sludge and the flowing water. The products of anaerobic decomposition are CO_2, CH_4, and H_2S, and if gas production is especially high, floating of bottom sludge may result, leading to an aesthetic problem as well as depletion of dissolved oxygen. Benthic oxygen demand, S_b^* [M/L^2T], is typically taken as a constant in most applications (Thomann, 1972), and the benthic flux of oxygen, S_b [M/L^3T], used in the advection–dispersion equation is derived from S_b^* using the relation

$$S_b = \frac{S_b^* A_s}{\forall} = \frac{S_b^*}{\overline{d}} \tag{4.70}$$

where A_s is the surface area of the bottom of the stream, $\forall$ is the volume of the stream section containing a benthic surface area, A_s, and $\bar{d}$ is the average depth of flow. Photosynthetic and respiration rates S_p and S_r are derived from S_p^* and S_r^* using relations similar to Equation 4.70. In many water-quality modeling studies, S_b^* values are obtained through calibration. Typical values of S_b^* at 20°C are given in Table 4.9, and calculated values of S_b at 20°C can be converted to other temperatures using the relation

$$S_{b_T} = S_{b_{20}}(1.065)^{T-20} \tag{4.71}$$

where S_{b_T} and $S_{b_{20}}$ are the values of S_b at temperature T and 20°C, respectively. At temperatures below 10°C, S_b declines faster than indicated by Equation 4.71, and in the range 5 to 0°C, S_b approaches zero (Chapra, 1997). The constant 1.065 in Equation 4.71 is sometimes taken to be in the range 1.05 to 1.06, not a specific value such as 1.065 (Novotny, 2003). There has been some debate about whether S_b should be taken as a constant at a given temperature, since it is almost certainly a function of the organic content of the sediments and the oxygen concentration in the overlying water, both of which vary with distance from the source.

The SOD includes oxygen demand from the following processes:

- Biological respiration and oxygen consumption of all living organisms residing in the upper aerobic benthic zone
- Chemical oxidation of reduced substances in the sediments, such as reduced iron, manganese, and sulfides after they come to oxygen-containing water
- Biochemical oxidation of methane and ammonia (simultaneous nitrification–denitrification) that evolve from the lower anaerobic sediments

Most of the SOD is attributed to biochemical oxidation of evolving methane from the lower anaerobic sediment in the upper oxygenated interstitial sediment layer. Generally, if the stream velocity is higher (>0.3 m/s), most of the fine sediments remain in suspension and the bed is composed mostly of sand and gravel, which have a very low organic content, and the bed will exhibit very low SOD. As more data become available, a functional approach to estimating the SOD will probably evolve as the preferred formulation; in the meantime, however, a paucity of field data would support taking S_b to be a constant.

TABLE 4.9   Typical Benthic Oxygen Demand Rates, S_b^*, at 20°C

Bottom Type	Range (g/m²·day)	Average Value (g/m²·day)
Filamentous bacteria (10 g/m²)	—	−7
Municipal sewage sludge near outfall	−2 to −10	−4
Municipal sewage sludge downstream of outfall (aged)	−1 to −2	−1.5
Estuarine mud	−1 to −2	−1.5
Sandy bottom	−0.2 to −1.0	−0.5
Mineral soils	−0.05 to −0.1	−0.07

Source: Thomann (1972).

Incorporating photosynthetic, respiratory, and benthic oxygen fluxes into the oxygen-sag model yields the following modified form of Equation 4.50:

$$V\frac{\partial c}{\partial x} = -k_1 L + k_2(c_s - c) + (S_p + S_r + S_b) \tag{4.72}$$

This equation yields a solution expressed in terms of the oxygen deficit, D $(= c_s - c)$, given by

$$\boxed{D(x) = D_{\text{SF}}(x) + \Delta D_S(x)} \tag{4.73}$$

where $D_{\text{SF}}(x)$ is the oxygen deficit predicted by the Streeter–Phelps equation (Equation 4.57), which assumes that the BOD in the river originates from the wastewater discharge and that the only source of oxygen is from atmospheric reaeration, and $\Delta D_S(x)$ is the additional oxygen deficit caused by the net effect of photosynthesis, respiration, and benthic oxygen demand and is given by (Thomann and Mueller, 1987)

$$\boxed{\Delta D_S(x) = -(S_p + S_r + S_b)\left[1 - \exp\left(-k_2 \frac{x}{V}\right)\right]} \tag{4.74}$$

Equations 4.73 and 4.74 leads to a critical point, x_c, with maximum oxygen deficit given by

$$\boxed{x_c = \frac{V}{k_2 - k_1} \ln\left[\frac{k_2}{k_1} - \frac{k_2 D_0(k_2 - k_1) + (S_p + S_r + S_b)(k_2 - k_1)}{k_1^2 L_0}\right]} \tag{4.75}$$

and a corresponding critical oxygen deficit, D_c, given by

$$\boxed{D_c = \frac{k_1}{k_2} L_0 \exp\left(-k_1 \frac{x_c}{V}\right) - \frac{S_p + S_r + S_b}{k_2}} \tag{4.76}$$

Example 4.12 An outfall discharges wastewater into a slow-moving river that has a mean velocity of 3 cm/s and an average depth of 3 m. After initial mixing, the dissolved-oxygen concentration in the river is 9.5 mg/L, the saturation concentration of oxygen is 10.1 mg/L, the ultimate BOD of the mixed river water is 20 mg/L, the rate constant for BOD is 0.48 day^{-1}, and the reaeration rate constant is 0.72 day^{-1}. During the night, algal respiration exerts an oxygen demand of 2 g/m^2·day, and sludge deposits downstream of the outfall exert a benthic oxygen demand of 4 g/m^2·day. Estimate the minimum dissolved oxygen and the critical location in the river.

SOLUTION   From the data given, the initial oxygen deficit, D_0, is $10.1 - 9.5 = 0.6$ mg/L, $k_1 = 0.48$ day^{-1}, $k_2 = 0.72$ day^{-1}, $S_r^* = -2$ g/m^2·day, and $S_b^* = -4$ g/m^2·day Since the average depth, $\bar{d}$, of the river is 3 m, the volumetric oxygen demand rates for respiration and benthic consumption, S_r and S_b, respectively, are

$$S_r = \frac{S_r^*}{\bar{d}} = -\frac{2}{3} = -0.667 \text{ g/m}^3 \cdot \text{day} = -0.667 \text{ mg/L} \cdot \text{day}$$

and

$$S_b = \frac{S_b^*}{d} = -\frac{4}{3} = -1.33 \, \text{g/m}^3 \cdot \text{day} = -1.33 \, \text{mg/L} \cdot \text{day}$$

Since $L_0 = 20 \, \text{mg/L}$ and $V = 3 \, \text{cm/s} = 2592 \, \text{m/day}$, Equation 4.75 gives the location, x_c, of the critical oxygen deficit as

$$x_c = \frac{V}{k_2 - k_1} \ln\left[\frac{k_2}{k_1} - \frac{k_2 D_0(k_2 - k_1) + (S_p + S_r + S_b)(k_2 - k_1)}{k_1^2 L_0}\right]$$

$$= \frac{2592}{0.72 - 0.48} \ln\left[\frac{0.72}{0.48} - \frac{0.72(0.6)(0.72 - 0.48) + (-0.667 - 1.33)(0.72 - 0.48)}{(0.48)^2(20)}\right]$$

$$= 4951 \, \text{m}$$

and Equation 4.76 gives the critical oxygen deficit, D_c, as

$$D_c = \frac{k_1}{k_2} L_0 \exp\left(-\frac{k_1 x_c}{V}\right) - \frac{S_p + S_r + S_b}{k_2}$$

$$= \frac{0.48}{0.72}(20) \exp\left(-\frac{0.48(4951)}{2592}\right) - \frac{-0.667 - 1.33}{0.72}$$

$$= 8.1 \, \text{mg/L}$$

Hence the minimum dissolved oxygen level in the stream is $10.1 - 8.1 = 2.2 \, \text{mg/L}$ and occurs 4951 m downstream of the outfall location.

Chapra and Di Toro (1991) considered the case of a stagnant water body in which respiration is the only oxygen sink and the oxygen-production (photosynthesis) rate is a function of the time of day, in which case the oxygen deficit, D, is given by

$$\frac{dD}{dt} + k_2 D = R - P(t) \tag{4.77}$$

where D is the oxygen deficit, k_2 is the reaeration coefficient, R is the (constant) respiration rate, and $P(t)$ is the primary production (photosynthesis) rate. Defining sunrise as $t = 0$ and the *photoperiod* (i.e., duration of sunlight) as f, the oxygen-production rate due to photosynthesis can be estimated by

$$P(t) = P_m \max\left(\sin\frac{\pi t}{f}, 0\right) \tag{4.78}$$

where P_m is the maximum instantaneous production rate. Equation 4.78 implies that solar noon is at $t = f/2$, and simple integration of Equation 4.78 over a 24-h period gives the daily average production rate as

$$P_{av} = \frac{2f}{\pi T} P_m \tag{4.79}$$

where $T = 24$ h. Chapra and Di Toro (1991) developed piecewise-periodic analytical solutions to Equations 4.77 and 4.78 by requiring that the dissolved-oxygen deficit at the beginning of the day is equal to its value at the end of the day, and by also requiring that the solutions be continuous at sunset. The analytical solution to Equations 4.77 and 4.78 for the photoperiod is

$$D_1(t) = \frac{R}{k_2} - \sigma \left[\sin\left(\frac{\pi t}{f} - \theta\right) + \gamma e^{k_2 t} \right], \qquad 0 \leq t \leq f \tag{4.80}$$

and for the dark period the solution is

$$D_2(t) = \frac{R}{k_2} - \sigma [\sin(\theta) + \gamma e^{-k_2 f}] e^{-k_2(t-f)}, \qquad f \leq t \leq T \tag{4.81}$$

where the parameter groups are

$$\theta = \tan^{-1} \frac{\pi}{k_2 f}, \qquad \gamma = \sin\theta \frac{1 + e^{-k_2(T-f)}}{1 - e^{-k_2 T}}, \qquad \sigma = \frac{P_m}{\sqrt{k_2^2 + (\pi/f)^2}} \tag{4.82}$$

These solutions satisfy the necessary periodic conditions that $D_1(f) = D_2(f)$ and $D_1(0) = D_2(T)$, θ and γ are dimensionless, and σ has units of M/L^3. A typical solution is shown in Figure 4.11, along with the instantaneous production and dissolved-oxygen profiles. Figure 4.11 shows that for the particular parameter set used ($k_2 = 7$ day^{-1}), the minimum and maximum dissolved-oxygen deficits both occur during the photoperiod, with the maximum deficit occurring shortly after sunrise and the minimum deficit occurring between solar noon and sunset. This is always true for this model, as can be demonstrated mathematically by differentiating Equations 4.80 and 4.81 with respect to t and seeking their extrema, of which D_1 has two and D_2 has none. The utility of Equations 4.80 and 4.81 is that the observed oxygen deficit as a function of time can be compared with these solutions, and the model parameters k_2, P_{av}, and R, can be extracted by the best fit between the data observed and the theoretical variation of the of the dissolved-oxygen deficit. The time of day, t_{min}, at which the dissolved-oxygen deficit is a minimum is obtained by setting the first derivative of Equation 4.80 to zero, which yields the following requirement:

$$\pi \cos\left[\pi\left(\frac{1}{2} + \frac{\phi}{f}\right) - \theta \right] - (k_2 f) \gamma e^{-k_2(\phi + f/2)} = 0, \qquad \phi > 0 \tag{4.83}$$

where ϕ is the time between solar noon and the time of minimum dissolved-oxygen deficit, given by the relation

$$\phi = t_{min} - \frac{f}{2} \tag{4.84}$$

Equation 4.83 shows that k_2 is expressed in terms of ϕ and f, which can both be estimated from observations of the dissolved oxygen versus time, as is apparent from Figure 4.11. To calculate the average production rate, P_{av}, using Equations 4.80 and 4.81, the following relation can be derived (Chapra and Di Toro, 1991):

$$\frac{\Delta}{P_{av}} = \frac{\pi \delta}{2\sqrt{(k_2 f)^2 + \pi^2}} \tag{4.85}$$

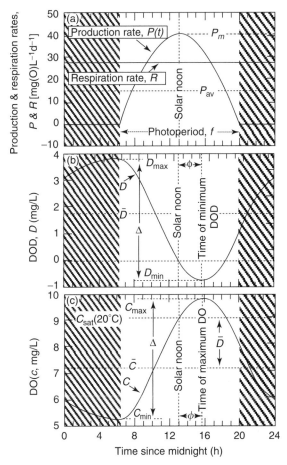

FIGURE 4.11   Dissolved oxygen versus time for a time-varying production rate. (From McBride and Chapra, 2005.)

where Δ is defined by the relation

$$\Delta = D_1(t)|_{t=t_{max}} - D_1(t)|_{t=t_{min}} \tag{4.86}$$

where t_{max} and t_{min} are derived by finding the extrema of D_1 in Equation 4.80, which require that

$$\pi \cos\left(\frac{\pi t_{min}}{f} - \theta\right) - (k_2 f)\gamma e^{-k_2 t_{min}} = 0, \qquad t_{min} > \frac{f}{2} \tag{4.87}$$

$$\pi \cos\left(\frac{\pi t_{max}}{f} - \theta\right) - (k_2 f)\gamma e^{-k_2 t_{max}} = 0, \qquad t_{max} < \frac{f}{2} \tag{4.88}$$

and δ (in Equation 4.85) is defined by the relation

$$\delta = -\left[\sin\left(\frac{\pi t_{max}}{f} - \theta\right) + \gamma e^{-k_2 t_{max}}\right] + \left[\sin\left(\frac{\pi t_{min}}{f} - \theta\right) + \gamma e^{-k_2 t_{min}}\right] \tag{4.89}$$

The respiration rate, R, can be estimated from k_2 and P_{av} by integrating Equations 4.80 and 4.81, which yields

$$R = P_{av} + k_2 \overline{D} \qquad (4.90)$$

where $\overline{D}$ is the diurnal average dissolved-oxygen deficit obtained from the dissolved-oxygen measurements. A typical dissolved-oxygen sensor and measurement procedure is shown in Figure 4.12. Utilization of Equations 4.83, 4.85, and 4.90 to estimate k_2, P_{av}, and R from dissolved-oxygen observations is called the *delta method*, and the development of this method is attributed to Chapra and Di Toro (1991). It is important to keep in mind that the delta method does not limit the lower bound of the dissolved oxygen, and parameters that yield negative dissolved-oxygen values should be discounted. Furthermore, in cases where observations indicate a minimum dissolved oxygen before sunrise and a maximum dissolved oxygen before noon, application of the Chapra and Di Toro (1991) dissolved-oxygen model is questionable.

Although modern programmable calculators can solve Equation 4.83 for k_2 with minimal effort, the following approximation was proposed by McBride and Chapra (2005) to estimate k_2 in lieu of using Equation 4.83,

$$k_2 = 7.5 \left(\frac{5.3\eta - \phi}{\eta\phi} \right)^{0.85} \qquad \text{where} \qquad \eta = \left(\frac{f}{14} \right)^{0.75} \qquad (4.91)$$

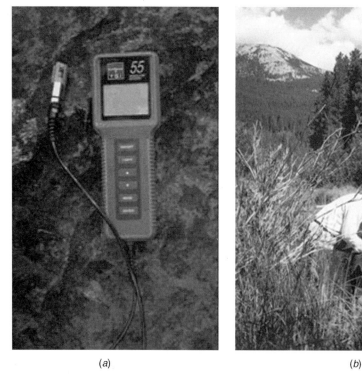

(a)	(b)

FIGURE 4.12 (*a*) Dissolved-oxygen sensor; (*b*) measurement technique. [(*a*) From YSI, 2005; (*b*) from USGS, 2005f.]

where ϕ and f are in hours and k_2 is in day^{-1}. The discrepancy between the exact (Equation 4.83) and approximate (Equation 4.91) relations are most pronounced for $f \geq 17$ h and $k_2 < 1$ day^{-1}. The average production rate, P_{av}, can be estimated using the following approximate relation in lieu of Equation 4.85 (McBride and Chapra, 2005):

$$\frac{\Delta}{P_{av}} = \frac{16}{\eta(33 + k_2^{1.5})} \tag{4.92}$$

Utilization of Equations 4.91 and 4.92 to estimate k_2 and P_{av} from dissolved-oxygen observations is called the *approximate delta method*, and development of this method is attributed to McBride and Chapra (1995). This method is mostly suited to streams with moderate reaeration ($k_2 < 10$ day^{-1}) and moderate photoperiods ($f = 10$ to 14 h).

Example 4.13 Measurements reported by Chapra and Di Toro (1991) taken on a slow-moving section of the Grand River (Michigan) characterized by profuse growths of aquatic plants showed that the diurnal dissolved-oxygen variations were characterized by: $\phi = 3.93$ h, $\Delta = 8.4$ mg/L, $\bar{D} = -0.272$ mg/L (i.e., average supersaturation), and $f = 13$ h (0700 to 2000 h). (a) Determine the values of k_2, P_{av}, and R corresponding to these data. (b) Compare these values with those obtained using the approximate relations.

SOLUTION From the data given, $f = 13$ h, $T = 24$ h, and Equation 4.82 gives

$$\theta = \tan^{-1}\frac{\pi}{k_2 f} = \tan^{-1}\frac{\pi}{13 k_2} \tag{4.93}$$

$$\gamma = \sin\theta\frac{1 + e^{-k_2(T-f)}}{1 - e^{-k_2 T}} = \sin\theta\frac{1 + e^{-11 k_2}}{1 - e^{-24 k_2}} \tag{4.94}$$

Also, from the given data, $\phi = 3.93$ h and Equation 4.83 give

$$\pi\cos\left[\pi\left(\frac{1}{2} + \frac{\phi}{f}\right) - \theta\right] - (k_2 f)\gamma e^{-k_2(\phi + f/2)} = 0$$

$$\pi\cos\left[\pi\left(\frac{1}{2} + \frac{3.93}{13}\right) - \theta\right] - (13 k_2)\gamma e^{-k_2(3.93 + 13/2)} = 0 \tag{4.95}$$

Simultaneous solution of Equations 4.93 to 4.95 yields

$$k_2 = 0.114\,\text{h}^{-1} = 2.7\,\text{day}^{-1}, \qquad \theta = 1.13\,\text{rad}, \qquad \gamma = 1.24$$

The times t_{min} and t_{max} when the oxygen deficit is a minimum and maximum, respectively, are given by Equations 4.87 and 4.88 as the solution to

$$\pi\cos\left(\frac{\pi t^*}{f} - \theta\right) - (k_2 f)\gamma e^{-k_2 t^*} = 0$$

$$\pi\cos\left(\frac{\pi t^*}{13} - 1.13\right) - (0.114)(13)(1.24)e^{-0.114 t^*} = 0 \tag{4.96}$$

where t^* represents the multiple solutions to Equation 4.96, and $t_{min} = t^*$ when $t^* > f/2$ and $t_{max} = t^*$ when $t^* < f/2$. Solutions to Equation 4.96 are $t^* = 10.4\,h$ and $0.575\,h$; hence, $t_{max} = 0.575\,h$ and $t_{min} = 10.4\,h$. Substituting into Equation 4.89 yields

$$\delta = -\left[\sin\left(\frac{\pi t_{max}}{f} - \theta\right) + \gamma e^{-k_2 t_{max}}\right] + \left[\sin\left(\frac{\pi t_{min}}{f} - \theta\right) + \gamma e^{-k_2 t_{min}}\right]$$

$$= -\left[\sin\left(\frac{\pi \times 0.575}{13} - 1.13\right) + 1.24 e^{-0.114 \times 0.575}\right]$$

$$+ \left[\sin\left(\frac{\pi \times 10.4}{13} - 1.13\right) + 1.24 e^{-0.114 \times 10.4}\right]$$

$$= 1.04$$

Substituting $\Delta = 8.4\,mg/L$, $\delta = 1.04$, $k_2 = 0.114\,h^{-1}$, and $f = 13\,h$ into Equation 4.85 gives

$$\frac{\Delta}{P_{av}} = \frac{\pi \delta}{2\sqrt{(k_2 f)^2 + \pi^2}}$$

$$\frac{8.4}{P_{av}} = \frac{\pi(1.04)}{2\sqrt{(0.114 \times 13)^2 + \pi^2}}$$

which yields

$$P_{av} = 17.9\,mg/L$$

Substituting $P_{av} = 17.9\,mg/L$, $k_2 = 0.114\,h^{-1} = 2.7\,day^{-1}$, and $\bar{D} = -0.272\,mg/L$ into Equation 4.90 gives

$$R = P_{av} + k_2 \bar{D} = 17.9 + 2.7(-0.272) = 17.2\,mg/L \cdot day$$

In summary, theoretical analysis of the measured diurnal variation in dissolved oxygen yields $k_2 = 2.7\,day^{-1}$, $P_{av} = 17.9\,mg/L \cdot day$, and $R = 17.2\,mg/L \cdot day$. These parameter values can be used in the Streeter–Phelps model to assess the impact of a continuous waste discharge into the river.

(b) As an (expedient) alternative to the delta method for calculating the values of k_2, P_{av}, and R, the approximate delta method may be used, since the preconditions $f < 17\,h$ and $k_2 > 1\,day^{-1}$ are met. Taking $f = 13\,h$ and $\phi = 3.93\,h$, Equation 4.91 gives

$$\eta = \left(\frac{f}{14}\right)^{0.75} = \left(\frac{13}{14}\right)^{0.75} = 0.946$$

$$k_2 = 7.5\left(\frac{5.3\eta - \phi}{\eta\phi}\right)^{0.85} = 7.5\left(\frac{5.3(0.946) - 3.93}{0.946(3.93)}\right)^{0.85} = 2.6\,day^{-1}$$

Taking $\Delta = 8.4\,mg/L$, $\eta = 0.946$, and $k_2 = 2.6\,day^{-1}$, Equation 4.92 yields

$$P_{av} = \frac{\eta(33 + k_2^{1.5})}{16}\Delta = \frac{0.946(33 + 2.6^{1.5})}{16}(8.4) = 18.5\,mg/L \cdot day$$

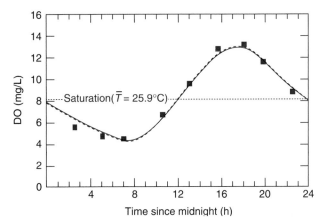

FIGURE 4.13   Dissolved oxygen versus time in the Grand River, Michigan. (From McBride and Chapra, 2005.)

Substituting $P_{av} = 18.5\,\text{mg/L}$, $k_1 = 2.6\,\text{day}^{-1}$, and $\overline{D} = -0.272\,\text{mg/L}$ into Equation 4.90 gives

$$R = P_{av} + k_2\overline{D} = 18.5 + 2.6(-0.272) = 17.8\,\text{mg/L} \cdot \text{day}$$

Comparing the parameter values obtained using the approximate delta method with the parameter values obtained using the full delta method indicates a 0.04% discrepancy in k_2 (2.6 day^{-1} versus 2.7 day^{-1}), a 3% discrepancy in P_{av} (18.5 mg/L versus 17.9 mg/L), and a 3% discrepancy in R (17.8 mg/L versus 17.2 mg/L). These results indicate that the approximate delta method gives results that are close to those of the full delta method.

The theoretical diurnal dissolved-oxygen curve is compared with the measurements in Figure 4.13, where the solid curve is based on the model-theoretical relations and the dashed curve is based on the approximate empirical relations (McBride and Chapra, 2005).

BOD Input Along the Channel   In addition to the discharge of municipal wastewater at discrete locations, a spatially distributed input of BOD is typically associated with runoff from agricultural areas, stormwater discharges, and highway runoff. Additional in-stream sources of distributed BOD include degradation products from sediment decomposition that subsequently diffuse from the sediment into the overlying water. In cases where there is a distributed BOD input, in addition to the discrete BOD input at a particular location, and atmospheric reaeration is the only source of oxygen, the oxygen deficit in the river is given by

$$D(x) = D_{SF}(x) + \Delta D_{BOD}(x) \tag{4.97}$$

where $D_{SF}(x)$ is the oxygen deficit predicted by the Streeter–Phelps equation (Equation 4.57), which assumes that the BOD in the river originates from a discrete discharge and the only source of oxygen is from atmospheric reaeration, and $\Delta D_{BOD}(x)$ is the additional oxygen deficit caused by the net distributed BOD and is given by (Thomann and Mueller, 1987)

$$\Delta D_{\text{BOD}}(x) = \frac{k_3 L_r}{k_1 k_2} \left[1 - \exp\left(-\frac{k_2 x}{V} \right) \right] - \frac{k_3 L_r}{(k_2 - k_1)k_1} \left[\exp\left(-\frac{k_1 x}{V} \right) - \exp\left(-\frac{k_2 x}{V} \right) \right]$$

(4.98)

where k_3 is the BOD decay coefficient (not including BOD removal by sedimentation of particulate BOD) (day^{-1}) and L_r is the ultimate BOD due to distributed sources (mg/L·day). To understand the difference between k_1 (the overall BOD removal coefficient, day^{-1}) and k_3, it should be recognized that k_1 is a composite of three processes: deoxygenation of BOD in free-flowing water, effect of benthic slimes on BOD adsorption and removal, and sedimentation of particulates. Hence, k_1 can be disaggregated as

$$k_1 = k_{1a} + k_{1b} - k_{1c}$$

(4.99)

where k_{1a} is the deoxygenation rate constant for free-flowing water approximately equaling the laboratory BOD bottle rate, k_{1b} is the BOD removal by sedimentation, and k_{1c} is the effect of absorption of BOD by benthic slimes, which results in oxygen demand. The magnitudes of k_{1a} have been reported in the range 0.1 to 0.6 day^{-1}, whereas the overall range of k_1 is typically between 0.1 and 5 day^{-1}, depending on the hydraulic and biological character of the stream and biodegradability of the organic pollution. All the processes represented in Equation 4.99 may not occur simultaneously. The constant k_3 used in Equation 4.98 is the BOD decay coefficient not including BOD removal by sedimentation of particulate BOD and is given by

$$k_3 = k_{1a} - k_{1c}$$

(4.100)

As a result of distributed BOD input, even in the absence of discrete wastewater discharges, there is usually a deficit in dissolved oxygen in streams and rivers. Background deficit concentrations are typically on the order of 0.5 to 2 mg/L and are typically assumed as a factor of safety is assessing the assimilative capacity of streams and rivers (Schnoor, 1996).

Example 4.14   A wastewater treatment plant discharges effluent into a river that has a mean velocity of 3 cm/s and a temperature of 20°C. After initial mixing across the river, the dissolved-oxygen concentration of the mixed river water is 6 mg/L, the ultimate BOD of this water is 20 mg/L, the laboratory rate constant for BOD decay in the river water is estimated to be 0.5 day^{-1}, and the reaeration constant is estimated to be 0.7 day^{-1}. Lateral distributed BOD sources extend 4 km downstream of the wastewater discharge, and these distributed sources are estimated to have an average ultimate BOD of 3 mg/L. BOD removal by sedimentation is estimated to increase the BOD decay rate by 20%, and adsorption by benthic slimes decreases the BOD decay rate by 5%. Assess the impact of the distributed BOD input on the oxygen concentration 4 km downstream of the wastewater discharge.

SOLUTION   At 20°C, the saturation concentration of dissolved oxygen is 9.1 mg/L (see Table 2.3), and from the data given, $L_0 = 20$ mg/L, $k_{1a} = 0.5$ day^{-1}, $k_{1b} = 0.20(0.5$ day$^{-1}) = 0.1$ day^{-1},

$k_{1c} = 0.05(0.5\ \text{day}^{-1}) = 0.025\ \text{day}^{-1}$, $k_2 = 0.7\ \text{day}^{-1}$, $V = 3\ \text{cm/s} = 2592\ \text{m/day}$, $x = 4\ \text{km} = 4000\ \text{m}$, and $D_0 = 9.1\ \text{mg/L} - 6\ \text{mg/L} = 3.1\ \text{mg/L}$. The BOD decay rates to be used in the analyses are

$$k_1 = k_{1a} + k_{1b} - k_{1c} = 0.5\ \text{day}^{-1} + 0.1\ \text{day}^{-1} - 0.025\ \text{day}^{-1} = 0.575\ \text{day}^{-1}$$
$$k_3 = k_{1a} - k_{1c} = 0.5\ \text{day}^{-1} - 0.025\ \text{day}^{-1} = 0.475\ \text{day}^{-1}$$

According to the Streeter–Phelps equation (Equation 4.57), in the absence of a distributed input of BOD, the dissolved-oxygen deficit 4 km downstream of the wastewater discharge is given by

$$D_{SF}(x) = \frac{k_1 L_0}{k_2 - k_1}\left[\exp\left(-\frac{k_1 x}{V}\right) - \exp\left(-\frac{k_2 x}{V}\right)\right] + D_0 \exp\left(-\frac{k_2 x}{V}\right)$$

$$D_{SF}(4000) = \frac{0.575(20)}{0.7 - 0.575}\left[\exp\left(-\frac{0.575(4000)}{2592}\right)\right.$$
$$\left. - \exp\left(-\frac{0.7(4000)}{2592}\right)\right] + 3.1\exp\left(-\frac{0.7(4000)}{2592}\right)$$
$$= 7.7\ \text{mg/L}$$

which corresponds to a dissolved-oxygen concentration of $9.1\ \text{mg/L} - 7.7\ \text{mg/L} = 1.4\ \text{mg/L}$. For a distributed source of BOD along the 4-km stretch of river, $k_3 = 0.475\ \text{day}^{-1}$, $L_r = 3\ \text{mg/L}$, and Equation 4.98 gives the contribution of the distributed load to the dissolved-oxygen deficit as

$$\Delta D_{BOD}(x) = \frac{k_3 L_r}{k_1 k_2}\left[1 - \exp\left(-\frac{k_2 x}{V}\right)\right] - \frac{k_3 L_r}{(k_2 - k_1)k_1}\left[\exp\left(-\frac{k_1 x}{V}\right) - \exp\left(-\frac{k_2 x}{V}\right)\right]$$

$$\Delta D_{BOD}(4000) = \frac{0.475(3)}{0.575(0.7)}\left[1 - \exp\left(-\frac{0.7(4000)}{2592}\right)\right]$$

$$- \frac{0.475(3)}{(0.7 - 0.575)(0.575)}\left[\exp\left(-\frac{0.575(4000)}{2592}\right) - \exp\left(-\frac{0.7(4000)}{2592}\right)\right]$$

$$= 0.9\ \text{mg/L}$$

Therefore, the distributed BOD load increases the dissolved-oxygen deficit by $0.9/7.7(100) = 12\%$. In the absence of the distributed BOD load, the dissolved-oxygen concentration 4 km downstream of the wastewater discharge is 1.4 mg/L, and with a distributed BOD load the dissolved-oxygen concentration is $1.4\ \text{mg/L} - 0.9\ \text{mg/L} = 0.5\ \text{mg/L}$. In either case, whether the distributed BOD load is present or not, the estimated dissolved-oxygen level would be devastating to the indigenous aquatic life, which typically requires about 5 mg/L of dissolved oxygen.

Sources and Sinks of Oxygen In cases where the river BOD originates from both a wastewater source and distributed sources of BOD along the channel, where the sources of oxygen are atmospheric reaeration and photosynthesis, and where the sinks of oxygen are the demand of the discharged wastewater plus respiration and benthic oxygen demand along the channel, the oxygen deficit, $D(x)$, along the channel is given by

$$D(x) = D_{SF}(x) + \Delta D_S(x) + \Delta D_{BOD}(x) \tag{4.101}$$

where $\Delta D_S(x)$ is given by Equation 4.74, and $\Delta D_{BOD}(x)$ is given by Equation 4.98. The form of Equation 4.101 explicitly gives the contributions of distributed oxygen sources and sinks on the oxygen deficit and provides a convenient form for assessing the relative impact of the various oxygen sources and sinks on the dissolved oxygen in rivers.

Example 4.15   A wastewater treatment plant discharges effluent into a river that has a mean velocity of 3 cm/s. Analysis of the waste discharge and distributed BOD input over a 500-m river stretch indicates that the dissolved oxygen 500 m downstream of the waste outfall is predicted to be 5 mg/L. In this analysis it is assumed that the reaeration constant is equal to 0.7 day^{-1} and the average temperature of the stream is 20°C. Photosynthesis within the river stretch is expected to generate 3 mg/L · day, plant respiration 1 mg/L · day, and sediment oxygen demand 0.5 mg/L · day. Estimate the expected fluctuation in dissolved oxygen 500 m downstream of the outfall.

SOLUTION   At 20°C the saturation concentration in the stream is 9.1 mg/L. From the data given, $D_{SF} + \Delta D_{BOD} = 9.1$ mg/L $-$ 5 mg/L $= 4.1$ mg/L, $k_2 = 0.7$ day^{-1}, $S_p = 3$ mg/L · day, $S_r = 1$ mg/L · day, and $S_b = 0.5$ mg/L · day. Since the average velocity in the river is 3 cm/s ($= 0.03$ m/s), the time, T, to travel 500 m is given by

$$T = \frac{500}{0.03} = 16{,}667\,\text{s} = 4.6\,\text{h}$$

Since the travel time is less than the photoperiod, the most critical condition for dissolved oxygen 500 m downstream of the outfall is when the waste travels for 4.6 h in darkness, photosynthesis is nonexistent, and both respiration and sediment oxygen demand are exerted. In this case, the dissolved-oxygen deficit is given by Equations 4.101 and 4.74 as

$$D(x) = [D_{SF}(x) + \Delta D_{BOD}(x)] + \Delta D_S(x) = [D_{SF}(x) + \Delta D_{BOD}(x)]$$

$$- (S_r + S_b)\left[1 - \exp\left(-k_2 \frac{x}{V}\right)\right]$$

$$D(500) = 4.1 - (-1 - 0.5)\left\{1 - \exp\left[-0.7\left(\frac{500}{0.03}\right)\right]\right\} = 5.6\,\text{mg/L}$$

which corresponds to a dissolved-oxygen concentration of 9.1 mg/L $-$ 5.6 mg/L $= 3.5$ mg/L.

When the 4.6 h travel time (over a distance of 500 m) occurs entirely during daylight hours, respiration is nonexistent and both photosynthesis and sediment oxygen demand occur. In this case, the dissolved-oxygen deficit is given by Equations 4.101 and 4.74 as

$$D(x) = [D_{SF}(x) + \Delta D_{BOD}(x)] + \Delta D_S(x)$$

$$= [D_{SF}(x) + \Delta D_{BOD}(x)] - (S_p + S_b)\left[1 - \exp\left(-k_2 \frac{x}{V}\right)\right]$$

$$D(500) = 4.1 - (3 - 0.5)\left\{1 - \exp\left[-0.7\left(\frac{500}{0.03}\right)\right]\right\} = 1.6\,\text{mg/L}$$

which corresponds to a dissolved-oxygen concentration of $9.1\,\text{mg/L} - 1.6\,\text{mg/L} = 7.5\,\text{mg/L}$. Therefore, the dissolved-oxygen concentration 500 m downstream of the waste discharge is expected to fluctuate over the range 3.5 to 7.5 mg/L within 24 h. As you might expect, this would cause significant stress to the indigenous aquatic life.

Dissolved Oxygen in Tidal Rivers and Estuaries The Streeter–Phelps model is used widely in practice, and an understanding of the assumptions incorporated in the model is essential for proper application. An important physical limitation of the Streeter–Phelps model is the neglect of longitudinal dispersion that can be important in some cases, particularly in tidal rivers and estuaries, where longitudinal dispersion causes both upstream and downstream effects on the dissolved oxygen. If longitudinal dispersion is considered, the steady-state advection–dispersion equation for dissolved oxygen is given by the simultaneous solution of the equations

$$K_L \frac{d^2 D}{dx^2} - V\frac{dD}{dx} - k_2 D + k_1 L = 0 \tag{4.102}$$

and

$$K_L \frac{d^2 L}{dx^2} - V\frac{dL}{dx} - k_1 L = 0 \tag{4.103}$$

where K_L is the longitudinal dispersion coefficient. Solving Equation 4.102 with the boundary conditions

$$D = 0 \text{ at } x = -\infty, \qquad D = D_0 \text{ at } x = 0, \qquad D = 0 \text{ at } x = +\infty \tag{4.104}$$

and solving Equation 4.103 with the boundary conditions

$$L = 0 \text{ at } x = -\infty, \qquad L = L_0 \text{ at } x = 0, \qquad L = 0 \text{ at } x = +\infty \tag{4.105}$$

and combining the results gives the following expression for the oxygen deficit (Thomann and Mueller, 1987; Schnoor, 1996):

$$D(x) = \begin{cases} \dfrac{L_0 k_1 m_1}{k_2 - k_1}\left\{\dfrac{1}{m_1}\exp\left[\dfrac{Vx}{2K_L}(1 - m_1)\right] - \dfrac{1}{m_2}\exp\left[\dfrac{Vx}{2K_L}(1 - m_2)\right]\right\}, & x \geq 0 \\[4mm] \dfrac{L_0 k_1 m_1}{k_2 - k_1}\left\{\dfrac{1}{m_1}\exp\left[\dfrac{Vx}{2K_L}(1 + m_1)\right] - \dfrac{1}{m_2}\exp\left[\dfrac{Vx}{2K_L}(1 + m_2)\right]\right\}, & x \leq 0 \end{cases} \tag{4.106}$$

where

$$m_1 = \sqrt{1 + \frac{4k_1 K_L}{V^2}} \quad \text{and} \quad m_2 = \sqrt{1 + \frac{4k_2 K_L}{V^2}} \tag{4.107}$$

Values of m_1 and m_2 contain important dimensionless numbers that measure the relative importance of dispersion and advection, where

$$\frac{k_1 K_L}{V^2}, \frac{k_2 K_L}{V^2} > 20, \qquad \text{dispersion predominates} \tag{4.108}$$

and

$$\frac{k_1 K_L}{V^2}, \frac{k_2 K_L}{V^2} < 0.05, \qquad \text{advection predominates} \tag{4.109}$$

Equation 4.106 requires prior estimation of the aeration constant, k_2, which can be done using field measurements (gas tracer method) or any of the appropriate formulas in Table 4.8. However, the appropriate velocity to be used in these formulas must be selected according to the following guidelines (Thomann and Mueller, 1987):

- When the net nontidal velocity, V_0, is greater than the average tidal velocity, V_T, use the net nontidal velocity.
- When the average tidal velocity, V_T, is greater than or equal to the net nontidal velocity, V_0, use the average tidal velocity.

It should be noted that the average tidal velocity, V_T, is given by

$$V_T = \frac{2}{T} \int_0^{T/2} v_1 \sin \frac{2\pi t}{T} dt = \frac{2}{\pi} v_1 \tag{4.110}$$

for a sinusoidally varying tidal velocity with period T and amplitude v_1.

Example 4.16   A tidal river has a mean velocity of 5 cm/s, a reaeration constant of 0.75 day^{-1}, and a longitudinal dispersion coefficient of 120 m^2/s. Wastewater is discharged into the river, and after initial mixing the river has an ultimate BOD of 10 mg/L and a BOD decay constant of 0.4 day^{-1}. If the temperature of the river is 20°C, determine the oxygen deficit 200 m downstream of the outfall. Assess whether consideration of longitudinal dispersion is important in predicting the effect of wastewater on dissolved oxygen levels in the river.

SOLUTION   From the data given, $V = 5$ cm/s $= 4320$ m/day, $k_1 = 0.4$ day^{-1}, $k_2 = 0.75$ day^{-1}, $K_L = 120$ m^2/s $= 1.04 \times 10^7$ m^2/day, and $L_0 = 10$ mg/L. Taking longitudinal dispersion into account, the dissolved-oxygen profile is given by Equation 4.106, where

$$m_1 = \sqrt{1 + \frac{4k_1 K_L}{V^2}} = \sqrt{1 + \frac{4(0.4)(1.04 \times 10^7)}{(4320)^2}} = 1.37$$

and

$$m_2 = \sqrt{1 + \frac{4k_2 K_L}{V^2}} = \sqrt{1 + \frac{4(0.75)(1.04 \times 10^7)}{(4320)^2}} = 1.63$$

Hence, Equation 4.106 gives the oxygen deficit 200 m downstream of the outfall as

$$
\begin{aligned}
D(200) = \frac{10(0.4)(1.37)}{0.75 - 0.4} &\left\{ \frac{1}{1.37} \exp\left[\frac{4320(200)}{2(1.04 \times 10^7)} (1 - 1.37) \right] \right. \\
&\left. - \frac{1}{1.63} \exp\left[\frac{4320(200)}{2(1.04 \times 10^7)} (1 - 1.63) \right] \right\} \\
= 1.9 \text{ mg/L}
\end{aligned}
$$

The dimensionless numbers given in Equations 4.108 and 4.109 are as follows:

$$\frac{k_1 K_L}{V^2} = \frac{0.4(1.04 \times 10^7)}{(4320)^2} = 0.22$$

$$\frac{k_2 K_L}{V^2} = \frac{0.75(1.04 \times 10^7)}{(4320)^2} = 0.42$$

Comparing these with the limits given in Equations 4.108 and 4.109 indicates that both advection and dispersion should be considered, but neither predominates.

Design criteria for wastewater discharges generally require that dissolved-oxygen levels on the boundary of a mixing zone not fall below specified water-quality standards. In these cases, the appropriate oxygen-deficit model can be used to estimate the allowable BOD in the wastewater discharge.

4.5   RESTORATION AND MANAGEMENT

The primary water-quality-related problems encountered in rivers and streams include (Novotny, 2003) (1) sedimentation and siltation of habitat; (2) low dissolved oxygen, resulting in fish kills and stresses; (3) overuse of riparian areas; (4) stream channelization; (5) drinking-water taste, odor, color, and organics problems; (6) toxic contamination of sediments; and (7) poor fishing. Techniques that can be used in river and stream restoration are divided into nonstructural and structural techniques. Nonstructural techniques are broadly defined as any method that does not require physical alteration of a watercourse or construction of a dam or other structure. Structural techniques range from "soft" approaches, such as the use of tree trunks and branches to slow water velocity, to "hard" engineering approaches, such as the use of gabion or riprap for bank stabilization.

4.5.1   Nonstructural Techniques

Nonstructural techniques for the restoration and management of rivers and streams typically include administrative or legislative policies and procedures that limit or regulate some activity. The most common nonstructural techniques are as follows:

1. *Flow regulation* consists of reserving or reclaiming flow for in-stream uses such as fishing, wildlife, and recreation.

2. *Plantings* create buffer zones that can be gradually reforested over time through planting of trees, brush, herbaceous vegetation, and grass. Strips of forest along both banks of a stream protect the stream from polluted runoff, and an example of a typical riparian forest is shown in Figure 4.14. Practical guidance on riparian reforestation in urban watersheds can be found in Herson (1992). Important considerations in reforestation of stream riparian zones include site assessment, soil preparation, species selection, planting techniques, and long-term maintenance.

3. *Pollution prevention techniques* include regulating activities in the stream, riparian zone, and surrounding watershed. For example, phasing construction to limit the amount of disturbed area at any given time greatly reduces downstream suspended-sediment levels.

FIGURE 4.14 Riparian forest. (From Stroud Water Research Center, 2005.)

FIGURE 4.15 Silt fence. (From State of North Carolina, 2005a.)

The utilization of silt fences, such as shown in Figure 4.15, to protect rivers from dislodged soil during construction is an effective pollution-prevention technique. Typically, silt fences are installed alongside water bodies but may also be used effectively anywhere there is a risk of erosion. Silt fences are temporary structures that can easily be knocked down or washed away by large surges of water or material that result from heavy rainfall. Therefore, they should be used as one tool available to prevent erosion and not considered as the only solution. They may be removed once a site is permanently stabilized and the threat of accelerated erosion no longer exists. Streams that are affected by nutrient loads from lawn fertilization can be improved by changing the type of fertilizer used or the frequency and timing of fertilization. Nonstructural pollution abatement activities also include use of fencing around streams and riparian zones, which is effective in excluding livestock and humans.

4. *Propagation facilities* are used to propagate aquatic species through incubation and spawning and is a common approach where it is desirable to maintain fishing in an area that otherwise would not have a sustainable sport-fish population. A fish hatchery on the Columbia River is shown in Figure 4.16. Hatchery-raised fish may not always survive or propagate in the wild; for example, salmon introduced into Lake Michigan have not been reproducing and must be restocked continuously.

5. *Land acquisition* can protect a watercourse through maintenance of buffer zones and prevention of potentially destructive land uses in the watershed. Land-acquisition approaches can lead to the establishment of greenways, buffer strips, and parks. These can be purchased by government or special foundations and trusts set up to provide such protection. For example, the state of Florida has created a fund by which riparian floodplain lands along the main state rivers (e.g., St. Johns River) are being acquired and converted to riparian wetlands and buffer zones. Similar programs exist in Wisconsin and Illinois.

6. *Land-use regulation* in the riparian zone and watershed is an effective legislative approach to controlling pollution sources.

7. *Biomanipulation* is a fish-management technique that involves direct manipulation of the fish community and other organisms that serve as prey, predators, or competitors

FIGURE 4.16   Fish hatchery on the Columbia River. (From USACE, 2005e.)

with the fish species of interest. Activities include game fish stocking, control of undesirable fish species, and prey enhancements to supplement food supplies.

4.5.2 Structural Techniques

Structural techniques for the restoration and management of rivers and streams are those that require some type of physical alteration of the watercourse and may include alterations to existing human-made structures, such as dams and levees. Structural stream-restoration techniques include bioengineering techniques, bank-armoring techniques, aquatic-habitat improvement methods, low-flow augmentation, in-stream and sidestream aeration, fish ladders, removal of river impoundments, and removal of contaminated sediment. Descriptions of several of these techniques follow.

1. *Bioengineering techniques* use plants to replace natural stream bank stabilization in situations where the stream bank has been eroded or lacks vegetation due to some destructive process. Planting can provide more ecological benefits than erosion control through addition of stream cover, shade, and improvement of bank-soil conditions. Native plant species are recommended because they are generally adapted to local environmental conditions.

2. *Bank armoring techniques* use rock, wood, steel, and other conventional construction materials to stabilize stream banks. An example of riprap (rock) stabilization of a stream bank is shown in Figure 4.17. Bank-armoring methods rarely provide significant

FIGURE 4.17   Riprap stabilization of a stream bank. (From NRCS, 2005a.)

FIGURE 4.18   Constructed riffle on the Brunswick River, Australia. (From Australia Waters and Rivers Commission, 2005.)

ecological benefits other than erosion control unless they are combined with bioengineering techniques.

3. *Aquatic habitat improvement methods* involve improving aquatic habitat through installation of certain in-stream structures. Disturbed streams often lack diverse morphological features, and habitat improvement structures can add gravel beds, structural complexity, restricted flow, and riffles and pools. These features are important to allow spawning and rearing areas for aquatic life. Selection of a particular technique or combination of techniques depends on current habitat deficiencies, watershed conditions, and the current morphology and hydrology of the watercourse. An example of a constructed riffle on the Brunswick River (Australia) is shown in Figure 4.18.

4. *Low-flow augmentation* provides cleaner diluting flow during times of water-quality emergencies and/or to sustain proper ecological conditions for fish and aquatic life. The source of the diluting flow may include upstream reservoirs, pumping from nearby water bodies, or recycling (pumping) flows of cleaner and more diluted downstream flows. Using highly treated effluents in effluent-dominated streams is also a feasible and acceptable source of base flow. As an example, in Milwaukee, Wisconsin, cleaner Lake Michigan water is pumped into harbor sections of the Milwaukee River during times of low oxygen levels in the harbor. The amount of dilution water needed can be estimated by water-quality models.

5. *In-stream and sidestream aeration* may be feasible for streams that exhibit low dissolved-oxygen levels. Aeration is also used to supplement oxygen in streams that have a low assimilative capacity to accept residual pollution by biodegradable organics, and the control of residual pollution would be ineffective and/or its cost would cause a widespread adverse socioeconomic impact. In-stream aeration is accomplished by turbine aeration in power plants or by installation of floating or submerged aerators. More natural in-stream cascades, spillways, and water falls may also provide additional oxygen. Sidestream elevated pool aeration has been used to supplement oxygen in Chicago waterways that

FIGURE 4.19   Sidestream aeration in Worth, Illinois. (From Chicago Public Library, 2005.)

receive an unusually high load of treated effluent. In sidestream aeration systems, water is pumped by Archimedes screw pumps into an elevated pool and cascades back into the water body (Butts, 1988; Butts et al, 1999). Both the pumps and cascading process supply oxygen. A sidestream aeration structure located in Worth, Illinois, is shown in Figure 4.19. Sidestream aeration appears to be more effective and aesthetic than in-stream aeration by floating aerators. Aeration by air diffusers located on the bottom of streams are subject to clogging.

Aeration is mostly a temporal measure that can be used (1) when summer low-oxygen levels drop for a short time (such as during the night and early morning hours) below the dissolved-oxygen limit for fish protection (4 to 6 mg/L depending on the type of fish); or (2) during winter, when stream aeration is reduced by ice cover. Aeration may be particularly useful for streams draining wetlands which exhibit dystrophic conditions (low dissolved oxygen levels caused by high rates of decomposition of organic matter in the wetlands) and as a supplement when the waste assimilative capacity of the water body is exhausted because the water body receives loads of biodegradable organics that cannot be controlled by removal at the source. The efficiency of aeration increases as the oxygen deficit increases.

The oxygen-concentration increase at in-stream aeration locations can be estimated using the relation (Novotny et al., 1989)

$$\Delta c = \frac{B_L D}{180 Q} K'$$ (4.111)

where Δc is the oxygen-concentration increase at the aeration location (mg/L), B_L is the aeration power input (kW), D is the average oxygen deficit (%), Q is the flow (m³/s), and K' is a coefficient representing the oxygen yield per kilowatthour at 50% average oxygen deficit, which is the mean of the upstream and downstream deficit at the point of aeration

TABLE 4.10   Aeration Coefficients

Aeration Mechanism	K' (kg O$_2$/kWh)
Cascades and rapids	1.5
Sharp-crested weir	0.6
Weirs and spillways	0.4
Turbine aeration	1.0
Surface aeration by floating aerators	0.5
Diffuse aeration	0.4

at 20°C. When accurate data are not available, K' can be estimated from Table 4.10. For weir and spillway aeration, the coefficient B_L in Equation 4.111 can be estimated by

$$B_L = gQh \tag{4.112}$$

where h is the fall height.

Butts (1988) developed the following empirical equation to predict the weir-aeration efficiencies for sidestream elevated pool aeration (SEPA) stations:

$$P_o = 0.32P_i + 4.13N + 2.62H + 54.78 \tag{4.113}$$

where P_o is the output dissolved oxygen (% saturation), P_i is the input dissolved oxygen (% saturation), N is the number of steps, and H is the total weir-system height (m). Equation 4.113 was as developed for the temperature range 15 and 25°C.

6. *Fish ladders* are cascades of small basins that have an elevation differential that migrating fish can easily overcome when moving upstream and also make it possible for the fish to rest in the basins. A typical fish ladder is illustrated in Figure 4.20. Fish ladders

FIGURE 4.20   Fish ladder on the Brooks River, Alaska. (From U.S. National Park Service, 2005a.)

FIGURE 4.21   Dam removal on the Ashuelot River, New Hampshire. (From U.S. Fish and Wildlife Service, 2005a.)

have been installed worldwide; however, experience with these installations on the Columbia River in the Pacific northwest (in the United States) is mixed and have proven insufficient to restore full salmon migration up the river.

7. *Removal of river impoundments* can be considered restoration techniques in addition to construction of new structures or modification of existing structures. In the United States many dams were built on streams for various purposes more than 100 years ago and, in Europe, building stream weirs (low-head dams) for providing head to mills dates back several centuries. Sediment has accumulated behind these dams from urban and rural diffuse sources, wastewater discharges, and from combined sewer overflows. Often, the impoundment today is filled with sediments, has ceased to function, and the sediments contain toxic pollutants. The removal of the McGoldrick dam on the Ashuelot River (New Hampshire) is shown in Figure 4.21.

8. *Removal of contaminanted sediment* is an important option in many cases. Some sites with contaminated sediments have been declared as hazardous contaminated sites that have to be cleaned up, and the most common cleanup methods are sediment removal and sediment capping. The removal of sediment by dredging is illustrated in Figure 4.22. Extreme caution must be taken when the sediment is dredged because of resuspension of the contaminants and the possibility of contaminated sediment movement downstream. In extreme cases, underwater dredging is not feasible and the river must be diverted from the contaminated site for the site to be dredged, sediment removed to suitable landfill sites, and the channel restored. Sediment capping is used in situations where the channel is not aggrading and there is a steady supply of clean sediment from upstream. The contaminated sediment is encapsulated by a clean cover with or without a geomembrane encapsulation.

A significant concern in many streams is accelerated stream bank erosion caused by human-induced changes in the watershed. Such erosion increases sediment load and inhibits aquatic life. Achieving natural stream stability plays an important role in minimizing stream bank erosion and resultant sediment pollution. Natural stream stability is achieved by allowing the stream to develop a stable profile and pattern and stable dimensions, so that the stream system neither aggrades nor degrades. Preservation of stream geomorphology is an important

FIGURE 4.22 Dredging of contaminated sediment. (From *Seattle Daily Journal of Commerce*, 2005.)

goal in the identification of stream reaches that require restoration, and aids in the development of sustainable stormwater-management programs. The specialty area dealing with the natural formation and equilibrium shape of rivers is *fluvial geomorphology*.

4.6 COMPUTER CODES

Several computer codes are available for simulating water quality in rivers. These codes typically provide numerical solutions to the advection–dispersion equation or some other form of the law of conservation of mass. Numerical solutions are produced at discrete locations and times for multiple interacting constituents, complex boundary conditions, spatially and temporally distributed contaminant sources and sinks, multiple fate processes, and variable flow and dispersion conditions. In engineering practice, the use of computer codes to apply the fundamental principles covered in this chapter is sometimes essential. In choosing a code for a particular application, there is usually a variety of codes to choose from. However, in doing work that is to be reviewed by regulatory agencies, or where professional liability is a concern, codes developed and maintained by the U.S. government have the greatest credibility and, perhaps more important, are almost universally acceptable in developing permit applications and defending design protocols on liability issues. Several of the more widely used codes that have been developed and endorsed by U.S. government agencies are described briefly here.

QUAL2E is the most widely used computer code for simulating streamwater quality. This code is capable of simulating several interacting water-quality constituents in connected (dendritic) streams that are well mixed across their cross section. The QUAL2E code solves the one-dimensional advection–dispersion equation in the form

$$\frac{\partial c}{\partial t} = \frac{1}{A_x}\frac{\partial}{\partial x}\left(A_x K_L \frac{\partial c}{\partial x}\right) - \frac{1}{A_x}\frac{\partial}{\partial x}A_x V c + kc + S_m \tag{4.114}$$

where c is the tracer concentration [M/L^3], A_x is the cross-sectional area of the channel [L^2], K_L is the longitudinal dispersion coefficient, V is the mean flow velocity [L/T], k is the first-order decay rate [$1/T$], and S_m is the internal mass influx [$M/T \cdot L^3$]. Equation 4.114 is solved numerically using a finite-difference scheme. The model can track several water-quality constituents, including dissolved oxygen, temperature, BOD, available nitrogen, available phosphorus, coliforms, and algae as chlorophyll a. Application is limited to simulating periods of constant flows and loads. The model allows specification of multiple discharges, withdrawals, and incremental inflow and outflow and has the capability of calculating required dilution flows to meet any specified dissolved-oxygen level. QUAL2E is currently maintained by the EPA Center for Water Quality Modeling in Athens, Georgia.

HSPF (Hydrological Simulation Program–Fortran) is the flagship model in the USEPA watershed-modeling system BASINS (see Chapter 9). The HSPF package is a series of computer codes designed to simulate: watershed hydrology, land-surface runoff, and fate and transport of pollutants in receiving waters. The hydrologic portion of the model includes a hydrology model, a nonpoint-source estimation model, and a stream-routing component based on the kinematic-wave approximation. The water-quality portion of the model includes sediment-transport simulation (settling–deposition–scouring) for three classes of particle sizes. Adsorption–desorption processes are calculated separately for each particle class in both the water column and the sediments. Degradation–transformation processes include hydrolysis, photolysis, oxidation, volatization, and biodegradation. Kinetic reactions are simulated as second-order processes. Chemical reaction simulation includes up to two secondary chemicals for each primary component. The model is limited in application to one-dimensional systems in which the pollutants are well mixed over the entire cross section. Zero-dimensional representation (lakes and other impoundments) implies that pollutants are completely mixed throughout the water body, and that the water body is not stratified. The model inputs are numerous and depend on the application. They may include one or more of the following groups:

- Time-series inputs of air temperature, rainfall, evapotranspiration, channel inflows, and wind
- Parameters such as channel geometry, soil moisture and vegetative conditions, infiltration and interflow data, overland sediment mobility, and wash-off potential
- Water-quality constituents in interflow and ground water
- Agrichemical quality constituents' data
- Reach/reservoir water-quality characteristics, coefficients, and kinetic rates

Typical outputs are pollutant concentrations. Yearly summaries include statistical analysis of time-varying contaminant concentrations that can be used for risk assessment.

WASP 6 (Water Quality Analysis Simulation Program 6) (USEPA, 1988) simulates contaminant transport in streams in one, two, or three dimensions. The contaminants that can be simulated include BOD, dissolved oxygen, and nutrients. WASP 6, one of the most versatile multidimensional water-quality models in the public domain, is used for applications to rivers, lakes, estuaries, and reservoirs. The model simulates constant and variable loading and flow conditions. The model is linked to the hydrodynamic model DYNHYD5, which can be used to determine flow conditions in any water body (Ambrose et al., 1992). The WASP 6 code consists of two subprograms: EUTRO and TOXI. EUTRO can simulate conventional pollutants (DO, BOD, nutrients, and eutrophication). TOXI describes sorption, volatization, ionization, photolysis, hydrolysis, biodegradation, and chemical oxidation and is used for simulation of the fate of toxic pollution (organic chemicals, metals) and the transport of sediments. TOXI also considers the transformation of pollutants in bottom sediments. Both submodels can be operated at different degrees of complexity and sophistication. The model can simulate up to three classes of sediments and three chemicals (each with up to five species and five physical forms) at a time. The model considers all biological, chemical, and physical transformations–reactions of interest in aquatic environments. The model was developed and is maintained by the U.S. Environmental Protection Agency.

SED2D simulates surface-water flow and cohesive sediment transport. It is a two-dimensional depth-averaged dynamic model. The modeling system consists of separate computer programs for hydrodynamic modeling, HYDRO2D, and cohesive sediment-transport modeling, CS2D. The SED2D modeling system may be used to predict both short-term (less than 1 year) and long-term scour and/or sedimentation rates in well-mixed bodies of water. CS2D solves the advection–dispersion equation for nodal depth-averaged concentrations of suspended sediment and bed-surface elevations. The processes of dispersion, aggregation, erosion, and deposition are simulated. A layered bed model is used in simulating bed formation and subsequent erosion. The model calculates flow and velocity fields, depth-averaged sediment concentrations, erosion and deposition of bed, and bed formation.

HEC-6 is a one-dimensional transient numerical code developed by the U.S. Army Corps of Engineers (USACE, 1991) that simulates sediment transport in river and reservoir systems with tributary inflows. This model can perform continuous simulation of aggradation and degradation in streams and deposition in reservoirs.

Only a few of the more widely used computer models have been cited here. Certainly, there are many other good models that are capable of performing the same tasks.

SUMMARY

Contaminants of concern in rivers and streams include suspended sediments, nutrients, pathogens, oxygen-depleting substances, and toxic chemicals. These contaminants are commonly discharged into streams from either point discharges or distributed (nonpoint) discharges. Water-quality engineers must be able to predict contaminant loadings that are consistent with a stream meeting prescribed water-quality standards. Such contaminant loadings are at the core of both mixing-zone delineation and total maximum daily load calculations that are an integral part of any restoration effort for impaired waters.

Mixing in streams occurs in all coordinate directions, with transverse mixing caused mostly by turbulent diffusion, and longitudinal mixing caused mostly by shear dispersion induced by transverse variations in the longitudinal velocity. A method is presented for calculating the downstream distance required for complete cross-sectional mixing of point discharges, and beyond this point further mixing can be described by the one-dimensional (along-stream) advection–dispersion equation. Methods are presented for estimating the contaminant concentration at the point of complete cross-sectional mixing, and for estimating the longitudinal dispersion coefficient from the stream-channel and flow characteristics. Solutions to the advection–dispersion equation are given for both instantaneous point discharges (spills) and continuous point discharges. In the case of instantaneous point discharges, the theoretical concentration distribution resulting from the instantaneous spill of a mass of contaminant that undergoes first-order decay is presented. This theoretical solution is fundamental to predicting the impact of spills on downstream locations and in extracting the longitudinal dispersion coefficients from tracer tests. For spills of volatile organic compounds, volatization is taken into account by using an effective first-order decay coefficient in the advection–dispersion equation.

Continuous discharges of contaminants from point sources occur frequently in the cases of treated domestic wastewater and other discharges of oxygen-demanding wastes. The classical theoretical analysis of such discharges is the Streeter–Phelps model, which accounts for the oxygen demand of the stream–wastewater mixture balanced by reaeration at the water surface. The Streeter–Phelps model can be used to estimate the minimum dissolved oxygen in a stream that results from the discharge of an oxygen-demanding contaminant. Oxygen-depleting processes not accounted for by the Streeter–Phelps model include nitrification, photosynthesis, respiration, benthic oxygen demand, and the distributed input of oxygen-demanding substances along the channel. Corrections to the Streeter–Phelps model to account for these additional processes are presented. Parameters required to describe these process include the reaeration rate, photosynthesis rate, and respiration rate; and these parameters can be estimated from dissolved-oxygen measurements using the delta method. In tidal rives and estuaries, the relative importance of dispersion to advection is increased, and an alternative method to the Streeter–Phelps model (which neglects dispersion) is presented.

Approaches to the restoration and management of water quality in streams include both nonstructural and structural approaches. Nonstructural approaches include flow regulation, pollution prevention, and land acquisition; structural approaches include bank armoring, stream aeration, and removal of contaminated sediment. Computer codes are commonly used to facilitate the analysis of fate and transport processes in rivers, and recommended codes include QUAL2E, HSPF, WASP 6, SED2D, and HEC-6.

PROBLEMS

4.1. A natural river has a top width of 18 m, a flow area of 75 m^2, a wetted perimeter of 25 m, and the roughness elements on the wetted perimeter have a characteristic height of 7 mm. If the flow rate in the river is 100 m^3/s and the temperature of the water is 20°C, estimate the friction factor and the turbulent diffusion coefficients in the vertical and transverse directions. [*Hint*: Use the hydraulic depth (= flow area/top width) as the characteristic depth of the channel.]

4.2. (a) If the river described in Problem 3.14 has a characteristic roughness of 8 mm, estimate the vertical and transverse mixing coefficients.

   (b) If the outfall discharges the effluent across the entire bottom width of 6 m, estimate the distance downstream to where the effluent is completely mixed across the channel cross section.

4.3. A single-port outfall is located on the side of a stream that is 15 m wide and 3 m deep, and the flow velocity in the stream is 2 m/s.

   (a) If the friction factor is estimated to be 0.035 (calculated using the Colebrook equation), how far downstream from the discharge location can the effluent be considered well mixed across the stream?

   (b) How is this mixing distance affected if the single-port outfall is replaced by a 5-m-long multiport outfall located in the center of the stream?

4.4. A regulatory mixing zone in a river extends 200 m downstream of a 5-m-long industrial multiport outfall. The river has an average depth of 3 m, an average width of

30 m, and an average velocity of 0.8 m/s; the outfall discharges $10 \, m^3/s$ of wastewater containing 5 mg/L of a toxic contaminant. It is estimated that the plume width on the downstream boundary of the mixing zone is 15 m. Estimate the plume dilution on the (downstream) boundary of the mixing zone.

4.5. A relatively clear river containing 5 mg/L of suspended solids has a temperature of 15°C and intersects a turbid river with a suspended-solids concentration of 35 mg/L and a temperature of 20°C. If the discharge in the clear river is $100 \, m^3/s$ and the discharge in the turbid river is $20 \, m^3/s$, estimate the suspended-solids concentration and temperature downstream of the confluence of the two rivers.

4.6. A river is 30 m wide, 3 m deep, and has a typical roughness height of 5 mm. If the average velocity in the river is 15 cm/s, determine the values of the longitudinal and transverse dispersion coefficients that should be used to make conservative estimates of mixing in the river.

4.7. Stream depths and vertically averaged velocities at 1-m intervals across a 10-m-wide stream are given in Table 4.11. If the friction factor is 0.04, estimate the longitudinal dispersion coefficient across the channel using the formulas in Table 4.1.

TABLE 4.11 Data for Problem 4.7

Distance from Side, y (m)	Depth, d (m)	Velocity, v (m/s)	Distance from Side, y (m)	Depth, d (m)	Velocity, v (m/s)
0	0.0	0.0	6	3.6	2.4
1	0.30	0.45	7	2.2	1.5
2	1.30	0.90	8	1.20	0.75
3	1.8	1.20	9	0.70	0.45
4	2.1	2.1	10	0.0	0.0
5	4.5	3.0			

4.8. Show that the maximum concentration at a distance x_0 downstream from an instantaneous spill in a river occurs at a time t_0 given by

$$t_0 = \frac{x_0}{V}\left[-\frac{K_L}{Vx_0} + \sqrt{\left(\frac{K_L}{Vx_0}\right)^2 + 1} \right]$$

where V is the mean velocity in the stream and K_L is the longitudinal dispersion coefficient. Determine the values of K_L/Vx_0 for which t_0 does not deviate by more than 1% from x_0/V.

4.9. Fifty kilograms of a conservative contaminant is spilled into a river that has an average velocity of 1 m/s, is 10 m deep and 52 m wide, and has a friction factor on the order of 0.04.

(a) Assuming that the contaminant is initially well mixed across the river, what maximum concentration is expected 150 m downstream of the spill?

(b) What value of K_L would you use in your calculations if the width of the river were 25 m?

4.10. Fifteen kilograms of a contaminant is spilled into a stream 4 m wide and 2 m deep.

(a) If the average velocity in the stream is 0.8 m/s, the longitudinal dispersion coefficient is 0.2 m^2/s, and the first-order decay constant of the contaminant is 0.05 h^{-1}, determine the maximum concentration in the stream at a drinking-water intake 1 km downstream from the spill.

(b) How would this concentration be affected if the decay constant is actually one-half of the estimated value?

4.11. The plume resulting from a spill of a conservative contaminant passes an observation point 1 km downstream of the spill location, where the concentration as a function of time is measured. If the river has a mean velocity of 25 cm/s and the concentration distribution is observed to be Gaussian with a maximum concentration of 5 mg/L, estimate the maximum concentration 1.5 km downstream of the spill location.

4.12. Derive Equation 4.35.

4.13. Dye is released instantaneously from a bridge across a river, and the dye concentrations as a function of time are measured at locations 500 and 1000 m downstream of the release location. The measured concentrations at the downstream locations are given in Table 4.12. Estimate the longitudinal dispersion coefficient in the river.

TABLE 4.12   Data for Problem 4.13

Time (min)	Concentration at $x = 500$ m (mg/L)	Concentration at $x = 1000$ m (mg/L)	Time (min)	Concentration at $x = 500$ m (mg/L)	Concentration at $x = 1000$ m (mg/L)
0	0	0	16	0.17	0.35
1	0	0	17	0.04	0.73
2	0	0	18	0.01	1.1
3	0	0	19	0	1.9
4	0	0	20	0	2.7
5	0	0	21	0	1.9
6	0.03	0	22	0	1.1
7	0.05	0	23	0	0.67
8	0.12	0	24	0	0.41
9	0.45	0	25	0	0.21
10	0.92	0	26	0	0.07
11	2.3	0	27	0	0.03
12	3.3	0	28	0	0.01
13	2.3	0.02	29	0	0
14	1.1	0.04	30	0	0
15	0.57	0.22			

4.14. If the river described in Problem 4.13 is 20 m wide and 5 m deep, estimate the mass of dye that was used in the tracer study. Verify that the assumption of a conservative dye is reasonable.

4.15. Henry's constant, H, is an important parameter for estimating the volatization of VOCs in streams. Use Equation 4.42 to estimate H for chloroethane, 1,2-dichloroethane, and vinyl chloride at 20°C. Compare these values with the values of H given in Table 4.5.

4.16. A river water contains ethylbenzene (C_8H_{10}) at a concentration of 22 mg/L. Express this concentration in mol/m^3. The molecular weight of ethylbenzene is 106.17.

4.17. A river has a mean velocity of 7 cm/s, a width of 20 m, and a depth of 1.3 m. The water and air temperatures are both 17°C, and the average wind speed is 8 m/s.

   (a) If the longitudinal dispersion coefficient in the river is 2.5 m^2/s, estimate the maximum concentration 10 km downstream of a location where 12 kg of tetrachloroethene has been spilled into the river.

   (b) How far downstream from the spill would you expect to find tetrachloroethene in the nonaqueous phase?

4.18. Drinking-water regulations in Florida require that TCE concentrations be less than 3 μg/L, and it can reasonably be assumed that water treatment plants remove at least 99% of the TCE in source waters. The intake of a water treatment plant is located in the river described in Problem 4.26, and you have been charged with developing a contingency plan for possible TCE spills upstream of the water-supply intake. Your plan is to be based on the scenario of a truck carrying 100 L of TCE running off the road into the river and spilling 75% of the TCE into the river. What is the maximum distance upstream of the intake that a spill will cause the intake to be closed temporarily? State your assumptions. How would you supply water to the public while the intake is closed?

4.19. Contaminants in rivers sometimes adsorb onto suspended solids in accordance with the equilibrium relation $c_0 = Kc$ where c_0 is the adsorbed concentration (= mass of contaminant per unit mass of suspended solids), K is an equilibrium constant [L^3/M], and c is the contaminant concentration in the water [M/L^3]. Show that sedimentation can be considered as a first-order decay process for contaminants in rivers. You have been retained as a consultant to assess the impact of a spill of 10 kg of chlorobenzene in the river described in Problem 4.25. The volatization coefficient is estimated to be 0.52 day^{-1}, the suspended-solids concentration is 7 mg/L, the adsorption constant, K, for chlorobenzene is 4.27 cm^3/g, and the average sedimentation velocity is 0.8 m/day. Assess the impact of the spill on the town's water supply intake 1 km downstream of the spill location. The drinking-water standard for chlorobenzene is 0.1 mg/L. Assess the relative importance of volatization and sedimentation on the expected concentration of chlorobenzene at the water-supply intake. If the city engineer decides to shut down the water-supply intake when the river water at the intake exceeds the drinking water standard, for approximately how long will the intake be shut down?

4.20. Estimate the reaeration rate in a stream that is 10 m wide and 2 m deep (on average), has a slope of 4×10^{-5}, and has a flow rate of 2 m^3/s. The stream temperature is 20°C and the dissolved-oxygen concentration is 7 mg/L. How would the reaeration rate be affected if the temperature in the river dropped to 15°C?

4.21. A small stream has a mean depth of 0.3 m, a mean velocity of 0.5 m/s, a discharge rate of 0.3 m^3/s, and a slope of 0.1%. Compare the values of the reaeration rate constants estimated using the formulas given in Table 4.8. Comment on your results.

4.22. After initial mixing of a wastewater discharge in a 5-m-deep river, the dissolved-oxygen concentration in the river is 7 mg/L and the temperature is 22°C. The average flow velocity in the river is 6 cm/s.

(a) If the ultimate BOD of the mixed river water is 15 mg/L, the rate constant for BOD at 20°C is 0.5 day^{-1}, and the reaeration rate constant at 20°C is 0.7 day^{-1}, estimate the minimum dissolved-oxygen concentration in the river.

(b) How far downstream of the outfall location will the minimum dissolved-oxygen concentration occur?

4.23. After initial mixing of the wastewater discharged in Problem 3.14, the 5-day BOD of the river water is 15 mg/L and the dissolved-oxygen concentration is 2 mg/L. If the temperature of the river water is 25°C, estimate the dissolved-oxygen concentration 500 m downstream of the discharge. Does the minimum dissolved-oxygen concentration occur within 500 m of the outfall?

4.24. Measurements in a river indicate that the BOD and reaeration rate constants are 0.3 and 0.5 day^{-1}, and the ultimate BOD of the river water after mixing with a wastewater discharge is 20 mg/L. If the average velocity in the river is 5 cm/s and the saturation concentration of dissolved oxygen is 12.8 mg/L, determine the initial dissolved-oxygen concentration of the mixed river water at which the minimum dissolved-oxygen concentration will occur at the outfall.

4.25. A river adjacent to a small town has an average width of 20 m, an average depth of 2.4 m, a discharge of 8.7 m³/s, and an average roughness height on the channel boundary of 25 cm. The town plans to construct a wastewater outfall in the river and discharge wastewater through two 6-m-long diffusers extending from both sides of the river. The diffusers are expected to induce complete vertical mixing. The wastewater discharge is expected to be 0.1 m³/s with a BOD$_5$ of 30 mg/L and a dissolved-oxygen concentration of 2 mg/L. The natural river has a BOD$_5$ of approximately 10 mg/L, and a dissolved-oxygen concentration of 8.1 mg/L, which is 2 mg/L below the saturation concentration of oxygen in the river.

(a) How far downstream of the discharge location will the wastewater be well mixed across the river? Call this location X.

(b) If the rate constants for BOD and reaeration are 0.1 and 0.7 day^{-1}, respectively, and it is assumed that the wastewater is well mixed across the river at the outfall location, estimate the 5-day BOD and dissolved oxygen at location X.

(c) What conclusions can you draw from the results of this problem?

4.26. A dairy processing plant is planning to discharge wastewater at a rate of 5 m³/s into a river that has a 7Q10 discharge of 30 m³/s. Available data indicate that the river has a bottom width of 20 m, side slopes of 2:1 (H:V), depth of flow of 3 m at a discharge of 35 m³/s, average roughness height of 1.5 cm, dissolved-oxygen concentration of 8.5 mg/L, 5-day BOD of 5 mg/L, temperature of 22°C, and average annual wind speed over the river of 5 m/s. The milk processing waste will have negligible dissolved oxygen and be at approximately the same temperature as the river. Laboratory tests on the combined river water and wastewater indicate that the BOD decay factor of the mixed water is 0.15 day^{-1} at 22°C. The dairy plant is to be located in a state where the dissolved oxygen in the stream is required by law to be greater than or equal to 5.0 mg/L, and the boundary of the mixing zone is located 800 m downstream of the wastewater discharge.

(a) Calculate the required length of the outfall such that complete cross-sectional mixing occurs at approximately 100 m downstream of the outfall. Does significant decay in BOD occur during cross-sectional mixing?

(b) Using the calculated outfall length, estimate the dissolved oxygen in the river 100 m downstream of the outfall, and find the maximum 5-day BOD of the dairy wastewater that will cause the river to meet the dissolved-oxygen criteria on the boundary of the mixing zone.

4.27. Treated domestic wastewater is discharged into a river with an average stream velocity of 20 cm/s and an estimated reaeration coefficient of 0.4 day^{-1}. The TKN of the mixed wastewater discharge is 0.9 mg/L and the nitrification rate constant is estimated to be 0.5 day^{-1}.

(a) Estimate the oxygen deficit due to nitrification 10 km downstream of the wastewater discharge.

(b) If the theoretical oxygen deficit neglecting nitrification is 5.5 mg/L, what is the percent error incurred by neglecting nitrification?

4.28. Repeat Problem 4.22 accounting for a respiration oxygen demand of 3 g/m$^2 \cdot$ day and a benthic oxygen demand of 5 g/m$^2 \cdot$ day. Determine the initial dissolved oxygen deficit that will cause the critical oxygen level to occur at the outfall.

4.29. A section of a slow-moving river has diurnal dissolved-oxygen variations where the minimum dissolved oxygen deficit occurs 3 h after solar noon, dissolved oxygen that varies over a range of 7.5 mg/L, and a daily mean oxygen deficit of 1.5 mg/L. If the photoperiod is 12 h, use the delta method to estimate the reaeration constant, mean photosynthesis rate, and mean respiration rate. How do these results compare with those obtained using the approximate delta method?

4.30. A waste effluent is discharged into a river that has a mean velocity of 5 cm/s and a temperature of 20°C. After initial mixing across the river, the dissolved-oxygen concentration of the mixed river water is 5 mg/L, the ultimate BOD is 25 mg/L, the laboratory rate constant for BOD decay in the river water is 0.4 day^{-1}, and the reaeration constant is estimated to be 0.8 day^{-1}. Lateral distributed BOD sources extend 5 km downstream of the discharge location, and the distributed BOD sources have an average ultimate BOD of 4 mg/L. BOD removal by sedimentation is estimated to increase the BOD decay rate by 15%, and adsorption by benthic slimes decreases the BOD decay rate by 10%. Assess the impact of the distributed BOD input on the oxygen concentration 5 km downstream of the discharge location.

4.31. A waste effluent is discharged into a river that has a mean velocity of 6 cm/s. Analysis of the waste discharge and distributed BOD input over a 1-km section of the river indicates that the dissolved oxygen 1 km downstream of the waste outfall is expected to be 8 mg/L, where it is assumed that the reaeration constant is equal to 0.8 day^{-1} and the average temperature of the stream is 20°C. Photosynthesis within the river section is expected to generate 7 mg/L · day, plant respiration 4 mg/L · day, and sediment oxygen demand 2 mg/L · day. Estimate the fluctuation in dissolved oxygen expected 1 km downstream of the outfall.

4.32. A wastewater is discharged into a large tidal river that has a mean velocity of 5 cm/s, a reaeration rate constant of 0.6 day^{-1}, and a longitudinal dispersion coefficient of 30 m^2/s. After initial mixing of the wastewater, the river has an ultimate BOD of 15 mg/L and a BOD decay constant of 0.3 day^{-1}. Account for longitudinal dispersion in determining the oxygen deficit 1 km downstream of the outfall. Assess whether it is important to consider longitudinal dispersion in this case.

CHAPTER 5

LAKES AND RESERVOIRS

5.1 INTRODUCTION

Lakes represent a large proportion of the world's readily available water supply, and the importance of lakes in a given region depends partly on their numbers and distribution. In Scandinavia, for example, lakes occupy almost 10% of the total land area, whereas lakes occupy less than 1% of the total land area in China and Argentina. It is estimated that there are about 3 to 6 million natural lakes in the world, covering an area of about 1.6 to 3 million square kilometers (Jørgensen et al., 2005). The study of lakes is an area of geophysics called *limnology*.

Lakes are large reservoirs of water in which currents are driven primarily by wind. Other factors that influence the distribution of currents in lakes include the bathymetry, density distribution, and inflow and outflow characteristics. A typical lake view is shown in Figure 5.1. Lakes that intersect the ground-water table and interact significantly with the ground water are called *seepage lakes*; the water levels in these lakes fluctuate as the ground-water table fluctuates. Lakes fed primarily by inflowing streams are called *drainage lakes*. Sources of water input to lakes in general include precipitation, stream inflow, and ground-water inflow; outflows from lakes include evaporation, outflow to ground water, and artificial withdrawals for such uses as water supply and irrigation. The quantification of ground-water flow into lakes is still an evolving area with a significant amount of uncertainty, and conventional practice is to use seepage meters to measure ground-water inflow to lakes (Schneider et al., 2005).

Freshwater systems are classified as *lentic* (standing) or *lotic* (flowing). Lentic systems such as lakes, reservoirs, and ponds are more susceptible to pollution than lotic systems because they act as sinks, retaining pollutants; lotic systems such as streams and rivers have more tendency to flush pollutants downstream. More specifically, lakes and reservoirs differ

FIGURE 5.1   Typical lake view. (From U.S. National Park Service, 2005b.)

from rivers and streams in several important ways (James, 1993): (1) lakes rarely receive discharges of organic matter large enough to cause serious oxygen depletion; (2) lakes have significantly longer retention times than most rivers; and (3) the principal water-quality gradients are in the vertical direction rather than in the longitudinal direction. *Ponds* are typically smaller and shallower than lakes, with sunlight typically penetrating to the bottom of a pond, where plants requiring photosynthesis can grow and the water temperature throughout the water column varies little.

Natural processes responsible for lake formation include tectonism, volcanism, landslides, glaciation, fluvial processes, and meteorites. Natural lakes occur most often in glaciated regions. Reservoirs are usually dammed rivers or stream valleys fed by a major tributary. A key operational difference between natural lakes and (human-made) reservoirs is that natural lakes tend to have uncontrolled outflows, whereas reservoirs have controlled outflows. Elongated and dendritic shapes are typical of reservoirs created by damming rivers, whereas natural lakes tend to be more circular. Lakes and reservoirs are both *impoundments*, and there is usually more water-level fluctuation and better mixing in reservoirs than in lakes. *Impounded water* (versus running water) is sometimes defined as a body of water having a detention time greater than or equal to 14 days (DeBarry, 2004). Pollutant loads to reservoirs are usually greater than for lakes located in drainage basins with similar land uses. This is due primarily to the fact that the drainage basins of reservoirs are generally larger than those of lakes. In a sample of lakes and reservoirs in the United States, the ratio of drainage basin to water body area for reservoirs was on average 14 times higher than for lakes. Lakes and reservoirs have a variety of uses, including recreation, water supply, hydropower, and flood control. Many features of natural lakes and reservoirs are similar, and the approaches for their use and management are very similar (Jørgensen et al., 2005).

Inland freshwater lakes and reservoirs provide the United States with 70% of its drinking water and supply water for industry, irrigation, and hydropower. Lake ecosystems support complex and important food web interactions and provide habitat needed to support numerous threatened and endangered species. In the United States, the most frequently reported pollutants affecting lakes, reservoirs, and ponds are nutrients, metals, and siltation, with major pollutant sources being agriculture, hydromodification, urban runoff, and storm sewers (USEPA, 1996c; USEPA, 2000a). Nutrients, primarily nitrogen and phosphorus, contribute to increased lake biomass, sometimes to undesirable levels. Significant nutrient sources include agricultural runoff, industrial and municipal wastewater discharges, and atmospheric deposition. Most reports of metal contamination in lakes are a result of the detection of mercury in fish tissue (Nazaroff and Alvarez-Cohen, 2001), and the major source of mercury contamination in lakes is thought to be atmospheric transport and deposition from electrical power plants. As an indication of the scope of the lake-pollution problem, 96% of the Great Lakes shoreline has been reported as impaired, due primarily to pollutants in fish tissue at levels that exceed standards to protect human health. Over geologic time, lakes naturally fill with sediment, which tends to deposit concentrically from the outer edges (where the velocities are the lowest) toward the middle.

Water-pollution problems in impoundments such as lakes and reservoirs can be quite persistent because of the long *hydraulic detention times* in these systems. The hydraulic detention time, t_d, is defined by the relation

$$t_d = \frac{V_L}{Q_o} \tag{5.1}$$

where V_L is the (average) volume of the impoundment and Q_o is the average outflow rate. The hydraulic detention time is sometimes called the *retention time* or *residence time*. To give an idea of typical time scales, detention times in large lakes are typically on the order of years, while the detention times for water in rivers and streams are typically on the order of days (Baumgartner, 1996). Detention times in lakes and reservoirs are typically classified as *short* if less than one year and *long* if they are more than one year. In lakes or reservoirs with short detention times, the degree of stratification depends on the detention time, whereas for long detention times (>1 year) stratification can be fully developed and independent of the detention time. In cases of *very short* detention time (<14 days), stratification does not develop (Jørgensen et al., 2005).

Example 5.1   The volumes and outflow rates for the Great Lakes are shown in Table 5.1. Determine the detention times for the Great Lakes.

SOLUTION   Using Equation 5.1 yields the result shown in Table 5.2. Based on these results, the detention times in the Great Lakes range from a low of 3 years in Lake Erie to a high of 179 years in Lake Superior.

TABLE 5.1   Data for Example 5.1

Lake	Volume (10^9 m³)	Outflow (10^9 m³/yr)
Superior	12,000	67
Michigan	4,900	36
Huron	3,500	161
Ontario	1,634	211
Erie	468	182

Source: Chapra and Reckhow, (1983).

TABLE 5.2 Results for Example 5.1

Lake	V_L (10^9 m^3)	Q_o (10^9 m^3/yr)	t_d (years)
Superior	12,000	67	179
Michigan	4,900	36	136
Huron	3,500	161	22
Ontario	1,634	211	8
Erie	468	182	3

Taking the surface area, A_L, of a lake as a measure of its volume, V_L, and taking the drainage area, A_D, as a measure of the average inflow and outflow, Q_o, from the lake, the detention time, t_d (days), of a natural lake can be estimated by the empirical relation (Bartsch and Gakstatter, 1978)

$$\log_{10} t_d = 4.077 - 1.177 \log_{10} \frac{A_D}{A_L}, \qquad 1 \text{ day} < t_d < 6000 \text{ days} \tag{5.2}$$

This equation applies only to natural lakes with uncontrolled discharges. It is important to note that long detention times do not necessarily correspond to large lakes, since small lakes with small outflow rates can have detention times comparable to those of large lakes with large outflow rates. Most natural lakes are fed by one or more streams and emptied by one outflow channel (Ramaswami et al., 2005).

Example 5.2 A natural lake has an estimated volume of 9×10^5 m^3, a surface area of 85,000 m^2, and an average uncontrolled outflow rate of 310 m^3/day. Estimate the hydraulic detention time in the lake and the drainage area that contributes runoff into the lake.

SOLUTION   From the data given, $V_L = 9 \times 10^5$ m^3, $Q_o = 310$ m^3/day, and Equation 5.1 gives the hydraulic detention time, t_d, as

$$t_d = \frac{V_L}{Q_o} = \frac{9 \times 10^5}{310} = 2900 \text{ days} = 8.0 \text{ years}$$

The detention time can be related empirically to the drainage area, A_D, by Equation 5.2. Taking $A_L = 85,000$ m^2, Equation 5.2 gives

$$\log_{10} t_d = 4.077 - 1.177 \log_{10} \frac{A_D}{A_L}$$

or

$$\log_{10} 2900 = 4.077 - 1.177 \log_{10} \frac{A_D}{85,000}$$

which leads to

$$A_D = 2.8 \times 10^5 \text{ m}^2$$

Hence, the estimated drainage area is about 3.3 times the size of the lake.

The reciprocal of the residence time is called the *flushing rate*. Impoundments with longer residence times (in months and years) typically have water-quality problems such as excessive biological productivity. The minimum residence time required for algae growth and development is at least several weeks (Novotny, 2003).

With low velocities and long detention times, most lake environments are favorable to sedimentation, resulting in most of the incoming sediments, and many of the organisms that grow and die in the lake, accumulating at the bottom of the lake. Over extended periods of time, sedimentation can change the character of a lake permanently, greatly increasing its organic content and ultimately converting it to a silted pond, swamp, marsh, or other type of wetland (Lamb, 1985; Novotny, 2003).

The *hydraulic loading* and *shape factor* are two parameters that are useful indicators of the biological productivity potential of lakes. Hydraulic loading, Q_s (in m/yr), is defined by the relation

$$Q_s = \frac{Q}{A} \tag{5.3}$$

where Q is the annual inflow to the lake (m^3/yr) and A is the surface area of the lake (m^2). The biological productivity potential of an impoundment is inversely proportional to the hydraulic loading, Q_s. The shape factor is defined as the length of the impoundment divided by its width. Elongated valley reservoirs (shape factors $\gg 1$) are less amenable to excessive biological productivity than are circular open lakes (shape factor ≈ 1).

The shallow water near the shore of an impoundment in which rooted (emergent) water plants (macrophytes) can grow is called the *littoral zone*. Littoral zones in lakes and reservoirs are essential for spawning and fish development, and therefore desirable lake depths should not be uniform and the bottom relief should provide a variety of landscapes. Deeper oxygenated zones are used for escape from summer warmer temperatures in the littoral zone. The extent of the littoral zone depends on the slope of the lake bottom.

5.2   NATURAL PROCESSES

Natural processes that typically dominate the water quality of lakes and (human-made) reservoirs are the same, and to facilitate discussion, these processes will be covered in the context of lakes but apply equally well to reservoirs. Several of the major processes affecting the water quality in lakes are described in the following sections.

5.2.1 Flow and Dispersion

Water movement in lakes influence the distribution of nutrients, microorganisms, and plankton and therefore affects biological productivity and the biota. Lake currents are driven primarily by wind, inflow/outflow, and the Coriolis force. For small shallow lakes, particularly long and narrow lakes, inflow/outflow characteristics are most important, and the predominant current is a steady-state flow through the lake. For very large lakes, wind is the primary generator of currents, and except for local effects, inflow/outflow have a relatively minor effect on lake circulation. The Coriolis effect, a deflecting force that is a function of the Earth's rotation, also plays a role in circulation in large lakes such as the Great Lakes.

Typical wind-induced circulation regimes in shallow and deep lakes are illustrated in Figure 5.2, where the main difference is that shallow lakes tend to have a single circulation

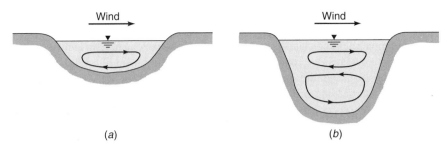

FIGURE 5.2 Wind-induced circulation in lakes: (*a*) shallow lake; (*b*) deep lake. (From Chin, David A., *Water-Resources Engineering.* Copyright © 2000. Reprinted by permission of Pearson Education, Inc., Upper Saddle River, NJ.)

cell, whereas deeper lakes tend to have more than one circulation cell. Wind-induced surface currents, called *wind drift*, are typically on the order of 2 to 3% of the wind speed. Lakes are typically classified as *shallow* if they are less than 7 to 10 m in depth, and *deep* if they are more than 10 m in depth; in some cases, deep lakes are defined as those with depths greater than 5 m (Novotny, 2003). However, the circulation pattern in a lake depends significantly on the surface area of the lake, since lakes with larger surface areas experience greater wind force and have a greater tendency to form a single circulation cell. Consequently, it is the ratio of the water body surface area to its depth that dictates whether a water body is shallow or deep.

Lake Okeechobee in Florida is an example of a large shallow lake where the currents are primarily wind driven. Lake Okeechobee has a surface area of approximately 1900 km^2 and an average depth of about 3 m. Chen and Sheng (2005) have demonstrated that wind-generated flows in the lake are sometimes sufficient to stir up a significant amount of bottom sediment. The phosphorus desorbed from the increased suspended sediment in the water column contributes significantly to phosphorus concentrations in the lake, which typically range from 50 to 100 µg/L (Schelske, 1989). This implication of this result is that adequate prediction of the phosphorus concentrations in Lake Okeechobee requires a circulation (hydrodynamic) model combined with both a sediment-transport model to describe the sediment fluidization and resuspension process and a water-quality model to describe the relationship between adsorbed and dissolved phosphorus. If Lake Okeechobee were a deep lake, wind-induced currents would probably not cause significant sediment suspension, and a simpler description of the fate and transport of phosphorus would be possible.

5.2.2 Light Penetration

Transmission of light through the water column influences primary productivity (growth of phytoplankton and macrophytes), distribution of organisms, and behavior of fish. The reduction of light penetration through the water column of a lake is a function of scattering and absorption. Light transmission is affected by the water-surface film, floatable and suspended particulates, turbidity, populations of algae and bacteria, and color. In a typical clear lake, 50% of the incident sunlight is absorbed in the upper 2 m, and very little light energy penetrates more than 10 m below the water surface (Wetzel, 1975).

An important measure based on the transmission of light is the depth to which photosynthetic activity is possible. The minimum light intensity required for photosynthesis has been established to be about 1% of the incident surface light (Cole, 1979). The portion of the lake from the surface to the depth at which the 1% intensity occurs is called the *euphotic zone*, the depth at which net photosynthesis is equal to zero is called the *compensation depth*, *compensation point*, or *compensation limit*. The lower limit of the euphotic zone occurs at approximately this location. Below the euphotic zone is the *aphotic zone*, where light penetration is negligible. Benthic plants do not exist in the aphotic zone (due to lack of light), and many lakes are sufficiently deep or turbid to prevent the development of benthic plants except in the immediate vicinity of the shoreline. If nutrients are abundant in the aphotic zone, production in the euphotic zone will subsequently increase if aphotic-zone water is mixed with euphotic-zone water, a process that occurs regularly in all aquatic systems (Laws, 2000).

5.2.3 Sedimentation

Deposition of sediment received from the surrounding watershed is an important physical process in lakes. Because of the low water velocities in lakes, sediments transported by inflowing waters tend to settle out. Sediment accumulation rates are strongly dependent both on the physiographic characteristics of the lake watershed and on various other characteristics of the lake. An extreme example of the result of excessive sedimentation is Kingman Lake shown in Figure 5.3. Kingman Lake is located in the District of Columbia. In general, prediction of sedimentation rates can be estimated by either periodic sediment surveys or estimation of watershed erosion and bed load. Accumulation of sediment in lakes can, over many years, reduce the life of the water body by reducing the water-storage capacity. Sediment flow into a lake also reduces light penetration, eliminates bottom habitat for many plants and animals, and carries with it absorbed chemicals and organic matter that settles to the bottom and can be harmful to the ecology of the lake. Where sediment accumulation is a major problem, proper watershed management including erosion and sediment control must be put into effect.

5.2.4 Eutrophication and Nutrient Recycling

Primary production refers to the photosynthetic generation of organic matter by algae, plants, and certain bacteria; and *secondary production* refers to the generation of organic matter by nonphotosynthetic organisms that consume the organic matter originating from

FIGURE 5.3   Kingman Lake. (From USEPA, 2005h.)

primary producers. Based on the level of (biological) productivity, water bodies such as lakes and rivers can be classified in terms of their trophic state as *oligotrophic* (poorly nourished), *mesotrophic* (moderately nourished), *eutrophic* (well nourished), and *hypereutrophic* (overnourished). These classes of lakes are described as follows:

1. *Oligotrophic lakes* have low biological productivity and are characterized by low algal concentrations and high water clarity. The water is clear enough so that the bottom can be seen at considerable depths. Lake Tahoe on the California–Nevada border, Crater Lake in Oregon, and the blue waters of Lake Superior are classic examples of oligotrophic lakes.

2. *Mesotrophic lakes* are intermediate between oligotrophic and eutrophic lakes. Although substantial depletion of oxygen may occur in the lake due to plant respiration and decomposition, the lake water remains aerobic. A mesotrophic lake condition is preferred for recreational, water-quality, and game-fishing reasons (DeBarry, 2004). Lake Ontario, Ice Lake in Minnesota, and Grindstone Lake in Minnesota are all examples of mesotrophic lakes.

3. *Eutrophic lakes* have high productivity because of an abundant supply of nutrients. Eutrophic lakes typically have undesirable high algal concentrations. Highly eutrophic lakes may also have large mats of floating algae that impart unpleasant tastes and odors to the water. As algae complete their life cycle and die off, their decomposition by bacteria and other organisms consumes oxygen and produces odors. The reduction in dissolved oxygen can be sufficient to cause fish kills. Lake Okeechobee in Florida, Halsted Bay of Lake Minnetonka in Minnesota, the Neuse River in North Carolina, and Lake Erie are all examples of eutrophic waters.

4. *Hypereutrophic lakes* are extremely eutrophic, with a high algal productivity level and intense algal blooms. They are often relatively shallow lakes with much accumulated organic sediment. They have extensive dense weed beds and often accumulations of filamentous algae. Recreational use of the waters in hypereutrophic lakes is often impaired. Examples of hypereutrophic lakes are Onondaga Lake in New York and Upper Klamath Lake in Oregon.

There is a natural progression from the oligotrophic state through the eutrophic state as part of the normal aging process that results from the recycling and accumulation of nutrients over a long period of time, typically over a time scale of centuries (Stefan, 1994). For example, nitrogen added to a lake is assimilated by algae; when the algae die, the bulk of the assimilated nitrogen is released and is available for assimilation by living algae, in addition to new nitrogen that is being added to the water body. Hence the nitrogen accumulates in the lake and increases the nourishment level. This natural aging process can be accelerated by several orders of magnitude as a result of large population densities and a predominance of agricultural land use in the lake catchment area. However, left untouched, many oligotrophic lakes have remained such since the last ice age (Davis and Masten, 2004). The process by which lakes become eutrophic is called *eutrophication*, and when this process is accelerated by input of organic wastes and/or nutrients from anthropogenic (human) sources, the process is called *cultural eutrophication*.

Eutrophication can have a number of deleterious effects, including (1) the excessive growth of floating plants that decrease water clarity, clog filters at water-treatment plants, and create odors; (2) significant fluctuations in oxygen and carbon dioxide levels associated with photosynthesis and respiration (in the euphotic zone), where low oxygen levels can cause the death of desirable fish species; (3) an increased sediment oxygen demand

FIGURE 5.4 Algae mat. (From Aber, 2002.)

(SOD) associated with the settling of aquatic plants, resulting in low dissolved oxygen levels near the bottom of the water body; and (4) loss of diversity in aquatic ecosystems. An example of an algae mat in a eutrophic lake is shown in Figure 5.4. Extensive mats of this type are readily visible on satellite images. Eutrophication is not synonymous with pollution; however, pollution can accelerate the rate of eutrophication.

A general trend that results from eutrophication is an increase in the numbers of organisms but a decrease in diversity of species, particularly among nonmotile species. Species associated with eutrophic systems are sometimes less desirable than species characteristic of oligotrophic systems. Cyanobacteria, which are frequently associated with organic nutrient enrichment, are a class of organisms frequently associated with undesirable water-quality conditions. Large-scale fish kills and the elimination of desirable species as a result of oxygen depletion may constitute a serious eutrophication consequence in some aquatic systems. High-elevation and high-latitude lakes can sustain only a few tolerant species of organisms and fish, and due to the cold water temperatures, they may not be desirable for primary contact recreation. Due to their aesthetic and natural values and due to the high sensitivity of the resident aquatic biota, such water bodies require a high degree of protection.

Lakes that receive a major portion of their nutrients from internal sources are called *autotrophic*, and those that receive a major portion of their nutrients from external sources are called *allotrophic*. In related terminology, *allochthonous* nutrients are those originating from the watershed contributing inflow to the lake, usually from nonpoint sources, while *autochthonous* nutrient sources include nutrients stored in the lake water and sediments.

Aquatic plants, including algae, that grow in surface waters can be broadly classified according to whether they move freely in the water or remain fixed in place, attached or rooted. Plants that move freely in water are called *phytoplankton* (e.g., free-floating algae); attached plants include *periphyton* (e.g., attached or benthic algae) and *macrophytes* (rooted, vascular aquatic plants). The types of fixed plant communities are varied and depend on the depth and clarity of the water. The level of eutrophication in lakes and reservoirs is usually measured by the amount of phytoplankton per unit volume of water, with commonly used measures including (1) total dry weight (g/L), (2) carbon contained in

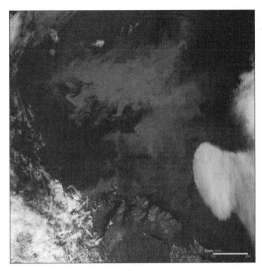

FIGURE 5.5   Satellite view of an algae bloom. (From NASA, 2002.)

phytoplankton (mg/L), (3) chlorophyll *a* contained in phytoplankton (μg/L), or (4) oxygen demand necessary to decompose the phytoplankton (mg O_2/L). Chlorophyll *a*, one of the green pigments involved in photosynthesis, is found in all algae, so it is used to distinguish the mass of algae in the water from other organic material such as bacteria. Chlorophyll *a* is a good indicator of algal concentrations and of nutrient enrichment. Excessive phytoplankton concentrations, as indicated by high chlorophyll *a* levels, cause adverse dissolved oxygen (DO) impacts such as wide diurnal variation in surface DO due to daytime photosynthesis and nighttime respiration and depletion of bottom DO through the decomposition of dead algae. Chlorophyll *a* and organic carbon are the most commonly used measures of plant concentration in lakes and reservoirs. When the concentration of phytoplankton algae during the late summer period exceeds a certain threshold nuisance value, the situation is called an *algae bloom*. A satellite view of an algae bloom resulting from excessive nutrients in surface runoff entering a lake is shown in Figure 5.5, where the algae bloom corresponds to the lighter area.

At least 19 elements are essential for life, and these basic elements are collectively called *nutrients*. Five of these nutrients are required in large amounts: carbon, hydrogen, oxygen, nitrogen, and phosphorus. The first three (C, H, O) are readily available in either water (H_2O) or dissolved carbon dioxide (CO_2) and are never long-term limiting factors for aquatic plant growth. However, the concentrations of nitrogen and phosphorus dissolved in natural waters are much lower, and it is usually one of these two elements that provides the limiting factor for aquatic plant growth. Consequently, nitrogen and/or phosphorus are usually considered responsible for eutrophication in natural water bodies. According to Liebig's law of the minimum, growth is generally limited by the essential nutrient that is in lowest supply. If all nutrients are available in adequate supply, massive algal and macrophyte blooms may occur, with severe consequences for the lake. Most commonly in lakes, phosphorus is the limiting nutrient for aquatic plant growth (Davis and Masten, 2004). In these situations, adequate control of phosphorus, particularly from anthropogenic sources, can control the growth of nuisance aquatic vegetation.

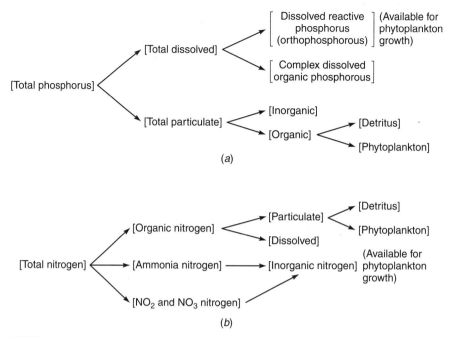

FIGURE 5.6   Components of total phosphorus and total nitrogen. (From Chin, David A., *Water-Resources Engineering.* Copyright © 2000. Reprinted by permission of Pearson Education, Inc., Upper Saddle River, NJ.)

Nitrogen and phosphorus are discharged into surface waters primarily via agricultural and urban runoff, municipal and industrial discharges, and combined sewer overflows. Phosphorus and nitrogen usually occur in a variety of combined forms, and it is common to quantify their concentrations in terms of *total phosphorus* (TP) and *total nitrogen* (TN). The combined forms of phosphorus and nitrogen that contribute to TP and TN are illustrated in Figure 5.6. Since not all components of TP and TN are available for uptake by phytoplankton, it is important to delineate the components that are present in the water. The two principal components of TP are the dissolved and particulate forms. Dissolved phosphorus includes several forms, of which dissolved reactive phosphorus (orthophosphate, PO_4^{3-}) is available for phytoplankton growth. Available phosphorus is commonly taken to be orthophosphate that passes a 0.45-μm membrane filter (Schnoor, 1996). The principal components of TN are organic, ammonia, nitrite (NO_2), and nitrate (NO_3) forms, of which ammonia, nitrite, and nitrate forms are utilized by phytoplankton for growth.

The design of systems to control eutrophication are based on identifying the limiting nutrient and determining the allowable nutrient levels to maintain a desirable concentration of plant biomass in the water. The limiting nutrient can be identified by comparing the ratio of nitrogen to phosphorus (N/P) in the water that is available for plant growth to the ratio of nitrogen to phosphorus required for plant growth. A consideration of the cell stoichiometry of phytoplankton shows a phosphorus content of 0.5 to 2.0 μg/μg chlorophyll *a* and a nitrogen content of 7 to 10 μg/μg chlorophyll *a*, indicating that a N/P ratio on the order of 10 is utilized in plant growth. Hence, if the ratio of available nitrogen to available phosphorus in the water significantly exceeds 10, phosphorus would be the limiting nutrient, whereas if the N/P ratio in the water is significantly less than 10, nitrogen would be the limiting nutrient. A relationship developed by Stumm and Morgan (1962) indicated that 1 atom of phosphorus and 16 atoms of nitrogen are needed to generate 154 molecules of oxygen during photosynthesis (this ratio of nitrogen to phosphorus is called the *Redfield ratio*), and Ryther and Dunstan (1971) reported that N/P ratio requirements for the growth of microalgae are in the range 3 to 30, with an average of 16. An alternative approach to determining whether nitrogen or phosphorus is the limiting nutrient is to plot the nitrogen concentrations versus phosphorus concentrations on an arithmetic plot. The straight line of best fit will intercept either the nitrogen or phosphorus abscissa, and the nutrient that is exhausted first (determined by the intercept on the abscissa) is the limiting nutrient. In a related approach, some studies have used the correlation of phosphorus and nitrogen levels to chlorophyll *a* concentrations to identify the limiting nutrient (Gin and Neo, 2005). The limiting nutrient is the one that is most correlated with chlorophyll *a*.

The control of eutrophication usually requires the specification of target biomass concentrations in terms of μg chlorophyll *a*/L (μg Chl*a*/L); typical target biomass concentrations in northern temperate lakes are in the range 1 to 4 μg Chl*a*/L for oligotrophic lakes and 5 to 10 μg Chl*a*/L for eutrophic lakes. Many lake models assume a priori that phosphorus is the limiting nutrient, in which case the biomass (chlorophyll *a*) concentration, c_b, in a water body can be estimated based on the TP concentration. This approach implicitly assumes a stable relationship between TP and available phosphorus (orthophosphate, PO_4^{3-}) and neglects other variables that affect algal growth, such as sedimentation, predation, nutrient recycling, and oxygen fluxes. Several empirical relations between c_b and TP are listed in Table 5.3, where c_b and TP are in μg/L. The variety of formulas in Table 5.3 reflect the seasonal, climatic, ecological, and hydrologic variations between lakes. The relations given in Table 5.3 should be used with caution since they are probably not valid for all values of TP. According to Jørgensen et al. (2005), with increasing concentrations of TP, the algal biomass first increases slowly, and then more rapidly, followed by an almost linear rise until some asymptotic value of biomass is reached at high concentrations of TP. A consequence of this nonlinear relation is that when the concentrations in a lake are very high prior to external phosphorus load reductions, even a very significant decrease in the phosphorus concentration does not necessarily result in a significant decrease in the chlorophyll *a* concentration. It has been estimated that the phosphorus concentration should be below 10 to 15 μg/L to limit algal blooms (Vollenweider, 1975). Total phosphorus concentrations of 10 and 20 μg/L are generally accepted as the boundaries between oligotrophy and mesotrophy and eutrophy, respectively.

In municipal wastewaters with activated-sludge treatment (without phosphorus removal), nitrogen tends to be the limiting nutrient; in mixed agricultural and urban runoff, phosphorus tends to be the limiting nutrient. The limiting nutrients that are typical for

TABLE 5.3 Empirical Relations Between Biomass and TP Concentrations

Formula	Reference
$\log_{10} c_b = 1.55 \log_{10} \text{TP} - 1.55 \log_{10} \dfrac{6.404}{0.0204\,(\text{TN/TP}) + 0.334}$	Smith and Shapiro (1981)
$\log_{10} c_b = 0.807 \log_{10} \text{TP} - 0.194$	Bartsch and Gakstatter (1978)
$\log_{10} c_b = 0.76 \log_{10} \text{TP} - 0.259$	Rast and Lee (1978)
$\log_{10} c_b = 1.449 \log_{10} \text{TP} - 1.136$	Dillon and Rigler (1974)

TABLE 5.4   Limiting Nutrients for Various Water Bodies

Nutrient Source	Typical N/P	Limiting Nutrient
Rivers and Streams		
Point-source dominated		
Without phosphorus removal	$\ll 10$	Nitrogen
With phosphorus removal	$\gg 10$	Phosphorus
Nonpoint-source dominated	$\gg 10$	Phosphorus
Lakes		
Large, nonpoint-source dominated	$\gg 10$	Phosphorus
Small, point-source dominated	$\ll 10$	Nitrogen

Source: Thomann and Mueller (1987).

TABLE 5.5   Ecological Zones in Lakes

Ecological Zone	Description
Marginal	The area immediately surrounding where land meets water
Littoral	The area from the shoreline lakeward to where rooted plants can no longer be supported
Pelagic	The zone of open water from the littoral zone to the center of the lake
Euphotic	The zone from the lake surface down to where light penetration decreases to 1 to 10% of near-surface light
Profundal	The zone of water and sediment occurring near the bottom of the lake below the euphotic zone
Benthic	The bottom stratum of the lake

various water bodies are shown in Table 5.4. According to the U.S. National Eutrophication Survey (USEPA, 1975), phosphorus is the limiting nutrient in most inland waters of the United States, where approximately 90% of the lakes studied showed phosphorus as the limiting nutrient. Studies of stormwater detention ponds have also shown phosphorus to be the limiting nutrient (Cullum, 1984). In marine systems, nitrogen is commonly found to be the limiting nutrient (Laws, 2000). More detailed analyses of the impact of nutrients on lake ecosystems generally take into consideration the various ecological zones listed in Table 5.5 and illustrated in Figure 5.7.

Example 5.3   (a) Water-quality measurements in a lake indicate that available phosphorus is on the order of 60 μg/L, and available nitrogen is about 2 mg/L. Is the lake nitrogen or phosphorus limited? (b) If the concentration of total phosphorus (TP) is 90 μg/L and the concentration of total nitrogen (TN) is 4 mg/L, estimate the biomass concentration and trophic state of the lake.

SOLUTION   (a) The ratio of available nitrogen to available phosphorus is $2 \times 10^3/60 = 33$. Since this ratio is greater than 10, the lake is phosphorus limited.

(b) From the data given, TP = 90 μg/L and TN = 4000 μg/L, and using the formulas in Table 5.1 we have the results shown in Table 5.6. Hence, the biomass concentration is

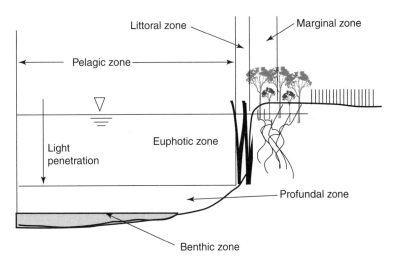

FIGURE 5.7   Ecological zones in lakes and reservoirs.

estimated to be in the range 17 to 52 μg/L. Based on this biomass concentration ($>10\,\mu$g/L) and the concentration of total phosphorus (>20 μg/L), the lake can be classified as eutrophic.

Measures of Eutrophication   It is important to note that eutrophication is not measured in a single, unique, unambiguous way (Stefan, 1994). Current methods measure symptoms (e.g., phytoplankton concentrations, μg Chl$a$/L) or causes (e.g., total phosphorus concentrations). Typical symptomatic criteria state that oligotrophic water bodies have chlorophyll a concentrations less 4 μg Chl$a$/L, mesotrophic water bodies have 4 to 10 μg Chl$a$/L, and eutrophic water bodies have more than 10 μg Chl$a$/L. Typical causal criteria are that oligotrophic water bodies have less than 10 μg TP/L, mesotrophic water bodies have TP concentrations in the range 10 to 20 μg TP/L, and eutrophic water bodies have more than 20 μg TP/L. Investigations of eutrophication problems in Wisconsin noted that algal blooms occurred when the concentration of inorganic nitrogen (NH_4^+, N_2^-, and NO^{2-}) exceeded 0.3 mg/L and the concentration of inorganic phosphorus exceeded 10 μg/L (Sawyer, 1947). It is important to keep in mind that algae uptake of nutrients is greatest during the productive summer period, causing lower nutrient concentrations; therefore, the critical nutrient concentrations should be measured during the winter or spring season, or whenever lake overturning occurs.

TABLE 5.6   Results for Example 5.3

Formula	c_b (μg/L)
Smith and Shapiro (1981)	52
Bartsch and Gakstatter (1978)	24
Rast and Lee (1978)	17
Dillon and Rigler (1974)	50

FIGURE 5.8   Secchi disk. (Courtesy of Bob Carlson.)

Secchi depth is frequently used as an indicator of algal abundance and general lake productivity. A *Secchi disk* is a circular plate divided into quarters painted alternately black and white. The disk is attached to a rope and lowered into the water until it is no longer visible; the depth of the Secchi disk at this point is the Secchi depth. An illustration of a Secchi disk being deployed is shown in Figure 5.8. Clarity is affected by algae, sediment particles, and other materials suspended in the water, and the *Secchi depth* is a measure of water clarity. Higher Secchi readings indicate clear water, and lower readings indicate turbid or colored water. Clear water lets light penetrate more deeply into the lake, and this light allows photosynthesis to occur and oxygen to be produced. A rule of thumb is that light can penetrate to a depth of 1.7 times the Secchi depth. Although it is only an indicator, Secchi depth is the simplest and one of the most effective tools for estimating lake productivity. For lakes that are phosphorus limited, the Secchi depth can be used to assess the trophic status using the *trophic status index* (TSI), defined by (Carlson, 1977)

$$\text{TSI} = 60 - 14.43 \ln(\text{SD}) \tag{5.4}$$

where SD is the Secchi depth in meters. Using correlations between the chlorophyll *a* concentrations, total phosphorus, and Secchi depth, the TSI can also be estimated by the relations

$$\text{TSI} = \begin{cases} 30.56 + 9.81 \ln(\text{Chl}a) & (5.5) \\ 4.14 + 14.43 \ln(\text{TP}) & (5.6) \end{cases}$$

TABLE 5.7   Trophic Status of Lakes

Water Quality	Oligotrophic	Mesotrophic	Eutrophic	Source
Total P (μg/L)	<10	10–20	>20	USEPA (1974)
	<10	10–30	>30	Nürnberg (1996)
Cholorophyll a (μg/L)	<4	4–10	>10	USEPA (1974)
	<3.5	3.5–9	>9	Nürnberg (1996)
Secchi disk depth (m)	>4	2–4	<2	USEPA (1974), Nürnberg (1996)
Hypolimnetic oxygen (% saturation)	>80	10–80	<10	USEPA (1974)
Phytoplankton production (g org. C/m^2 · day)	7–25	75–250	350–700	Mason (1991)
Trophic status index	<40	35–45	>45	Carlson (1977)

where Chla is the concentration of chlorophyll a in μg/L and TP is the concentration of total phosphorus in μg/L. In assessing the efficacy of using Equations 5.4, 5.5, or 5.6 as indicators of trophic status, it is important to note that the best indicator of trophic status varies from lake to lake, and Secchi depth values may be erroneous in lakes where turbidity is caused by factors other than algae. It has been reported that the TSI works best as a trophic-state indicator in northern temperate lakes (Osgood, 1982) and performs poorly in lakes with excessive weed problems (North American Lake Management Society, 1990). Based on observations in several northern lakes, most oligotrophic lakes had a TSI below 40, mesotrophic lakes had a TSI between 35 and 45, most eutrophic lakes had a TSI greater than 45, and hypereutrophic lakes have a TSI greater than 60 (Sloey and Spangler, 1978; Krenkel and Novotny, 1980). In cases where the TSI is estimated from independent values of SD, Chla, and TP using Equations 5.4 to 5.6, respectively, the average TSI given by these three equations is used to measure the trophic state (Jørgensen et al., 2005).

A summary of the criteria used to assess the trophic status of a lake are given in Table 5.7. The lack of a precise definition of trophic status makes it difficult to develop an accurate engineering tool that would enable estimation of the stage of the eutrophication process of a given water body.

5.2.5 Thermal Stratification

Thermal stratification in lakes can have a pronounced effect on water quality, since temperature has a significant influence on the rates of chemical and biological reactions, and strong temperature gradients can significantly limit the diffusion of dissolved oxygen from the water surface to the bottom of a lake. Temperature and its distribution within lakes and reservoirs affect not only the water quality within the lake but also the thermal regime and quality of a river system downstream of the lake.

Three distinct classes of lakes are commonly identified: strongly stratified, weakly stratified, and nonstratified. Strongly stratified lakes are typically deep and characterized by horizontal isotherms, weakly stratified lakes are characterized by isotherms that are tilted along the longitudinal axis of the lake, and nonstratified lakes are characterized by isotherms that are essentially vertical, in which case the temperature distribution at any location is roughly uniform with depth. Density stratification in lakes is due primarily to temperature differences, although salinity and suspended solids concentrations may also affect density. A simple approximation for the relationship between density of water, ρ_w, temperature, T, and salinity, S, is given by (Cowley, 1968)

$$\rho_w = 1 + \{10^{-3}[(28.14 - 0.0735T - 0.00469T^2) + (0.802 - 0.002T)(S - 35)]\} \quad\text{g/cm}^3 \tag{5.7}$$

where T is in degrees Celsius and S is in parts per thousand. The aquatic community present in a lake is highly dependent on thermal structure. The fundamental processes that influence the thermal stratification of lakes are heat and momentum transfer across the lake surface and gravity forces acting on density differences within the lake.

In warm summer weather, heat added at the surface of deep ($>10\,$m) lakes is concentrated in the top few meters, resulting in a warm, less dense layer that is well mixed, overlaying a distinctly colder and weakly mixed lower layer. The well-mixed surface layer is called the *epilimnion*; the weakly mixed lower layer is called the *hypolimnion*. The warmer epilimnion is separated from the colder hypolimnion by a thin layer with a sharp temperature gradient called the *metalimnion* or *mesolimnion*. The sharp temperature gradient in the mesolimnion is called the *thermocline*, and in freshwater lakes, the thermocline is defined as having a minimum temperature gradient of 1°C/m. When a thermocline does not exist, the epilimnion and hypolimnion are not defined (French et al., 1999; Davis and Masten, 2004). In shallow lakes or shallow portions of deep lakes the thermocline eventually intercepts the lake bottom so that no hypolimnion exists. The turbidity of lake waters has a strong influence on the thickness of the epilimnion since surface heat attenuates rapidly in turbid waters. The depth of the epilimnion is related to the size of the lake, where it can be as shallow as 1 m in small lakes and as deep as 20 m or more in large lakes. Waters in the epilimnion tend to be well oxygenated, while waters in the hypolimnion tend to be low in oxygen.

In cool fall weather, the surface layers of deep lakes begin to cool and become more dense than the underlying water, leading to a gravity circulation, supplemented by the wind, that causes the lake waters to overturn and become better mixed. During winter, lakes are usually unstratified except at higher latitudes where further cooling of the lake below 4°C[1] under cold winter conditions causes the surface layers of the lake to become colder than 4°C, making them less dense than underlying water and the lake is again stratified. As the temperature warms during the spring, the surface waters warm to 4°C, becoming more dense than the underlying water and causing a turnover in the lake waters. As warm summer weather returns, the lake tends to again become stratified, and the seasonal cycle is complete. A typical example of the lake stratification cycle in a temperate lake is illustrated in Figure 5.9, where the lake is stably stratified in the summer and winter, with seasonal overturning in between.

Lakes can be classified on the basis of their annual pattern of overturning. These classifications are as follows:

1. *Amictic:* lakes that never overturn and are permanently covered with ice, found in the Antarctic and very high mountains.
2. *Holomictic:* lakes that mix from top to bottom as a result of wind-driven circulation. Several subcategories are defined:

---

[1]Recall that the maximum density of water occurs at 3.94°C.

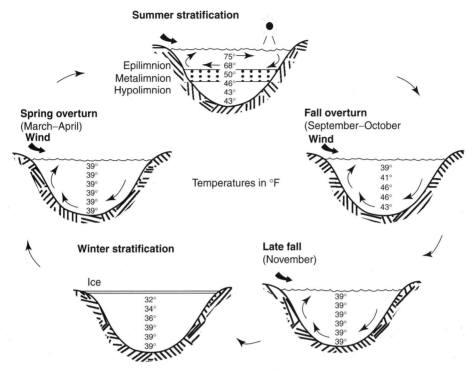

FIGURE 5.9   Lake stratification cycle. (From State of Wisconsin, 2005.)

a. *Oligomictic:* lakes characterized by overturning that is unusual, irregular, and short in duration; generally small to medium tropical lakes or very deep lakes.

b. *Monomictic:* lakes that undergo one regular overturn per year. Warm lakes in which the winter temperature never drops below 4°C may be monomictic, as may be cold lakes in which the summer water temperature never rises above 4°C. Lakes in tropical regions and as far north as 40° latitude are generally monomictic. Cold monomictic lakes are connected with high latitude and altitude and are frozen over most of the year.

c. *Dimictic:* lakes that overturn twice a year, in the spring and fall, one of the most common types of annual mixing in cool temperate regions such as central and eastern North America. Most lakes in temperate climates in which the summer water temperature is above 4°C and the winter temperature is below 4°C are dimictic.

d. *Polymictic:* lakes that circulate frequently or continuously, cold lakes that are continually near or slightly above 4°C, or warm equatorial lakes where air temperature changes very little.

3. *Meromictic:* lakes that do not circulate throughout the entire water column. The lower water stratum is perennially stagnant.

It is important to keep in mind that lakes in the tropics are characterized by a diurnal climate (day/night variability), compared with a seasonal climate that is characteristic of temperate zones. Consequently, polymictic lakes are found more widely in tropical regions. On a practical note, inaccurate conclusions are sometimes made from water-quality measurements taken in shallow tropical waters during morning hours when the water is mixed, as it may become stratified and anoxic during the day. Lake and reservoir water-quality monitoring in tropical regions must consider these daily differences much more strictly than in the temperate regions, where differences in some variables such as pH, oxygen, and nutrient concentrations are not very pronounced over a period of a few hours (Jørgensen et al., 2005).

During stratification, the steep temperature gradient (i.e., thermocline) in the metalimnion suppresses many of the mass-transport phenomena that are otherwise responsible for the vertical transport of water-quality constituents within a lake. In strongly stratified lakes, the exchange of water and dissolved constituents between the epilimnion and hypolimnion can be reduced to molecular diffusion rates.

Retardation of mass transport between the epilimnion and the hypolimnion results in sharply differentiated water quality and biology between the lake strata, and one of the most important differences between the layers is often dissolved oxygen. As oxygen is depleted from the hypolimnion without being replenished, life functions of many organisms are impaired, and the biology and biologically mediated reactions fundamental to water quality are altered. Since light is limited in the hypolimnion, algae that settle into this zone can only respire, imposing an additional demand on the dissolved oxygen which, combined with the benthic oxygen demand, causes the hypolimnion to become *anoxic* (i.e., devoid of oxygen). Anoxic conditions in the hypolimnion trigger reducing chemical reactions, which convert some chemical compounds from their more oxidized states into reduced ones. Typically, reduced states of chemicals are more soluble in water. Anoxic conditions in the bottom sediments cause dissolution of some metals, such as iron and manganese, which interfere with potable uses of water. Some toxic compounds, such as mercury, can undergo microbiological methylation in the bottom sediment, which makes them far more toxic than their other forms. Similarly, phosphates are released from the sediments during conditions of anoxia in the hypolimnion and in the sediments. Anoxic conditions in the hypolimnion are typically most severe in meromictic lakes and less severe in holomictic lakes.

Nutrients released from the bottom sediments during stratified conditions are not available to phytoplankton in the epilimnion. However, during overturn periods, vertical mixing distributes the nutrients through the water column. The high nutrient availability is short-lived because the soluble reduced forms are rapidly oxidized to insoluble forms that precipitate out and settle to the bottom. Phosphorus and nitrogen are also deposited on the bottom of the lake through sorption to particles that settle to the bottom and as dead plant material that is added to the sediments.

Thermal stratification is common in lakes located in temperate climates with distinct warm and cold seasons. From a water-quality viewpoint, lake turnover brings nutrients from lower layers into the surface layers, which have higher oxygen levels and more sunlight, thereby stimulating biological productivity. Depth and wind have significant influences on the thermal structure of lakes, with shallow lakes (<10 m) rarely stratifying for long periods, and very deep lakes (>30 m) generally remaining stratified on a long-term basis, either permanently in tropical climates or seasonally outside the tropics (James, 1993). The stability of a lake can be measured by a *densimetric Froude number*, Fr_D, defined by

$$Fr_D = \frac{V}{\sqrt{(\Delta\rho/\rho_0)gd}}$$

(5.8)

where V is the average velocity in the lake, $\Delta\rho$ is the change in density over the depth, ρ_0 is the average density of the water, g is the acceleration due to gravity, and d is the average depth of the lake. Measurements by Long (1962) in laboratory channels indicated that if $\mathrm{Fr}_D \ll 1/\pi = 0.32$, the impoundment may become stratified; if $\mathrm{Fr}_D \approx 1/\pi$, the impoundment is weakly stratified; and if $\mathrm{Fr}_D \gg 1/\pi$, the impoundment may be vertically mixed. According to Tchobanoglous and Schroeder (1985), well-stratified lakes have $\mathrm{Fr}_D \ll 0.1$, weakly stratified lakes have $0.1 < \mathrm{Fr}_D < 1$, and fully mixed lakes have $\mathrm{Fr}_D > 1$. The average velocity, V, and the change in density, $\Delta\rho$, are sometimes expressed in the forms

$$V = \frac{QL}{V_L} \tag{5.9}$$

$$\Delta\rho = \beta d \tag{5.10}$$

where Q is the volumetric discharge through the impoundment (lake or reservoir), L is the length of the impoundment, V_L is the volume of the impoundment, and β is the average density gradient, defined as the change in density per unit depth. For freshwater impoundments, β and ρ_0 are commonly approximated by 10^{-3} and $1000\,\mathrm{kg/m^3}$, respectively.

Example 5.4   Measurements in a 10-m-deep lake show a mean velocity of 10 cm/s and a density difference between the top and bottom of the lake of 4.1 kg/m³. If the mean density of the lake water can be taken as 998 kg/m³, estimate the strength of the stratification.

SOLUTION   From the data given, $d = 10\,\mathrm{m}, V = 10\,\mathrm{cm/s} = 0.1\,\mathrm{m/s}, \Delta\rho = 4.1\,\mathrm{kg/m^3}, \rho_0 = 998\,\mathrm{kg/m^3}$, and hence the densimetric Froude number is given by Equation 5.8 as

$$\mathrm{Fr}_D = \frac{V}{\sqrt{(\Delta\rho/\rho_0)gd}} = \frac{0.1}{\sqrt{(4.1/998)(9.81)(10)}} = 0.16$$

Since $0.1 < \mathrm{Fr}_D < 1$, the lake should be classified as weakly stratified.

Overturning in the water column serves two highly important functions: The downward mixing of oxygen-rich surface waters below the thermocline introduces oxygen into the bottom waters of aquatic systems and recharging surface waters with nutrients trapped below the thermocline. The lower depths of a water body will become anoxic if the consumption of oxygen by biological or chemical processes exceeds the rate of resupply by vertical mixing and diffusion. Since virtually all aquatic organisms require oxygen for respiration, it is generally considered desirable for all parts of the water column to remain oxygenated. The development of low oxygen concentrations below the thermocline in any aquatic system is frequently associated with undesirable changes in the type and abundance of organisms living in the water. Shallow systems that are mixed to the bottom at all times do not develop seasonal oxygen depletion problems.

Water flowing into a lake will travel through the lake at a depth having the same density as the inflow until it becomes mixed. If the incoming water has a lower temperature and higher density than the lake water, when the incoming velocity subsides, the incoming water will "plunge" beneath the surface, with extensive mixing possible. If the incoming water is nutrient laden, these nutrients will be mixed with the lake water. If it is high in organic matter, which would settle in the hypolimnion, the microbial metabolism could

FIGURE 5.10 Low-level discharge from a lake.

deplete the oxygen supply. Conversely, the discharge of the lake, whether from a surface spillway or outlet pipe at the bottom of the lake, will have a major influence on the temperature of the lake, the depth of the metalimnion, the water temperature of the discharge, and the receiving stream. An example of a low-level discharge from a lake is shown in Figure 5.10. Such lake discharges could affect water quality and oxygen content and, in turn, the aquatic species makeup of the receiving stream. Releases of cold, low-dissolved-oxygen hypolimnetic water downstream cause reductions of BOD removal rates and decreases in reaeration rates, resulting in reduced oxygen levels and an overall reduction in the waste assimilative capacity. In addition, the release of cold hypolimnetic water may affect primary contact recreation such as swimming—the water is simply too cold to swim in.

Extreme depletion of dissolved oxygen (DO) may occur in ice- and snow-covered lakes in which light is insufficient for photosynthesis. If depletion of DO is great enough, fish kills may result.

5.3 WATER-QUALITY MODELS

Water-quality models are commonly utilized to assess various lake management techniques. Ideally, the modeling of lake water quality simulates lake processes and their interconnected and independent relationships. Several water-quality models that are commonly used in lake environments are described below.

5.3.1 Zero-Dimensional (Completely Mixed) Model

The response of lakes to the input of contaminants can sometimes be estimated by assuming that the lake is well mixed. This approximation is justified (1) when wind-induced circulation is strong and (2) when the time scale of the analysis is sufficiently long (on the order of a year) that seasonal mixing processes yield a completely mixed lake. Completely mixed models are frequently called *zero-dimensional models*, since they do not have any spatial dimension. Contaminant mass fluxes into a lake can come from a variety of sources, including municipal and industrial waste discharges, inflows from polluted rivers, direct surface runoff, contaminant releases from sediments, and contaminants contained in rainfall (atmospheric sources). Denoting the rate of contaminant mass inflow to a lake by $\dot{M}$, assuming that the lake is well mixed, and assuming that the contaminant undergoes first-order decay with a decay factor, k, the law of conservation of contaminant mass requires that

$$\frac{d}{dt}V_L c = \dot{M} - Q_o c - kV_L c \tag{5.11}$$

where V_L is the volume of the lake, c is the average contaminant concentration in the lake, and Q_o is the average outflow rate. The first-order decay factor, k, is the sum of the decay factors of all first-order decay processes by which the contaminant is removed from the water, including settling, chemical, and biological transformations. Equation 5.11 states that the rate of change of contaminant mass in the lake, $d(V_L c)/dt$, is equal to the mass influx, $\dot{M}$, minus the mass outflow rate, $Q_o c$, minus the rate at which mass is removed by first-order decay, $kV_L c$. It is emphasized here that the mass influx, $\dot{M}$, includes the contaminant influx from all sources, including direct discharges from outfalls and releases from sediments. Equation 5.11 is sometimes referred to as a *Vollenweider model* after Vollenweider (1968, 1975, 1976). Assuming that the volume of the lake, V_L, remains constant, Equation 5.11 can be put in the form

$$V_L \frac{dc}{dt} + (Q_o + kV_L)c = \dot{M} \tag{5.12}$$

which simplifies to the following differential equation that describes the contaminant concentration in the lake as a function of time:

$$\frac{dc}{dt} + \left(\frac{Q_o}{V_L} + k\right)c = \frac{\dot{M}}{V_L} \tag{5.13}$$

Taking the mass inflow rate, $\dot{M}$, to be constant and beginning at $t = 0$, and taking the initial condition as

$$c = c_0 \quad \text{at} \quad t = 0 \tag{5.14}$$

yields the following solution to Equation 5.13 (Thomann and Mueller, 1987):

$$\boxed{c(t) = \frac{\dot{M}}{Q_o + kV_L}\left\{1 - \exp\left[-\left(\frac{Q_o}{V_L} + k\right)t\right]\right\} + c_0 \exp\left[-\left(\frac{Q_o}{V_L} + k\right)t\right]} \tag{5.15}$$

where the first term on the right-hand side of Equation 5.15 gives the buildup of concentration due to the continuous mass input, $\dot{M}$, and the second term accounts for the dieaway of the initial concentration, c_0. The mass inflow rate, $\dot{M}$, and contaminant concentration, c, in Equation 5.15 as a function of time are illustrated in Figure 5.11 for cases in which the initial concentration, c_0, is less than and greater than the asymptotic concentration

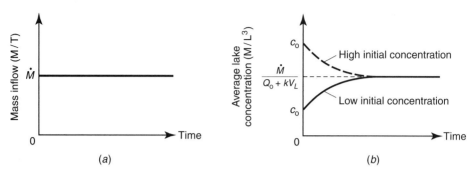

FIGURE 5.11 Response of a well-mixed lake to a constant contaminant inflow: (a) mass inflow; (b) lake response. (From Chin, David A., *Water-Resources Engineering*. Copyright © 2000. Reprinted by permission of Pearson Education, Inc., Upper Saddle River, NJ.)

given by Equation 5.15. Taking $t \rightarrow \infty$ in Equation 5.15 gives the asymptotic concentration, c_∞, as

$$c_\infty = \frac{\dot{M}}{Q_o + kV_L} \tag{5.16}$$

Equation 5.15 can also be used to calculate the lake response to a mass inflow over a finite interval, Δt. This case is illustrated in Figure 5.12(a) for a mass inflow rate, $\dot{M}$, over an interval Δt. The response is described by Equation 5.15 up to $t = \Delta t$, beyond which the response is described by

$$c(t) = c_1 \exp\left[-\left(\frac{Q_o}{V_L} + k\right)(t - \Delta t)\right] \tag{5.17}$$

which derived from Equation 5.15 for a contaminant mass inflow of zero and an initial concentration of c_1. In the case of a variable mass inflow illustrated in Figure 5.12(b), where the contaminant mass inflow rate is equal to $\dot{M}_1$ up to $t = \Delta t$, and equal to $\dot{M}_2$ thereafter, the response of the lake is described by Equation 5.15 up to $t = \Delta t$, beyond which the concentration in the lake is described by

$$c(t) = \frac{\dot{M}_2}{Q_o + kV_L}\left\{1 - \exp\left[-\left(\frac{Q_o}{V_L} + k\right)(t - \Delta t)\right]\right\}$$
$$+ c_1 \exp\left[-\left(\frac{Q_o}{V_L} + k\right)(t - \Delta t)\right] \tag{5.18}$$

which is derived from Equation 5.15 with a mass inflow rate, $\dot{M}_2$, beginning at $t = \Delta t$ with an initial concentration of c_1.

The analyses described here can also be applied to cases where the contaminants are removed by the settling of suspended solids in the lake. In this case, where the contaminants are adsorbed onto suspended solids, the removal rate due to sedimentation can be described by a settling velocity, v_s, and the conservation of mass equation can be written as

$$\frac{d}{dt}(V_L c) = \dot{M} - Q_o c - kV_L c - v_s A_L c \tag{5.19}$$

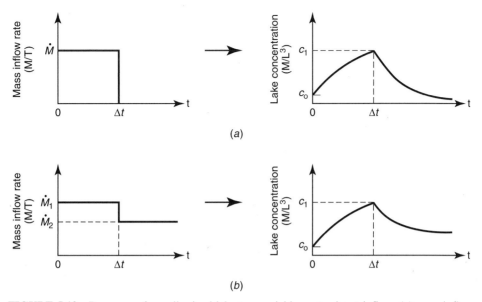

FIGURE 5.12   Response of a well-mixed lake to a variable contaminant inflow: (*a*) mass inflow over a finite interval; (*b*) variable mass inflow. (From Chin, David A., *Water-Resources Engineering*. Copyright © 2000. Reprinted by permission of Pearson Education, Inc., Upper Saddle River, NJ.)

where A_L is the surface area of the lake over which settling occurs. Equation 5.19 can be slightly rearranged into the form

$$\frac{d}{dt}(V_L c) = \dot{M} - Q_o c - k' V_L c \tag{5.20}$$

where

$$\boxed{k' = k + \frac{v_s A_L}{V_L}} \tag{5.21}$$

Since Equation 5.20 is identical with the conservation equation that neglects sedimentation as a removal process (Equation 5.11), with the decay coefficient, k, replaced by an effective decay coefficient, k', all the previous results are applicable provided that k is replaced by k'. Several variations in this model have been used in practice. In cases where there is significant vertical stratification, the lake can be considered as a well-mixed epilimnion overlying a well-mixed hypolimnion, with limited interaction between the two zones. Also, in large lakes ($A_L > 50$ to $100\,\text{km}^2$) it may be necessary to subdivide the lake into a number of well-mixed smaller lakes.

Example 5.5   The average concentration of total phosphorus in a lake is $30\,\mu\text{g/L}$, and an attempt is to be made to reduce the phosphorus level in the lake by reducing phosphorus inflows into the lake. The target phosphorus concentration is $15\,\mu\text{g/L}$. (a) If the discharge from the lake averages $0.09\,\text{m}^3/\text{s}$, the first-order decay rate for phosphorus is 0.01 day^{-1}, and the volume of the lake is $300,000\,\text{m}^3$, estimate the maximum allowable phosphorus inflow in kg/yr. (b) If this loading is maintained for 3 years but suddenly doubles in the fourth year, estimate the phosphorus concentration in the lake one month into the fourth year.

SOLUTION   (a) From the data given, $Q_o = 0.09$ m³/s $= 7776$ m³/day, $k = 0.01$ day⁻¹, and $V_L = 300{,}000$ m³. For an ultimate concentration, c_∞, of 15 μg/L ($= 15 \times 10^{-6}$ kg/m³), Equation 5.16 gives

$$c_\infty = \frac{\dot{M}}{Q_o + kV_L}$$

which rearranges to

$$\dot{M} = (Q_o + kV_L)c_\infty = [7776 + 0.01(300{,}000)](15 \times 10^{-6}) = 0.16 \text{ kg/day} = 59 \text{ kg/yr}$$

(b) Maintaining a mass loading of 0.16 kg/day for 3 years ($= 1095$ days), with $c_0 = 30$ μg/L $= 30 \times 10^{-6}$ kg/m³ yields a concentration at time t given by Equation 5.15 as

$$c(t) = \frac{\dot{M}}{Q_o + kV_L}\left\{1 - \exp\left[-\left(\frac{Q_o}{V_L} + k\right)t\right]\right\} + c_0 \exp\left[-\left(\frac{Q_o}{V_L} + k\right)t\right]$$

which for $t = 1095$ days gives

$$c(1095) = \frac{0.16}{7776 + 0.01(300{,}000)}\left\{1 - \exp\left[-\left(\frac{7776}{300{,}000} + 0.01\right)(1095)\right]\right\}$$

$$+ 30 \times 10^{-6} \exp\left[-\left(\frac{7776}{300{,}000} + 0.01\right)(1095)\right]$$

$$= 15 \text{ μg/L}$$

Hence after 3 years the phosphorus level has already decreased to the target level of 15 μg/L. In the fourth year, the mass flux doubles to $\dot{M}_2 = 2 \times 0.16 = 0.32$ kg/day, and the concentration as a function of time is given by Equation 5.18 as

$$c(t) = \frac{\dot{M}_2}{Q_o + kV_L}\left\{1 - \exp\left[-\left(\frac{Q_o}{V_L} + k\right)(t - \Delta t)\right]\right\} + c_1 \exp\left[-\left(\frac{Q_o}{V_L} + k\right)(t - \Delta t)\right]$$

where $c_1 = 15$ μg/L $= 15 \times 10^{-6}$ kg/m³, $\Delta t = 1095$ days, and after 1 month ($= 30$ days) $t = 1095 + 30 = 1125$ days. The concentration in the lake is then given by

$$c(1125) = \frac{0.32}{7776 + 0.01(300{,}000)}\left\{1 - \exp\left[-\left(\frac{7776}{300{,}000} + 0.01\right)(1125 - 1095)\right]\right\}$$

$$+ 15 \times 10^{-6} \exp\left[-\left(\frac{7776}{300{,}000} + 0.01\right)(1125 - 1095)\right]$$

$$= 25 \text{ μg/L}$$

Hence, the lake concentration rebounds to almost the original concentration within one month. This is a reflection of the relatively short detention time in the lake.

Although the completely mixed model cannot predict the specific change of water quality at individual locations within lakes and reservoirs, this type of model is particularly useful in estimating the behavior and water-quality trends if the temporal scale of the simulations is sufficiently long, such as months or years (Kuo and Yang, 2002). In cases where a water body has been loaded with phosphorus for a number of years, large quantities of phosphorus might have accumulated in the lake bottom sediments, adding years to the estimated time to reach an equilibrium concentration (Jørgensen et al., 2005)

5.3.2 One-Dimensional (Vertical) Models

In cases where there is significant variability in water quality over the depth of a lake, a one-dimensional (vertical) model is frequently used to simulate the fate and transport of water-quality constituents. Such vertical variations in water quality are commonly a result of thermal stratification of the lake or reservoir. One-dimensional water-quality models in lakes typically discretize the water body into homogeneous (completely mixed) layers, in which case the governing advection–dispersion equation can be put in the form

$$V\frac{\partial c}{\partial t} = AK_z\frac{\partial^2 c}{\partial z^2}\Delta z + Q_z\frac{\partial c}{\partial z}\Delta z + Q_{in}c_{in} - Q_{out}c \pm VS_m \qquad (5.22)$$

where V is the volume of a layer [L^3], c is the tracer concentration [M/L^3], Δz is the thickness of a layer [L], A is the cross-sectional area of a layer [L^2], K_z is the vertical dispersion coefficient [L^2/T], Q_{in} in the volumetric inflow rate [L^3/T], Q_{out} is the volumetric outflow rate [L^3/T], Q_z is a vertical advection coefficient [L^3/T], c_{in} is the concentration of the tracer in the inflow [M/L^3], and S_m are the sources and/or sinks of tracer mass [M/L^3].

Estimation of the Vertical Diffusion Coefficient   In many one-dimensional lake models, it is necessary to estimate the vertical diffusion coefficient from measured data. Vertical mixing is generally a function of the density profile, which, in lakes, can be related directly to the temperature profile. A widely used method to estimate vertical diffusion coefficients in lakes from temperature data is the *flux-gradient method*, first proposed by Jassby and Powell (1975). Recognizing that the thermal structure in lakes results from the interaction of solar heating and wind stress on the surface of the lake, and assuming horizontal homogeneity, the diffusion equation describing the vertical transport of heat in the water column is given by

$$\frac{\partial T}{\partial t} = \frac{\partial}{\partial z}\left(K_z\frac{\partial T}{\partial z}\right) + S_T \qquad (5.23)$$

where T is temperature and S_T is a heat source or sink in the water column. An example of a heat source is solar radiation. The first term on the right-hand side of Equation 5.23 represents the diffusion transport of heat in the vertical direction, and the diffusion coefficient K_z is a function of the thermal and current structure of the lake. The heat source term, S_T, can be omitted in the water column except at the surface because light extinction usually limits penetration of solar radiation into deep water. Integration of Equation 5.23 with respect to z from the bottom of the lake to depth z yields

$$K_z \frac{\partial T}{\partial z} = \int_{-D}^{z} \frac{\partial T}{\partial t} dz \tag{5.24}$$

where the heat flux into the sediments and the radiation absorbed by the sediments at $z = -D$ is not included. Equation 5.24 can be rearranged to give the following expression for the vertical mixing coefficient:

$$K_z(z) = \frac{\displaystyle\int_{-D}^{z} \frac{\partial T}{\partial t} dz}{\partial T/\partial z} \tag{5.25}$$

where the numerator represents the accumulated rate of change of stored heat between z and the bottom of the lake, and the denominator is the temperature gradient at depth z. It should be noted that Equation 5.25 does not apply at the surface because solar radiation at $z = 0$ is not included. Accurate temperature readings are essential to successful application of the flux gradient method, and negative values of K_z, which do not have physical meaning, can occur in computations due to errors in the measured temperature gradient. Typical values of K_z in thermally stratified lakes are in the range 10^{-9} to $10^{-6}\,\text{m}^2/\text{s}$.

The flux gradient method yields an infinite diffusion coefficient when $\partial T/\partial z$ approaches zero. In this case, another method developed by Sundaram et al. (1969) can be used. As an alternative and more simplistic approach, the vertical diffusion coefficient is sometimes expressed as a function of the stability effects in terms of the Richardson number, where

$$K_z(z) = K_0(1 + \sigma_1 R_i)^{-1} \tag{5.26}$$

where K_0 is the vertical diffusion coefficient without stratification ($R_i = 0$) σ_1 is an empirical constant, and R_i is the Richardson number, which is defined as

$$R_i = -\alpha_v g z^2 \frac{\partial T/\partial z}{u_*^2} \tag{5.27}$$

where α_v is the coefficient of volumetric thermal expansion of water and u_* is the shear velocity. Advantages of using Equations 5.26 and 5.27 to estimate K_z are that it is easy to use and the computation procedure is straightforward.

5.3.3   Two-Dimensional Models

Two-dimensional water-quality models have been developed for long-deep reservoirs in which significant vertical water-quality gradients are coupled with horizontal water-quality gradients. Typically, two-dimensional models solve the advection–diffusion equation in a vertical longitudinal plane through the reservoir. These models are mostly used to predict the two-dimensional temperature structure of deep reservoirs through the annual stratification cycle. Depth-integrated two-dimensional models have also been developed for shallow wide lakes, where these models are typically driven by wind shear and include no stratification effects. A simplified two-dimensional model is the near-shore mixing model, which is described in the following section.

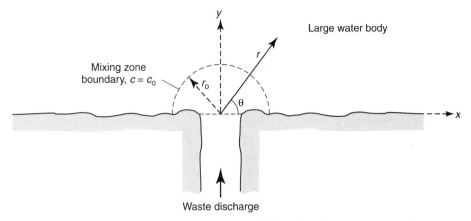

FIGURE 5.13   Wastewater discharge into a lake. (From Chin, David A., *Water-Resources Engineering.* Copyright © 2000. Reprinted by permission of Pearson Education, Inc., Upper Saddle River, NJ.)

Near-Shore Mixing Model   Near-shore mixing models are concerned with the distribution of contaminants in the vicinity of waste discharges into large bodies of water such as lakes. Consider the wastewater discharge illustrated in Figure 5.13, where the advective currents are negligible and the steady-state advection–dispersion equation with first-order decay is given by

$$D\left(\frac{\partial^2 c}{\partial x^2} + \frac{\partial^2 c}{\partial y^2}\right) - kc = 0 \tag{5.28}$$

where D is the dispersion coefficient, c is the contaminant concentration, and k is the first-order decay constant. Equation 5.28 can be written in polar (r, θ) coordinates as

$$\frac{\partial^2 c}{\partial r^2} + \frac{1}{r}\frac{\partial c}{\partial r} + \frac{1}{r^2}\frac{\partial^2 c}{\partial \theta^2} - \frac{k}{D}c = 0 \tag{5.29}$$

For a radially symmetric concentration distribution, $\partial c/\partial \theta$ and $\partial^2 c/\partial \theta^2$ are both equal to zero and Equation 5.29 reduces to

$$\frac{d^2 c}{dr^2} + \frac{1}{r}\frac{dc}{dr} - \frac{k}{D}c = 0 \tag{5.30}$$

which has the general solution

$$c(r) = AI_0\left(\sqrt{\frac{kr^2}{D}}\right) + BK_0\left(\sqrt{\frac{kr^2}{D}}\right) \tag{5.31}$$

where A and B are constants and I_0 and K_0 are modified Bessel functions of the first and second kind, respectively (see Appendix E.2). Taking the boundary conditions as

$$c(r_0) = c_0 \tag{5.32}$$

$$c(\infty) = 0 \tag{5.33}$$

requires that the concentration is equal c_0 on the boundary of a mixing zone at a distance r_0 from the discharge location and that the pollutant concentration decays to zero at a large distance from the discharge location. Imposing the boundary conditions given by Equations 5.32 and 5.33 on Equation 5.31 gives the concentration distribution in the lake as (O'Connor, 1962)

$$c = \frac{K_0(\sqrt{kr^2/D})}{K_0(\sqrt{kr_0^2/D})} c_0 \qquad (5.34)$$

The U.S. Environmental Protection Agency (USEPA) defines a mixing zone as an area within a water body where pollutants discharged from pipes are allowed to mix at high concentrations before entering the surrounding water in safe concentrations. The common practice of delineating mixing zones surrounding outfall pipes assumes that contaminants become diluted as they enter the surrounding waters and are dispersed over a larger area, justifying less stringent discharge standards within the mixing zone. Most states within the United States have policies allowing the use of mixing zones, although the spatial extent of mixing zones are strictly limited. There is currently a ban on mixing zones within the Great Lakes that applies to discharges of certain persistent, highly toxic chemicals that become more threatening to public health and the environment over time as they bioaccumulate in the food chain. The ban on mixing zones in the Great Lakes applies to the most toxic of these chemicals, including mercury, dioxins, polychlorinated biphenyls, and various pesticides.

Example 5.6   An industrial plant discharges wastewater through a single-port outfall into the shoreline region of a lake. Field measurements indicate that the dispersion coefficient in the lake is $1\,\text{m}^2/\text{s}$, the decay rate of the contaminant is $0.1\,\text{day}^{-1}$, and the effluent dilution 30 m from the outfall is 12. Estimate the distance from the outfall required to achieve a dilution of 100.

SOLUTION   Dilution is defined as the initial concentration divided by the final concentration; hence, if c_{30} is the contaminant concentration 30 m from the outfall and c_i is the concentration at the discharge location,

$$\frac{c_i}{c_{30}} = 12$$

and Equation 5.34 gives the concentration at a distance r from the outfall as

$$c_r = \frac{K_0(\sqrt{kr^2/D})}{K_0(\sqrt{k \cdot 30^2/D})} c_{30} = \frac{K_0(\sqrt{kr^2/D})}{K_0(\sqrt{k \cdot 30^2/D})} \frac{c_i}{12}$$

Hence the dilution, S_r, at a distance r from the outfall is given by

$$S_r = \frac{c_i}{c_r} = 12 \frac{K_0(\sqrt{k \cdot 30^2/D})}{K_0(\sqrt{kr^2/D})}$$

Since $S_r = 100$, $k = 0.1\,\text{day}^{-1}$, and $D = 1\,\text{m}^2/\text{s} = 8.64 \times 10^4\,\text{m}^2/\text{day}$, we are looking for r such that

$$100 = 12 \frac{K_0\sqrt{0.1(30)^2/8.64 \times 10^4}}{K_0\sqrt{0.1r^2/8.64 \times 10^4}}$$

or

$$\frac{K_0(0.0323)}{K_0(0.00108r)} = 8.33$$

Solving for r gives $r = 918\,m$. Hence, a dilution of 100 is achieved 918 m from the outfall. It is somewhat doubtful that this formulation would be applicable out to 918 m from the shoreline.

5.4 RESTORATION AND MANAGEMENT

The best shoreline for a lake is a natural one. Natural shoreline protection, such as native-species woody growth and scrub–shrub species, provides habitat for terrestrial and aquatic species, and also provides a buffer to filter pollutants as they get washed over the vegetation during rainfall–runoff events. The capability of natural vegetation to withstand shoreline erosion is also an important consideration since the root system of natural vegetation tends to give the soil a supporting structure that holds the shoreline together. This support does not exist with shallow-rooted vegetation such as grass lawns, which are considered by some to be the worst lake shoreline (DeBarry, 2004). Ideally, lakes should be surrounded by an undisturbed buffer, however, the desirability of lakefront properties is often given a higher priority.

5.4.1 Control of Eutrophication

The consequences of eutrophication are weed-choked shallow areas, algal blooms, oxygen-depleted deep waters, degradation of potable water supplies, limitations on recreational water use, degraded fisheries, reduced storage capacity, and disruption of downstream biological communities. The prevention and control of eutrophication is an important issue of concern in most lakes, particularly those that supply drinking water. Most techniques to control eutrophication are designed to manage the inflow of nutrients and sediment into the lake, and it is widely recognized that sediment deposited on the bottom of lakes can play a decisive role as a nutrient source. Benthic sediments can continually release nutrients into the water by diffusion, even if aerobic conditions are present at the sediment–water interface (Bernhardt, 1994). Phosphorus is typically the limiting nutrient affecting eutrophication, and measures to control eutrophication usually fall into the following categories: (1) control of point sources, (2) control of non-point sources, (3) limitation of phosphorus content in the lake, (4) limitation of internal loading, and (5) limitation of algal development in the lake, without changing the phosphorus budget.

Control of Point Sources   Point sources of phosphorus typically consist of domestic wastewater discharges, which usually contain on the order of 5 to 10 mg/L P and 20 to 40 mg/L N. An example of a point-source discharge is shown in Figure 5.14. Advanced wastewater treatment with precipitation, sedimentation, and/or filtration are conventional approaches for phosphorus removal from domestic wastewater.

Control of Nonpoint Sources   Nonpoint sources of phosphorus are commonly associated with agricultural operations. Other nonpoint sources of phosphorus to lakes include

FIGURE 5.14   Point-source discharge. (From United Nations Environment Programme, 2005.)

wastewater from populated areas not connected to a sewage system (e.g., septic tanks), effluents from isolated farmhouses, water birds, and open-air bathing facilities. Nonpoint sediment and nutrient loading is illustrated in Figure 5.15, where the lighter area is associated with sediment- and nutrient-laden runoff entering the lake.

Best management practices (BMPs) are typically used to control the input of nutrients. Control measures in agricultural areas include spreading of animal manure in a way that is compatible with plant growth and lake protection, evergreen strips along tributaries, and the use of crops that are compatible with the ground slope, condition of the soil, and the distance of the crop area from the lake. For example, maize planted on sloping fields is particularly disadvantageous because catastrophic erosion can take place during heavy rainfall.

FIGURE 5.15   Nonpoint-source discharge. (From University of Nevada, 2005.)

Small inflowing streams can be cleansed using seepage trenches, seepage filter basins with artificial soil filters, and using biological measures in pre-reservoirs and bioreactors. Seepage trenches are channels in porous soils where the water in the trench enters an adjacent lake by percolation through vadose zone. Phosphorus is removed from the water as it passes through the subsoil, and the process is effective in direct proportion to the amount of fine-grained sandy silt in the soil. Seepage trenches are typically limited to inflow rates of less than 100 L/s and soils with a high adsorption capacity (e.g., sandy silt). Seepage filter basins with artificial soil filters are used in cases where the native soil does not have sufficient filter capacity. Seepage filter basins are typically filled with fine particulate materials (<0.5 mm diameter) which contain a high proportion of aluminum and iron oxides, which guarantee a high capacity for phosphate adsorption. Subsurface hydraulic conductivities in excess of 10 m/day are required for seepage trenches and filter basins to function adequately (Bernhardt, 1994). The use of pre-reservoirs as bioreactors is based on the growth of microorganisms that fix the phosphorus in their biomass. Following sedimentation of the biomass, the phosphorus remains in the sediment, provided that conditions are sufficiently aerobic. Anaerobic conditions at the bottom of the impoundment should be avoided.

Chemical Treatments for Phosphorus   Aluminum salts such as aluminum sulfate (alum, $AlSO_4$) and sodium aluminate ($Na_2Al_2O_4$) or ferrous chloride ($FeCl_2$) have a strong affinity to adsorb and absorb inorganic phosphorus and remove phosphorus-containing particulate matter from the water column as part of the floc (loose precipitate) that forms. The result, after the floc settles, is not only a reduction of phosphorus availability but also a substantial increase in water clarity. Adverse effects may occur if the dosage of alum is too high. Particularly in softer more acidic waters, excessive input of aluminum salts can decrease the lake pH and result in concentrations of dissolved aluminum in the water column that are toxic to fish and other biota. For this reason, some experts no longer recommend the application of alum in lakes (Jørgensen et al., 2005). Alum treatment of Mohegan Lake (New York) with a specialized pontoon barge is shown in Figure 5.16. The pontoon barge shown in Figure 5.16 was equipped with a fathometer and speedometer to regulate

FIGURE 5.16   Alum treatment. (From Allied Biological, 2005.)

delivery rates according to bathymetric contours and the speed of the vessel. The chemical was sprayed directly on the surface via boomless spray nozzles out of twin polyethylene tanks with a combined capacity of 1600 L. Dose rates were based on previous jar tests designed to pinpoint optimum floc formation. The treatment of this 42 ha lake was completed in 3 days. Based on water-quality sampling conducted before and after the application, a number of improvements were observed at Lake Mohegan. Orthophosphorus declined from 25 µg/L to between 10 and 15 µg/L. Algal populations were reduced from 5570 organisms per milliliter to 380 organisms. Clarity (Secchi depth) improved from about 1 m before treatment to about 3 m one week posttreatment.

Limitation of Internal Loading   Sediments on the lake bottom are the primary source of internal phosphorus loading. Conventional measures that can be taken to control internal loading from benthic sediments include covering, dredging, removal of hypolimnetic water, and artificial destratification. Sheeting is sometimes used to cover the bottom sediment, but the sheeting typically has to be weighted to prevent it from floating. An example of sheeting used for sediment cover and lake shading is shown in Figure 5.17. Sand and gravel cannot be used to keep the sheeting in place because they provide a good substrate for macrophyte seedlings after 1 to 2 years. The sheeting should be permeable to gasses to prevent the sheeting from balooning due to the release of gases from the sediment. Alternatives to covering the bottom of the lake with sheeting include covering the bottom with foil, clay, crushed bricks, or other inert materials (Jørgensen et al., 2005).

Dredging can only be used to remove the sediment from small and shallow lakes. Major considerations are the treatment and deposition of the material removed and the need to pay particular attention to the amount of toxic metals present. The practice of removing sediment from lakes and from shallow areas is more effective than covering it with synthetic sheeting, because the nutrients are actually removed from the lake. It does, however, involve more technical problems and higher costs, and it is feasible only if a location is available for the deposition of the sediment removed (Bernhardt, 1994). The primary advantage of dredging is its relatively long-lasting effect. However, dredging operations can cause extensive damage to the benthic community, which may be an important food

FIGURE 5.17   Application of sediment cover and lake shading. (From State of Washington, 2005.)

source for fish and may disturb fish spawning habits if not carefully designed and implemented.

Removal of hypolimnetic water removes water that is within the hypolimnion of a lake, and this water is usually low in oxygen content and rich in nutrients. Withdrawing the hypolimnetic-nutrient-rich waters selectively using a siphon or deepwater outlet in a dam can decrease the quantity of nutrients stored and recycled in the water body. However, hypolimnetic withdrawals may also trigger thermal instability and lake turnover. In addition, the discharge of these nutrient-rich, often anaerobic waters from the hypolimnion of eutrophic lakes may cause adverse effects in downstream receiving waters.

Increasing the oxygen content in the lower portions of the lake reduces the release of phosphorus from the benthic sediments. However, although phosphorus is released from benthic sediments at considerably higher rates under anaerobic conditions, considerable amounts of phosphates and substances stimulating algal growth are still released from the sediment under aerobic conditions (Bernhardt, 1994). Phosphorus release from benthic sediments typically occurs when the oxygen concentration in the overlying water becomes less than 5 mg/L, and anaerobic conditions are typically associated with oxygen concentrations below 2 mg/L. The reason why phosphorus release from benthic sediments is greater under anaerobic conditions than under aerobic conditions is that oxygenated forms of iron and manganese in natural waters form an insoluble precipitate with phosphorus, thereby limiting the amount of dissolved phosphorus under aerobic conditions.

Artificial destratification of a stratified lake can change the composition of the algal population, reduce the number of algae and the algal growth rate, and aerate the lake to compensate for the oxygen deficit resulting from metabolic activity. With artificial circulation of the lake, the oxygen content of the water is increased at all depths. Specifically, at the sediment–water interface aerobic conditions are created, suppressing the release of nutrients from the sediment.

In general, over the long term, in-lake treatments will only be effective if accompanied by efforts to reduce external nutrient loads. In-lake treatments typically require significant expenditures for equipment, chemicals, and labor. Case studies, as well as additional information on the methods, costs, and potential negative side effects of each approach can be found in Olem and Flock (1990).

Limitation of Algal Development Control measures used to limit algal development in a lake without changing the phosphorus budget include artificial mixing, manipulation of the pelagic food web (biomanipulation), reduction of the residence time of water in the lake, and the use of nontoxic natural products to control algal growth. Commonly used substances to control algae (algicides) include copper sulfate pentahydrate and other chelated copper compounds. Potassium permanganate has also been found to be effective for algal control in some cases. Copper sulfate application methods and dosages will vary depending on the lake or reservoir conditions, and caution is required because copper sulfate addition can have a detrimental effect on fish and other aquatic life. For this reason, some experts recommend against the use of copper sulfate (Jørgensen et al., 2005) except in emergency circumstances. Typical effective copper sulfate concentrations are 1 to 2 mg/L. Application methods include dissolving the copper sulfate crystals using porous bags pulled by a boat and using specifically designed boats with either an application hopper that feeds copper sulfate crystals directly to the surface of the water body, or with a spray pump through which dissolved copper sulfate is sprayed along the water surface. An example of copper sulfate application to a lake is shown in Figure 5.18. Timing of application is important.

FIGURE 5.18   Copper sulfate application. (From USDA, 2005b. Photo by Jonathan House.)

Regular monitoring of algal numbers or chlorophyll *a* to determine when algal blooms begin will allow the application of the algicide before an algal bloom develops.

5.4.2 Control of Dissolved-Oxygen Levels

Low levels of dissolved oxygen may occur in lakes as a result of natural conditions as well as cultural eutrophication. The lowest concentrations of dissolved oxygen tend to occur in the deeper waters of the hypolimnion during thermal stratification in late summer, during long periods of snow and ice cover in winter, or in dense macrophyte beds at night or following long periods of cloud cover. One important option to consider for lakes that have problems with low dissolved oxygen is to manage the fisheries for species able to tolerate relatively low levels of oxygen or that do not inhabit areas of the lakes (such as the hypolimnion) that experience oxygen depletion. Problems with low dissolved oxygen can also be alleviated by one or more of the following methods:

1. Decreasing the quantity of organic matter decomposing in the lake (the major oxygen consuming process) by adopting one or more of the following practices:
   a. Limiting the export of organic materials from the watershed to the lake, in particular excessive exports associated with human activities, such as runoff from feedlots or direct discharges of sewage wastewaters
   b. Dredging to remove organic-rich sediments
   c. Decreasing in-lake productivity by reducing nutrient loads and nutrient availability
2. Increasing photosynthesis (an oxygen-generating process), especially during critical times (e.g., winter) and in critical locations (e.g., deeper waters of the hypolimnion) subjected to oxygen depletion, primarily by increasing light penetration
3. Destratifying the lake by bringing low-oxygen waters in the hypolimnion in contact with the lake surface and the well-oxygenated waters of the epilimnion (artificial circulation)
4. Direct aeration

Lake destratification by aeration mechanisms during summer oxygen depletion may sometimes be very damaging. When the low-oxygen hypolimnetic water is brought to the surface, it may result in fish kills, and lake aeration devices may also cause an undesirable shift in algal population (Novotny, 2003).

The approaches and optimal design criteria for lake aeration mechanisms vary between systems installed to alleviate problems with winterkill (oxygen depletion during winter as a result of ice cover) as opposed to low levels of dissolved oxygen in the hypolimnion during summer. During winter, the goal is not to aerate the entire water body but to create an oxygen-rich refuge area for fish near the lake surface. Major design concerns include problems with equipment ice-up and the need (for safety reasons) to minimize the loss or weakening of the lake's ice cover. Hypolimnetic aeration systems must deal with the more difficult problem of aerating waters at greater depths. Where the objective is to establish or maintain a cold-water fishery, hypolimnetic aeration must be achieved without disturbing the lake's thermal stratification. Otherwise, low levels of dissolved oxygen may be avoided by preventing thermal stratification through artificial circulation of the water column. In both cases, during both winter and summer, maximum reliability at minimal cost are important design objectives. Aeration systems in common use are described below (Novotny, 2003).

Pump and Baffle Aeration System   Using this method, oxygen-poor water is extracted from a nearshore area of the lake, pumped to the top of a chute located on shore, and then allowed to cascade over a set of baffles (constructed of wooden boards). An example of such a cascade is shown in Figure 5.19. The turbulence created as the water passes over the baffles helps to reaerate the water. The reoxygenated water is then returned to a different part of the lake, away from the intake area, creating a zone of oxygen-rich water. Generally, approximately 10% of the lake's volume should be aerated. Pump and baffle systems have several major advantages relative to other aeration techniques. In particular, when properly operated, only a small area of the lake's ice cover is opened. Open areas and thin ice are safety hazards for which the operator of an aeration system is liable. All of the major pieces of equipment are on shore. In addition, the chute can be mounted on a trailer and moved from one lake to another or to different areas of the lake as needed. Generally, to prevent winterkill, aeration will be required for about two months, depending on winter conditions. By monitoring dissolved oxygen levels in the lake, the system can be operated only during those times when needed. Pump and baffle systems have been built by lake associations or may be purchased as a unit from a number of manufacturers.

Artificial Circulation   Artificial circulation eliminates thermal stratification, or prevents its formation, either by mechanical pumping or through the injection of compressed air from a pipe or ceramic diffuser at the lake's bottom. An example of a ceramic-diffuser system is illustrated in Figure 5.20, where two operational diffusers releasing air bubbles are shown in Figure 5.20(*a*) and the controller surrounded by diffuser units is shown in the foreground in Figure 5.20(*b*). If sufficiently powered, the rising column of bubbles will produce lakewide mixing. As a result, the conditions that create hypolimnetic oxygen depletion (isolation of the deeper waters from the atmosphere with little to no primary production in these deeper, darker waters) are eliminated. Artificial circulation is one of the most commonly used lake restoration techniques. Examples of its utility for improving fisheries yields include the Parvin Reservoir in Colorado and Corbett Lake in British Columbia. The technique is best used in lakes that are not nutrient limited; nutrient concentrations are often higher in the hypolimnion, and as a result, mixing can stimulate

FIGURE 5.19   Water cascade into a lake. (From Malaysia University, 2003 Photo by Suzana Mohkeri.)

increased algal growth. In addition, artificial circulation is not a feasible option for cold-water fish species, which use the hypolimnion as a thermal refuge during summer. Aeration–destratification must begin at the time of vernal overturn and be continuous. Regulatory agencies are often reluctant to permit such systems.

Water Fountains   Water fountains pump water from the surface layers lakes into the air and, in the process, aerate lake waters. Water fountains have a limited effect in aerating lake water since they tend to pump water from the upper layers of the lake that are already well aerated. Several designs of water fountains are available, and several of the more popular designs are shown in Figure 5.21. In selecting a water fountain for a particular lake, aesthetics are usually the primary motivation, and aeration a secondary benefit.

Hypolimnetic Aeration   Hypolimnetic aeration is used in cases of high hypolimnetic oxygen deficits, taste and odor problems, and increased concentrations of manganese and iron (Prepas and Burke, 1997). Hypolimnetic aerators may be used to increase oxygen levels in the hypolimnion without disturbing the lake's thermal stratification. An airlift device is used to bring cold hypolimnetic water to the surface. The water is aerated, by contact with the atmosphere; gases such as methane, hydrogen sulfide, and carbon dioxide, which may accumulate under anaerobic conditions, are lost, and then the water is returned to the hypolimnion. Hypolimnetic aerators require a large hypolimnion to work properly and are generally ineffective in shallow lakes and reservoirs. Costs depend on the amount of

(a)   (b)

FIGURE 5.20   Diffused-air circulation system. (From Vertex Water Features, 2005.)

compressed air needed, which is a function, in turn, of the area of the hypolimnion, the rate of oxygen consumption in the lake, and the degree of thermal stratification. Example applications include Waccabuc Lake in New York, Larson and Mirror Lakes in Wisconsin, and Tory Lake in Ontario.

Oxygen Injection   Studies have shown that it is often more cost-effective and practical to inject pure oxygen into the hypolimnion, as opposed to air injections or aerating the hypolimnion via airlift systems. At Richard B. Russell Reservoir in Georgia, oxygen levels in the hypolimnion have been increased from less than 3 mg/L to more than 9 mg/L, with an oxygen transfer efficiency of about 75%. Liquid oxygen is stored in tanks on site and connected to several supply heads submerged and anchored in the reservoir. Flexible membrane diffusers mounted on the supply heads are used to maximize absorption efficiency and minimize maintenance requirements. Flexible membrane systems should last 2 to 6 years (or 10 years or more if operated less than six months per year); the compressor and distribution system should last substantially longer (estimated 30-year life).

The Calleguas Municipal Water District (CMWD) in California treats water stored in Lake Bard at the Lake Bard Water Filtration Plant (LBWFP). Each spring, the lake stratifies and by early summer, the available dissolved oxygen (DO) in the lake hypolimnion is depleted. Without oxygen, reduced nutrients and metals released from the lake bottom degrade water quality, making it harder to treat. Because ozone generators convert only about 6% of liquid oxygen (LOX) to ozone, the CMWD significantly increases the value of LOX by using off-gas to oxygenate the hypolimnion of Lake Bard. The system takes oxygen-rich off-gas from the ozone contactor and oxygenates the lake hypolimnion;

FIGURE 5.21   Water fountains. (From Vertex Water Features, 2005.)

pipelines efficiently convey and disperse oxygen throughout the lake. The liquid oxygen tanks used at the CMWD plant and the Outlet Tower bridge which is used to support the system that injects the oxygen-rich off-gas into the hypolimnion of Lake Bard are shown in Figure 5.22. This lake-oxygenation system is an example of a "green solution" that takes a waste product and uses it to benefit operations and the environment.

Snow Removal to Increase Light Penetration   Snow removal from the lake surface to increase light penetration and photosynthesis (oxygen generation) under the ice is a low-tech, low-cost alternative to aerators that may be sufficient to prevent winterkill in

FIGURE 5.22 Lake Bard hypolimnion oxygenation system. (From American Academy of Environmental Engineers, 2005 Project and photo by Kennedy/Jerks Consultants.)

lakes with marginal levels of dissolved oxygen. A typical snow-removal tractor is shown in Figure 5.23. Snow is a much more effective absorber of light than is ice. While 85% of the available light will penetrate 12.5 cm of clear ice, 5 cm of snow over 7.5 cm of ice will block out almost all light. Even thin layers of snow can greatly decrease light penetration, decreasing primary productivity, and thus leading to oxygen depletion and winterkill.

In selecting a method to improve dissolved-oxygen levels in lakes, the most cost-effective approach for increasing levels of dissolved oxygen depends on the size (area and depth) of the lake, nature and causes of the problem, and fisheries-management objectives.

FIGURE 5.23 Snow removal tractor. (From Shaver Lake Power Center, 2005.)

5.4.3 Control of Toxic Contaminants

Toxic contamination of lakes is caused by metals, pesticides, oils, and other pollutants contained in agricultural, industrial, and urban runoff. Contaminated sediments are sometimes categorized as *legacy pollution*, since they resulted from past practices that are no longer in effect. As an example of this type of pollution, lake associations years ago used arsenic or copper salts to control algae and weeds, and these pollutants are now deposited in the sediments long after these practices were terminated. Elevated levels of these substances can degrade water quality and affect aquatic organisms. Some of these pollutants may bioaccumulate in the fish tissue, limiting the suitability of fish for human consumption. Possible corrective actions include:

- Elimination or reducing the source of the contaminants by applying best management practices
- Dredging and removing contaminated lake sediments
- Isolating (capping) contaminants concentrated in the bottom sediments from the overlying water column by covering the sediments with a relatively impermeable layer, such as bentonite (a form of clay) or a plastic liner
- On-site water treatment, such as diluting the contaminated water with clean water pumped from other sources, or withdrawing and treating the lake water and then returning the treated water to the lake
- Addition of chemicals such as alum (aluminum sulfate) to the lake, which may accelerate the precipitation of toxic substances out of the water column into the bottom sediments
- Deepwater aeration for contaminants (e.g., some metals and ammonia) that precipitate or become nontoxic in the presence of dissolved oxygen
- Controlling changes in water level if the exposure and suspension of contaminated sediments tend to increase the solubility and mobilization of the toxic substance
- Biomanipulation if the potential for human exposure to bioaccumulated toxics can be reduced by altering the food chain or target fish species for fisheries management (e.g., avoiding game fish such as trout that are top predators and have high levels of body fat)

Relatively few field tests have been conducted to evaluate the long-term effectiveness of most of the foregoing techniques; in addition, some of these techniques can have potentially serious negative side effects (e.g., dredging operations may resuspend toxic contaminants and actually increase bioavailability). Often, no action is the most environmentally sound and cost-effective approach, allowing natural processes such as sedimentation to gradually reduce the concentration and availability of toxic substances after the source of contaminants has been eliminated.

5.4.4 Control of Acidity

Lake waters may naturally be acidic: for example, in regions with naturally acidic soils or when a lake is part of wetland system. Anthropogenic lake acidity is caused by large inputs of organic acids, acid mine drainage, or acidic atmospheric deposition. Acid atmospheric deposition (acid rain) is presently confined mainly to industrial regions of the northern hemisphere, such as the northeastern United States and adjacent parts of Canada,

Scandinavia, parts of Great Britain and Ireland, and small areas in central Europe and Asia. The low pH/high acidity is caused by the low buffering capacity (low alkalinity) of the lake to neutralize these acidic inputs. The main water-quality problem of acidified lakes is the increased concentration of heavy metals in the water.

Extensive research has been conducted to refine and test methods for neutralizing lake waters using a variety of neutralizing agents as well as application techniques. There are five basic approaches to treating acidic lakes:

1. *Limestone addition to the lake surface.* Small limestone particles, limestone powder, or a limestone slurry are dispersed via boat, plane, or helicopter over the lake. The dispersal of limestone into a lake by boat is illustrated in Figure 5.24. During winter, the limestone may be spread on the ice by truck, entering the lake in the spring as the ice melts. Direct addition of limestone to the lake surface is the most commonly employed method for decreasing lake acidity. Because limestone is used for agricultural liming, it is usually available at low cost. However, the cost of the limestone dispersal can be significant, particularly for remote lakes without road access. Repeated applications are usually needed, and lakes with short water retention times may need to be treated annually.

2. *Injection of base materials into the lake sediment.* Limestone, hydrated lime, or sodium carbonate can be injected into the lake sediments, resulting in a gradual decrease in lake acidity. This technique is largely experimental, however, and limited to small, shallow lakes with soft organic sediments and road access for transport of the application equipment. The treatment may remain effective substantially longer than surface applications, but at the same time the lake's benthic community is disturbed, turbidity may increase, and the costs are higher.

3. *Mechanical stream doser.* Lake acidity may be decreased by neutralizing the acidic waters in upstream tributaries. Mechanical dosers are automated devices that release dry powder or slurried limestone directly into the stream, with the quantity of material added controlled by monitors of stream flow or stream chemistry. The treatment is continuous,

FIGURE 5.24 Limestone addition to a lake. (From State of Virginia, 2005.)

FIGURE 5.25   Limestone placed in the St. Mary's River. (From U.S. Forest Service, 2005.)

expensive, and generally not recommended for lakes unless all other alternatives are not feasible.

4. *Limestone addition to the watershed.* Limestone is spread on all or parts of the lake's watershed, decreasing the acidity of runoff and shallow ground-water flow into the lake. Although the costs of one application are higher, the overall costs may be lower than for limestone applications to surface waters, because the effects are much more long lasting. Watershed liming may be especially appropriate for lakes with short retention times (less than six months). Liming of soils also increases their fertility and pollutant-retention capacity. An example of watershed liming is in the St. Mary's Wilderness, where the U.S. Forest Service placed crushed limestone in the St. Mary's River, shown in Figure 5.25, to decrease the acidity of receiving lakes within the wilderness area.

5. *Pumping of alkaline ground water.* Where abundant supplies of alkaline ground water are available, these waters may be discharged directly into lakes or lake tributaries, decreasing acidity. Applications of this method have been limited.

Several books have been published on methods for liming lakes and streams, including Olem (1991), Olem et al. (1991), and Brocksen et al. (1992).

5.4.5 Control of Aquatic Plants

In lakes and reservoirs where thick beds of macrophytes cover a high proportion of the lake bottom, an aquatic-plant control program may be needed to improve yields of large, predatory game fish and increase the growth rates of panfish species, such as bluegill and white and black crappie. For lake uses such as swimming and boating, minimizing macrophyte beds are desired. For fisheries management, on the other hand, moderate growths of aquatic plants enhance the fisheries. The complete elimination of macrophyte beds may be as harmful to fisheries as are excessive plant growths.

The objective of aquatic plant management is to provide the appropriate amount of aquatic plants, taking into account the effects of macrophytes on fish communities, other

FIGURES 5.26   Lake dredging. (From Delta Contracting Company of New Jersey, 2005.)

lake uses (e.g., swimming and boating), nutrient cycles, and aesthetics. Macrophytes and terrestrial vegetation also help to stabilize the lake bed and shoreline, reducing problems with lakeshore erosion and high turbidity.

Excessive plant growth as a result of eutrophication or the inadvertent introduction of an exotic macrophyte species is a common lake problem. Approaches to controlling nuisance plant growths are as follows:

1. *Sediment removal and sediment tilling.* Lakes can be dredged to remove sediments and deepen the lake so that less of the lake bottom receives adequate light for macrophyte growth. Portions of lakes are frequently dewatered prior to dredging, as shown in Figure 5.26. The maximum depth at which macrophytes are able to grow depends on water transparency and the plant species. Hydrilla, a nuisance exotic plant in southern waters, can grow at lower light intensities than can native plants, limiting the effectiveness of lake deepening. Reductions in nutrient loads, to control eutrophication, can increase lake transparency, increasing the depth at which macrophytes can grow and countering the effectiveness of dredging to reduce macrophyte growth. Sediment removal and tilling (e.g., rototilling using cultivation equipment) can also be used to disturb the lake bottom, tearing out plant roots for short-term macrophyte control. Both dredging and tilling can have negative side effects, including destruction of the benthic community and an increase in turbidity and siltation. ·

2. *Water-level drawdown.* In lakes where water levels can be controlled, lake levels can be lowered to expose macrophytes in the littoral zone to prolonged drying and/or freezing. Some species of plants are permanently damaged by these conditions, killing the entire plant, including roots and seeds, with exposures of 2 to 4 weeks. Other plant species are unaffected or even increase. An example of water-level drawdown to control aquatic plants is shown in Figure 5.27. Water-level drawdowns in the winter tend to be more effective than during the summer (Jørgensen et al., 2005).

3. *Shading and sediment covers.* Covers can be placed on the water or sediment surface as a physical barrier to plant growth or to block light. Sediment covers made of polypropylene, fiberglass, or a similar material can effectively prevent growth in small areas such as near docks and swimming areas, but are generally too expensive to install over large areas. Applications of silt, sand, clay, or gravel have also been used, but plants eventually root in them. Shading to reduce growth rates can be provided by floating sheets of polyethylene or by planting evergreen trees along the lakeshore. An example of shading is illustrated in Figure 5.17.

FIGURE 5.27   Lake drawdown to control aquatic plants.

4. *Introduction of grass carp.* Grass carp is an exotic fish species that feeds on macrophytes; however, grass carp do not consume all aquatic plant species. Generally, they avoid alligatorweed, water hyacinth, cattails, spatterdock, and water lily. The fish prefer plant species that include elodea, pondweeds (*Potamogeton* spp.), and hydrilla. A grass carp is shown in Figure 5.28. Low stocking densities of grass carp can produce selective grazing on the preferred plant species. The use of grass carp for aquatic plant management is allowed only in certain states.

5. *Introduction of insects that infest macrophytes.* Several exotic insect species have been imported to the United States and approved by the U.S. Department of Agriculture for use in macrophyte control. Each insect species grows and feeds on only select target plant species. In particular, insects have been used in southern waters to aid in the control of alligatorweed and water hyacinth. Because insect populations tend to grow more slowly than the plants, insects work best when used in conjunction with another plant control technique (e.g., harvesting or herbicides). No significant negative side effects from insect infestations have been documented.

6. *Mechanical harvesting.* Mechanical harvesters constructed on low-draft barges can be used to cut and remove rooted plants and floating water hyacinths. A typical mechanical harvester is shown in Figure 5.29. Cutting rates range from 0.1 to 0.3 ha/h, depending on machine size. Harvesters can effectively clear an area of vegetation, although the benefits are only temporary. Rates of plant regrowth can be rapid (within weeks) but can be slowed if the cutter blade is lowered into the upper sediment layer. Cut plants are removed from the lake, eliminating an internal source of nutrients and organics with potential long-term benefits. However, some plant species, such as milfoil, may be fragmented and dispersed and actually increase in abundance after harvesting operations. Also, small fish can be caught and killed by mechanical harvesters. Harvesting operations should precede spawning periods and avoid important spawning and nursery areas.

7. *Herbicides.* Herbicides used to kill aquatic plants include Diquat, endothall, 2,4-D, glyphosate, and fluridone. Although herbicide treatments can reduce macrophyte growths rapidly, the benefits are short term and the potential for negative side effects is high. Plants are left in the lake to die and decompose, releasing plant nutrients and in some cases causing

FIGURE 5.28 Grass carp.

oxygen depletion and algal blooms. Plants generally regrow after several weeks or months, or may be replaced by other, more tolerant macrophyte species. Most chemicals currently approved are toxic to aquatic organisms and humans only at relatively high doses. Little information is available, however, on the long-term ecological consequences of herbicide use. Herbicide applicators must be licensed, have adequate insurance, wear protective gear, and use only USEPA-approved chemicals, following label directions exactly. A typical herbicide application boat is shown Figure 5.30. Generally, because herbicides do not remove nutrients or organics from the lake or address the cause of the aquatic plant problem, herbicides should be used only where other techniques are unacceptable or ineffective.

FIGURE 5.29 Mechanical harvester.

FIGURE 5.30   Herbicide application boat.

5.4.6   Attainability of Lake Uses

The most critical water-quality indicators for assessing the attainability of various lake uses are dissolved oxygen, nutrients, chlorophyll a, and toxicants. In performing use-attainability analyses, the relative importance of three forms of oxygen demand should be considered: respiratory demand of phytoplankton and macrophytes, water-column demand, and benthic demand. If use impairment is occurring, assessments of the significance of each oxygen sink can be useful in evaluating the feasibility of achieving sufficient pollution control or in implementing the best management practices to attain a desired use. The presence of toxics such as pesticides, herbicides, and heavy metals in sediments or the water column are also important considerations in evaluating use attainability. These pollutants may prevent the attainment of uses related to fish propagation that would otherwise be supported by the water-quality criteria for dissolved oxygen and other parameters.

5.5   COMPUTER CODES

Several computer codes are available for simulating the water quality in lakes. These codes typically provide numerical solutions to the advection–dispersion equation, or some other form of the law of conservation of mass, at discrete locations and times for multiple interacting constituents, complex boundary conditions, spatially and temporally distributed contaminant sources and sinks, multiple fate processes, and variable flow and dispersion conditions. In choosing a code for a particular application, there is usually a variety of codes to choose from. However, in doing work that is to be reviewed by regulatory agencies, or where professional liability is a concern, codes developed and maintained by the U.S. government have the greatest credibility and, perhaps more important, are almost universally acceptable in developing permit applications and defending design protocols on liability issues. Several of the more widely used models that have been developed and endorsed by U.S. government agencies are described briefly here.

CE-QUAL-R1 (USAWES, 1986) is a one-dimensional (in vertical) horizontally averaged continuous simulation code. This code conceptualizes the lake as a vertical sequence of horizontal layers, where contaminants are uniformly distributed in each layer. Density and wind-driven vertical mixing of constituents is simulated through the series of horizontal layers. The code simulates inflows, outflows, vertical diffusion, and interactions of a number of water-quality constituents. Inflows and outflows are added directly to or removed directly from the appropriate layers. This code is not suitable for cases where lateral or longitudinal variations in water quality are important. Water-quality constituents that can be modeled include temperature, algae, phosphorus, nitrogen, dissolved oxygen, metals, suspended sediments, dissolved solids, pH, and several other water-quality parameters.

CE-QUAL-W2 (Cole and Buchak, 1995) is a two-dimensional water-quality and hydrodynamic code supported U.S. Army Waterways Experiment Station. The code has been widely applied to stratified surface-water systems such as lakes, reservoirs, and estuaries and computes water levels, horizontal and vertical velocities, temperature, and 21 other water-quality parameters (such as dissolved oxygen, nutrients, organic matter, algae, pH, the carbonate cycle, bacteria, and dissolved and suspended solids).

WASP 7 (Water Quality Analysis Simulation Program 7) (USEPA, 1988) simulates contaminant transport in streams and lakes in one, two, or three dimensions. The contaminants that can be simulated include BOD, dissolved oxygen, and nutrients. WASP 7, one of the most versatile multidimensional water quality codes in the public domain, is used for applications in rivers, lakes, and reservoirs. This code helps users interpret and predict water-quality responses to natural phenomena and human-made pollution for various pollution management decisions. WASP 7 is a dynamic compartment-modeling program for aquatic systems, including both the water column and the underlying benthos. WASP 7 also can be linked with hydrodynamic and sediment transport models that can provide flows, depths velocities, temperature, salinity, and sediment fluxes. WASP 7 has been used to examine eutrophication of Tampa Bay (Florida); phosphorus loading to Lake Okeechobee (Florida); eutrophication of the Neuse River estuary (North Carolina); eutrophication of the Coosa River and reservoirs (Alabama); PCB pollution of the Great Lakes, eutrophication of the Potomac estuary, kepone pollution of the James River estuary, volatile organic pollution of the Delaware estuary, heavy metal pollution of the Deep River (North Carolina) and mercury in the Savannah River (Georgia).

Only a few of the more widely used computer codes have been cited here. Certainly, there are many other good models that are capable of performing the same tasks.

SUMMARY

Lakes differ from rivers in that they typically have much lower velocities and longer detention times, and the water-quality gradients of concern are usually in the vertical direction. Natural processes that affect the water quality in lakes are flow and dispersion, light penetration, sedimentation, eutrophication, and thermal stratification. A major cause for concern is that lakes tend to accumulate nutrients and ultimately become eutrophic, in which case the respiration and photosynthesis of phytoplankton can cause significant and deleterious effects on the dissolved oxygen in the lake. The design of systems to control eutrophication involve the identification of the limiting nutrient, usually phosphorus or nitrogen, and control of nutrient fluxes into the lake. The effect of contaminant inputs on the water quality in lakes can sometimes be analyzed adequately using a completely mixed model, which is an appropriate formulation when the wind-induced circulation is strong and the time scale of the analysis is sufficiently long that seasonal mixing processes yield a completely mixed lake. In some cases, one- and two-dimensional water-quality models are necessary to adequately describe the fate and transport of pollutants in lakes. In cases where water-quality problems exist, proven techniques are available to control eutrophication, dissolved oxygen, toxic contaminants, acidity, and aquatic plants.

PROBLEMS

5.1. A rectangular flood-control lake is $100\,\text{m} \times 70\,\text{m}$ and has an average depth of $5\,\text{m}$.

  (a) If the average inflow and outflow rate is $0.05\,\text{m}^3/\text{s}$, estimate the detention time in the lake.

  (b) Compare this detention time with a larger lake that is $200\,\text{m} \times 140\,\text{m}$ and $10\,\text{m}$ deep, with an average inflow/outflow rate of $0.1\,\text{m}^3/\text{s}$.

5.2. A large natural lake has a surface area of $2.5 \times 10^6\,\text{m}^2$ and the catchment for the lake has an area of $2.0 \times 10^7\,\text{m}^2$. Estimate the hydraulic detention time of the lake.

5.3. The orthophosphate concentration in a lake is measured as $30\,\mu\text{g/L}$ and the available nitrogen is measured at $0.2\,\text{mg/L}$. Determine the limiting nutrient for algal growth. If the biomass concentration in the lake is too high, suggest a method to limit the growth of biomass.

5.4. The total phosphorus concentration in a lake is measured as $15\,\mu\text{g/L}$, and the concentration of total nitrogen is estimated as $0.17\,\text{mg/L}$. Estimate the biomass concentration and trophic state of the lake.

5.5. The density difference between the top and bottom of a 7-m-deep lake is measured as $3.5\,\text{kg/m}^3$, and the mean density of the lake water is $998\,\text{kg/m}^3$. If the velocity in the lake is typically on the order of 2% of the wind speed, estimate the maximum wind speed for the lake to remain strongly stratified.

5.6. The currents in a 7-m-deep lake are on the order of $5\,\text{cm/s}$. If the mean density of the water is $998\,\text{kg/m}^3$, estimate the density difference between the top and the bottom of the lake for the lake to be strongly stratified. What temperature difference could be responsible for such a density variation?

5.7. You have been appointed to be the project engineer to direct the cleanup of a polluted lake in an urban development. Congratulations! The lake is approximately circular with a radius of $100\,\text{m}$ and an average depth of $5\,\text{m}$. The target biomass concentration in the lake is $5\,\mu\text{g Chl}a/\text{L}$, and the current biomass concentration is estimated to be $15\,\mu\text{g Chl}a/\text{L}$. If the average inflow and outflow from the lake is $5\,\text{L/s}$, estimate the allowable concentration of total phosphorus in the lake inflow such that the target biomass concentration is reached in six months. The decay rate for phosphorus can be taken as $0.01\,\text{day}^{-1}$.

5.8. The average concentration of total phosphorus (TP) in a flood-control lake is $25\,\mu g/L$, and daily flows into and out of the lake are typically on the order of $0.13\,m^3/s$.

    (a) If the decay rate of TP is 0.2 day^{-1} and the volume of the lake is $2.8 \times 10^5\,m^3$, estimate the mass loading that must be maintained to ultimately bring the TP concentration down to $15\,\mu g/L$.

    (b) If this mass loading is maintained for one week, estimate the TP concentration at the end of each day.

5.9. A eutrophic lake is estimated to have a mean total phosphorous concentration of $47\,\mu g/L$ that fluctuates by $\pm 50\%$ during the year.

    (a) Estimate the corresponding percentage fluctuation in biomass in the lake.

    (b) If the lake is $200\,m$ long $\times 100\,m$ wide $\times 3\,m$ deep, the average inflow (and outflow) rate is $0.1\,m^3/s$, and the decay rate for phosphorus is 0.03 day^{-1}, compare the detention time in the lake with the time scale for phosphorus decay. What can you infer from this result regarding the fate of phosphorus in the lake?

    (c) If the inflow concentration of phosphorus is reduced to $20\,\mu g/L$, what is the concentration of phosphorus that can ultimately be expected in the lake?

    (d) How long will it take for the mean phosphorus concentration to be decreased by one-half?

5.10. If the mass loading of TP in Problem 5.8 suddenly drops to zero at the end of the first week, estimate the daily TP concentration during the second week.

5.11. If the mass loading of TP in Problem 5.8 doubles at the end of the first week, estimate the daily TP concentration during the second week.

5.12. A recreational lake is approximately circular with a diameter of $700\,m$ and an average depth of $3\,m$. Cultural eutrophication has caused the lake to have an excessive concentration of algae. The total phosphorus concentration is measured at $40\,\mu g/L$, and it is estimated that the first-order decay factor for phosphorus in the lake is 0.008 day^{-1}. Average annual inflow into the lake is $37.6\,L/s$, annual rainfall is $150\,cm$, and annual evaporation is $130\,cm$.

    (a) Estimate the current mass loading of phosphorus in kilograms per year, and the mass loading that is required to reduce the algae concentration in the lake by one-half.

    (b) With the reduced mass loading, how long will it take for the phosphorus concentration in the lake to be within 5% of the equilibrium concentration of phosphorus?

5.13. The surface area of the lake described in Problem 5.8 is $28,000\,m^2$. If the effective settling velocity of TP is $0.1\,m/day$, repeat Problem 5.8 to assess the effect of sedimentation on the TP concentrations.

5.14. Temperature profiles measured in White Lake, Michigan, on May 21 (T_1) and June 6, 1974 (T_2) are given in Table 5.8. Estimate the variation of the vertical diffusion coefficient over the depth of the lake on May 21 and June 6, 1974.

TABLE 5.8 Data for Problem 5.14

Depth (m)	T_1 (°C)	T_2 (°C)	Depth (m)	T_1 (°C)	T_2 (°C)
15	11.0	12.1	7	12.8	16.8
14	12.1	12.3	6	13.0	18.0
13	12.1	12.4	5	13.1	18.0
12	12.1	12.7	4	13.1	18.2
11	12.1	13.3	3	13.2	18.2
10	12.2	13.7	2	13.3	18.2
9	12.4	14.7	1	13.7	18.2
8	12.6	16.4			

Source: Lung (2001).

5.15. A near-shore discharge of a contaminant from a single-port outfall into a lake results in a contaminant concentration of 10 mg/L at a distance of 10 m from the discharge location. The lake currents are negligible, the dispersion coefficient in the lake is 5 m²/s, and the decay rate of the contaminant is 0.01 day⁻¹. Estimate the contaminant concentration 100 m from the discharge location.

5.16. Measurements in the vicinity of a near-shore outfall in a lake indicate that the contaminant concentration 20 m from the outfall is approximately twice the concentration at a distance 40 m from the outfall. Observations also indicate that the currents in the lake are negligible, and the decay constant of the contaminant is 0.05 day⁻¹. Estimate the dispersion coefficient.

CHAPTER 6

WETLANDS

6.1 INTRODUCTION

Wetlands are areas where water covers the ground surface or is present at or near the ground surface for varying periods of time during the year. The prolonged presence of water creates conditions that favor the growth of specially adapted plants (hydrophytes) and promote the development of characteristic wetland (hydric) soils. Although wetlands are often wet, a wetland might not be wet year-round. In fact, some of the most important wetlands are only seasonally wet. Wetlands occur at the interface between land and water and are transition zones where the flow of water, the cycling of nutrients, and the energy of the sun meet to produce a unique ecosystem. The wetlands ecosystem is characterized by unique hydrology, soils, and vegetation that make these areas important features of a watershed. Wetlands are among the most productive ecosystems in the world, comparable to rain forests and coral reefs, and they perform a variety of ecological functions. The combination of shallow water, high levels of nutrients, and primary productivity is ideal for the development of organisms that form the base of the food web and feed many species of fish, amphibians, shellfish, and insects. Wetlands are points of ground-water recharge, absorb water and airborne pollutants, attenuate floodwater, control erosion, cycle minerals such as nitrogen, produce organic matter through carbon fixation, and provide feeding refuge and reproductive habitats for a wide variety of fish and wildlife. More than one-third of the threatened and endangered species in the United States live only in wetlands, and nearly half use wetlands at some point in their lives. Wetlands are found on every continent in the world except Antarctica, and constitute about 4 to 6% of the land surface on Earth (Mitsch and Gosselink, 2000). Wetlands are found at the interface of terrestrial ecosystems, such as upland forests and grasslands, and aquatic ecosystems in rivers, lakes, and oceans. Wetlands are also found in seemingly isolated situations, where

Water-Quality Engineering in Natural Systems, by David A. Chin
Copyright © 2006 John Wiley & Sons, Inc.

the nearby aquatic system is often a ground-water aquifer. Approximately half of the state of Alaska is classified as wetlands (70 to 80 million hectares), with Florida (4.5 million hectares), Louisiana (3.6 million hectares), Minnesota (3.5 million hectares), and Texas (3.1 million hectares) having the next largest areas of wetlands. Wetlands are being lost in the United States at a rate of about 40,000 ha/yr (USEPA, 2000a), and over half of the wetlands globally have been destroyed over the past 150 years (Thom et al., 2001). It has been estimated that more than one-half of the approximately 1 million square kilometers of wetlands in the lower 48 U.S. states have been lost between the arrival of the first settlers and the present (Novotny, 2003).

Some states (in the United States) have created wetland replacement banks funded by fees that developers pay for a permit to drain a wetland on developed property. The fee is used to create a wetland that is the same size or larger in a place where it would be most beneficial.

Until recently in the United States and throughout the world, wetlands were considered as a source of disease (malaria) and an obstacle to human use of land resources for growth, agriculture, and economic development. In reality, in addition to attenuating surface runoff (flood control) and ecological function, wetlands have an important role in controlling the quality of surface waters, and because of their role in removing many anthropogenic pollutants and nutrients in surface waters, wetlands are sometimes called "the kidneys of the landscape." The most important uses and ecological functions of wetlands are:

1. *Flood storage and conveyance.* Riverine wetlands and adjacent floodplains form natural corridors in which floodwater is stored and lateral inflow is attenuated.

2. *Erosion reduction and sediment control.* Wetlands adjacent to rivers and lakes contain vegetation that slows runoff and flood flow, thus mitigating shoreline erosion and enhancing sedimentation. Vegetation serves also to filter sediments from water, and the roots of the vegetation then bind and stabilize the deposited silt.

3. *Ground-water recharge and discharge.* Wetlands and ground water are interconnected (unless they are artificial wetlands designed to treat wastewater and stormwater runoff); however, the connection is poor since the very existence of the wetland usually implies highly impervious substrate soils. If shallow ground water is discharging into a wetland, it can bring nutrients (primarily nitrate nitrogen) and minerals, which are then used by the wetland vegetation. If the wetland is recharging, it is a natural barrier preventing some mobile pollutants from entering the ground water.

4. *Pollution control.* Wetland vegetation and microbes residing on the stems and roots of vegetation and in sediments can filter, absorb, and decompose suspended and dissolved organic matter and convert organic-nitrogen and ammonia-nitrogen to nitrate (nitrification) with subsequent denitrification of nitrate to nitrogen gas. Some phosphorus can be used by the vegetation and/or absorbed onto sediments. The accumulated residual organic matter can also immobilize toxic pollutants and bury them without harm to the biota.

5. *Buffers for pollution control.* Wetlands have very high assimilative capacity for nitrate-nitrogen and hydrophobic toxic compounds, including metals. The assimilative capacity for toxic compounds is related to the primary productivity that is stimulated by nutrient inputs. Wetlands are also efficient in removing turbidity and BOD.

6. *Wildlife habitats.* The land–water interface is among the richest wildlife habitats in the world. This is because of the abundance of water, nutrients, and shelter provided by the wetlands and their vegetation. Many endangered species rely on wetlands.

Wetlands are broadly classified as either natural wetlands or constructed wetlands. Natural wetlands are formed by nature with limited human influence, whereas constructed wetlands are engineered systems that are designed to function like natural wetlands. Constructed wetlands are used to treat municipal and industrial wastewater, agricultural runoff, and urban runoff.

6.2 NATURAL WETLANDS

Natural wetlands can be broadly classified as either coastal wetlands or inland wetlands, where coastal wetlands are influenced by alternate floods and ebbs of tides, and inland wetlands are nontidal. Coastal wetlands in the United States are found along the Atlantic, Pacific, Alaskan, and Gulf coasts, and there are a wide variety of estuarine and marine fish, shellfish, birds, and mammals that must have coastal wetlands to survive. Coastal wetlands may be further categorized as either marine or estuarine wetlands (Chang, 2002), depending on whether they are adjacent to an open ocean or adjacent to an estuary. In contrast to coastal wetlands, inland wetlands are most common on floodplains along rivers and streams (*riverine wetlands*), in isolated depressions surrounded by dry land (*palustrine wetlands*), along littoral zones of lakes and ponds (*lacustrine wetlands*), and in other low-lying areas where the ground water intercepts the soil surface or where precipitation sufficiently saturates the soil. Riverine and lacustrine wetlands are sometimes collectively referred to as *riparian wetlands*. Wetlands are typically classified as marshes, swamps, bogs, or fens.

6.2.1 Marshes

Marshes are wetlands that are frequently or continually inundated with water and characterized by emergent soft-stemmed vegetation that is adapted to saturated-soil conditions. A typical marsh is illustrated in Figure 6.1. There are many different kinds of marshes,

FIGURE 6.1   Typical marsh. (Courtesy of David A. Chin.)

ranging from *prairie potholes* in the central and northern United States to the vast Everglades in Florida. Marshes are found in coastal and inland areas, and can contain fresh or salt water. Marshes receive most of their water from surface runoff, and many marshes are directly connected to the ground water. Marshes typically recharge the ground water and moderate stream flow by providing water to streams, which is an especially important function during periods of drought. The presence of marshes in a watershed reduces damage caused by floods because they slow and store floodwater. Marshes are very important in preserving the quality of surface waters. As water moves slowly through a marsh, sediment and other pollutants settle to the bottom of the marsh. Vegetation and microorganisms found in marshes utilize excess nutrients that can otherwise pollute surface water.

Freshwater marshes have larger and more diverse plant populations than saltwater or tidal marshes, and irregularly flooded salt marshes exhibit the fewest species of plants (DeBarry, 2004). Salt marshes are typically composed mainly of rushes, sedges, and grasses. Animals seek refuge from predators in the thick marsh vegetation. After marsh plants die, microorganisms break the plants down into detritus, which serves as a food source for many small animals.

6.2.2 Swamps

A *swamp* is any wetland dominated by wood plants, and are characterized by saturated soils during the growing season and standing water during portions of the year. Swamps serve vital roles in flood protection, nutrient removal, and sediment removal. Swamps are divided into two major categories, depending on the type of vegetation present: forested swamps and shrub swamps.

Forested Swamps   *Forested swamps* are found throughout the United States and are often inundated with floodwater from nearby rivers and streams. A typical forested swamp is illustrated in Figure 6.2. In very dry years they may represent the only shallow water for

FIGURE 6.2   Typical forested swamp.

kilometers, and their presence is critical to the survival of wetland-dependent species such as wood ducks (*Aix sponsa*), river otters (*Lutra canadensis*), and cottonmouth snakes (*Agkistrodon piscivorus*). Common species of trees found in forested wetlands are red maple and pin oak (*Quercus palustris*) in the northern United States, overcup oak (*Quercus lyrata*) and cypress (*Taxodium* spp.) in the south, and willows (*Salix* spp.) and western hemlock (*Tsuga* spp.) in the northwest. *Bottomland hardwood swamp* is a type of forested swamp found in the south-central United States.

Bottomland hardwood forests are found along rivers and streams in the southeast and south-central United States, generally in broad floodplains. These forested wetlands are typically made up of different species of gum (*Nyssa* spp.), oak (*Quercus* spp.), and bald cypress (*Taxodium distichum*); these trees have the ability to survive in areas that are either seasonally flooded or covered with water much of the year. Bottomland hardwoods serve a critical role in a watershed by reducing the risk and severity of flooding to downstream communities by providing areas to store floodwater. In addition, hardwood forests improve water quality by filtering and flushing nutrients, processing organic wastes, and attenuating sediment before it reaches open water.

Shrub Swamps   *Shrub swamps* are similar to forested swamps except that shrubby vegetation such as buttonbush, willow, dogwood (*Cornus* spp.), and swamp rose (*Rosa palustris*) predominates. Mangrove swamps are a common type of shrub swamp dominated by mangroves, as illustrated in Figure 6.3. Most mangrove swamps are denser than shown in Figure 6.3(*a*), and Figure 6.3(*b*) shows a close-up view of a typical dense mangrove swamp. Mangrove swamps are coastal wetlands found in tropical and subtropical regions. The word *mangrove* refers to both the wetland itself and to the salt-tolerant trees that dominate these wetlands. Florida's southwest coast supports one of the largest mangrove swamps in the world. In the continental United States, only three species of mangrove grow: red, black, and white mangroves. Red mangrove (*Rhizophora mangle*) is easily recognized by its distinctive arching roots. Black mangrove (*Avicennia* spp.), which often grows more inland, has root projections (pneumatophores), which help to supply the plant with air in submerged soils. White mangroves (*Laguncularia racemosa*) grow farther inland with no outstanding root structures. In Florida, mangrove swamps are dominated by

(*a*)   (*b*)

FIGURE 6.3   Mangrove swamp.

red and black mangroves. Mangrove swamps are constantly replenished with nutrients transported by freshwater runoff from the land and flushed by the ebb and flow of tides. Mangroves perform a variety of important functions, such as buffering the shoreline against heavy wave action, filtering sediment- and pollutant-laden water, preventing shoreline erosion, and serving as rookeries for a variety of wading birds. Mangrove trees support a significant fish habitat for many juvenile fish species, including barracuda and snapper.

6.2.3 Bogs

Bogs are one of North America's most distinctive types of wetland. They are characterized by spongy peat deposits, acidic waters, and a floor covered by a thick carpet of sphagnum moss. A typical bog (the Männikjärve bog in Estonia) is illustrated in Figure 6.4. Bogs receive all or most of their water from precipitation rather than from surface runoff, ground water, or streams. Consequently, bogs are low in the nutrients needed for plant growth, a condition that is enhanced by acid-forming peat mossses. The unique and demanding physical and chemical characteristics of bogs result in the presence of plant and animal communities that demonstrate many adaptations to low nutrient levels, waterlogged conditions, and acidic waters, such as carnivorous plants. In the United States, bogs are found in the glaciated northeast and Great Lakes regions, where they are referred to as *northern bogs*, and bogs are also found in the southeast, where they are referred to as *pocosins*. Bogs serve an important ecological function in preserving downstream flooding by absorbing precipitation, and bogs have been recognized for their role in regulating the global climate by storing large amounts of carbon in peat deposits.

6.2.4 Fens

Fens are peat-forming wetlands that receive nutrients from sources other than precipitation, usually from upslope sources through surface runoff over surrounding mineral soils

FIGURE 6.4   Aerial view of a bog wetland. (From Aber, 2001. Copyright: J. S. Aber.)

FIGURE 6.5   Typical fen. (From McGlynn, 2005. Photo by Steve Cook.)

and from ground-water recharge. A typical fen is illustrated in Figure 6.5, where the sloping land above the fen is shown clearly in the background. Fens differ from bogs in that they are less acidic and have higher nutrient levels, which make them able to support a much more diverse plant and animal community. Fens are often covered by grasses, sedges, rushes, and wildflowers. Fens are mostly a northern hemisphere phenomenon, occurring in the northeastern United States, Great Lakes region, the Rocky Mountains, and much of Canada. Fens are generally associated with low temperatures and short growing seasons, where ample precipitation and high humidity cause excessive amounts of moisture to accumulate. Fens provide important benefits in a watershed, including preventing or reducing flood risks, improving water quality, and providing habitat for unique plant and animal communities.

6.3   DELINEATION OF WETLANDS

A variety of definitions are used around the world to delineate wetlands. In the United States, the U.S. Army Corps of Engineers' definition of a wetland carries particular weight, since the Corps is responsible for delineating wetlands for regulatory purposes and issuing dredge-and-fill permits required by the Clean Water Act. The Corps of Engineers defines wetlands as "those areas that are inundated or saturated by surface or ground water at a frequency and duration sufficient to support, and that under normal circumstances do support, a prevalence of vegetation typically adapted for life in saturated soil conditions" (US ACE, 1987). Wetlands delineated for regulatory purposes are called *jurisdictional wetlands*. According to the *U.S. Army Corps of Engineers Wetlands Delineation Manual* (US ACE, 1987), three factors must generally be considered in delineating a wetland: vegetation, soils, and hydrology.

6.3.1 Vegetation

Wetland vegetation is defined as macrophytes that are adapted to inundated or saturated conditions. For purposes of delineation, plants can be grouped into five categories: obligate wetland plants (OBLs), facultative wetland plants (FACWs), facultative plants (FACs), facultative upland plants (FACUs), and obligate upland plants (UPLs). The characteristics of the five plant categories are given in Table 6.1. To meet the wetland vegetation requirement, more than 50% of the dominant species in a delineated wetland area must be OBLs, FACWs, or FACs. Species lists of plants in these categories can be found in USACE (1987). Marsh grass (*Spartina alterniflora*) and bald cypress (*Taxodium distichum*), both shown in Figure 6.6, are plant species that are almost always found in wetlands.

TABLE 6.1 Plant Categories Used in Wetland Delineation

Category	Symbol	Definition
Obligate wetland plant	OBL	Plants that occur almost always (estimated probability >99%) in wetlands under natural conditions but which may also occur rarely (estimated probability <1%) in nonwetlands. Examples: *Spartina alterniflora, Taxodium distichum.*
Facultative wetland plant	FACW	Plants that usually occur in wetlands (estimated probability >67 to 99%) but also occur in nonwetlands (estimated probability >1 to 33%). Examples: *Fraxinus pennsylvanica, Cornus stolonifera.*
Facultative plant	FAC	Plants with a similar likelihood (estimated probability 33 to 67%) of occurring in both wetlands and nonwetlands. Examples: *Gleditsia triaconthos, Smilax rotundifolia.*
Facultative upland plant	FACU	Plants that sometimes occur in wetlands (estimated probability 1 to 33%) but occur more often in nonwetlands (estimated probability >67 to 99%). Examples: *Quercus rubra, Potentilla arguta.*
Obligate upland plant	UPL	Plants that rarely occur in wetlands (estimated probability <1%) but almost always occur in nonwetlands (estimated probability >99%) under natural conditions. Examples: *Pinus echinata, Bromus mollis.*

(a)   (b)

FIGURE 6.6   Wetland plants: (*a*) marsh grass (*Spartina alterniflora*); (*b*) bald cypress (*Taxodium distichum*). (From NOAA, 2005b.)

6.3.2 Soils

Wetland soils, called *hydric soils*, are defined as soils that are saturated, flooded, or ponded long enough during the growing season to develop anaerobic conditions in the upper soil layers that favor the growth and regeneration of hydrophytic vegetation. Most soils, including hydric soils, are composed predominantly of minerals such as quartz, feldspars, and clay minerals. However, hydric soils commonly have a buildup of organic matter at the soil surface, which can make the surface horizon dark colored. If the organic matter content (measured as organic carbon) is greater than 20 to 30% of the soil's weight and this organic-rich layer is more than 40 cm thick, it is considered an organic soil. *Peat* is composed mostly of recognizable plant fragments that are only partly decomposed; *muck* contains highly decomposed organic matter. When drained of excess water and carefully managed, muck soils are among the most important vegetable-producing soils in the eastern United States.

A property unique to hydric soils is their color or color patterns. Besides the dark shading from the presence of organic matter, iron compounds are the most important coloring agents in soils. Hydric soils tend to exhibit gray or blue-gray colors (known as *gleying* or *gleyed colors*), especially just beneath the topsoil or surface horizon, as shown in Figure 6.7. This results from the chemically reduced oxidation state of iron compounds as opposed to the rusty red (oxidized) and brown colors of drier, nonhydric soils. Where shallow water tables fluctuate, gray, yellow, and red colors can also occur as small splotches, or as thread-like or network patterns, created by accumulations or depletions of iron and manganese. Because they result from processes of reduction and oxidation, these color indicators of wetness are collectively termed *redoximorphic features*.

When a hydric soil is drained, it may no longer be referred to as a hydric soil, unless the supported vegetation is hydrophytic, and indicators of hydrology support the designation as a hydric soil.

6.3.3 Hydrology

Hydrologic factors that determine whether an area is a wetland are the frequency, timing, and duration of inundation or soil saturation, as shown in Table 6.2 for nontidal areas. Zone I is

FIGURE 6.7   Hydric soil. (From Suffolk County Soil and Water Conservation District, 2005.)

TABLE 6.2   Hydrologic Zones for Nontidal Areas

Zone	Name	Percent of Time Inundated	Comments
I	Permanently inundated	100	Inundation >2 m mean water depth. Aquatic, not wetlands.
II	Semipermanently to nearly inundated or saturated	75–100	Inundation defined as <2 m mean water depth.
III	Regularly inundated or saturated	25–75	
IV	Seasonally inundated or saturated	12.5–25	
V	Irregularly inundated or saturated	5–12.5	Many areas having these hydrologic characteristics are not wetlands.
VI	Intermittently or never inundated or saturated	<5	Areas with these hydrologic characteristics are not wetlands.

Source: USACE (1987).

aquatic, zones II, III, and IV are wetlands, and zone VI is upland. Zone V may or may not be considered a wetland, depending on the other indicators. Recorded water-level data from adjacent streams and lakes, as well as model predictions, can be used to establish the hydrologic condition of an area.

If an area meets any of the wetland criteria regarding vegetation, soils, and hydrology, the area is classified by the Corps of Engineers as a *jurisdictional wetland.*

6.4 WETLAND HYDROLOGY

Wetlands are part of a landscape mosaic that provides several watershed functions, and removal or alteration of wetlands through alteration of hydrology significantly affects the health and function of the broader landscape. The hydrology of a wetland creates the unique physicochemical conditions that make such an ecosystem different from both well-drained terrestrial systems and deepwater aquatic systems. Hydrologic pathways such as rainfall, surface runoff, ground-water flow, tides, and flooding rivers transport energy and nutrients to and from wetlands. Water depth, flow patterns, and duration and frequency of flooding are the result of hydrologic inputs and outputs, and these factors have a direct influence on the biochemistry of soils and the biota of wetlands. Except in nutrient-poor wetlands such as bogs, water inputs are the major source of nutrients to wetlands. When hydrologic conditions in wetlands change, even slightly, the biota may respond with massive changes in species composition, richness, and ecosystem productivity. Several animals are particularly noted for their contributions to hydrologic modifications and subsequent changes in wetlands. Beavers are noted for building dams, muskrats for burrowing, and geese (particularly Canada geese, shown in Figure 6.8) for consuming excessive amounts of wetland vegetation.

FIGURE 6.8   Canada geese. (From Jackson Bottom Wetlands Preserve, 2005.)

The *hydroperiod* is the seasonal pattern of water level in a wetland and can be considered a hydrologic signature of each wetland type. Factors that affect the hydroperiod include (1) the balance between inflows and outflows, (2) the surface contours of the landscape, and (3) the subsurface soil, geology, and ground-water conditions. The balance between inflows and outflows defines the water budget of the wetland, whereas the surface contours and subsurface conditions define the capacity of the wetland to store water. For wetlands that are not subtidal or permanently flooded, the amount of time that a wetland contains standing water is called the *flood duration*, and the average number of times that a wetland is flooded in a given period is known as the *flood frequency*. Each depth of inundation has a range of possible flood durations, and each duration has a corresponding flood frequency.

The hydrologic budget of a wetland is given by the relation

$$\frac{\Delta V}{\Delta t} = P_n + S_n + G_n - \text{ET} \qquad\qquad (6.1)$$

where ΔV is the volume of water added to the wetland in time increment Δt, P_n is the incremental net precipitation, S_n is the incremental net surface-water inflow, including flooding streams, G_n is the incremental net ground-water inflow, and ET is the incremental evapotranspiration. The incremental net precipitation is equal to the incremental precipitation minus the amount of this precipitation intercepted by the vegetation. The components of the wetland water budget are discussed below.

6.4.1 Net Surface-Water Inflow

Wetlands receive surface inflows in many forms. *Overland flow* is a nonchannelized sheet flow that usually occurs during and immediately following rainfall or a spring thaw, or as tides rise in coastal wetlands. Channelization of flow, which is usually associated with urbanization, has a significant impact on the functioning wetlands with significant overland inflow. In addition, roads can block or severely alter the outflow dynamics of the system, and increased flow rates through culverts can be a major impediment to wetland functioning as well as fish migration. A special case of surface flow occurs in riparian wetlands that are in floodplains adjacent to rivers or streams and are occasionally flooded by those rivers or streams. Examples of riparian wetlands are the delta marshes of the Mississippi River (Gosselink, 1984) and the hardwood swamps of the southeastern United States (Wharton et al., 1982). Some wetlands can be more isolated, receiving only intermittent surface-water input. These systems include *vernal pools* in California (Zedler, 1987), *prairie pothole wetlands* in the midwest (Kantrud et al., 1989), and *playa lakes* in the southern high plains.

6.4.2 Net Ground-Water Inflow

Wetlands can recharge ground water or can be located in areas where ground water is discharged to the wetland. Movement of ground water into or out of a wetland is a function of the permeability of the soils, which is partially affected by vegetation and soil type. The drawdown of the water table caused by urbanization of nearby areas can have a deleterious effect on wetland function, and such an effect has been documented in the forested wetlands of Florida (Mortellaro et al., 1995). The effects of water-table drawdown on wetlands include reduced hydroperiod, lower water levels, a shift to drier, nonwetland plant species, death of animal species, loss of fish and amphipods, reduced use by birds and wildlife, and increased fire damage.

6.4.3 Evapotranspiration

Evapotranspiration (ET) typically accounts for loss of water ranging from 20 to 80% of the annual water budget in most wetland systems (Mitsch and Gosselink, 2000). Evapotranspiration plays a role in the attenuation of floodwaters as well as maintenance of the soil-water-redox conditions in a wetland. Changing the water input, loss of plants, and changes in soil conditions will affect ET and in turn change the functioning of the wetland. It is noteworthy that transpiration can exceed open-water evaporation losses in prairie-pothole systems (LaBaugh et al., 1998).

As a general rule, natural wetlands are more prevalent in cool or wet climates than in hot or dry climates, and steep terrain tends to have fewer wetlands than gently sloping landscapes. Wetlands occur most extensively in regions where precipitation is in excess of losses such as evapotranspiration and surface runoff. The dividing line between precipitation excess in the eastern United States and precipitation deficit in the western United States is the Mississippi River.

6.5 CASE STUDY: THE EVERGLADES AND BIG CYPRESS SWAMP

Southern Florida, from Lake Okeechobee southward to Florida Bay, contains several unique natural wetlands. Within a 34,000-km^2 area are three major types of wetlands: the Everglades, the Big Cypress Swamp, and the coastal mangroves of Florida Bay. The relative locations of these wetland systems are shown in Figure 6.9.

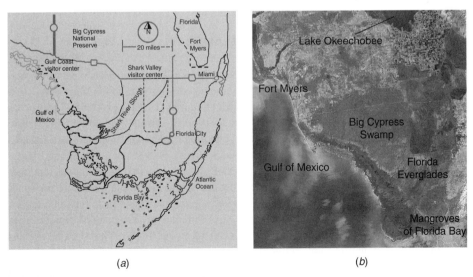

(a)   (b)

FIGURE 6.9   Wetland systems in south Florida: (*a*) map of south Florida; (b) satellite view. (From NASA, 2005a.)

Water flowing through the Everglades from Lake Okeechobee to Florida Bay is typically only a few centimeters deep and about 80 km wide. For this reason, the Everglades has been affectionately called the "river of grass" (Douglas, 1947). The Everglades is dominated by sawgrass (*Cladium jamaicense*), which is actually a sedge, not a grass. In addition to sheet flow, the Everglades contains deeper water sloughs and tree islands, or *hammocks*, that support a vast diversity of plants, including hardwood trees, palms, orchids, and other air plants. Surface water in the Everglades flows in a generally southerly or southwesterly direction through the grasses and other macrophytes at relatively low velocities that generally range from 0.5 to 2 cm/s (Ricassi and Schaffranek, 2003). Under most conditions, flow conditions are in the laminar regime (Lee et al., 2004). Tracer experiments in the Everglades have indicated longitudinal dispersion coefficients on the order of 5×10^{-5} m^2/s in regions where the average velocity is on the order of 0.3 cm/s (Harvey et al., 2005).

About half of the original Everglades has been lost to agriculture and urban development, and the health of the Everglades has been affected significantly by changes in the quantity, quality, timing, and distribution of water entering the Everglades. For example, natural phosphorus levels in the Everglades are on the order of 10 μg/L, and where the phosphorus levels rise from 10 μg/L to 50 μg/L an invasion of cattails and other foreign species significantly changes the local wetland ecology and affects the habitat for many endangered species. To address these and other impacts, the Everglades is currently the site of the largest wetland-restoration effort in the United States. The comprehensive restoration blueprint includes plans for improving the water quality as it enters the Everglades from agricultural areas to the north and for modifying the hydrology to conserve and restore habitat for declining populations of wading birds such as the wood stork and the white ibis.

The Big Cypress Swamp is dominated by cypress trees interspersed with pine flatwoods and wet prairie. The Big Cypress Swamp receives about 125 cm of rainfall per year but does not receive major inputs of overland flow as the Everglades does. The third major wetland type, mangroves, forms impenetrable thickets where the sawgrass and cypress swamps meet the saline waters of the Florida coastline.

6.6 CONSTRUCTED TREATMENT WETLANDS

Constructed treatment wetlands are engineered systems that utilize natural processes involving wetland vegetation, soils, and their associated microbial assemblages to assist, at least partly, in treating an effluent or other water source (USEPA, 2000a). Constructed treatment wetlands were initially applied around 1960 in Europe and North America to exploit the biodegradation ability of plants. Constructed treatment wetlands are used to treat wastewater from municipal and industrial sources, as well as contaminated water from agricultural and stormwater sources. The advantages of using constructed treatment wetlands include low construction and operating costs, and they are appropriate for small communities and as a final-stage treatment in large municipal systems. Constructed wetlands for treating municipal wastewater are an excellent option for small communities where inexpensive land is available and skilled operators are hard to find. In arid regions and communities reaching their limits of water availability, water reuse using these systems is an attractive option that can help achieve water conservation and wildlife habitat goals. A disadvantage of constructed treatment wetlands is their relatively slow rate of operation in comparison to conventional wastewater-treatment technology. Constructed wetlands have proven effective in removing metals from waste streams; however, if a large volume of waste is involved, a sizable expanse of wetlands and rigorous rotation of areas in the treatment process may be required (Stone, 1999). Also, once the plants take up and concentrate the metals, they themselves become a hazardous waste that must be disposed of properly. As of 2000, there were more than 600 active constructed-treatment wetland projects in the United States (USEPA, 2000a). As of 2004, the largest constructed wetland in the world was the 6700-ha stormwater-treatment wetland (locally designated as STA 3/4) located in the Everglades Agricultural Area of Florida (Bufe, 2005). Although constructed wetlands mimic natural wetlands, they represent created ecosystems that are not part of the original wetland resources of a region, and therefore continuous flooding with polluted water is more acceptable.

There are two types of constructed wetlands: surface-flow wetlands and subsurface-flow wetlands. Constructed *surface-flow wetlands*, sometimes called *free water surface wetlands*, are similar to natural wetlands with predominant overland flow. In constructed *subsurface-flow wetlands*, sometimes called *vegetated submerged beds*, water exits the wetland through subsurface drains. Schematic diagrams of natural and constructed wetlands used in water-quality control are illustrated in Figure 6.10. Both surface- and subsurface-flow wetlands are used to treat municipal wastewaters. However, surface-flow wetlands that receive less than secondary treated wastewater are potential health hazards and odor sources and are not well suited for use at parks, playgrounds, and similar public facilities; gravel-bed subsurface-flow systems often dominate these applications. In using constructed treatment wetlands to treat urban runoff, agricultural runoff, and livestock and poultry wastes, surface-flow wetlands are most commonly used. Wetland treatment of landfill wastes use both surface- and subsurface-flow systems. Overall, subsurface-flow systems are more effective than surface-flow systems at removing pollutants at high application rates. However, overloading, surface flooding and media clogging can result in reduced efficiency of subsurface systems (Shutes, 2001).

6.6.1 Surface-Flow Wetlands

A constructed surface-flow (SF) wetland attempts to reproduce a natural surface-flow wetland. A constructed surface-flow wetland typically consists of basins or channels with a subsurface barrier of clay or impervious geotechnical material (lining) to prevent seepage. The basins are then filled with soils to support emergent vegetation. If the soil is brought from an existing wetland, wetland vegetation may emerge without seeding; however, seeding and planting of vegetation is part of the construction process.

The fully vegetated zones of most SF systems used to treat domestic wastewater are anaerobic. This anaerobic condition is caused by the significant oxygen demand of primary or lagoon-treated influent, and the secondary invasion of duckweed, which almost always accompanies the detritus from emergent species to effectively inhibit atmospheric reaeration of the water column. Under anaerobic conditions, biological reactions are extremely slow, even though there is considerable biological growth attached to the dense stand of emergent plants. The attached-growth microorganisms are responsible for much of the biological treatment, however, the shade provided by the emergent plants keeps the water column cooler than it would be in open areas, and the lack of sunlight penetration limits the natural pathogen kill that occurs in open-water zones. The dominant forces removing pollutants from wastewater in fully vegetated zones are physical. If the detention time in a fully vegetated zone is about 2.5 days, the flocculation and high-rate sedimentation resulting

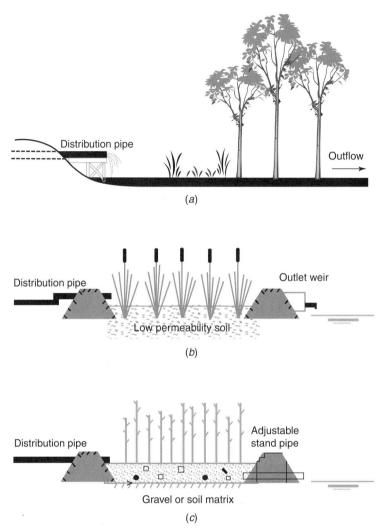

FIGURE 6.10 Wetland treatment systems: (*a*) natural; (*b*) surface-flow; (*c*) subsurface-flow. (Copyright © 1996 from *Treatment Wetlands* by R. H. Kadlec and R. L. Knight. Reproduced by permission of Routledge/Taylor & Francis Group, LLC.)

from tortuous travel through dense plant stems results in efficient removal of settleable (>100 μm), supracolloidal (between 1 and 100 μm), and colloidal (<1 μm) solid fractions of total suspended solids (TSS), BOD, heavy metals, and organic fractions of total nitrogen and phosphorus. In addition to the fully vegetated zone, which can produce effluent meeting secondary treatment standards, surface-flow wetlands may also contain open-water zones to meet higher treatment standards. In the open-water zone, where aerobic conditions prevail because of the oxygen produced by submerged vegetation and atmospheric reaeration, more rapid aerobic biochemical reactions oxidize soluble BOD and nitrify ammonia-nitrogen. In addition, solar radiation increases pathogen kills. One key to preventing algae blooms in open-water zones is to make the detention time small, usually

FIGURE 6.11 Constructed surface-flow wetland. (From North Carolina State University, 2005.)

2 to 3 days, depending on temperature. The final zone of a constructed surface-flow wetland should be another fully vegetated zone to flocculate and settle biomass created by aerobic reactions in the open-water zone, denitrify the nitrate-nitrogen produced, and keep wildlife (particularly birds) away from the effluent collection system to minimize their effect on effluent quality. A typical constructed surface-flow wetland is shown in Figure 6.11, where the inflow pipe and sedimentation area (forebay) is readily apparent at the head of the wetland.

The water depth in constructed SF wetlands typically range from 5 to 80 cm (0.2 to 2.6 ft), with design flows ranging from less than 4 m^3/day (1000 gpd) to more than 75,000 m^3/day (20 mgd). Annual phosphorus removal in surface-flow constructed wetlands is typically minimal (Kreissl and Brown, 2000). Phosphorus uptake by plants is mostly returned to the water during the senescent period, making the maximum phosphorus removal obtainable in the range of less than 1.5 mg/L in systems that are (lightly) loaded at less than 1.5 kg/ha · day.

To develop a wetland that is a low-maintenance site, the natural successional process must be allowed to proceed. Often, this may mean some initial period of invasion by undesirable plant species, but if proper hydrologic and nutrient loads are maintained, this invasion is usually temporary.

6.6.2 Subsurface-Flow Wetlands

Constructed subsurface-flow (SSF) wetlands are essentially horizontal gravel filters that provide treatment similar to the first, fully vegetated zone of the constructed surface-flow (SF) wetland. Because of their reduced size and inherent barriers to prevent human contact, SSF wetlands have been used as on-site treatment facilities following septic tanks and, usually, before soil absorption. Constructed SSF systems have also been used for domestic wastewater treatment after primary settling with good success in meeting secondary effluent standards (TSS, BOD ≤ 30 mg/L).

FIGURE 6.12 Constructed subsurface-flow wetland. (From Nova Scotia Agricultural College, 2005. Photo by Steven Tattrie.)

Subsurface-flow wetlands are typically about 50 cm to 1.0 m deep from top to bottom and contain soil, gravel, or rock media with emergent aquatic plants rooted in the media. In the United States, the most common medium is gravel, but sand and soil have been used in Europe. Most controlled studies indicate that these systems perform as well without plants as with plants (Kreissl and Brown, 2000). The water surface in SSF wetlands is maintained beneath the upper surface of the underlying porous medium. In operational systems in the United States, the porous medium depth ranges from 30 to 90 cm (1 to 3 ft), and design flows range from less than 190 m^3/day (50,000 gpd) to 13,000 m^3/day (3.5 mgd). The SSF system is underlain with an impermeable layer of clay or synthetic liner, and the system is built with a slight inclination of 1 to 3% between the inlet and the outlet. A typical subsurface-flow wetland is shown in Figure 6.12, where the surface vegetation is in the early stage of development.

Problems have arisen when SSF systems were used to remove nitrogen or perform other functions for which they are not physically capable. In both surface- and subsurface-flow constructed wetlands, physical pollutant removal may be temporary because chemical and microbiological degradation in anaerobic areas can return some settled organic and inorganic matter to the water column by transforming particulate matter to other, more soluble, matter.

Constructed subsurface-flow wetlands have been used extensively in Europe under the names *root zone method, hydrobotanical system, soil filter trench, biological macrophytic and marsh bed*, and *vegetated submerged bed*.

6.6.3 Wetland Regulations in the United States

Under the Clean Water Act (CWA), states and tribes must adopt water-quality standards for all natural water bodies. Water-quality standards include specification of a "designated use" for each water body, criteria to protect each designated use, and an antidegradation policy (CWA Section 303). Natural (jurisdictional) wetlands are classified as *waters of the United States*, and water-quality standards must be adopted for all natural wetlands. Such water-quality standards usually preclude using natural wetlands for water treatment. As a consequence, constructed treatment wetlands are typically located in uplands, outside waters of the United States. As a general rule, municipal wastewater effluent must be treated to at least secondary levels before it enters waters of the United States (CWA Section 301), such as rivers and oceans. Circumstances that would allow a constructed treatment wetland to be located in an existing wetland or other waters of the United States include (1) the source water meets all applicable water-quality standards and criteria; (2) use of a constructed wetland would result in a net environmental benefit to the aquatic system's natural functions and values; and (3) a constructed treatment wetland would help restore the aquatic system to its historic, natural condition.

If a constructed treatment wetland is not itself a water of the United States but discharges pollutants into a water of the United States, the discharge requires a National Pollutant Discharge Elimination System (NPDES) permit under CWA Section 402. The Section 402 program is administered at the federal level by the USEPA, and over 40 states are authorized by the USEPA to administer the NPDES permitting program within their state.

Filling of natural wetlands is a common occurence in land development. If construction activities involve the placement of dredged or fill material to the waters of the United States, including wetlands, permit authorization under CWA Section 404 is required. At the federal level, the USEPA and the U.S. Army Corps of Engineers jointly administer the Section 404 program, with the U.S. Fish and Wildlife Service, the National Marine Fisheries Service, and state resource agencies having advisory roles. An *individual permit* is usually required for potentially significant dredge and fill impacts; however, for most dredge and fill activities that will have only minimal adverse effects, the U.S. Army Corps of Engineers often grants *general permits*. These general permits may be issued on a nationwide, regional, or state basis for particular categories of activities, such as minor road crossings, utility line backfill, and bedding. An example of a dredge and fill operation is shown in Figure 6.13. Section 404 regulations require that new wetlands be created if

FIGURE 6.13   Dredge-and-fill operation. (From USGS, 2005g.)

natural wetlands are filled. However, natural wetlands typically provide much better ecosystem services than do human-made wetlands (DeBarry, 2004).

In general, wetlands constructed or restored for the primary purpose of treating wastewater will not be recognized as compensatory mitigation to offset wetland losses authorized under federal regulatory programs (USEPA, 2000a). In some cases, however, components of constructed wetland treatment systems that provide wetland functions beyond what is needed for treatment purposes may be used for compensatory mitigation. For example, project sponsors may be eligible to receive mitigation credit for using treated effluent as part of a constructed treatment wetland system that restores or creates additional wetland area beyond the area need for treatment purposes.

6.6.4 Basic Principles for Wetland Restoration and Creation

Wetland restoration and creation is an area of practice that is sometimes considered a specialty area within ecological engineering. A key philosophy in ecological engineering is to design systems that are as close to the natural features of the landscape as possible and require a minimum amount of maintenance (Mitsch and Jørgensen, 1989). Specific principles to be considered in restoring and creating wetlands are (Novotny, 2003):

- Design the system for minimum maintenance. The system of plants, animals, microbes, substrate, and water flows should be developed for self-maintenance and self-design.
- Design a system that utilizes natural energies, such as the potential energy of streams.
- Develop a system considering the landscape. The best sites are locations where wetlands existed previously or where nearby wetlands still exist.
- Design the system as an *ecotone*. This means including a buffer strip around the site but also means that the wetland site itself needs to be viewed as a buffer between the upland and the aquatic system to be protected.
- Take into consideration the uses of the surrounding lands. For example, in agricultural zones, planned idling of land for soil-erosion control may obviate the need for a wetland to control pollution by nonpoint runoff.
- Hydrologic conditions are paramount. Without water for at least part of the growing season, a wetland is impossible.
- Give the system time. Wetlands are not created overnight.
- The soils should be surveyed. Highly permeable soils do not support healthy wetland systems.
- Consider the chemical composition of influent waters, including ground-water discharge to the wetland. This affects biological productivity and/or bioaccumulation of toxic materials.

6.6.5 Design of Constructed Treatment Wetlands

Guidance for the design of wetlands to treat municipal wastewaters can be found in the USEPA manual, *Constructed Wetlands Treatment of Municipal Wastewaters* (EPA/625/R-99/010), and the Water Environment Federation (WEF) Manual of Practice, *Natural Systems for Wastewater Treatment* (FD-16). Design goals for wetlands range

FIGURE 6.14   Constructed wetland components. (From Earth Action Partnership, 2005. Photo by Pamela Feagler.)

from an exclusive commitment to treatment functions to systems that provide advanced treatment combined with enhanced wildlife habitat and public recreational opportunities. The size of constructed wetlands range from small on-site units designed to treat septic tank effluent from a single-family dwelling to 16,200 ha (40,000 acres) of wetlands being designed in South Florida for treating phosphorus in agricultural stormwater runoff. A typical constructed wetland for stormwater treatment is shown in Figure 6.14. Storm water enters the first bay from the large drain pipe shown in Figure 6.14(*a*) and slows down immediately, which allows the sedimentation of pollutants. The cleaner water then flows gently through the meandering wetland channel shown in Figure 6.14(*b*), which covers about 0.4 ha of ground. The channels are planted with wetland grasses and plants which help to further clean the water. Figure 6.14(*c*) shows the end of the channel at the last bay, where the water level is controlled by an outlet-pipe structure.

Hydrology is the most important variable in wetland design. If the proper hydrologic conditions are developed, chemical and biological conditions will respond accordingly. Parameters used to characterize the hydrologic conditions of treatment wetlands include hydroperiod, seasonal pulses, hydraulic loading rate, and detention time. Other important variables in wetland design include basin morphology, chemical loading, soil composition, vegetation, and management after construction. Typical design parameters for constructed wetlands treating wastewater effluents are listed in Table 6.3, and these parameters are covered in more detail in the following sections.

TABLE 6.3 Design Parameters for Constructed Wetlands

Design Parameter	Surface-Flow Systems	Subsurface-Flow Systems	Natural Systems
Minimum size requirement (ha/1000 m³ · day)	2–4	1.2–1.7	5–10
Hydraulic loading (cm/day)	2.5–5	5.8–8.3	1–2
Maximum water depth (cm)	50	Water level below ground surface	50; depends on native vegetation
Bed depth (cm)	—	30–90	—
Minimum aspect ratio	2:1	—	1.4:1
Minimum hydraulic residence time (days)	5–10	5–10	14
Minimum pretreatment	Primary; secondary is optional	Primary	Primary; secondary; nitrification; TP reduction
Configuration	Multiple cells in parallel and series	Multiple beds in parallel	Multiple discharge sites
Distribution	Swale, perforated pipe	Inlet zone (>0.5 m wide) of large gravel	Swale, perforated pipe
Maximum loading (kg/ha · day)			
BOD₅	100–110	80–120	4
Suspended solids	Up to 175	Up to 150	—
TKN	7.5	30	3
Phosphorus	0.12–1.5	3.6–17	—
Additional considerations	Mosquito control with mosquito fish; remove vegetation	Allow flooding capability for weed control	Natural hydroperiod should be >50%; no vegetation harvest

Source: Water Pollution Control Federation (1992).

Location   The ideal site for a constructed wetland is located a reasonable distance from the wastewater source, at an elevation that permits gravity flow to the wetland, between the wetland cells, and to the final discharge point. The site should be available at a reasonable cost, should not require extensive clearing or earthwork for construction, should have a deep nonsensitive ground-water table, and contain subsoils that provide a suitable liner when compacted.

Preapplication Treatment   Some form of preliminary treatment precedes all wetland treatment systems in the United States, ranging from primary levels to tertiary levels. The required level of preapplication treatment depends on the functional intent of the wetland, on the level of public exposure expected, and on the need to protect habitat values. The minimal preliminary treatment for municipal wastewater would be the equivalent of primary treatment, accomplished using septic tanks or Imhoff tanks for small systems or pond units with deep zones for sludge accumulation for large systems. Providing the equivalent of secondary treatment is considered prudent before allowing public access to the wetland or developing specific habitats that encourage birds and other wildlife. This level of treatment can be accomplished at a first-stage wetland unit where public access is restricted and habitat values are minimized. Tertiary treatment with nutrient removal may be necessary before discharging to natural wetlands where preservation of the existing habitat and ecosystem is desired.

Hydroperiod   The hydroperiod is defined as the pattern of water depth fluctuation over time. Wetlands that have a seasonal fluctuation of water depth have the most potential for developing a diversity of plants, animals, and biogeochemical processes. Alternate flooding and aeration of soils promote nitrification–denitrification, and deepwater areas, devoid of emerging vegetation, offer habitats for fish (e.g., *Gambusia affinis*, the mosquito-eating fish). Fluctuating water levels can provide needed oxidation of organic sediments and can, in some cases, rejuvenate a system to higher levels of chemical retention. Water levels are typically controlled by inflow and outflow structures such as feed pumps, gates, and weirs. A typical outflow weir protected by a trash rack is shown in Figure 6.15. During the startup period, low water levels are needed to avoid flooding newly emerged plants. Startup periods for the establishment of plants may take 2 to 3 years, and the development of an adequate litter sediment compartment may take another 2 to 3 years (Mitsch and Gosselink, 2000).

Seasonal Pulses   Storms and seasonal patterns of floods can significantly affect the performance of wetlands designed for the control of nonpoint surface runoff. Highest nutrient loadings from agricultural sources occur during the first storms after fertilizer application, and a good wetland design should take advantage of these pulses for system replenishment, and provide for excess wet-weather storage if nutrient retention is a primary

FIGURE 6.15   Wetland outlet weir. (From Emmons and Oliver Resources, 2005).

objective. Infrequent floods and droughts are important for dispersing biological species to the wetland and adjusting resident species composition.

Wetland Hydraulics The Manning equation is widely accepted as a model for the flow of water through surface-flow wetlands (Water Environment Federation, 2001). The form of the Manning equation that is most frequently used is given by

$$v = \frac{1}{n} y^{2/3} S^{1/2} \tag{6.2}$$

where v is the average flow velocity (m/s), n is the Manning roughness coefficient (dimensionless), y is the depth of flow (meters), and S is the hydraulic gradient (dimensionless). In most applications of the Manning equation, resistance to flow occurs only on the bottom and submerged sides of an open channel, and published values of n coefficients for various conditions are available (e.g., Chin, 2006). However, in surface-flow wetlands, resistance to flow extends through the entire depth of the water because of the presence of emergent vegetation and litter. Manning's n in wetlands can be estimated in terms of the depth of flow, y (in meters), by the relation

$$n = \frac{a}{y^{1/2}} \tag{6.3}$$

where a is called the *resistance factor*, which can be estimated using the guidance given in Table 6.4. The head differential between the water surface at the inlet and at the outlet of a wetland provides the energy required to overcome the frictional resistance. Although constructing a wetland with a sloping bottom can provide some of the head differential, the preferred approach is to construct a bottom with minimal slope that still allows complete drainage when needed and to provide outlet structures that allow the water level to be adjusted to compensate for the resistance that may increase with time.

Overland flow in wetlands has also been described using the following more generalized relationship between discharge per unit width, flow depth, hydraulic gradient, and conductance of the wetland:

$$q = K_w y^{\beta} S^{\alpha} \tag{6.4}$$

where q is the discharge per unit width (m²/day), y is the depth of flow (meters), S is the hydraulic gradient (dimensionless), K_w is the hydraulic conductance coefficient for overland

TABLE 6.4 Resistance Factor in Estimating Manning's n

Wetland Description	Resistance Factor, a $(\text{s} \cdot \text{m}^{1/6})$
Sparse, low standing vegetation, $y > 0.4$ m	0.4
Moderately dense vegetation, $y \geq 0.3$ m	1.6
Very dense vegetation and litter, $y < 0.3$ m	6.4

Source: Reed et al. (1995).

TABLE 6.5   Experimental Values of Parameters in Kadlec's Equation

Source	Location	K_w (m²/day · m$^\beta$)	α	β
Kadlec (1990)	Houghton Lake, MI	16×10^9	2.5	0.7
Kadlec et al. (1981)	Houghton Lake, MI	0.39×10^9	3.0	1.0
Hammer and Kadlec (1986)	Houghton Lake, MI	0.66×10^9	3.0	1.0
Chescheir et al. (1987)	Eastern NC	2.6×10^9	3.0	1.0
Chen (1976)	Laboratory (Kentucky Blue grass)	182×10^9	3.75	0.50
	Laboratory (Bermuda grass)	52.5×10^9	3.75	0.39

FIGURE 6.16   Aerial view of a wetland flow. (From State of Victoria, Australia 2005.)

flow (m²/day · m), β is an exponent related to the vegetation microtopography and the stem density–depth distribution, and α is an exponent that reflects the degree of laminar or turbulent flow conditions. Equation 6.4 is sometimes referred to as *Kadlec's equation* (after Kadlec, 1990), and values of K_w, α, and β for various wetland environments found in Michigan and eastern North Carolina are given in Table 6.5. Kadlec's equation is the basis for simulating overland flow in the Wetland Package, a computer code developed by Restrepo et al. (1998) to be used in conjunction with the very popular MODFLOW code (Harbaugh and McDonald, 1996) for simulation ground-water and surface-water flows, particularly in the wetland areas of southern Florida. An aerial view of overland flow in a wetland is shown in Figure 6.16. It is noteworthy that flows through wetlands such as that shown in Figure 6.16 are seldom in the form of uniform sheet flow only but occur mostly as a combination of sheet flow and channel flow.

Hydraulic Loading Rate   The *hydraulic loading rate*, q, is defined by the relation

$$q = \frac{Q}{A} \qquad\qquad (6.5)$$

where Q is the average flow rate through the wetland [L^3/T], and A is the surface area of the wetland [L^2]. Knight (1990) reviewed data from a variety of wetlands constructed for domestic wastewater treatment and found that loading rates varied between 0.7 and 5 cm/day. Knight (1990) recommends a rate of 2.5 to 5 cm/day for surface-flow (constructed) wetlands and 8 cm/day for subsurface-flow wetlands. According to Mitsch and Gosselink (2000), these loading rates are probably too restrictive for wetlands used for the control of nonpoint pollution and stormwater runoff, but few studies have been undertaken to determine the optimum design rates for such wetlands.

Detention Time   The *detention time*, t, is defined by the relation

$$t = \frac{Vn}{Q} \tag{6.6}$$

where V is the volume of water in the wetland (surface-flow wetland) or the volume of the medium through which the water flows (subsurface-flow wetland), n is the porosity of the medium, and Q is the average flow rate through the wetland. The porosity, n, is defined by the relation

$$n = \frac{\text{water volume}}{\text{total volume}} \tag{6.7}$$

where $n = 0.9$ to 1 for surface-flow wetlands, depending on the growth density of vegetation, and n is the void fraction of the substrate for subsurface-flow wetlands. Optimum detention time ranges from 5 to 14 days for treatment of municipal wastewater. Florida wetland regulations require that the volume in the permanent pools of the wetland must provide for a residence time of at least 14 days. The detention time is sometimes called the *retention time* (Novotny, 2003); however, this terminology is somewhat of a misnomer since the water in the wetland is not retained but is detained.

Basin Morphology   Several aspects of the morphology of constructed wetland basins need to be considered when designing wetlands. A typical wetland design is shown in Figure 6.17. Constructed treatment wetland designs should avoid rectangular basins, rigid structures, and straight channels whenever possible (USEPA, 2000a). A flat littoral zone maximizes the area of appropriate water depth for emergent plants, thus allowing more wetland plants to develop more quickly and allowing wider bands of different plant communities. Plants will also have room to move uphill if water levels are raised in the basins due to flows being higher than predicted. Florida regulations for the Orlando area require littoral zones to have a slope of 6:1 or flatter to a point 60 to 77 cm below the water surface. Slopes of 10:1 or flatter are even better (Mitsch and Gosselink, 2000). Bottom slopes of less than 1% are recommended for wetlands used to control runoff, and slopes less than 0.5% are recommended for wetlands used to treat wastewater.

In cases where nutrient and sediment retention are desired objectives, flow conditions should be such that the entire wetland is effective. This may require several inflow locations to avoid channelization of flows. Steiner and Freeman (1989) suggest a maximum length/width ratio of 10:1, with a minimum ratio of 2:1 suggested by Mitsch and Gosselink (2000).

Providing a variety of deep and shallow areas is optimum. Deep areas (> 50 cm) offer habitat for fish, increase the capacity of the wetland to retain sediments, enhance

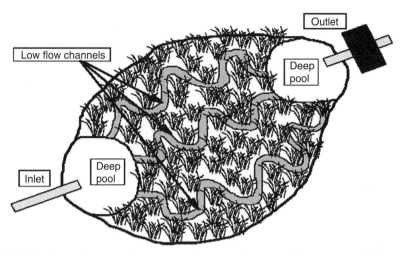

FIGURE 6.17   Typical wetland design to maximize the flow path. (From Ontario Ministry of the Environment, 2003.)

nitrification, and provide low-velocity areas where water flow can be redistributed. Shallow depths ($< 50\,cm$) provide maximum soil–water contact for certain chemical reactions such as denitrification and can accommodate a greater variety of emergent vascular plants.

Constituent Loadings   The effectiveness of constructed wetlands to remove BOD, suspended solids, nitrogen, phosphorus, and metals is usually of concern in the design of constructed wetlands. Water-constituent loadings on wetlands are usually expressed in terms of mass (of constituent) per unit area (of wetland) per unit time, with typical units being $g/m^2 \cdot year$. Typical removal efficiencies for several pollutants in constructed and natural wetlands are given in Table 6.6. Removal efficiencies of wetlands tend to decrease as constituent loadings increase; however, this relationship is very site-specific and sometimes not readily apparent. Hydraulic loading, detention time, and pollutant uptake of the system are all factors that affect removal efficiency. The data shown in Table 6.6 show that when properly designed, constructed wetlands can perform better than natural wetlands for pollutants such as phosphorus, ammonia, and some heavy metals. The efficiencies given in Table 6.6 can be applied to constructed treatment wetlands for both domestic wastewater treatment and stormwater runoff. Specific removal efficiencies are discussed in more detail below.

BOD Removal   The removal of particulate BOD in wetlands occurs rapidly through settling and by entrapment in the submerged plant parts and litter in surface-flow (SF) wetlands

TABLE 6.6   Typical Pollutant Removal Efficiencies of Constructed and Natural Wetlands (Percent)

Wetland Type	Suspended Solids	Total Phosphorus	NH_3	Lead	Zinc
Constructed	80	58	44	83	42
Natural	76	5	25	69	62

Source: Strecker et al. (1992).

or in the media in subsurface-flow (SSF) systems. The trapped material and soluble BOD are oxidized primarily by periphytic microorganisms in the system. A unique aspect of wetland systems is the concurrent production of BOD caused by the internal decomposition of plant litter and other naturally occurring organic materials. Thus, some irreducible BOD concentration is always present in the final wetland effluent, making these systems incapable of complete BOD removal. This residual background BOD may range from 2 to 10 mg/L.

Suspended-Solids Removal Wetland systems produce residual background concentrations of total suspended solids (TSS) which are a by-product of decomposing vegetation and other natural organics in the wetlands. Background concentrations range from 2 to 10 mg/L. In general, the removal of TSS is more rapid than the removal of BOD. The general consensus in the literature is that if a wetland is sized for either BOD or total nitrogen compliance and is designed and operated properly, it should produce TSS concentrations close to background concentrations.

Nitrogen Removal The form of nitrogen entering a wetland system is a function of the type of pretreatment used. Typically, most of the wetland influent nitrogen will be a combination of organic nitrogen and ammonia nitrogen. This combination is typically characterized as total Kjeldahl nitrogen (TKN). After BOD, TSS, and coliforms, ammonia removal is the most common requirement imposed on wetland treatment systems. In most cases, the requirement is related to the oxygen demand in the receiving stream, and the concern is the presence of the ionized ammonia fraction (NH_4^+). Biological nitrification is the only significant long-term mechanism available for ammonia removal in wetlands. Nitrification is an aerobic process; however, the environment in a wetland is not generally aerobic, except near the water surface and on submerged plant and root surfaces. In theory, at least 4.6 g of oxygen is required to oxidize 1 g of ammonia. The available oxygen sources in surface- and subsurface-flow wetlands do not provide the conditions necessary for rapid nitrification. Thus, although BOD and TSS removal may be substantially accomplished within a few days, ammonia removal (nitrification) may require a few weeks to reach low levels without supplemental oxygen sources. In contrast to ammonia removal, nitrate removal in wetlands can be rapid and effective because the anoxic conditions and carbon sources necessary to support the treatment reactions are naturally present in surface- and subsurface-flow wetlands. Nitrification reactions required for ammonia removal are the limiting factor for nitrogen removal in wetlands. A rule of thumb is that when not overloaded, wetlands, can consistently retain nitrogen in the amounts of 10 to 20 g N/m$^2 \cdot$ year.

Phosphorus Removal Phosphorus removal in natural treatment systems occurs because of plant uptake, adsorption, complexation, and precipitation. Direct settling can account for removal of any influent phosphorus associated with particulate matter. Available data indicate that 30 to 50% phosphorus removal is a reasonable expectation for wetland systems with residence times of less than 10 days (Water Environment Federation, 2001). A rule of thumb is that when not overloaded, wetlands can consistently retain phosphorus in the amounts of 1 to 2 g P/m$^2 \cdot$ year.

Pathogen Removal Natural sources of fecal coliforms in wetlands create an irreducible background concentration of fecal coliforms that is likely to be present in the wetland effluent. This background level is estimated to be 300 to 500 fecal coliform units (cfu) per 100 mL. Virus removal is comparable to fecal coliform removal in these systems. Available data indicate that a 2 to 3 log reduction in fecal coliforms can be expected with a 5- to

10-day hydraulic residence time in a wetland system (Water Environment Federation, 2001). In most cases, this reduction is not enough to meet a typical discharge limit of 200 cfu per 100 mL. Supplemental disinfection is typically required when wetland systems have a fecal coliform limit for discharge compliance.

Metals Removal   The ultimate sink for metals in wetlands is in the organic benthic sediments. Fortunately, most constructed treatment wetlands generate fresh organic material continuously in the form of plant litter, but in the long term or with high loadings, they may reach toxic levels for the wildlife inhabiting the wetland. With typical municipal wastewater, the useful life of wetlands with respect to metals accumulation is on the order of hundreds of years or more (Water Environment Federation, 2001).

Design Models   Design models for wetland systems can be broadly grouped into two types: areal loading models and volumetric models. A design model developed by Kadlec and Knight (1996) is an example of an areal loading model; wetland models developed by Reed et al. (1995) and Crites and Tchobanoglous (1998) are examples of volumetric models. When considering alternative designs for a project, it is advisable to compare the results from these models and adopt those best suited to the particular project.

Kadlec–Knight Model   Kadlec and Knight (1996) proposed using the following mass-balance equation to describe the removal of various constituents in wetlands:

$$q \frac{dC}{dx^*} = k_A(C - C^*) \tag{6.8}$$

where q is the hydraulic loading rate [L/T], C is the constituent concentration [M/L^3], x^* is the fractional distance from the inlet to the outlet (dimensionless), k_A is the areal removal rate constant [L/T], and C^* is the residual or background constituent concentration [M/L^3]. Integrating Equation 6.8 over the length of a wetland yields

$$\boxed{\frac{C_o - C^*}{C_i - C^*} = \exp\left(-\frac{k_A}{q}\right)} \tag{6.9}$$

where C_i and C_o are the inflow and outflow concentrations, respectively. Application of Equation 6.9 requires estimates of the parameters C^* and k_A, which are given in Table 6.7. Equation 6.9 is particularly useful for estimating the area of wetland necessary to achieve a given removal. Rearranging Equation 6.9 gives

$$A = \frac{Q}{k_A} \ln \frac{C_i - C^*}{C_o - C^*} \tag{6.10}$$

where Q is the flow rate through the wetland.

Example 6.1   Domestic wastewater from a rural town is to be treated using a constructed wetland. The typical flow rate into the wetland is 14,000 m^3/day and the average level of

TABLE 6.7 Parameters in the Kadlec–Knight Model

Wetland Type and Constitutent	k_A (m/yr)	C^* (mg/L)
Surface flow		
BOD	34	$3.5 + 0.053C_i$
Suspended solids	1000	$5.1 + 0.16C_i$
Total phosphorus	12	0.02
Total nitrogen	22	1.5
Ammonia nitrogen	18	0
Nitrate nitrogen	35	0
Subsurface flow		
BOD	180	$3.5 + 0.053C_i$
Total phosphorus	12	0.02
Total nitrogen	27	1.5
Ammonia nitrogen	34	0
Nitrate nitrogen	50	0

Source: Kadlec and Knight (1996).

nitrate nitrogen in the inflow is 100 mg/L. What area of wetland is required to produce an outflow with an average nitrate nitrogen concentration of 10 mg/L?

SOLUTION   From the data given, $Q = 14{,}000 \, \text{m}^3/\text{day} = 5.11 \times 10^6 \, \text{m}^3/\text{yr}$, $C_i = 100 \, \text{mg/L}$, $C_o = 10 \, \text{mg/L}$, and Table 6.7 gives $k_A = 35 \, \text{m/yr}$ and $C^* = 0 \, \text{mg/L}$. The required wetland area is given by Equation 6.10 as

$$A = \frac{Q}{k_A} \ln\left(\frac{C_i - C^*}{C_o - C^*}\right)$$

$$= \frac{5.11 \times 10^6}{35} \ln \frac{100 - 0}{10 - 0} = 336{,}000 \, \text{m}^2 = 33.6 \, \text{ha}$$

Therefore, to adequately remove nitrate nitrogen from the wastewater, about 34 ha of wetland is required. This calculation should be repeated for all the constituents of concern, and the maximum calculated wetland area should be used in the final design.

In utilizing the Kadlec and Knight (1996) model the following limitations should be noted: (1) the model only takes into account the input wastewater volume, Q, and does not take into account water gains and losses within the wetland; (2) the database used to develop this model includes a large number of lightly loaded surface-flow wetland systems, which tends to produce low rate constants, k_A, that might, in turn, result in unnecessarily large wetland system designs; and (3) the incorporation of a background concentration, C^*, might result in excessive wetland sizes to achieve low concentrations.

Reed Model   Reed et al. (1995) proposed the following wetland design model:

$$\boxed{\frac{C_o}{C_i} = \exp(-K_T t)} \qquad (6.11)$$

TABLE 6.8   Typical Parameters Used in Reed et al. (1995) Model

Wetland Type and Constituent	K_{20} (day^{-1})	θ	Background Concentration (mg/L)
Surface flow			
BOD	0.678	1.06	6
Suspended solids	—	—	6
Ammonia nitrogen	0.2187	1.048	0.2
Nitrate nitrogen	1.000	1.15	0.2
Total phosphorus	a	—	0.05
Fecal coliforms[b]	2.6	1.19	2000 cfu/100 mL
Subsurface flow			
BOD	1.104	1.06	6
Suspended solids	—	—	6
Ammonia nitrogen	c	1.048	0.2
Nitrate nitrogen	1.000	1.15	0.2
Total phosphorus	a	—	0.05
Fecal coliforms[b]	2.6	1.19	2000 cfu/100 mL

Source: Water Environment Federation (2001).

[a] Total phosphorus removal does not depend on temperature and can be estimated using the relation $C_o/C_i = \exp(-K_p/q)$, where K_p is typically equal to 27.3 mm/day, and q is the average hydraulic loading rate in cm/day.
[b] Fecal coliform removal can be estimated using the relation $C_o/C_i = [1 + K_T(t/x)]^x$, where K_T is estimated using Equation 6.12, t is the hydraulic residence time, and x is the number of wetland cells in series.
[c] Can be estimated using the relation $K_{20} = 0.1854 + 0.3922(rz)^2 (0.6077)$, where rz is the fraction of the subsurface-flow wetland occupied by plant roots. A typical value of rz is 0.5.

where K_T is a rate constant at the wetland temperature T, and t is the hydraulic residence time in the wetland. The rate constant, K_T, is estimated from a standardized value at 20°C, K_{20}, using the relation

$$K_T = K_{20}\theta^{T-20} \tag{6.12}$$

where typical values of K_{20} and θ are given in Table 6.8. The required wetland area corresponding to the Reed et al. (1995) model is given by

$$A = Q_A \frac{\ln(C_i/C_o)}{K_T yn} \tag{6.13}$$

where Q_A is the average flow in the wetland (= average of inflow and outflow), y is the average depth of flow in the wetland, and n is the porosity of the wetland. Typical values of n are in the range 0.65 to 0.75 for surface-flow wetlands and 0.35 to 0.45 for subsurface-flow wetlands, and typical values of y are in the range 15 to 60 cm in surface-flow wetlands and 30 to 60 cm in subsurface-flow wetlands.

Crites–Tchobanoglous Model   Crites and Tchobanoglous (1998) estimate that K_{20} can be taken as 0.1070 day^{-1}. The water-quality parameter that requires the greatest treatment area for removal is the limiting design factor, and that area should be selected for the specific project. Crites and Tchobanoglous (1998) proposed using the same equations as those proposed by Reed et al. (1995) (Equations 6.11 to 6.13), with some modifications in parameter estimates and formulations. Crites and Tchobanoglous (1998) estimated the background concentration of BOD in the range 5 to 6 mg/L and estimated the total suspended solids (TSS) removal by the following relations:

Surface-flow wetland:

$$\frac{C_o}{C_i} = 0.1139 + 0.000213q \tag{6.14}$$

Subsurface-flow wetland:

$$\frac{C_o}{C_i} = 0.1058 + 0.00011q \tag{6.15}$$

where q is the hydraulic loading rate in mm/day.

In applying wetland models for design, a safety factor is applied after the wetland size has been determined. The safety factor typically ranges from 15 to 25%, depending on the uncertainty of the available data and on the stringency of performance expectations.

Empirical Models Empirical equations are sometimes used to relate the inflow and outflow concentrations of particular constituents. Several of these equations are given in Table 6.9, where C_i is the inflow concentration in mg/L, C_o is the outflow concentration in mg/L, A is the area of the wetland in hectares, Q is the wetland inflow in m^3/day, and q ($= Q/A$) is the hydraulic loading rate in cm/day.

TABLE 6.9 Empirical Equations to Predict Wetland Removal Rates

Wetland Type and Constituent	Empirical Equation	Correlation Coefficient, r^2	Number of Wetlands
Surface flow			
BOD	$C_o = 4.7 + 0.173C_i$	0.62	440
Suspended solids	$C_o = 5.1 + 0.158C_i$	0.23	1582
Ammonia nitrogen	$A = 0.01Q/\exp(1.527\ln C_o - 1.05\ln C_i + 1.69]$	—	—
	$C_o = 0.336C_i^{0.728}q^{0.456}$	0.44	542[a]
Nitrate nitrogen	$C_o = 0.093C_i^{0.474}q^{0.745}$	0.35	553[a]
Total nitrogen	$C_o = 0.409C_i + 0.122q$	0.48	408[a]
Total phosphorus	$C_o = 0.195C_i^{0.91}q^{0.53}$	0.77	373[a]
	$C_o = 0.37C_i^{0.70}q^{0.53}$	0.33	166[b]
Subsurface flow			
Suspended solids	$C_o = 4.7 + 0.09C_i$	0.67	77
Ammonia nitrogen	$C_o = 3.3 + 0.46C_i$	0.63	92
Nitrate nitrogen	$C_o = 0.62C_i$	0.80	95
Total nitrogen	$C_o = 2.6 + 0.46Ci + 0.124q$	0.45	135
Total phosphorus	$C_o = 0.51C_i^{1.10}$	0.64	90

Source: Kadlec and Knight (1996).

[a]Surface-flow marshes.

[b]Surface-flow swamps.

Example 6.2 Contaminated water flows into a 33.6-ha wetland at a rate of $14,000\,\text{m}^3/\text{day}$ and with a nitrate-nitrogen concentration of $100\,\text{mg/L}$. Estimate the nitrate-nitrogen concentration in the water leaving the wetland and the removal efficiency of the wetland.

SOLUTION   From the data given, $A = 33.6\,\text{ha} = 3.36 \times 10^5\,\text{m}^2$, $Q = 14,000\,\text{m}^3/\text{day}$, and $C_i = 100\,\text{mg/L}$. The hydraulic loading rate, q, is derived from the given data by the relation

$$q = \frac{Q}{A} = \frac{14,000}{3.36 \times 10^5} = 0.0417\,\text{m/day} = 4.17\,\text{cm/day}$$

From Table 6.9, the concentration in the wetland effluent can be estimated by

$$C_o = 0.093 C_i^{0.474}\, q^{0.745}$$
$$= 0.093(100)^{0.474}(4.17)^{0.745} = 2.39\,\text{mg/L}$$

The removal efficiency of the wetland is given by

$$\text{removal efficiency} = \frac{100 - 2.39}{100} \times 100 = 98\%$$

Substrate The substrate is the primary medium supporting rooted vegetation and mostly consists of soils. The soil or subsoil must have a permeability low enough to retain standing water. If a wetland is designed to improve water quality, the substrate plays an important role in its ability to retain certain constituents. The substrate is not as significant for the retention of suspended organic matter and solids (along with the constituents adsorbed to sediment particles), because their retention is based primarily on net deposition that results from the slow velocities characteristic of wetlands. Characteristics of substrates that affect the design of surface-flow wetlands include organic content, texture, nutrient content, and iron and aluminum content.

The organic content of wetland soils has a significant impact on the retention of various water constituents in the wetland. Organic soils generally have a higher cation-exchange capacity than mineral soils and therefore can more effectively remove certain metals through ion exchange. The substrate should be composed mainly of clay with some organic matter for plant growth and a cation-exchange capacity of 15 or greater. Organic soils can also enhance nitrogen removal by providing an energy source and anaerobic conditions appropriate for denitrification. Organic matter in wetland soils generally varies between 5 and 75%, with higher concentrations in peat-building systems such as bogs, and lower concentrations in riparian wetlands. In constructed wetlands, organic soils are sometimes avoided because they are low in nutrients, can cause low pH, and often provide inadequate support for rooted aquatic plants. For optimal nitrogen removal, the pH should be between 6.5 and 8.5. The wetland substrate shown in Figure 6.18 contains leafy organic material and fine material; the pencil is included for scale.

Local soils underlain with impermeable clay or bentonite, to prevent downward percolation, is often the best design for constructed wetlands. Loam soils are typically preferable, while sandy soils, although generally low in nutrients, anchor plants adequately and readily allow water to reach the plant roots.

FIGURE 6.18   Wetland substrate. (From University of British Columbia, 2005.)

Low nutrient levels that are characteristic of organic, clay, and sandy soils can cause problems for initial plant growth. Fertilization may be necessary in some cases to establish plants and enhance growth; however, fertilization should be avoided if possible in wetlands that will eventually be used as sinks for macronutrients.

When soils are submerged and anaerobic conditions occur, iron is reduced from the ferric (Fe^{3+}) to the ferrous (Fe^{2+}) state, releasing phosphorus that was previously held as insoluble ferric phosphate compounds. The Fe–P compound can be a significant source of phosphorus to overlying and interstitial waters after flooding and anaerobic conditions occur, particularly if the wetland is constructed on former agricultural land (Mitsch and Gosselink, 2000). After an initial pulse of released phosphorus in such constructed wetlands, the iron and aluminum contents of a wetland soil exert significant influence on the ability of that wetland to retain phosphorus. All things being equal, soils with higher aluminum and iron concentrations are more desirable because their affinity for phosphorus is higher.

Vegetation   The vegetation in constructed wetlands used for water treatment usually differs from the vegetation in natural wetlands: Treatment wetlands are constructed with the main goal of improving water quality, whereas a goal in creating or restoring natural wetlands is to develop a diverse vegetation cover and habitat. Constructed (treatment) wetlands generally have higher concentrations of various constituents in the water, which limits the number of plants that will survive in those wetlands. Relatively few plants thrive in high-nutrient, high-BOD, high-sediment waters; among these plants are cattails (*Typha* spp.), bulrushes (*Schoenoplectus* spp., *Scirpus* spp.), and reed grass (*Phragmites australis*). A typical growth of cattails is shown in Figure 6.19. When water is deeper than 30 cm, emergent plants often have difficulty growing, and surface-flow wetlands can become covered with duckweed (*Lemma* spp.) in temperate zones and with water hyacinths (*Eichhornia crassipes*) in the tropics. There are only a few studies on the best types of vegetation for treatment wetlands. Esry and Cairns (1989) investigated the best

FIGURE 6.19   Cattails (*Typha* spp.). (From USACE, 2005*a*.)

types of vegetation for wetlands that treat stormwater runoff and found that sawgrass and pickerelweed perform adequately, whereas bulrush did not perform well.

Management of Constructed Wetlands   Development of wildlife in wetlands is usually desirable. However, animals such as beaver and muskrat can create obstructions to flow, destroy vegetation, and burrow into dikes. Canada geese grazing on emergent vegetation can cause mass destruction, and deeper wetlands often become havens for undesirable fish such as carp that can cause excessive turbidity and uproot vegetation. Hazing or wildlife exclusion devices, such as noise-making devices or netting and fencing, should be used if the effluent or other water source being treated is toxic or presents a significant threat to wildlife. An example of a fence for wildlife exclusion is shown in Figure 6.20. Typical fences are built of high-tension fencing material woven into a tight pattern to form a barrier against intrusion. The bottom of the fence is usually set into the ground 30 cm to prevent burrowing animals from gaining access. Power sources vary but the main types are solar and battery powered.

Wetlands with significant areas of stagnant water are breeding grounds for mosquitoes, and sometimes, chemical or biological control, such as by mosquito fish (*Gambusia affinis*), stickleback, bats, and purple martins are desirable.

Nutrients removed from the inflowing water in constructed (treatment) wetlands are sequestered in the sediments, released to the atmosphere (nitrogen), or taken up into plant

FIGURE 6.20   Fence for wildlife exclusion. (From U.S. Fish and Wildlife Service, 2005b.)

roots and aboveground shoots and foliage. The harvesting of plants generally does not result in the removal of a large quantity of contaminants, such as nutrients, unless the plants are harvested several times during the growing season. Drawdowns of water levels in a wetland followed by burning is sometimes used to control the invasion of woody vegetation if that invasion is considered undesirable. When controlled burning of wetlands is used, the impact on wildlife and the potential introduction of inorganic nutrients to the water column from the ash and sediments should be taken into account.

The water level in surface-flow treatment wetlands is the key to both water-quality improvement and vegetation success. Wetlands constructed to treat municipal wastewater typically have little control on the inflow unless authorities have given permission to bypass the wetland if needed, or if a wastewater storage lagoon is built upstream of the wetland. Most constructed wetlands have an outflow control structure, such as a flume or gate, that can control the depth in the wetland. Water depths of about 30 cm are optimum for many herbaceous macrophytes used in treatment wetlands (Mitsch and Gosselink, 2000). High water levels favor the treatment of phosphorus that is associated with sedimentation and similar processes, while shallow water levels lead to closer proximity of sediments to overlying water, causing anaerobic or near-anaerobic conditions in the aquatic system during the growing season. Low water levels favor the reduction of nitrate nitrogen through denitrification.

6.6.6   Wetlands for Treating Roadway Runoff

Constructed wetlands can be a highly efficient means of removing pollutants from highway and urban runoff (FHWA, 1996). Surface-flow wetlands are the predominant type of wetland used for treating stormwater runoff, and an ideal design of the wetland as a water quality control would include the creation of a detention basin upstream of the wetland, sometimes called a *forebay*. The detention basin or forebay provides an area where heavy particulate matter can settle out, thereby minimizing disturbance of the wetland soils and vegetation. The removal efficiencies for several pollutants in constructed wetlands used to treat stormwater runoff were listed in Table 6.6. In designing wetlands for treating road-way runoff, the following guidelines are suggested:

- The wetland site must have sufficient inflow to maintain a constant water pool. If infiltration is excessive relative to the inflow, a clay or plastic liner should be used.
- Approximately 75% of the wetland pond area should have water depths less than 30 cm. This provides the optimal growth depth for most wetland plant species.
- A perimeter area of approximately 3 to 6 m beyond the constant pool of the wetlands, which is periodically flooded during runoff events, should be established for aquatic emergents.
- Wetland basins should have a detention time of at least 24 hours for a 1-year storm.
- At least five different wetland species should be established. This minimizes the risk that the wetland will not become established successfully.

Once a wetland has stabilized, there are few maintenance requirements. The removal of accumulated sediment is usually the primary concern, and the sediment forebay should be cleaned out every 5 to 10 years upon reaching a specified sediment depth.

Stormwater treatment efficiencies associated with wetlands vary with the type and age of wetland. For example, wet prairie, broadleaf emergent marsh, open-water slough, willow/shrub marsh, cypress swamp, and mixed hardwood swamp are all defined as wetlands; however, each has its own unique ecosystem, which may provide different treatment efficiencies. Several studies have shown that although wetlands appeared to provide good pollutant removals on an individual-storm basis, long-term effects such as seasonality of vegetation may actually produce a net export of nutrients (Whalen and Cullum, 1988). Harper et al. (1986) made several recommendations based on an extensive stormwater-runoff wetlands treatment monitoring project. The results of this study indicated that wet-land systems best suited to stormwater management are those that already exhibit relatively long hydroperiods (e.g., hardwood hammocks, cypress domes, and marshes). Further, hardwood wetlands have the ability to evapotranspire much larger quantities of hydrologic inputs. Consequently, smaller hardwood-wetland areas may be able to accept relatively larger volumetric inputs than other wetland types.

Natural pristine wetlands should not be used for treating roadway runoff, since disruptions to the natural ecology are likely to be deleterious. However, damaged natural wetlands that have been ditched and drained, and where upland plants are invading and replacing wet-land plants, are good candidates for revitalization for wetland treatment. In the United States, real-estate developers that are issued permits to fill in and build on natural wetlands are required by federal law to create a wetland in a nearby area and monitor it for 5 years. This practice is called *wetland mitigation*. Ongoing research is currently aimed at determining whether such constructed wetlands ultimately function as well as the natural wetlands that are destroyed in the process (*Water Environment and Technology*, 2000).

SUMMARY

Wetlands are areas where water covers the ground surface or is present at or near the ground surface for sufficient periods of time during the year to cause hydrophytic vegetation and hydric soils to dominate the area. Wetlands are classified as either natural wetlands or constructed wetlands. Natural wetlands are formed by nature with limited human influence, while constructed wetlands are engineered systems that are designed to function like natural wetlands. Natural wetlands are further categorized as marshes, swamps, bogs, and fens depending on the types of vegetation and local hydrology. For purposes of regulation, guidelines have been developed by the U.S. Army Corps of Engineers to delineate natural wetlands based on specific vegetation, soil, and hydrology measures. Constructed treatment wetlands are engineered systems that utilize natural processes involving wetland vegetation, soils, and their associated microbial assemblages to assist, at least partly, in treating an effluent or other water source. Constructed treatment wetlands are further categorized as surface- or subsurface-flow wetlands. Detailed guidelines are provided for the design and management of constructed treatment wetlands.

PROBLEMS

6.1. The design flow rate into a constructed wetland is 10,000 m³/day with a suspended solids concentration in the inflow of 100 mg/L. Estimate the area of wetland required to produce an outflow with a suspended solids concentration of 10 mg/L.

6.2. Water flows into a 50-ha wetland at a rate of 15,000 m³/day and with a BOD of 150 mg/L.

(**a**) Estimate the BOD of the water leaving the wetland.

(**b**) What is the removal efficiency of the wetland for BOD?

6.3. The Wakodahatchee Wetland in south Florida currently treats secondary wastewater effluent and has an average inflow rate of 5.26 m³/s (120 mgd), total phosphorus concentration of 2.12 mg/L, and total nitrogen concentration of 28.03 mg/L. The effluent from the wetland currently has a total phosphorus concentration of 1.46 mg/L and a total nitrogen concentration of 8.43 mg/L. Outflow from the wetland is currently discharged by deep-well injection. This constructed wetland is to be upgraded such that the effluent from the wetland can be discharged into the L-30 canal, which has total phosphorus and total nitrogen concentrations of 0.24 and 1.9 mg/L, respectively. TMDL permitting requirements mandate that the effluent from the wetland cannot have total nitrogen and total phosphorus levels in excess of L-30 background concentrations.

(**a**) Estimate the upgraded wetland area required to achieve the water-quality goals.

(**b**) If the existing wetland has an area of 15.8 ha (39 acres), and the total site area available for the wetland system is 22.7 ha (56 acres), comment on the feasibility of modifying the existing wetland to meet the effluent water-quality goals.

(**c**) If space is a limiting factor, what alternative treatment modifications could be explored to meet the TMDL goals?

CHAPTER 7

GROUND WATER

7.1 INTRODUCTION

Ground water is a major source of drinking water in the United States, supplying approximately 40% of public water utilities and accounting for almost all of the water supply to rural households (USGS, 1998). It has been estimated that approximately 50% of the U.S. population relies on ground-water sources for drinking water (Solley et al., 1988). The direct use of ground water for drinking is the reason why drinking-water standards are usually applied to ground water, and the reason why ground-water contamination is such a sensitive issue. A typical ground water contamination scenario is shown in Figure 7.1, where the contaminant source is located on the ground surface and the contaminant plume is migrating toward a water-supply well. Ground water frequently contributes the base flow to rivers and streams, a condition that occurs when the river stage is lower than the adjacent water table. Under these conditions, ground-water inflow can contaminate rivers and streams.

Regulations associated with ground-water and wellhead protection programs require engineers to predict the fate and transport of contaminants released either directly into the ground water or on land surfaces above the ground water. These quantitative predictions are used to assess the impact of existing or potential contaminant sources on ground-water quality, to design systems to mitigate any deleterious effects, and to design systems to remediate contaminated ground water.

7.2 NATURAL GROUND-WATER QUALITY

The chemical constituents that occur naturally in ground water enter the aquifer with rainwater through the recharge area, as leachate from the upper soil layer, and from the

Water-Quality Engineering in Natural Systems, by David A. Chin
Copyright © 2006 John Wiley & Sons, Inc.

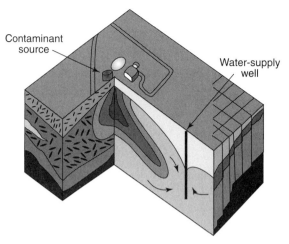

FIGURE 7.1 Contaminant dispersion in groundwater. (From Scientific Software Group, 2005.)

dissolution of minerals as the ground water flows through the porous medium. Mineral salts and dissolved (ionized) minerals are the most common sources of natural groundwater quality. The quality of ground water is important since it becomes the base flow of most perennial rivers and streams. The dissolved chemical composition of ground water includes positively charged ions (cations) and negatively charged ions (anions). The most abundant cations in ground water include calcium (Ca^{2+}), iron (Fe^{2+} or Fe^{3+}), magnesium (Mg^{2+}), sodium (Na^+), and hydrogen ion acidity (H^+). The most abundant negatively charged ions (anions) in ground water include sulfate (SO_4^{2-}), nitrate (NO_3^-), chloride (Cl^-), and bicarbonate (HCO_3^-). The cations and anions in ground water are balanced based on their equivalent weight.

Water entering the ground water from atmospheric precipitation is generally acidic (pH < 7). The normal pH of unpolluted precipitation is 5.6, which is estimated based on the normal saturation vapor pressure of carbon dioxide (CO_2) in the atmosphere. Significant atmospheric emissions of acid-forming oxides (SO_2 and NO_x) from power plants, industries, and automobiles can cause the pH of rainfall to be as low as 3. *Acid precipitation* refers to precipitation with a pH below 5.6. Production of CO_2 in the soil by bacteria can result in pH values of soil and ground water of less than 5. Acidic water entering well-aerated soil reduces some metals from their less soluble forms (e.g., Fe^{3+}) to their more soluble form (e.g., Fe^{2+}). The bicarbonate (HCO_3^-) content of water in the upper ground-water zone originates primarily from the dissolution of limestone ($CaCO_3 \cdot nH_2O$) and dolomite ($Ca \cdot MgCO_3 \cdot nH_2O$) carbonate minerals as described by the following equilibrium equations:

$$H_2O + CO_2 \rightleftharpoons H_2CO_3 \rightleftharpoons H^+ + HCO_3^- \tag{7.1}$$

and

$$CaCO_3 + H_2CO_3 \rightleftharpoons Ca^{2+} + 2HCO_3^- \tag{7.2}$$

In more acidic water (pH < 4.5) the following reaction is more typical:

$$CaCO_3 + 2H^+ \rightleftharpoons Ca^{2+} + H_2O + CO_2 \tag{7.3}$$

These equations demonstrate the buffering capacity of carbonate materials, which is particularly important in protecting ground water in areas where acid rain occurs. Equations 7.2 and 7.3 further demonstrate that the removal of acidity by limestone and dolomite will increase the hardness of the ground water. *Hardness* is defined as the content of polyvalent cations, such as Ca^{2+}, Mg^{2+}, Fe^{2+}, and Sr^{2+} expressed on a $CaCO_3$ equivalent basis.

Dissolution of minerals is the most important process controlling the quality of ground water. The solubility of minerals and solids is determined by the dissolution–precipitation reaction and their equilibria. A carbonate mineral such as limestone will partially dissolve into its ions as follows:

$$CaCO_3 \rightleftharpoons Ca^{2+} + CO_3^{2-} \tag{7.4}$$

and the equilibrium in this dissolution–precipitation reaction is given by the *reaction equilibrium coefficient*, K, given by

$$K = \frac{[Ca^{2+}][CO_3^{2-}]}{[CaCO_3]} \tag{7.5}$$

where the brackets indicate molar concentrations. The magnitude of the equilibrium coefficient for carbonate minerals indicates a very low solubility in the neutral pH range; however, other minerals, such as salt (NaCl) or gypsum ($CaSO_4$), have relatively high solubility, so that when ground water encounters these layers, high concentrations of Na^+, Cl^- or Ca^{2+}, and SO_4^{2-} are common.

Several substances that are normally considered as contaminants at elevated levels occur naturally in ground water and, in some rare instances, the natural levels exceed ground water standards. For example, minerals such as galena may cause elevated levels of lead (Pb). As a further example, Klusman and Edwards (1977) measured toxic metals in ground water from the mineral belt of Colorado and found that drinking-water standards were violated in 14% of the samples for cadmium (Cd) and 9% for zinc (Zn). Other metals that occur naturally in ground water include antimony, arsenic, beryllium, chromium, copper, lead, mercury, nickel, selenium, silver, and thallium. Table 7.1 lists the natural constituents of ground water according to their relative abundance.

7.3 CONTAMINANT SOURCES

The most common sources of ground-water contamination are septic tanks, leaking underground storage tanks (LUSTs), land application of wastewater, irrigation and irrigation return flow, solid-waste disposal sites (i.e., landfills), waste-disposal injection wells, and hazardous chemicals in agriculture (USEPA, 2000a). The most frequently reported contaminants in ground water are petroleum products, volatile organic compounds, nitrates, pesticides, and metals (USEPA, 1990b).

TABLE 7.1 Typical Natural Constituents of Groundwater

Major Constituents (> 5 mg/L)

Bicarbonate	Chloride	Sodium
Carbonic acid	Magnesium	Sulfate
Calcium	Silicon	

Minor Constituents (0.1 to 10 mg/L)

Boron	Iron	Potassium
Carbonate	Nitrate	Strontium
Fluoride		

Trace Constituents (< 0.1 mg/L)

Aluminum	Copper	Selenium
Antimony	Lead	Silver
Arsenic	Manganese	Thallium
Barium	Nickel	Thorium
Beryllium	Phosphate	Uranium
Cadmium	Radium	Vanadium
Chromium	Radon	Zinc
Cobalt		

Organic Compounds (Shallow Aquifers)

Amino acids	Humic acids	Tannins
Carbohydrates	Hydrocarbons	Total organic
Fluvic acids	Lignins	carbon (TOC)

Organic Compounds (Deep Aquifers)

Acetate	Propionate

Source: Davis and DeWiest (1966).

7.3.1 Septic Tanks

Septic tanks discharge pathogenic microorganisms, synthetic organic chemicals, nutrients (such as nitrogen and phosphorus), and other contaminants directly into the ground water and can cause serious problems if drinking water sources are too close to the septic tanks. A typical two-chamber septic tank is shown in Figure 7.2. In addition to siting concerns, for septic tanks to work properly it is important that a zone of unsaturated soil exist between the leach bed and the water table so that the effluent from the septic tank does not enter the ground water directly. In the United States, approximately 29% of the population disposes of sewage by individual (on-site) septic systems. Typically, lots using on-site septic systems are rural in nature and utilize their own wells for water supply. Typically, approximately 10% of the water withdrawn from the wells is lost between the well withdrawal and the septic systems through evapotranspiration and consumptive uses such as car washing.

The discharge from septic-tank systems is commonly estimated as 280 L/capita · day, and this effluent typically contains 40 to 80 mg/L of nitrogen, 10 to 30 mg/L of phosphorus,

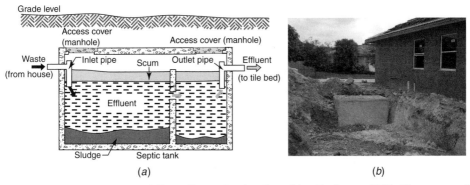

FIGURE 7.2   Septic tank. [(*a*) From Carson Dunlop Consulting Engineers, 2005; (*b*) courtesy of David A. Chin.]

and 200 to 400 mg/L of BOD_5 (Sikora et al., 1976; Canter et al., 1987). Based on reported efficiencies of soil absorption systems, the following concentrations are typical of septic-tank effluent entering the ground water (Canter and Knox, 1985): BOD_5 of 30 to 80 mg/L, COD of 60 to 140 mg/L, ammonia nitrogen of 19 to 80 mg/L, and total phosphorus of 5 to 10 mg/L. Other constituents of concern in septic-tank effluent include bacteria, viruses, nitrates, synthetic organics, and toxic metals. Organic matter, BOD_5, pathogenic microorganisms, and phosphorus are effectively removed by most properly designed and permitted septic-tank systems, and these contaminants are rarely found more that 1.5 m below the level of discharge or beyond the immediate vicinity of the seepage field (Reneau and Petry, 1976; Brown et al., 1979). The nitrification process is typically completed in septic-tank drainfields located in well-drained soils, and the mobile nitrate-nitrogen enters the ground water.

Septic tanks are most likely to contribute to ground-water contamination in areas where (1) there is a high density of homes with septic tanks, (2) the soil layer over permeable bedrock is thin, (3) the soil is extremely permeable, or (4) the water table is less than 1 m below the ground surface.

7.3.2 Leaking Underground Storage Tanks

Underground tanks store gasoline at service stations and are widely used by industry, agriculture, and homes to store oil, hazardous chemicals, and chemical waste products. A leaking underground storage tank is shown in Figure 7.3. Because there are so many underground storage tanks and only a small portion of them are corrosion resistant, the problem of leaking underground storage tanks is a major source of diffuse pollution (Novotny, 2003).

7.3.3 Land Application of Wastewater

Land application of waste sludges and treated wastewater are significant sources of heavy metals, toxic chemicals, and pathogenic microorganisms. A typical wastewater infiltration basin is illustrated in Figure 7.4. There are three types of land application of wastewater: (1) slow-rate systems, (2) overland-flow systems, and (3) rapid infiltration systems. These wastewater systems are described below.

FIGURE 7.3   Leaking underground storage tank. (From USEPA, 2005e.)

FIGURE 7.4   Wastewater infiltration basin. (From NEWater, 2005.)

Slow-rate systems (SRSs) are most common in the United States for treatment of municipal wastewater and effluent reuse in arid areas; in Europe, these systems have been in use for centuries. The hydraulic loading rate for these systems is mostly matched to the irrigation and nutrient requirements for crops and soil permeability. In arid regions, the hydraulic loading is related to the irrigation requirement and prevention of salt buildup in soils. These systems are essentially irrigation systems and have problems similar to those of irrigation return flow and its impact on ground water and base flow. Of the three types of land-application systems, SRSs exhibit the highest nutrient removal due to the combined effect of nutrient uptake by crops and attenuation by soils. The disadvantage of low-rate application systems is the large area requirement, typically 30 ha of land per 1000 m^3/day of treated sewage.

In *overland-flow systems* (OFSs) wastewater is treated as it moves in graded and maintained grassed and vegetated sloped areas, and the treated effluent is collected as residual runoff at the bottom of the slope. Percolation of wastewater is not desirable and should be minimized by selection of low-permeability soils, soil compaction, and/or locating these systems over an impermeable subsurface stratum. Under these conditions, the impact of OFSs on ground-water resources should be minimal. OFSs are similar to grassed buffer strips used for treatment of urban and agricultural runoff. Nitrogen removal is accomplished by nitrification–denitrification processes and depends on the BOD/nitrogen ratio. If the nitrogen in the influent is primarily in nitrate form, the removal is minimal (Reed et al., 1995).

Rapid infiltration systems (RISs) rely on infiltration and filtration of wastewater in permeable soils. If subsoils are permeable, the effluent will reach the ground water and if designed improperly, may become a cause for ground-water contamination. Removal of contaminants in the upper soil layer is accomplished by physical–chemical interaction (adsorption) and biochemical degradation (both aerobic and anaerobic). Vegetation and its nutrient uptake is not considered. If most nitrogen is in nitrate form, removal efficiency is greatly reduced.

Problems associated with land application of wastewater are similar to those for septic tanks; however, much greater volumes of wastewater are concentrated in a smaller area. Mobile pollutants such as nitrates are of greatest concern; other contaminants (BOD, pathogenic microorganisms, and phosphates) remain near the area of application. Bacteria and viruses die off quite rapidly as wastewater passes through the soil material. The portion of the aquifer that is recharged by treated wastewater effluents should not be used as a source of drinking water and access should be restricted; water should be withdrawn at some safe distance from the recharge area.

Sludge generated by wastewater treatment facilities is commonly applied to agricultural lands as a fertilizer and soil conditioner. The effect of land application of sludge on ground-water quality depends on the transformation that occurs within the topsoil horizon. Although most of the toxic metals will be retained by the topsoil, the toxic metal content of sludge is of concern. Concentrations of toxic metals in wastewater sludge are much higher than those in raw wastewater.

7.3.4 Irrigation and Irrigation Return Flow

Using water that is high in dissolved solids to irrigate an area causes a portion of the irrigation water to be returned to the atmosphere by evapotranspiration, and since evapotranspired water has no salt content, there is a subsequent salt and contaminant buildup in soils. The portion returned to the atmosphere may range from less than 20% in high-rate application systems in humid climatic conditions to almost 100% in low-rate application systems in arid and semiarid climates. Figure 7.5 illustrates irrigation using ground water as a source.

To maintain acceptable salt content in soils and to sustain crop growth and fertility of the soils, excess irrigation water must be applied if natural precipitation is not sufficient to control salt buildup in soils. The excess irrigation water, containing increased salinity and

FIGURE 7.5 Irrigation from groundwater.

leachate from soils, is either collected by subsurface drainage systems or percolates directly into the ground water. The irrigation water collected by subsurface drainage systems or leached into the ground water is called *irrigation return flow* and represents one of the more serious problems associated with diffuse pollution of ground water. The concentration of salts in water percolating through the soil-root level into irrigation return flow or to ground water can be computed using the relation

$$c_i Q_i = c_{aq} (Q_i - Q_e) \qquad (7.6)$$

where c_i is the salt or contaminant concentration in the water or wastewater used for irrigation, Q_i is the amount of irrigation water also including precipitation that is not lost as surface runoff (= effective precipitation), c_{aq} is the salt or contaminant concentration of water percolating from the root zone downward, and Q_e is the amount of water released from the soil by evapotranspiration. The amount of excess irrigation water that has to be applied to control salt or contaminant buildup in soil depends on the tolerance of a crop to salt in the soil water, the salt content of the irrigation water, evapotranspiration rate, crop uptake, and other losses from the system. The *leaching ratio*, Q_i/Q_e, is derived from Equation 7.6:

$$\frac{Q_i}{Q_e} = \frac{c_{aq}}{c_{aq} - c_i} \qquad (7.7)$$

The salinity of irrigation water is usually expressed as conductivity in microsiemens per centimeter [$1000\,\mu S/cm \approx 640\,mg/L$ of total dissolved solids (TDS)]. The salt tolerance of crops ranges from less than $500\,\mu S/cm$ for salt-sensitive crops such as most fruit trees and some vegetables (celery, strawberries, or beans) to more than $1500\,\mu S/cm$ for salt-tolerant crops such as cotton, beets, barley, and asparagus. Most common grain crops and vegetables have medium tolerance (500 to $1500\,\mu S/cm$) to salts. The *leaching requirement* (LR) is defined by

$$LR = \frac{EC_i}{EC_{aq}} = \frac{c_i}{c_{aq}} \tag{7.8}$$

where EC_i is the electric conductivity of irrigation water and EC_{aq} is the electric conductivity of drainage water. Combining Equations 7.7 and 7.8, the *leaching ratio* is

$$\frac{Q_i}{Q_e} = \frac{1}{1 - LR} \tag{7.9}$$

Although irrigation return flow has been recognized as a significant water-quality problem, the Clean Water Act in the United States specifically excludes agricultural runoff and irrigation return flows from the definition of pollution.

Example 7.1   An avocado crop can tolerate water in the root zone with a total dissolved solids (TDS) concentration of up to $300\,mg/L$, and avocados require $10\,cm$ of water to support growth during the spring planting season. Available irrigation water and effective rainfall combined has a TDS content of $60\,mg/L$, soil evaporation during the spring planting season is $30\,cm$, and the effective rainfall is $25\,cm$. (a) Estimate the amount of irrigation water required, and the expected TDS concentration in the root zone. (b) Determine the leaching requirement, leaching ratio, and maximum requirement for irrigation plus rainfall to avoid excessive TDS in the root zone.

SOLUTION   (a) The irrigation requirement is determined on a volumetric basis according to the relation

$$irrigation\ requirement = crop\ requirement + evaporation - rainfall$$

$$= 10\,cm + 30\,cm - 25\,cm$$

$$= 15\,cm$$

From the data given, $Q_i =$ rainfall + irrigation $= 25\,cm + 15\,cm = 40\,cm$, $c_i = 60\,mg/L$, and $Q_e = 30\,cm$. Equation 7.6 gives the resulting TDS in the root zone, c_{aq}, as

$$c_{aq} = \frac{c_i Q_i}{Q_i - Q_e} = \frac{60(40)}{40 - 30} = 240\,mg/L$$

Therefore, the TDS in the root zone is expected to be $240\,mg/L$. This is less than the allowable maximum for avocados of $300\,mg/L$. In the event that the root-zone TDS concentration turned out to be greater than $300\,mg/L$, the irrigation requirement would need to be increased beyond the volumetric requirement of $15\,cm$, such that $c_{aq} \leq 300\,mg/L$.

(b) The leaching requirement, LR, is given by Equation 7.8 as

$$\text{LR} = \frac{c_i}{c_{aq}} = \frac{60}{300} = 0.20$$

and the leaching ratio is given by Equation 7.9 as

$$\frac{Q_i}{Q_e} = \frac{1}{1 - \text{LR}} = \frac{1}{1 - 0.20} = 1.25$$

This result indicates that the minimum irrigation plus effective rainfall required the keep the root-zone TDS concentration less than 300 mg/L is $1.25(Q_e) = 1.25(30 \text{ cm}) = 37.5 \text{ cm}$. In this case, the irrigation plus effective rainfall is 40 cm (≥ 37.5 cm) and yields an adequate root-zone TDS concentration.

7.3.5 Solid-Waste Disposal Sites

Solid-waste disposal sites are commonly called *landfills*. Modern landfills are constructed with leachate-collection and treatment systems, but most older landfills are simply large holes in the ground filled with waste and covered with dirt (Bedient et al., 1994). Leaking liquids and leachate from older landfills can be a significant source of ground-water contamination. Typical landfills are shown in Figure 7.6. Modern landfills are sophisticated engineering operations employing resource recovery (collection of methane and subsequent conversion to energy), leachate collection and subsequent treatment, and daily covering of wastes with soil. After ceasing operation, a landfill site can be reclaimed.

For each well-designed and well-operated landfill there are hundreds of abandoned unsanitary dumps of refuse and toxic chemicals that cause ground-water contamination problems. During the decade of 1970–1980, a large number of landfills were developed, including some receiving radioactive wastes. Stored and decomposing wastes are leaching from disintegrating drums left on these sites and will represent a serious problem for decades. In the United States such sites have been inventoried, and if severe problems have occurred, they were classified by the EPA as Superfund sites. Although solid-waste

(a)   (b)

FIGURE 7.6 Landfill: (*a*) closed, far view; (*b*) open, near view. [(*a*) From Energy Information Administration, 2005; (*b*) from Prince William Conservation Alliance, 2005.]

**TABLE 7.2   Leachate Characteristics from Municipal Solid Waste
Disposal Sites**

Component	Median Value (mg/L)	Ranges of All Values (mg/L)
Alkalinity (as $CaCO_3$)	3050	0–20,850
Biochemical oxygen demand (BOD_5)	5700	81–33,360
Chemical oxygen demand (COD)	8100	40–89,520
Copper (Cu)	0.5	0–9.9
Lead (Pb)	0.75	0–2.0
Zinc (Zn)	5.8	3.7–8.5
Chloride (Cl^-)	700	4.7–2500
Sodium (Na^+)	767	0–7700
Total dissolved solids (TDS)	8955	584–44,900
Ammoniacal nitrogen (NH_4^+)	218	0–1106
Total phosphate (PO_4^{3+})	10	0–30
Iron (Fe)	94	0–2820
Manganese (Mn)	0.22	0.05–125
pH	5.8	3.7–8.5

Source: USEPA (1977).

disposal sites are considered point sources of pollution, leachate from unsanitary landfills and dumps may have polluted large portions of ground water and appear as contaminated base flow in rivers and streams. Dangerous toxic compounds are commonly part of the overall composition of landfill leachate, especially when the landfill is used for the disposal of toxic chemicals. Table 7.2 shows the ranges in concentration for various chemical constituents of typical leachate from municipal solid-waste disposal sites. In countries that use coal for household heating, the composition of leachate may be quite different from that typical for U.S. conditions (Johansen and Carlson, 1976).

There are several methods for managing leachate: natural attenuation by soils, prevention of leachate formation, collection and treatment, pretreatment to reduce volume and solubility, and detoxification of hazardous wastes prior to landfilling. Leachate undergoes natural attenuation by various chemical, physical, and biological processes as it migrates through soil. Whether natural attenuation will be adequate to prevent ground-water contamination should be evaluated for each site. The generation of leachate can be minimized by restricting rainwater from infiltrating the landfill. This is accomplished by providing appropriate surface drainage and/or placement of an impermeable liner over the daily accumulation of refuse. Another method of controlling leachate is to collect it at the bottom of the landfill and treat it before discharging it into surface water or land. In most cases, leachate collected must be pretreated before discharge into sewers by an anaerobic biological treatment unit. The high BOD strength of the leachate makes it difficult to treat in conventional aerobic treatment units, and without pretreatment, conventional biological treatment plants could become overloaded. Newly constructed landfills require a clay and geomembrane lining and suitable low-permeability (clay) substratum to virtually eliminate potential seepage of leachate into ground water.

Most regulations recommend or require that landfill sites be developed on uplands rather than in floodplains and on low-permeability soils. Geologically, such sites are difficult to find, and these sites must also be socially and politically acceptable. If a landfill receives

FIGURE 7.7 Injection well. (From Healthgate Resources, 2005.)

hazardous (toxic) waste, a TCLP extraction toxicity analysis must be performed. The solid-waste disposal site is considered hazardous (toxic) if the TCLP extract from a representative sample of waste contains any of the regulated toxic compounds in concentrations that exceed their allowable limit. Key indicators of leachate presence in ground water are elevated levels of specific conductance, temperature, chloride ion, color, turbidity, and COD.

7.3.6 Waste-Disposal Injection Wells

Waste-disposal injection wells are used to inject contaminated water, surface runoff, and hazardous wastes deep into the ground and away from drinking-water sources, but poor well design, faulty construction, inadequate understanding of the subsurface geology, and deteriorated well casings can all cause contaminants to be introduced into drinking-water sources. The wellhead of an injection well is shown in Figure 7.7.

7.3.7 Agricultural Operations

The uses of pesticides and fertilizers in agricultural practice are significant sources of synthetic organic chemicals and nutrients in ground water. The impact of agricultural practices on ground-water quality are discussed extensively in Section 9.5.

7.4 FATE AND TRANSPORT MODELS

Contaminants in ground water undergo a variety of fate and transport processes. The fate processes that are most often considered include sorption onto the solid matrix and first-order decay, both of which affect the amount of contaminant mass in ground water. Transport processes include advection at the mean (large-scale) ground-water seepage velocity and mixing caused by small-scale variations in the seepage velocity associated with spatial variability in the hydraulic conductivity. Typical velocities of ground water

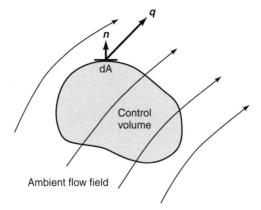

FIGURE 7.8 Control volume in a porous medium. (From Chin, David A., *Water-Resources Engineering.* Copyright © 2000. Reprinted by permission of Pearson Education, Inc., Upper Saddle River, NJ.)

may range from less than 1 cm/yr in tight clays to more than 100 m/yr in permeable sand and gravel. The normal range for ground-water velocities is 1 to 10 m/yr.

The dispersion of dissolved tracers in ground water differs from the dispersion of dissolved tracers in surface water due to the presence of the solid matrix. In ground water, the diffusive flux, q_i^d [M/L^{2}T], of a tracer is expressed in the modified Fickian form

$$\mathbf{q}^d = -\mathbf{D}\,\nabla(nc) \tag{7.10}$$

where $\mathbf{D}$ is a dispersion coefficient that includes the effect of the solid matrix, c is the tracer concentration in the ground water, n is the porosity of the porous medium, and nc is the mass of tracer per unit volume of porous medium. The mass flux associated with the larger-scale (advective) fluid motions is given by

$$\mathbf{q}^a = n\mathbf{V}c \tag{7.11}$$

where $\mathbf{q}^a$ is the advective tracer mass flux [M/L^{2}T], and $\mathbf{V}$ is the mean seepage velocity. The total flux of a tracer, $\mathbf{q}$, within a fluid is the sum of the advective and diffusive fluxes and is given by

$$\mathbf{q} = \mathbf{q}^a + \mathbf{q}^d = n\mathbf{V}c - \mathbf{D}\,\nabla(nc) \tag{7.12}$$

Consider the finite control volume shown in Figure 7.8, where this control volume is contained within the porous medium. The law of conservation of mass requires that the net flux [M/T] of tracer mass into the control volume is equal to the rate of change of tracer mass [M/T] within the control volume. This relation is given by

$$\frac{\partial}{\partial t}\int_V cn\,dV + \int_S \mathbf{q}\cdot\mathbf{n}\,dA = \int_V S_m n\,dV \tag{7.13}$$

where V is the volume of the control volume, S is the surface area of the control volume, $\mathbf{q}$ is the flux vector given by Equation 7.12, $\mathbf{n}$ is the unit normal pointing out of the control volume, and S_m is the tracer mass flux per unit volume of ground water originating within the control volume. Equation 7.13 can be simplified using the divergence theorem, which relates a surface integral to a volume integral by the relation

$$\int_S \mathbf{q} \cdot \mathbf{n} \, dA = \int_V \nabla \cdot \mathbf{q} \, dV \qquad (7.14)$$

Combining Equations 7.13 and 7.14 leads to the result

$$\frac{\partial}{\partial t} \int_V cn \, dV + \int_V \nabla \cdot \mathbf{q} \, dV = \int_V S_m n \, dV \qquad (7.15)$$

Since the control volume is fixed in space and time, the derivative of the volume integral with respect to time is equal to the volume integral of the derivative with respect to time, and Equation 7.15 can be written in the form

$$\int_V \left(\frac{\partial cn}{\partial t} + \nabla \cdot \mathbf{q} - S_m n \right) dV = 0 \qquad (7.16)$$

This equation requires that the integral of the quantity in parentheses must be equal to zero for any arbitrary control volume, and this can be true only if the integrand itself is equal to zero. Following this logic, Equation 7.16 requires that

$$\frac{\partial cn}{\partial t} + \nabla \cdot \mathbf{q} - S_m n = 0 \qquad (7.17)$$

This equation can be combined with the expression for the mass flux given by Equation 7.12 and written in the expanded form

$$\frac{\partial cn}{\partial t} + \nabla \cdot (n\mathbf{V}c - \mathbf{D}\nabla nc) = S_m n \qquad (7.18)$$

Assuming that the porosity, n, is invariant in space and time, Equation 7.18 simplifies to

$$\frac{\partial c}{\partial t} + \mathbf{V} \cdot \nabla c + c(\nabla \cdot \mathbf{V}) = \mathbf{D}\nabla^2 c + S_m \qquad (7.19)$$

In the case of incompressible fluids, conservation of fluid mass requires that

$$\nabla \cdot \mathbf{V} = 0 \qquad (7.20)$$

and combining Equations 7.19 and 7.20 yields the following diffusion equation:

$$\frac{\partial c}{\partial t} + \mathbf{V} \cdot \nabla c = \mathbf{D}\nabla^2 c + S_m \qquad (7.21)$$

The dispersion coefficient, **D**, in porous media is generally anisotropic, and denoting the principal components of the dispersion coefficient as D_i, the advection–dispersion equation can be expressed in the form

$$\frac{\partial c}{\partial t} + \sum_{i=1}^{3} V_i \frac{\partial c}{\partial x_i} = \sum_{i=1}^{3} D_i \frac{\partial^2 c}{\partial x_i^2} + S_m \qquad (7.22)$$

where x_i are the principal directions of the dispersion coefficient tensor. This form of the advection–dispersion equation, which is appropriate for flow through porous media, is identical to the form of the advection–dispersion equation used in surface waters.

Solutions to the advection–dispersion equation for specified initial and boundary conditions are commonly referred to as *dispersion models* or *fate and transport models* and can be either analytic or numerical. Numerical models provide discrete solutions to the advection–dispersion equation in space and time and are most useful in cases of complex geology and irregular boundary conditions. Analytic models provide continuous solutions to the advection–dispersion equation in space and time and are most useful in cases of simple geology and simple boundary conditions. Most ground-water contamination problems can be analyzed assuming steady-state flow conditions, which implies that the flow velocity and dispersion characteristics remain constant with time. Several useful analytic dispersion models are described in the following sections.

7.4.1 Instantaneous Point Source

In the case where a mass, M, of conservative contaminant is injected instantaneously over a depth, H, of a uniform aquifer with mean seepage velocity, V, the resulting concentration distribution, $c(x, y, t)$, is given by the fundamental solution to the advection–dispersion equation, which can be written in the form

$$c(x, y, t) = \frac{M}{4\pi t H n \sqrt{D_L D_T}} \exp\left[-\frac{(x - Vt)^2}{4D_L t} - \frac{y^2}{4D_T t} \right] \qquad (7.23)$$

where t is the time since the injection of the contaminant, n is the porosity, x is the coordinate measured in the direction of the seepage velocity, y is the transverse (horizontal) coordinate, the contaminant source is located at the origin of the coordinate system, and D_L and D_T are the longitudinal and transverse dispersion coefficients. Equation 7.23 is more commonly applied in cases where a contaminant is initially mixed over a depth, H, of the aquifer, not the entire depth of the aquifer, and vertical dispersion is negligible compared with longitudinal and horizontal-transverse dispersion. Contaminants are seldom released instantaneously into the ground water. However, if the duration of release is short compared to the time of interest, and if the volume spilled is small enough not to influence the ground-water flow pattern significantly near the release point, the instantaneous release assumption is justified. A contaminant mass cannot be added realistically at a point over a depth H. If the contaminant mass is added over an area A_0 and the initial concentration is c_0, the following substitution into Equation 7.23 is appropriate:

$$\frac{M}{Hn} = c_0 A_0 \tag{7.24}$$

and Equation 7.23 can be applied provided that

$$4\pi t \sqrt{D_L D_T} \gg A_0 \tag{7.25}$$

This relation is based on the requirement that the size of the contaminated area is much greater than the size of the initial spill area.

Example 7.2   Ten kilograms of a contaminant is spilled over the top 2 m of an aquifer. The longitudinal and horizontal-transverse dispersion coefficients are 1 and $0.1 \, \text{m}^2/\text{day}$, respectively; vertical mixing is negligible; the porosity is 0.2; and the mean seepage velocity is 0.6 m/day. (a) Estimate the maximum contaminant concentrations in the ground water 1 day, 1 week, 1 month, and 1 year after the spill. (b) What is the contaminant concentration at the spill location after 1 week?

SOLUTION   From the data given, $M = 10 \, \text{kg}$, $H = 2 \, \text{m}$, $D_L = 1 \, \text{m}^2/\text{day}$, $D_T = 0.1 \, \text{m}^2/\text{day}$, $n = 0.2$, and $V = 0.6 \, \text{m/day}$. According to Equation 7.23, the maximum concentration, c_{max}, occurs at $x = Vt$ and $y = 0 \, \text{m}$; hence,

$$c_{max}(t) = \frac{M}{4\pi t H n \sqrt{D_L D_T}}$$

Substituting the given parameters gives

$$c_{max}(t) = \frac{10}{4\pi t (2)(0.2) \sqrt{1 \times 0.1}} = \frac{6.30}{t} \, \text{kg/m}^3 = \frac{6300}{t} \, \text{mg/L}$$

which yields the results shown in Table 7.3.

(b) The concentration at the spill location ($x = 0 \, \text{m}$, $y = 0 \, \text{m}$) as a function of time is given by Equation 7.23 as

$$c(0, 0, t) = \frac{M}{4\pi t H n \sqrt{D_L D_T}} \exp\left[-\frac{(Vt)^2}{4D_L t}\right]$$

TABLE 7.3   Results for Example 7.2

t (days)	c_{max} (t) (mg/L)
1	6300
7	900
30	210
365	17.5

which at $t = 7$ days gives

$$c(0, 0, 7) = \frac{10}{4\pi(7)(2)(0.2)\sqrt{1 \times 0.1}} \exp\left[-\frac{(0.6 \times 7)^2}{4(1)(7)}\right] = 0.48 \, \text{kg/m}^3 = 480 \, \text{mg/L}$$

Hence, after 7 days, the concentration at the site of the spill is approximately 53% of the maximum concentration of 900 mg/L.

7.4.2 Continuous Point Source

In the case where a conservative contaminant of initial concentration c_0 is injected continuously at a rate Q [L^3/T] into a uniform aquifer of depth H and mean seepage velocity V, the concentration distribution downstream of the source, $c(x, y, t)$, is given by (Fried, 1975)

$$c(x, y, t) = \frac{Qc_0}{4\pi H \sqrt{D_L D_T}} \exp\left(\frac{Vx}{2D_L}\right) [W(0, B) - W(t, B)] \tag{7.26}$$

where the x coordinate is in the direction of the seepage velocity; y is the transverse (horizontal) coordinate; the source is located at the origin of the coordinate system; D_L and D_T are the longitudinal and transverse dispersion coefficients, respectively; $W(\alpha, \beta)$ is defined as

$$W(\alpha, \beta) = \int_\alpha^\infty \frac{1}{y} \exp\left(-y - \frac{\beta^2}{4y}\right) dy \tag{7.27}$$

and B is defined by

$$B = \left[\frac{(Vx)^2}{4D_L^2} + \frac{(Vy)^2}{4D_L D_T}\right]^{1/2} \tag{7.28}$$

$W(\alpha, \beta)$ is identical to the well function for a leaky aquifer that is used in ground-water hydrology. To facilitate the evaluation of Equation 7.26, values of $W(\alpha, \beta)$ are tabulated in Table 7.4. As $t \to \infty$, the concentration distribution given by Equation 7.26 approaches the steady-state solution (Bear, 1972)

$$c(x, y) = \frac{Qc_0}{2\pi H \sqrt{D_L D_T}} \exp\left(\frac{Vx}{2D_L}\right) K_0\left\{\left[\frac{V^2}{4D_L}\left(\frac{x^2}{D_L} + \frac{y^2}{D_T}\right)\right]^{1/2}\right\} \tag{7.29}$$

where K_0 is the modified Bessel function of the second kind of order zero (described in Appendix E.2).

Example 7.3   A conservative contaminant is injected continuously through a 4-m-deep perforated well into an aquifer with a mean seepage velocity of 0.8 m/day and longitudinal and transverse dispersion coefficients of 2 and 0.2 m^2/day, respectively. If the injection rate of the contaminated water is 0.7 m^3/day, with a contaminant concentration of 100 mg/L, estimate the steady-state contaminant concentrations at locations 1, 10, 100, and 1000 m downstream of the injection well. Neglect vertical diffusion.

TABLE 7.4 Well Function for Leaky Aquifer, W (α, β)

α \ β	0.00	0.002	0.004	0.007	0.01	0.02	0.04	0.06	0.08	0.10
0.00		12.6611	11.2748	10.1557	9.4425	8.0569	6.6731	5.8456	5.2950	4.8541
1×10^{-6}	13.2383	12.4417	11.2711	10.1557						
2×10^{-6}	12.5451	12.1013	11.2259	10.1554						
5×10^{-6}	11.6289	11.4384	10.9642	10.1290	9.4425					
8×10^{-6}	11.1589	11.0377	10.7151	10.0602	9.4313					
1×10^{-5}	10.9357	10.8382	10.5725	10.0034	9.4176	8.0569				
2×10^{-5}	10.2426	10.1932	10.0522	9.7126	9.2961	8.0558				
5×10^{-5}	9.3263	9.3064	9.2480	9.0957	8.8827	8.0080	6.6730			
7×10^{-5}	8.9899	8.9756	8.9336	8.8224	8.6625	7.9456	6.6726			
1×10^{-4}	8.6332	8.6233	8.5937	8.5145	8.3983	7.8375	6.6693	5.8658	5.2950	4.8541
2×10^{-4}	7.9402	7.9352	7.9203	7.8800	7.8192	7.4472	6.6242	5.8637	5.2949	4.8541
5×10^{-4}	7.0242	7.0222	7.0163	6.9999	6.9750	6.8346	6.3626	5.8011	5.2848	4.8530
7×10^{-4}	6.6879	6.6865	6.6823	6.6706	6.6527	6.5508	6.1917	5.7274	5.2618	4.8478
1×10^{-3}	6.3315	6.3305	6.3276	6.3194	6.3069	6.2347	5.9711	5.6058	5.2087	4.8292
2×10^{-3}	5.6394	5.6389	5.6374	5.6334	5.6271	5.5907	5.4516	5.2411	4.9848	4.7079
5×10^{-3}	4.7261	4.7259	4.7253	4.7237	4.7212	4.7068	4.6499	4.5590	4.4389	4.2990
7×10^{-3}	4.3916	4.3915	4.3910	4.3899	4.3882	4.3779	4.3374	4.2719	4.1839	4.0771
1×10^{-2}	4.0379	4.0378	4.0375	4.0368	4.0351	4.0285	4.0003	3.9544	3.8920	3.8190
2×10^{-2}	3.3547	3.3547	3.3545	3.3542	3.3536	3.3502	3.3365	3.3141	3.2832	3.2442
5×10^{-2}	2.4679	2.4679	2.4678	2.4677	2.4675	2.4662	2.4613	2.4531	2.4416	2.4271
7×10^{-2}	2.1508	2.1508	2.1508	2.1507	2.1506	2.1497	2.1464	2.1408	2.1331	2.1232
1×10^{-1}	1.8229	1.8229	1.8229	1.8228	1.8227	1.8222	1.8220	1.8164	1.8114	1.8050
2×10^{-1}	1.2227	1.2226	1.2226	1.2226	1.2226	1.2224	1.2215	1.2201	1.2181	1.2155
5×10^{-1}	0.5598	0.5598	0.5598	0.5598	0.5598	0.5597	0.5595	0.5592	0.5587	0.5581
7×10^{-1}	0.3738	0.3738	0.3738	0.3738	0.3738	0.3737	0.3736	0.3734	0.3732	0.3729
1.00	0.2194	0.2194	0.2194	0.2194	0.2194	0.2194	0.2193	0.2192	0.2191	0.2190
2.00	0.0489	0.0489	0.0489	0.0489	0.0489	0.0489	0.0489	0.0489	0.0489	0.0488
5.00	0.0011	0.0011	0.0011	0.0011	0.0011	0.0011	0.0011	0.0011	0.0011	0.0011
7.00	0.0001	0.0001	0.0001	0.0001	0.0001	0.0001	0.0001	0.0001	0.0001	0.0001
8.00	0.0000	0.0000	0.0000	0.0000	0.0000	0.0000	0.0000	0.0000	0.0000	0.0000

TABLE 7.5   Results for Example 7.3

x (m)	c(x, 0) (kg/m³)	c(x, 0) (mg/L)
1	0.0094	9.4
10	0.0037	3.7
100	0.0012	1.2
1000	0.00039	0.39

SOLUTION   From the data given, $H = 4$ m, $V = 0.8$ m/day, $D_L = 2$ m²/day, $D_T = 0.2$ m²/day, $Q = 0.7$ m³/day, and $c_0 = 100$ mg/L $= 0.1$ kg/m³. The steady-state concentration is given by Equation 7.29 as

$$c(x, y) = \left[\frac{Qc_0}{2\pi H\sqrt{D_L D_T}}\right] \exp\left(\frac{Vx}{2D_L}\right) K_0 \left\{\left[\frac{V^2}{4D_L}\left(\frac{x^2}{D_L} + \frac{y^2}{D_T}\right)\right]^{1/2}\right\}$$

which yields

$$c(x, 0) = \left[\frac{0.7(0.1)}{2\pi(4)\sqrt{2 \times 0.2}}\right] \exp\left[\frac{(0.8)x}{2(2)}\right] K_0 \left\{\left[\frac{0.8^2}{4 \times 2}\left(\frac{x^2}{2}\right)\right]^{1/2}\right\}$$

$$= 0.00440 \exp(0.2x) K_0 (0.2x)$$

The steady-state downstream concentrations are given listed in Table 7.5.

7.4.3   Continuous Plane Source

The case where a conservative contaminant of concentration c_0 is continuously released from a plane source of dimension $Y \times Z$ is illustrated in Figure 7.9. The resulting concentration distribution, $c(x, y, z, t)$, is given by (Domenico and Robbins, 1985)

$$c(x, y, z, t) = \frac{c_0}{8} \operatorname{erfc}\left[\frac{x - Vt}{2(\alpha_x Vt)^{1/2}}\right] \left\{\operatorname{erf}\left[\frac{y + Y/2}{2(\alpha_y x)^{1/2}}\right] - \operatorname{erf}\left[\frac{y - Y/2}{2(\alpha_y x)^{1/2}}\right]\right\}$$

$$\left\{\operatorname{erf}\left[\frac{z + Z}{2(\alpha_z x)^{1/2}}\right] - \operatorname{erf}\left[\frac{z - Z}{2(\alpha_z x)^{1/2}}\right]\right\} \tag{7.30}$$

where V is the mean seepage velocity, and α_x, α_y, and α_z are the dispersivities in the coordinate directions. The dispersivity in a porous medium is defined as the dispersion coefficient divided by the mean seepage velocity, where

$$\alpha_x = \frac{D_x}{V}, \qquad \alpha_y = \frac{D_y}{V}, \qquad \alpha_z = \frac{D_z}{V} \tag{7.31}$$

and D_x, D_y, and D_z are the dispersion coefficients in the x, y, and z directions, respectively. In Equation 7.30, x is the longitudinal (flow) direction, y is the horizontal-transverse direction, and z is the vertical-transverse direction. If there is no spreading in the vertical, z, direction, the error functions containing the z terms in Equation 7.30 are ignored and $c_0/8$

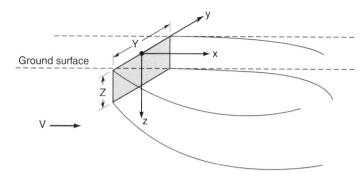

FIGURE 7.9 Dispersion from a continuous plane source. (From Chin, David A., *Water-Resources Engineering*. Copyright © 2000. Reprinted by permission of Pearson Education, Inc., Upper Saddle River, NJ.)

becomes $c_0/4$ (Domenico and Schwartz, 1990). The distance, x_0, from the source to the location where the contaminant plume is well mixed over the aquifer thickness, H, can be estimated by (Domenico and Palciauskas, 1982)

$$x_0 = \frac{(H - Z)^2}{\alpha_z} \tag{7.32}$$

For distances of less than x_0, Equation 7.30 is applicable; for distances greater than x_0, the distance x in the denominator of the error function of the z term is replaced by x_0, prohibiting further spreading for $x > x_0$. Domenico (1987) showed that for contaminants that undergo first-order decay with a decay factor λ, Equation 7.30 becomes

$$
\begin{aligned}
c(x, y, z, t) = \frac{c_0}{8} &\exp\left\{\frac{x}{2\alpha_x}\left[1 - \left(1 + \frac{4\lambda\alpha_x}{V}\right)^{1/2}\right]\right\} \mathrm{erfc}\left[\frac{x - Vt(1 + 4\lambda\alpha_x/V)^{1/2}}{2(\alpha_x Vt)^{1/2}}\right] \\
&\cdot \left\{\mathrm{erf}\left[\frac{y + Y/2}{2(\alpha_y x)^{1/2}}\right] - \mathrm{erf}\left[\frac{y - Y/2}{2(\alpha_y x)^{1/2}}\right]\right\} \left\{\mathrm{erf}\left[\frac{z + Z}{2(\alpha_z x)^{1/2}}\right]\right. \\
&\left. - \mathrm{erf}\left[\frac{z - Z}{2(\alpha_z x)^{1/2}}\right]\right\}
\end{aligned}
\tag{7.33}
$$

Example 7.4 A continuous contaminant source is 3 m wide × 2 m deep and contains a contaminant at a concentration of 100 mg/L. The mean seepage velocity in the aquifer is 0.4 m/day, the aquifer is 7 m deep, and the longitudinal, horizontal-transverse, and vertical-transverse dispersivities are 3, 0.3, and 0.03 m, respectively. (a) Assuming that the contaminant is conservative, determine the downstream location at which the contaminant plume will be fully mixed over the depth of the aquifer. (b) Estimate the contaminant concentrations at the water table at locations 10, 100, and 1000 m downstream of the source after 10 years. (c) If the contaminant undergoes biodegradation with a decay rate of 0.01 day^{-1}, estimate the effect on the concentrations downstream of the source.

SOLUTION   (a) From the data given, $Y = 3$ m, $Z = 2$ m, $c_0 = 100$ mg/L $= 0.1$ kg/m^3, $V = 0.4$ m/day, $H = 7$ m, $\alpha_x = 3$ m, $\alpha_y = 0.3$ m, and $\alpha_z = 0.03$ m. The contaminant plume

TABLE 7.6   Results for Example 7.4(b)

x(m)	$c(x, 0, 0, 3650)$ (kg/m^3)	$c(x, 0, 0, 3650)$ (mg/L)
10	0.046	46
100	0.0090	9.0

becomes well mixed at a distance x_0 downstream, where x_0 is given by Equation 7.32 as

$$x_0 = \frac{(H-Z)^2}{\alpha_z} = \frac{(7-2)^2}{0.03} = 833 \text{ m}$$

(b) The concentration along the line $y = 0$ m, $z = 0$ m is given by Equation 7.30 as

$$c(x, 0, 0, t) = \frac{c_0}{8} \text{erfc}\left[\frac{x - Vt}{2(\alpha_x Vt)^{1/2}}\right] \left\{ \text{erf}\left[\frac{Y/2}{2(\alpha_y x)^{1/2}}\right] - \text{erf}\left[\frac{-Y/2}{2(\alpha_y x)^{1/2}}\right] \right\}$$

$$\cdot \left\{ \text{erf}\left[\frac{Z}{2(\alpha_z x)^{1/2}}\right] - \text{erf}\left[\frac{-Z}{2(\alpha_z x)^{1/2}}\right] \right\}$$

and therefore at $t = 10$ years $= 3650$ days,

$$c(x, 0, 0, 3650) = \frac{0.1}{8} \text{erfc}\left[\frac{x - 0.4(3650)}{2[3(0.4)(3650)]^{1/2}}\right] \left\{ \text{erf}\left[\frac{3/2}{2(0.3x)^{1/2}}\right] \right.$$

$$\left. - \text{erf}\left[\frac{-3/2}{2(0.3x)^{1/2}}\right] \right\} \left\{ \text{erf}\left[\frac{2}{2(0.03x)^{1/2}}\right] - \text{erf}\left[\frac{-2}{2(0.03x)^{1/2}}\right] \right\}$$

which simplifies to[1]

$$c(x, 0, 0, 3650) = 0.05 \text{ erfc}\left(\frac{x - 1460}{132}\right) \text{erf}\left(\frac{1.37}{\sqrt{x}}\right) \text{erf}\left(\frac{5.77}{\sqrt{x}}\right)$$

Since the contaminant becomes well mixed at $x_0 = 833$ m, this formulation can only be used for calculating the concentrations at $x \le 833$ m. At $x = 10$ m and $x = 100$ m, the equation yields the results shown in Table 7.6.

At $x = 1000$ m, the plume is well mixed over the vertical, and the contaminant concentration is calculated by replacing x by x_0 ($= 833$ m) in the denominator of the error function in the z term to yield

$$c(x, 0, 0, 3650) = 0.05 \text{ erfc}\left(\frac{x - 1460}{132}\right) \text{erf}\left(\frac{1.37}{\sqrt{x}}\right) \text{erf}\left(\frac{5.77}{\sqrt{833}}\right)$$

$$= 0.0111 \text{ erfc}\left(\frac{x - 1460}{132}\right) \text{erf}\left(\frac{1.37}{\sqrt{x}}\right)$$

which gives $c(1000, 0, 0, 3650) = 0.0011$ kg/m^3 $= 1.1$ mg/L.

---

[1]This simplification uses the identity $\text{erf}(-x) = -\text{erf}(x)$.

(c) If the contaminant undergoes first-order decay with $\lambda = 0.01$ day^{-1}, the concentration profile along the line $y = 0$ m, $z = 0$ m, is given by Equation 7.33 as

$$c(x,0,0,3650) = \frac{0.1}{8} \exp\left\{\frac{x}{2(3)}\left[1 - \left(1 + \frac{4(0.01)(3)}{0.4}\right)^{1/2}\right]\right\}$$

$$\cdot \mathrm{erfc}\left[\frac{x - 0.4(3650)[1 + 4(0.01)(3/0.4)]^{1/2}}{2[3(0.4)(3650)]^{1/2}}\right]$$

$$\cdot \left\{\mathrm{erf}\left[\frac{3/2}{2(0.3x)^{1/2}}\right] - \mathrm{erf}\left[\frac{-3/2}{2(0.3x)^{1/2}}\right]\right\}\left\{\mathrm{erf}\left[\frac{2}{2(0.03x)^{1/2}}\right]\right.$$

$$\left. - \mathrm{erf}\left[\frac{-2}{2(0.03x)^{1/2}}\right]\right\}$$

which simplifies to

$$c(x,0,0,3650) = 0.05 \exp(-0.0234x)\, \mathrm{erfc}\left(\frac{x - 1665}{132}\right) \mathrm{erf}\left(\frac{1.37}{\sqrt{x}}\right) \mathrm{erf}\left(\frac{5.77}{\sqrt{x}}\right)$$

and at $x = 10$ m and $x = 100$ m yields the results shown in Table 7.7. Replacing x by x_0 ($= 833$ m) in the z term yields $c(1000, 0, 0, 3650) = 7.48 \times 10^{-14}$ kg/m$^3 \approx 0$ mg/L.

The results of this example show that biodegradation will have a significant effect on the contaminant concentrations downstream of the source. Beyond $x = 100$ m, the biodegraded contaminant concentrations are negligible.

7.5   TRANSPORT PROCESSES

Dispersion of contaminants in ground water is caused by spatial variations in hydraulic conductivity and, to a much smaller extent, by pore-scale mixing and molecular diffusion. Pore-scale mixing results from the differential movement of ground water through pores of various sizes and shapes, a process called *mechanical dispersion*; the combination of mechanical dispersion and molecular diffusion is called *hydrodynamic dispersion*. Dispersion caused by large-scale variations in hydraulic conductivity is called *macrodispersion*. Consider a porous medium in which several samples of characteristic size L are tested for their hydraulic conductivity, K. The hydraulic conductivity (K) is then a random

TABLE 7.7   Results for Example 7.4(c)

x (m)	$c(x, 0, 0, 3650)$ (kg/m^3)	$c(x, 0, 0, 3650)$ (mg/L)
10	0.015	15
100	1.23×10^{-7}	1.23×10^{-4}

TABLE 7.8 Hydraulic Conductivity Statistics

Formation	$\langle Y \rangle = \langle \ln K \rangle$ (K in m/day)	K_G (m/day)	σ_Y
Sandstone	-2.0	0.13	0.92
	-0.98	0.38	0.46
Sand and gravel	—	—	1.01
	—	—	1.24
	—	—	1.66
Silty clay	-0.15	0.86	2.14
Loamy sand	0.59	1.81	1.98

Source: Freeze (1975).

space function (RSF) with support scale L. Assuming that K is log normally distributed, it is convenient to work with the variable Y defined as

$$Y = \ln K \tag{7.34}$$

where Y is a normally distributed random space function, characterized by a mean, $\langle Y \rangle$; variance, σ_Y^2; and correlation length scales, λ_i, in the x_i-coordinate directions. The geometric mean hydraulic conductivity, K_G, is related to $\langle Y \rangle$ by

$$K_G = e^{\langle Y \rangle} \tag{7.35}$$

Freeze (1975) analyzed data from a variety of geologic cores, and the statistics of the measured hydraulic conductivities are tabulated in Table 7.8. These data indicate relatively high values of σ_Y, which reflect a significant degree of variability about the mean hydraulic conductivity. The variance of the hydraulic conductivity is inversely proportional to the magnitude of the support scale, with larger support scales resulting in smaller variances in the hydraulic conductivity. Consequently, whenever values of σ_Y are cited, it is sound practice also to state the corresponding support scale. The support scale of the data shown in Table 7.8 is on the order of 10 cm. The spatial covariance of Y must also be associated with a stated support scale, since both σ_Y and the correlation length scale, λ_i, depend on the support scale. Larger support scales generally yield larger correlation length scales. Porous media in which the correlation length scales of the hydraulic conductivity in the principal directions differ from each other are called *anisotropic media*, and porous media where the correlation length scales of the hydraulic conductivity in the principal directions are all equal are called *isotropic media*. Detailed discussions of dispersion in both isotropic and anisotropic media can be found in Dagan (1989), Chin and Wang (1992), Gelhar (1993), and Chin (1997). The mean seepage velocity, V_i, in isotropic porous media is given the the Darcy equation,

$$V_i = -\frac{K_{\text{eff}}}{n_e} J_i \tag{7.36}$$

where K_{eff} is the effective hydraulic conductivity, n_e is the effective porosity, and J_i is the slope of the piezometric surface in the i-direction. The effective hydraulic conductivity in isotropic media can be expressed in terms of the statistics of the hydraulic conductivity field by the relations (Dagan, 1989)

$$
\begin{aligned}
&\text{One-dimensional flow:} && K_{\text{eff}} = K_G\left(1 - \frac{\sigma_Y^2}{2}\right) \\[2ex]
&\text{Two-dimensional flow:} && K_{\text{eff}} = K_G && \text{(7.37)} \\[2ex]
&\text{Three-dimensional flow:} && K_{\text{eff}} = K_G\left(1 + \frac{\sigma_Y^2}{6}\right)
\end{aligned}
$$

The dispersion coefficient in porous media can be stated generally as a tensor quantity, D_{ij}, which is typically expressed in terms of the magnitude of the mean seepage velocity, V, by the relation (Bear, 1979)

$$ D_{ij} = \alpha_{ij} V \qquad (7.38) $$

where α_{ij} is the *dispersivity* of the porous medium. In general porous media, α_{ij} is a symmetric tensor with six independent components and can be written in the form

$$ \alpha_{ij} = \begin{bmatrix} \alpha_{11} & \alpha_{12} & \alpha_{13} \\ \alpha_{21} & \alpha_{22} & \alpha_{23} \\ \alpha_{31} & \alpha_{32} & \alpha_{33} \end{bmatrix} \qquad (7.39) $$

where $\alpha_{ij} = \alpha_{ji}$. In cases where the flow direction coincides with one of the principal directions of the hydraulic conductivity, the off-diagonal terms in the dispersivity tensor are equal to zero, and α_{ij} can be written in the form

$$ \alpha_{ij} = \begin{bmatrix} \alpha_{11} & 0 & 0 \\ 0 & \alpha_{22} & 0 \\ 0 & 0 & \alpha_{33} \end{bmatrix} \qquad (7.40) $$

where α_{11} is generally taken as the dispersivity in the flow direction, and α_{22} and α_{33} are the dispersivities in the horizontal and vertical transverse principal directions of the hydraulic conductivity. The component of the dispersivity in the direction of flow is called the *longitudinal dispersivity*, and the other components of the dispersivity are called the *transverse dispersivities*.

The dispersivites used to describe the transport of contaminants in porous media cannot be taken as constant unless the contaminant cloud has traversed several correlation length scales of the hydraulic conductivity, or the contaminant cloud is sufficiently large to encompass several correlation length scales. If either of these conditions is violated, the dispersivity increases as the contaminant cloud moves through the porous medium, includes an expanding range of hydraulic conductivity variations and ultimately approaches a constant value called the *asymptotic macrodispersivity* or simply the *macrodispersivity*. In isotropic media, the correlation length scale, λ, of the hydraulic conductivity is the same in all directions, and the components of the macrodispersivity can be estimated using the approximate relations (Dagan, 1989; Chin and Wang, 1992)

$$ \alpha_{11} = \sigma_Y^2 \lambda, \qquad \alpha_{22} = \alpha_{33} = 0 \qquad (7.41) $$

where it is interesting to note that the heterogeneous structure of the porous medium does not create transverse macrodispersion. Derivation of Equation 7.41 assumes that the local-mean seepage velocity is statistically homogeneous, and the spatial correlation of the hydraulic conductivity can be represented by an exponential function of spatial separation. According to Chin and Wang (1992), assumptions in the theoretical approximations used in deriving Equation 7.41 can be taken as valid up to $\sigma_Y = 1.5$.

In cases where the porous medium is stratified, isotropic in the horizontal plane, and anisotropic in the vertical plane, the correlation length scale of the hydraulic conductivity in the horizontal plane can be denoted by λ_h, and the correlation length scale in the vertical direction denoted by λ_v. The *anisotropy ratio*, e, is then defined by

$$e = \frac{\lambda_v}{\lambda_h}$$

(7.42)

and is typically on the order of 0.1 in most stratified media. Gelhar and Axness (1983) have derived approximate relations to estimate the components of the macrodispersivity in the case that the flow is in the plane of isotropy. In this case, the longitudinal and transverse components of the macrodispersivity tensor can be estimated by

$$\boxed{\alpha_{11} = \sigma_Y^2 \lambda_h, \qquad \alpha_{22} = \alpha_{33} = 0}$$

(7.43)

The relationships given in Equation 7.43 are approximately valid for $\sigma_Y < 1$, but the exact range of validity has not been established (Chin, 1997). Typical values of σ_Y, λ_h, and λ_v in several formations are listed in Table 7.9. It is important to note that even though the hydraulic-conductivity statistics given in Table 7.9 depend on the support scale of the samples used to derive the statistics, the macrodispersivities calculated using these statistics are (theoretically) independent of the support scale of the samples. In estimating the (total) dispersivity in porous media, the macrodispersivities calculated using either Equation 7.41

TABLE 7.9   Variances and Correlation Length Scales of Hydraulic Conductivity

Formation	σ_Y	λ_h (m)	λ_v (m)	Reference
Sandstone	1.5–2.2	—	0.3–1.0	Bakr (1976)
	0.4	8	3	Goggin et al. (1988)
Sand	0.9	>3	0.1	Byers and Stephens (1983)
	0.6	3	0.12	Sudicky (1986)
	0.5	5	0.26	Hess (1989)
	0.4	8	0.34	Woodbury and Sudicky (1991)
	0.4	4	0.2	Robin et al. (1991)
	0.2	5	0.21	Woodbury and Sudicky (1991)
Sand and	5	12	1.5	Boggs et al. (1990)
gravel	2.1	13	1.5	Rehfeldt et al. (1989)
	1.9	20	0.5	Hufschmied (1986)
	0.8	5	0.4	Smith (1978, 1981)

or 7.43 are additive to the dispersivities associated with hydrodynamic dispersion, which result from pore-scale mixing and molecular diffusion.

Example 7.5 Several hydraulic conductivity measurements in an isotropic aquifer indicate that the spatial covariance, C_Y, of the log-hydraulic conductivity can be approximated by the equation

$$C_Y = \sigma_Y^2 \exp\left(-\frac{r_1^2}{\lambda^2} - \frac{r_2^2}{\lambda^2} - \frac{r_3^2}{\lambda^2}\right)$$

where $\sigma_Y = 0.5$, $\lambda = 5$ m, the spatial lags r_1 and r_2 are measured in the horizontal plane, and r_3 is measured in the vertical plane. The mean hydraulic gradient is 0.001, the effective porosity is 0.2, and the mean log-hydraulic conductivity is 2.5 (where the hydraulic conductivity is in m/day). Estimate the effective hydraulic conductivity and the macrodispersion coefficient in the aquifer.

SOLUTION   From the data given, the hydraulic conductivity field is described statistically by $\langle Y \rangle = 2.5$, $\sigma_Y = 0.5$, and $\lambda = 5$ m. The geometric mean hydraulic conductivity, K_G, is given by Equation 7.35 as

$$K_G = e^{\langle Y \rangle} = e^{2.5} = 12 \text{ m/day}$$

and the effective hydraulic conductivity, for three-dimensional flow, is given by Equation 7.37 as

$$K_{\text{eff}} = K_G\left(1 + \frac{\sigma_Y^2}{6}\right) = 12\left(1 + \frac{0.5^2}{6}\right) = 12.5 \text{ m/day}$$

The mean seepage velocity, V, in the aquifer is given by Equation 7.36 as

$$V = -\frac{K_{\text{eff}}}{n_e} J$$

where $J = -0.001$ and $n_e = 0.2$; hence,

$$V = -\frac{12.5}{0.2}(-0.001) = 0.063 \text{ m/day}$$

Since $\sigma_Y = 0.5$ and $\lambda = 5$ m, the longitudinal macrodispersivity, α_{11}, can be estimated by Equation 7.41 as

$$\alpha_{11} = \sigma_Y^2 \lambda = (0.5)^2(5) = 1.25 \text{ m}$$

and, according to Equation 7.41, the theoretical transverse macrodispersivities are both zero. The longitudinal dispersion coefficient, D_{11}, is given by

$$D_{11} = \alpha_{11} V = 1.25(0.063) = 0.079 \text{ m}^2/\text{day}$$

The relative importance of advective transport to dispersive transport can be measured by the *Peclet number*, Pe, defined as

$$\boxed{Pe = \frac{VL}{D_L}} \tag{7.44}$$

where V is the mean seepage velocity, L is the characteristic length scale, and D_L is the characteristic longitudinal dispersion coefficient. For Pe > 10 advection dominates, for Pe < 0.1 dispersion dominates, and for $0.1 \leq Pe \leq 10$ both advection and dispersion are important. In municipal well fields, values of Pe within several meters of the well tend to be high, indicating that contaminant transport is advection-dominated and dispersion effects are relatively small. A Peclet number, Pe_m, can be defined based on the molecular diffusion coefficient, where

$$Pe_m = \frac{Vd}{D_m} \tag{7.45}$$

where d is the characteristic pore size and D_m is the molecular diffusion coefficient. Previous investigations have shown that the pore-scale longitudinal dispersion coefficient is much greater than the molecular diffusion coefficient when $Pe_m > 10$, and the pore-scale transverse dispersion coefficient is much greater than the molecular diffusion coefficient when $Pe_m > 100$ (Perkins and Johnson, 1963).

Example 7.6 The mean seepage velocity in an aquifer is 1 m/day, the mean pore size is 1 mm, and the molecular diffusion coefficient of a certain toxic contaminant in water is 10^{-9} m²/s . Determine whether molecular diffusion should be considered in a pore-scale contaminant transport model.

SOLUTION From the data given, $V = 1$ m/day, $d = 1$ mm = 0.001 m, and $D_m = 10^{-9}$ m²/s = 8.64×10^{-5} m²/day. The Peclet number, Pe_m, is given by

$$Pe_m = \frac{Vd}{D_m} = \frac{1(0.001)}{8.64 \times 10^{-5}} = 12$$

Since $Pe_m > 10$, molecular diffusion has a negligible contribution to longitudinal dispersion, but since $Pe_m < 100$, molecular diffusion will contribute significantly to transverse dispersion.

In most practical cases longitudinal dispersion is dominated by macrodispersion, horizontal-transverse dispersion is influenced significantly by temporal variations in the seepage velocity, and vertical-transverse dispersion is dominated by small-scale hydrodynamic dispersion (Rehfeldt and Gelhar, 1992). Field studies indicate that horizontal-transverse dispersivities can be related to longitudinal dispersivities using a ratio of longitudinal to horizontal-transverse dispersivity in the range 6 to 20 (Anderson, 1979; Klotz et al., 1980). The horizontal-transverse dispersivity is usually much larger than the vertical-transverse dispersivity. Common practice is to estimate the longitudinal dispersivity using a theoretical or empirical relation such as Equation 7.41, estimate the horizontal-transverse dispersivity as one-tenth of the longitudinal dispersivity, and estimate the vertical-transverse

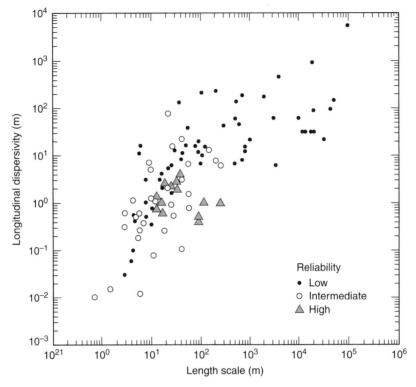

FIGURE 7.10 Longitudinal dispersivity versus length scale in groundwater.

dispersivity as one-hundredth of the longitudinal dispersivity (Zheng and Bennett, 1995). These estimates are conservative relative to those suggested by the U.S. Environmental Protection Agency (1985a), where the horizontal-transverse dispersivity is one-third of the longitudinal dispersivity and the vertical-transverse dispersivity is one-twentieth of the longitudinal dispersivity.

Longitudinal dispersivities derived from 59 sites around the world have been collated by Gelhar and colleagues (1992) and are shown in Figure 7.10. Data from these sites yielded 106 values of longitudinal dispersivity, ranging from 0.01 to 5500 m at scales of 0.75 m to 100 km. Based on the results shown in Figure 7.10, it is clear that the longitudinal dispersivity increases with the distance traveled by the contaminant cloud (i.e., scale), indicating that field formations are seldom homogeneous and that the variability in hydraulic conductivity increases with scale. Of all the experiments reviewed by Gelhar and colleagues (1992), only 14 studies were considered to provide highly reliable estimates of the dispersivity, a further 31 values were considered of intermediate reliability, and the most reliable dispersivity estimates were at the lower end of the length scale. In the absence of field measurements of the hydraulic conductivity, from which the spatial statistics are parameterized by $\langle Y \rangle$, σ_Y, and λ, Figure 7.10 provides a useful basis for estimating the dispersivity in porous formations. The length scale, L, in Figure 7.10 can be taken as either the distance travelled by a tracer released from a point or the length scale measuring the size of the tracer cloud. In either case, the length scale, L, measures the spatial extent of seepage velocity variations experienced by the tracer cloud. Analyses by

Neuman (1990) indicate that the longitudinal macrodispersivity, α_{11}, can be related to the travel distance, L, by the relation

$$\alpha_{11} = 0.0175 L^{1.46}, \qquad 100\,\text{m} < L < 3500\,\text{m} \qquad (7.46)$$

where α_{11} and L are both measured in meters. For $L < 100\,\text{m}$, a better match with field data was obtained using

$$\alpha_{11} = 0.0169 L^{1.53}, \qquad L < 100\,\text{m} \qquad (7.47)$$

Analyses by Al-Suwaiyan (1998) demonstrate that the observed macrodispersivities reported by Gelhar and colleagues (1992) are scattered about the mean (approximated by Equations 7.46 and 7.47), with the upper limit of the scatter at about five times the mean and the lower limit at about one-fifth of the mean. These uncertainty limits should be accounted for whenever Equations 7.46 and 7.47 are used in contaminant-transport predictions. For very small scales, on the order of the pore size, dispersion is caused primarily by pore-scale mechanical dispersion and molecular diffusion, where the longitudinal and transverse dispersivities can be estimated by the relations

$$\begin{aligned} \alpha_L &= \alpha_L^* + \frac{D_m}{\tau V} \\ \alpha_T &= \alpha_T^* + \frac{D_m}{\tau V} \end{aligned} \qquad (7.48)$$

where α_L^* and α_T^* are pore-scale longitudinal and transverse dispersivities respectively, D_m is the molecular diffusion coefficient in water, τ is the tortuosity (which accounts for the effect of the solid matrix on diffusion), and V is the mean seepage velocity. The molecular diffusion coefficient divided by the tortuosity represents the effective molecular diffusion coefficient in porous media and is sometimes called the *bulk diffusion coefficient*. Values of α_L^* are typically on the order of the pore size of the porous medium, α_T^* is typically on the order of 0.1 to $0.01\alpha_L^*$ (Delleur, 1998), τ is typically in the range 2 to 100 (lower values are associated with coarse material such as sands; higher values are associated with finer material such as clays), and typical values of the molecular diffusion coefficient are in the range 10^{-5} to $10^{-3}\,\text{m}^2/\text{day}$ at 25°C (Fetter, 1999).

For travel distances longer than 3500 m, the longitudinal dispersivity tends to asymptote to an upper limit that is consistent with a finite variability in the hydraulic conductivity. In cases where the dispersivity increases with travel distance, the Fickian assumption of a constant dispersion coefficient is not supported and the dispersion is termed *non-Fickian*. However, a Fickian approximation to the mixing process is obtained by adjusting the dispersion coefficient with length scale, and the advection–dispersion equation can be used to approximate the dispersion process. Typical values of the longitutinal dispersivity for various ranges of length scales are shown in Table 7.10.

Example 7.7 A contaminant plume in an aquifer is approximately 50 m long, 10 m wide, and 3 m deep. The characteristic pore size in the aquifer is 3 mm, the molecular diffusion

TABLE 7.10 **Typical Longitudinal Dispersivities for Various Length Scales**

	Longitudinal Dispersivity (m)	
Scale (m)	Average	Range
<1	0.001–0.01	0.0001–0.01
1–10	0.1–1.0	0.001–1.0
10–100	25	1–100

Source: Schnoor (1996).

coefficient is $2 \times 10^{-9} \, \text{m}^2/\text{s}$, the tortuosity is 1.5, and the mean seepage velocity is 0.5 m/day. Estimate the components of the dispersion coefficient.

SOLUTION   From the data given, $L_x = 50 \, \text{m}$, $L_y = 10 \, \text{m}$, $L_z = 3 \, \text{m}$, $D_m = 2 \times 10^{-9} \, \text{m}^2/\text{s} = 1.73 \times 10^{-4} \, \text{m}^2/\text{day}$, $\tau = 1.5$, and $V = 0.5 \, \text{m/day}$. The length scale, L, of the contaminant plume can be approximated by the relation

$$L = \sqrt{L_x L_y} = \sqrt{50(10)} = 22 \, \text{m}$$

and since $L < 100 \, \text{m}$, the longitudinal macrodispersivity can be estimated by Equation 7.47 as

$$\alpha_{11} = 0.0169 L^{1.53} = 0.0169(22)^{1.53} = 1.9 \, \text{m}$$

The horizontal-transverse macrodispersivity, α_{22}, can be estimated as $0.1\alpha_{11}$, which gives

$$\alpha_{22} = 0.1\alpha_{11} = 0.1(1.9) = 0.19 \, \text{m}$$

and the vertical-transverse macrodispersivity, α_{33}, can be estimated as $0.01\alpha_{11}$, which gives

$$\alpha_{33} = 0.01\alpha_{11} = 0.01(1.9) = 0.019 \, \text{m}$$

The local longitudinal dispersivity, α_L, is given by Equation 7.48, where α_L^* is on the order of the pore size (0.003 m), and hence

$$\alpha_L = \alpha_L^* + \frac{D_m}{\tau V} = 0.003 + \frac{1.73 \times 10^{-4}}{1.5(0.5)} = 0.0032 \, \text{m}$$

and taking α_T^* as $0.1\alpha_L^* = 0.1(0.003) = 0.0003 \, \text{m}$, Equation 7.48 gives

$$\alpha_T = \alpha_T^* + \frac{D_m}{\tau V} = 0.0003 + \frac{1.73 \times 10^{-4}}{1.5(0.5)} = 0.00053 \, \text{m}$$

The principal components of the dispersion coefficient are then given by

$$D_{11} = (\alpha_{11} + \alpha_L)V = (1.9 + 0.0032)(0.5) = 0.95 \, \text{m}^2/\text{day}$$

$$D_{22} = (\alpha_{22} + \alpha_T)V = (0.19 + 0.00053)(0.5) = 0.095 \, \text{m}^2/\text{day}$$

$$D_{33} = (\alpha_{33} + \alpha_T)V = (0.019 + 0.00053)(0.5) = 0.0097 \, \text{m}^2/\text{day}$$

It must be emphasized that these estimates of the dispersion coefficient are order-of-magnitude estimates only, and it would be entirely appropriate to take $D_{11} = 1 \, \text{m}^2/\text{day}$, $D_{22} = 0.1 \, \text{m}^2/\text{day}$, and $D_{33} = 0.01 \, \text{m}^2/\text{day}$. The dominance of macrodispersion over pore-scale mechanical dispersion and molecular diffusion in the longitudinal and horizontal-transverse directions are evident.

Field experiments to estimate local values of dispersivity at a particular site typically consist of releasing dye from an injection well, measuring the breakthrough dye concentrations at a downstream well, and estimating the dispersivity based on the rate of growth of variance of the dye cloud or matching the measured concentrations to a solution to the advection–dispersion equation. These tests can be either *natural-gradient tests* or *forced-gradient tests*. Natural-gradient tests are conducted under natural flow conditions, and forced-gradient tests are conducted under artificial (pumping) stress conditions. Examples of forced-gradient conditions include converging radial flow, and flows induced by placing an injection well upstream of a pumping well. It is important to keep in mind that the dispersivities estimated under different stress conditions tend to be different. Results reported by Tiedeman and Hsieh (2004) show that among forced-gradient tests, a converging radial-flow test tends to yield the smallest longitudinal dispersivity (α_{11}), an equal-strength two-well test tends to yield the largest α_{11}, and an unequal strength two-well test tends to yield an intermediate value of α_{11}. Tiedeman and Hsieh (2004) also showed that values of α_{11} estimated under forced-gradient conditions can significantly underestimate α_{11} under natural-gradient flow conditions. In support of this result, Chao et al. (2000) reported that based on numerical simulations using point sources, α_{11} from radial-flow tests were 5 to 10 times smaller than α_{11} for natural-gradient tests.

7.6   FATE PROCESSES

Fate processes include all mechanisms that remove tracer mass from the water environment. These processes include chemical reactions, decay, and sorption. *Sorption* processes include *adsorption*, *chemisorption*, and *absorption*, where adsorption is the process by which a solute attaches itself to a solid surface, chemisorption occurs when the solute is incorporated onto a solid surface by ion exchange, and absorption occurs when the solute diffuses into the solid matrix and is sorbed onto the interior surfaces. Fate processes in the environment are complex and difficult to study at the field scale, and usually are studied under idealized laboratory conditions, with the results fitted to idealized models. The most commonly considered fate processes in ground water are sorption and decay.

7.6.1 Sorption

Models that describe the partitioning of dissolved mass onto solid surfaces are called *sorption isotherms*, since they describe sorption at a constant temperature. The most widely used sorption isotherm is the *Freundlich isotherm*, given by

$$F = K_F c_{aq}^n \tag{7.49}$$

where F is the mass of tracer sorbed per unit mass of solid phase, c_{aq} is the concentration of the tracer dissolved in the water (aqueous concentration), and K_F and n are constants. The constant, n, is typically in the range 0.7 to 1.2 (Domenico and Schwartz, 1998), and for many contaminants at low concentrations, the constant n is approximately equal to unity. In this case, the Freundlich isotherm, Equation 7.49, is linear and is written in the form

$$F = K_d c_{aq} \tag{7.50}$$

where K_d is called the *distribution coefficient* [L^3/M] and is defined as the ratio of the sorbed tracer mass per unit mass of solid matrix to the aqueous concentration. Although the Freundlich isotherm is widely used in practice, it has an important limitation in that it places no limit on the sorption capacity of the solid matrix. The *Langmuir isotherm* allows for a maximum sorption capacity on the solid matrix and is defined by

$$F = \frac{K_l \bar{S} c_{aq}}{1 + K_l c_{aq}} \tag{7.51}$$

where K_l is the *Langmuir constant* and $\bar{S}$ is the maximum sorption capacity. At low solute concentrations, when $K_l c_{aq} \ll 1$, the Langmuir isotherm is similar to the linear Freundlich isotherm, with F linearly proportional to c_{aq}, while at high concentrations, when $K_l c_{aq} \gg 1$, F approaches the limiting value of $\bar{S}$. Despite the apparent advantages of using the Langmuir isotherm, the linear Freundlich isotherm remains the most widely used.

In applying linear isotherms such as Equation 7.50, care should be taken since the actual isotherm is usually *piecewise linear*, and therefore extrapolation beyond the range of experimental conditions used to estimate K_d is not recommended (Fetter, 1992). Values of K_d in Equation 7.50 range from near zero to 10^3 cm^3/g or greater. In the case of organic compounds, the mass of the organic compound sorbed per unit mass of solid matrix has been observed to depend primarily on the amount of organic carbon in the solid matrix (Karickhoff et al., 1979), and it is more appropriate to deal with the *organic carbon sorption coefficient* K_{oc}, which is defined as the ratio of sorbed mass of organic compound per unit mass of organic carbon to the aqueous concentration. Therefore, the distribution coefficient, K_d, is related to the organic carbon sorption coefficient, K_{oc}, by

$$K_d = f_{oc} K_{oc} \tag{7.52}$$

where f_{oc} is the fraction of organic carbon in the porous medium [M/M]. Values of f_{oc} typically range from 0.02 to 3% (Domenico and Schwartz, 1990), and suggested values of f_{oc} for several soil textures are given in Table 7.11. The mass fraction of natural organic

TABLE 7.11 Typical Values of Carbon Content in Soils

Soil	f_{oc}
Silty clay	0.01–0.16
Sandy loam	0.10
Silty loam	0.01–0.02
Unstratified silts, sands, gravels	0.001–0.006
Medium to fine sand	0.0002
Sand	0.0003–0.10
Sand and gravel	0.00008–0.0075
Coarse gravel	0.0011

Source: Prakash (2004).

material in soil can be represented using two different notations: f_{om} represents the mass fraction of organic matter present in the soil, and f_{oc} represents the mass fraction of organic carbon in soil. Mass fraction f_{om} considers the mass of the entire organic molecule, while f_{oc} considers only the mass of carbon present in the organic matter. Using the rule of thumb that the weight of organic matter is roughly double the weight or organic carbon, a useful approximation is $f_{om} \approx 2 f_{oc}$.

The formulation described by Equation 7.52 is appropriate whenever the organic fraction exceeds 0.1% (Schwarzenbach and Westall, 1981; Banerjee et al., 1985), and values of K_{oc} for several organic compounds typically found in contaminated ground water are given in Appendix B.2. When the organic fraction is less than 0.1%, sorption of organic compounds on nonorganic solids can become significant, and it is not automatic that the soil or aquifer organic carbon will be the primary surface onto which the organic compounds will partition (Fetter, 1993). The octanol–water partition coefficient, K_{ow}, is a widely available and easily measured parameter that gives the distribution of a chemical between n-octanol and water in contact with each other. This coefficient is defined by

$$K_{ow} = \frac{c_o}{c_w} \tag{7.53}$$

where c_o is the concentration in the octanol and c_w is the concentration in the water. Octanol serves as a generalized surrogate for organic media, and the reason for using octanol is historical: During the early years of pharmaceutical research, researchers found that octanol served as an inexpensive surrogate for human tissue and, as a result, pharmaceutical studies often involved partitioning tests using octanol as an index for drug uptake by organisms. K_{ow} is now commonly reported for most synthetic organic chemicals. Studies with a variety of organic contaminants have generally shown a linear relationship between log K_{oc} and log K_{ow}, with nonlinearities caused by sorption to the mineral portion of soils. Several proposed empirical relationships are shown in Table 7.12, where K_{oc} is in cm^3/g and K_{ow} is dimensionless. Clearly, there is no universal relation for deriving K_{oc} from the easily measured K_{ow}, although Fetter (1993) has shown that estimates of K_{oc} derived using equations in Table 7.12 are likely to fall within one standard deviation of the geometric mean of K_{oc} estimated from the combined predictions of all the equations listed in Table 7.12. Values of log K_{ow} for selected organic compounds are listed in Table 7.13, and

TABLE 7.12   Empirical Relationships Between K_{oc} and K_{ow}

Equation	Chemicals	Reference
$\log K_{oc} = 1.00 \log K_{ow} - 0.21$	10 polyaromatic hydrocarbons	Karickhoff et al. (1979)
$\log K_{oc} = 1.00 \log K_{ow} - 0.201$	Miscellaneous organics	Karickhoff et al. (1979)
$\log K_{oc} = 0.544 \log K_{ow} + 1.377$	45 organics, mostly pesticides	Kenaga and Goring (1980)
$\log K_{oc} = 1.029 \log K_{ow} - 0.18$	13 pesticides	Rao and Davidson (1980)
$\log K_{oc} = 0.94 \log K_{ow} + 0.22$	s-Trizines and dinitroanalines	Rao and Davidson (1980)
$\log K_{oc} = 0.989 \log K_{ow} - 0.346$	5 polyaromatic hydrocarbons	Karickhoff (1981)
$\log K_{oc} = 0.937 \log K_{ow} - 0.006$	Aromatics, polyaromatics, triazines	Brown and Flagg (1981)
$\log K_{oc} = 1.00 \log K_{ow} - 0.317$	DDT, tetrachlorobiphenyl, lindane, 2,4-D, and dichloropropane	McCall et al. (1983)
$\log K_{oc} = 0.72 \log K_{ow} + 0.49$	Methylated and chlorinated benzenes	Schwarzenbach and Westall (1981)
$\log K_{oc} = 1.00 \log K_{ow} - 0.317$	22 polynuclear aromatics	Hassett et al. (1980)
$\log K_{oc} = 0.524 \log K_{ow} + 0.855$	Substituted phenylureas and alkyl-N-phenylcarbamates	Briggs (1973)

it is best to estimate K_{oc} from these values using an empirical relation derived for similar chemicals. Values of K_{ow} range over many orders of magnitude (typically, 10^1 to 10^7), and therefore K_{ow} is usually reported as $\log K_{ow}$ (typically, 1 to 7). The higher the value of K_{ow}, the greater the tendency of the compound to partition from the water into the organic phase.

The amount of sorbed mass per unit volume of the porous medium, c_s, is related to the mass of tracer sorbed per unit mass of solid phase, F, by the expression

$$c_s = \rho_b F \tag{7.54}$$

where ρ_b is the bulk density of the solid matrix (= oven-dried mass of a soil sample divided by the sample volume). Typical values of the bulk density of several porous media are given in Table 7.14. Combining Equations 7.50 and 7.54 leads to the following linear relationship between the concentration of the sorbed mass, c_s, and the concentration of the dissolved mass, c_{aq}:

$$\boxed{c_s = \beta c_{aq}} \tag{7.55}$$

where β is a dimensionless constant given by

$$\boxed{\beta = \rho_b K_d} \tag{7.56}$$

The fate term, S_m, in the dispersion equation, Equation 7.22, is equal to the rate at which tracer mass is added to the water in a unit volume of water. In the case of sorption, the rate

TABLE 7.13 Values of K_{ow} for Selected Organic Compounds

Compound	log K_{ow}	Reference
Acetone	−0.24	Schwarzenbach et al. (1993)
Atrazine	2.56	Schwarzenbach et al. (1993)
Benzene	2.01–2.13[a]	Hansch and Leo (1979), MacKay (1991)
Carbon tetrachloride	2.64–2.83	Chou and Jurs (1979), Hansch and Leo (1979), Schnoor (1996)
Chlorobenzene	2.49–2.84[a]	Hansch and Leo (1979), MacKay (1991)
Chloroform	1.95–1.97[a]	Hansch and Leo (1979)
DDT	3.98–7.92	Pontolillo and Eganhouse (2001)
Dieldrin	5.48	Schwarzenbach et al. (1993)
Lindane	3.78	Schwarzenbach et al. (1993)
Malathion	2.89	Schwarzenbach et al. (1993)
Naphthalene	3.29–3.35[a]	MacKay (1991), Schnoor (1996)
n-Octane	5.18	Schwarzenbach et al. (1993)
Parathion	3.81	Schwarzenbach et al. (1993)
Phenol	1.46[a]–1.49	Hansch and Leo (1979), MacKay (1991)
Polychlorinated biphenyls	4.09–8.23	Schwarzenbach et al. (1993)
2,3,7,8-Tetrachlorodibenzo-p-dioxin	6.64	Schwarzenbach et al. (1993)
Toluene	2.69[a]	MacKay (1991)
1,1,1-Trichloroethane	2.47[a]–2.51	MacKay (1991), Schnoor (1996)
Trichloroethylene	2.29[a]	MacKay (1991)
p-Xylene	3.12–3.18	Schwarzenbach et al. (1993)

[a]At 25°C.

at which tracer mass is added to the aqueous phase is equal to the rate at which tracer mass is lost from the solid phase. Therefore, the rate at which tracer mass is added to the water per unit volume of water, S_m, is given by

$$S_m = -\frac{1}{n}\frac{\partial c_s}{\partial t} = -\frac{\beta}{n}\frac{\partial c_{aq}}{\partial t} \tag{7.57}$$

TABLE 7.14 Typical Values of Bulk Density and Porosity in Porous Media

Porous Media	Bulk Density (kg/m^3)	Porosity
Limestone and shale	2780	0.01–0.20
Sandstone	2130	0.10–0.20
Gravel and sand	1920	0.30–0.35
Gravel	1870	0.30–0.40
Fine to medium mixed sand	1850	0.30–0.35
Uniform sand	1650	0.30–0.40
Medium to coarse mixed sand	1530	0.35–0.40
Silt	1280	0.40–0.50
Clay	1220	0.45–0.55

Source: Tindall and Kunkel (1999).

In applications to flow in porous media, the aqueous contaminant concentration, c_{aq}, is commonly denoted by c, and hence Equation 7.57 can be written as

$$S_m = -\frac{\beta}{n}\frac{\partial c}{\partial t} \qquad (7.58)$$

Substituting this sorption model into the advection–dispersion equation, Equation 7.22, yields

$$\left(1 + \frac{\beta}{n}\right)\frac{\partial c}{\partial t} + \sum_{i=1}^{3} V_i \frac{\partial c}{\partial x_i} = \sum_{i=1}^{3} D_i \frac{\partial^2 c}{\partial x_i^2} \qquad (7.59)$$

The term $1 + \beta/n$ is commonly referred to as the *retardation factor*, R_d, where

$$\boxed{R_d = 1 + \frac{\beta}{n}} \qquad (7.60)$$

Dividing both sides of Equation 7.59 by R_d yields the following form of the dispersion equation:

$$\boxed{\frac{\partial c}{\partial t} + \sum_{i=1}^{3} \frac{V_i}{R_d}\frac{\partial c}{\partial x_i} = \sum_{i=1}^{3} \frac{D_i}{R_d}\frac{\partial^2 c}{\partial x_i^2}} \qquad (7.61)$$

Comparing Equation 7.61, which accounts for the sorbing of contaminants onto porous media, to the advection–dispersion equation for conservative contaminants, Equation 7.22, it is clear that both equations have the same form, with sorption being accounted for by reducing the mean seepage velocity and dispersion coefficients by a factor $1/R_d$. In other words, the fate and transport of a sorbing tracer can be modeled by neglecting sorption but reducing the mean velocity and dispersion coefficients by a factor $1/R_d$. Contaminants that have a higher partitioning coefficient ($K_d > 10^3$ cm^3/g) will move at a very slow rate, if at all. On a cautionary note, the use of the retardation coefficient assumes that partitioning reactions are very fast relative to the rate of ground-water flow and that equilibrium is achieved between the aqueous and adsorbed phases of the contaminant. In cases where this is not true, contaminant plumes will migrate faster than predicted by assuming equilibrium between the aqueous and adsorbed phases of the contaminant.

Example 7.8   One kilogram of a contaminant is spilled over a 1-m depth of ground water and spreads laterally as the ground water moves with an average seepage velocity of 0.1 m/day. The longitudinal and transverse dispersion coefficients are 0.03 and 0.003 m^2/day, respectively; the porosity is 0.2; the density of the aquifer matrix is 2.65 g/cm^3; log K_{oc} is 1.72 (K_{oc} in cm^3/g); and the organic fraction in the aquifer is 5%. (a) Calculate the concentration at the spill location after 1 hour, 1 day, and 1 week. (b) Compare these values with the concentration obtained by neglecting sorption.

SOLUTION   (a) The distribution coefficient, K_d, is given by

$$K_d = f_{oc} K_{oc}$$

where $f_{oc} = 0.05$ and $K_{oc} = 10^{1.72} = 52.5\,\text{cm}^3/\text{g}$. Therefore,

$$K_d = (0.05)(52.5) = 2.63\,\text{cm}^3/\text{g}$$

The dimensionless constant β is given by

$$\beta = \rho_b K_d = (1 - n)\rho_s K_d$$

where $n = 0.2$, $\rho_s = 2.65\,\text{g/cm}^3$, and therefore

$$\beta = (1 - 0.2)(2.65)(2.63) = 5.58$$

The retardation factor, R_d, is then given by

$$R_d = 1 + \frac{\beta}{n} = 1 + \frac{5.58}{0.2} = 29$$

For an instantaneous release, the resulting concentration distribution is given by

$$c(x, y, t) = \frac{M}{4\pi H n t \sqrt{D_L D_T}} \exp\left[-\frac{(x - Vt)^2}{4 D_L t} - \frac{y^2}{4 D_T t} \right]$$

where $M = 1\,\text{kg}$, $n = 0.2$, $H = 1\,\text{m}$, $D_L = 0.03/R_d\,\text{m}^2/\text{day}$, $D_T = 0.003/R_d\,\text{m}^2/\text{day}$, $V = 0.1/R_d\,\text{m/day}$, $x = 0\,\text{m}$, and $y = 0\,\text{m}$. Substituting these values into the expression above for the concentration distribution yields

$$c(0, 0, t) = \frac{(1) R_d}{4\pi (1)(0.2) t \sqrt{(0.03)(0.003)}} \exp\left[-\frac{(-0.1/R_d t)^2}{4(0.03/R_d) t} \right]$$

$$= \frac{41.95 R_d}{t} \exp\left(-\frac{0.083 t}{R_d} \right) \quad \text{kg/m}^3$$

In the absence of sorption, $R_d = 1$, for a sorbing contaminant $R_d = 29$, and the concentrations at $t = 1$ hour, 1 day, and 1 week are given in Table 7.15.

TABLE 7.15   Results for Example 7.8

Time	Without Sorption ($R_d = 1$) (kg/m^3)	With Sorption ($R_d = 29$) (kg/m^3)
1 hour	199	28,960
1 day	7.7	1,215
1 week	0.67	170

(b) Sorption results in higher contaminant concentrations in the ground water near the spill. This is a result of the requirement that higher water concentrations are necessary to maintain an equilibrium with the sorbed mass. Unrealistically high concentrations calculated at early times are a result of the model assumption that the spill occurs over an infinitesimally small volume. To be realistic, the calculated concentrations must be less than the solubility of the contaminant.

The sorption characteristics of metals and radionuclides are more difficult to predict than for organic compounds. Metals usually exist as cations in the aqueous phase, and the degree to which metals partition onto the solid matrix is determined by the *cation-exchange capacity* of the solid matrix and the presence of other cations that compete for exchange sites. The cation-exchange capacity is greatest in matrices with high clay content and organic matter. Several metals can exist in several oxidation states and are often complexed with *ligands* that are present in the aqueous phase. The mobility of metals depends on both the oxidation state and speciation. As a general rule, clays will have the largest K_d values for specific inorganic solutes, cations are more strongly adsorbed than anions, and divalent cations will be adsorbed more readily than those of monovalent species (Fetter, 2001). Thibault et al. (1990) estimated K_d values for metals in soils based on soil texture, where soils containing greater than 70% sand-sized particles were classed as sands, those containing more than 35% clay-sized particles were classed as clays, loam soils had an even distribution of sand-, clay-, and silt-sized particles or consisted of up to 80% silt-sized particles, and organic soils contained more than 30% organic matter. The geometric mean K_d values for several metals and other elements in various soils are shown in Table 7.16.

The retardation factor, R_d, is most often used as a reduction factor to be applied to the mean velocity (as described previously); however, this factor is also useful as a measure of

TABLE 7.16 Values of K_d (cm³/g) for Selected Elements

Element	Sand	Silt	Clay	Organic
Am	1,900	9,600	8,400	112,000
C	5	20	1	70
Cd	80	40	560	800
Co	60	1,300	550	1,000
Cr	70	30	1,500	270
Cs	280	4,600	1,900	270
I	1	5	1	25
Mn	50	750	180	150
Mo	10	125	90	25
Ni	400	300	650	1,100
Np	5	25	55	1,200
Pb	270	16,000	550	22,000
Pu	550	1,200	5,100	1,900
Ra	500	36,000	9,100	2,400
Se	150	500	740	1,800
Sr	15	20	110	150
Tc	0.1	0.1	1	1
Th	3,200	3,300	5,800	89,000
U	35	15	1,600	410
Zn	200	1,300	2,400	1,600

Source: Thibault et al. (1990).

the fraction of contaminant that is in the pore water. To see this clearly, the retardation factor defined by Equation 7.60 can be expressed in the form

$$R_d = 1 + \frac{K_d \rho_b}{n} = \frac{Vnc + K_d \rho_b Vnc}{Vnc} = \frac{\text{pollutant mass in soil and water}}{\text{pollutant mass in water alone}} \qquad (7.62)$$

where V is a bulk volume and c is the contaminant concentration in the pore water. According to Equation 7.62, the inverse of the retardation factor represents the fraction of contaminant that is present in the water. To illustrate the application of the retardation factor in this context, consider the case where a mass M of contaminant is contained within the aquifer and the contaminated water is to be extracted using a pumping well. After the first pore volume is extracted, the mass of contaminant remaining in the aquifer is $M(1 - R_d^{-1})$. Similarly, after the second pore volume is extracted, the mass remaining is $M(1 - R_d^{-1})^2$, and after j pore volumes have been extracted, the contaminant mass remaining is equal to $M(1 - R_d^{-1})^j$. Therefore, the fraction of initial mass, M, flushed after j pore volumes, F_j, is given by

$$\boxed{F_j = \frac{M - M(1 - R_d^{-1})^j}{M} = 1 - (1 - R_d^{-1})^j} \qquad (7.63)$$

This relation is very useful is determining the number of pore volumes that must be removed from an aquifer to obtain a given level of site remediation. Equation 7.63 assumes that soil-water sorption–desorption equilibrium occurs instantaneously. When pollutant desorption from soil occurs much slower than the flow rate of water through the contaminated zone, contaminant concentrations will be lower than the predicted equilibrium, resulting in even lower mass fractions removed. In cases where the sorption–desorption process is relatively slow, contaminant concentration distributions in ground water show a heavy tail, in contrast to a near-Gaussian distribution.

Example 7.9   A contaminated aquifer is estimated to contain 30 kg of contaminant spread over a 100-m^3 volume of aquifer. The porosity of the aquifer is 0.2 and the retardation factor of the contaminant is 5. Estimate the volume of pore water that must be removed to reduce the mass of contaminant in the aquifer by 90%.

SOLUTION   From the data given, $M = 30$ kg, $V = 100$ m^3, $n = 0.2$, $R_d = 5$, and $F_j = 0.9$. The number of pore volumes, j, that must be removed from the aquifer to extract 90% of the contaminant mass is given by Equation 7.63, where

$$F_j = 1 - (1 - R_d^{-1})^j$$
$$0.90 = 1 - (1 - 5^{-1})^j$$

which yields

$$j = 10.3$$

and

$$\text{pore-water extraction} = jnV = 10.3(0.2)(100 \text{ m}^3) = 206 \text{ m}^3$$

Therefore, 10.3 pore volumes or 206 m³ must be extracted to yield a 90% reduction in contaminant mass in the aquifer.

It is important to keep in mind that the retardation factor is only a constant if the sorption isotherm is linear and can be described by a single parameter, the distribution coefficient. In cases where the sorption isotherm is nonlinear, the retardation factor will depend on the aqueous concentration, with increasing retardation coefficients corresponding to increasing aqueous concentrations (Mojid and Vereecken, 2005).

7.6.2 First-Order Decay

Many chemical compounds in the environment ultimately decompose into other compounds, usually through chemical reactions such as hydrolysis or biodegradation. Biodegradation rates in ground water are much less than biodegradation rates in soils, due primarily to the much lower density of microorganisms in ground water. The most frequently used model of decomposition is the following first-order decay model:

$$S_m = -\lambda c \tag{7.64}$$

where S_m is the rate at which tracer mass is added to ground water per unit volume of ground water, c is the concentration of tracer in the ground water, and λ is the first-order decay coefficient. Substituting this decay model into the dispersion equation, Equation 7.22, leads to

$$\frac{\partial c}{\partial t} + \sum_{i=1}^{3} V_i \frac{\partial c}{\partial x_i} = \sum_{i=1}^{3} D_i \frac{\partial^2 c}{\partial x_i^2} - \lambda c \tag{7.65}$$

which can be modified by changing variables from c to c^*, where

$$c = c^* e^{-\lambda t} \tag{7.66}$$

Substituting Equation 7.66 into Equation 7.65 and dividing both sides by $e^{-\lambda t}$ yields

$$\frac{\partial c^*}{\partial t} + \sum_{i=1}^{3} V_i \frac{\partial c^*}{\partial x_i} = \sum_{i=1}^{3} D_i \frac{\partial^2 c^*}{\partial x_i^2} \tag{7.67}$$

which is exactly the same as the advection–dispersion equation for a conservative tracer. The practical implication of this result is that the fate and transport of a tracer undergoing first-order decay is the same as if the tracer is initially assumed to be conservative, and the resulting concentration distribution reduced by a factor $e^{-\lambda t}$, where t is the time since release of the tracer mass.

The first-order decay coefficient, λ, is frequently expressed in terms of the *half-life*, T_{50}, which is the time required for 50% of the initial mass to decay and is related to the first-order decay coefficient by

$$T_{50} = \frac{\ln 2}{\lambda} \tag{7.68}$$

The half-lives of several organic compounds in soils have been compiled by Howard et al. (1991), and these results are summarized in Table 7.17. The variability of the degradation

TABLE 7.17   First-Order Decay Rates of Selected Organic Compounds in Soil

Compound	Half-Life, T_{50} (days)	First-Order Decay Rate, λ (day^{-1})
Acetone	2–14	0.050–0.35
Benzene	10–730	0.00095–0.069
Bis(2-ethylhexyl)phthalate	10–389	0.00178–0.069
Carbon tetrachloride	7–365	0.0019–0.099
Chloroethane	14–56	0.0124–0.0495
Chloroform	56–1800	0.000385–0.0124
1,1-Dichloroethane	64–154	0.00450–0.0108
1,2-Dichloroethane	100–365	0.00190–0.00693
Ethylbenzene	6–228	0.00304–0.116
Methyl *tert*-butyl ether	56–365	0.00190–0.0124
Methylene chloride	14–56	0.0124–0.0495
Naphthalene	1–258	0.00269–0.693
Phenol	0.5–7	0.099–1.39
Toluene	7–28	0.0248–0.099
1,1,1-Trichloroethane	140–546	0.00127–0.00495
Trichloroethene	321–1650	0.000420–0.00216
Vinyl chloride	56–2880	0.000241–0.0124
Xylenes	14–365	0.00190–0.0495

Source: Howard et al. (1991).

rates shown in the table reflect both the error in assuming that biodegradation is described by a simple first-order decay model as well as the variability in environmental conditions that affect biodegradation, principally the amount and type of bacteria present in the subsurface, geologic and hydraulic characteristics, temperature, and concentration of dissolved oxygen in the ground water.

Example 7.10   Ten kilograms of a contaminant is spilled into the ground water and is well mixed over a 1-m depth. The mean seepage velocity in the aquifer is 0.5 m/day, the porosity is 0.2, the longitudinal dispersion coefficient is 1 m^2/day, the horizontal-transverse dispersion coefficient is 0.1 m^2/day, vertical mixing is negligible, and the first-order decay constant of the contaminant is 0.01 day^{-1}. (a) Determine the maximum concentration in the ground water after 1,10,100, and 1000 days. (b) Compare these concentrations to the maximum concentration without decay.

SOLUTION   (a) From the data given, $M = 10$ kg, $H = 1$ m, $V = 0.5$ m/day, $n = 0.2$, $D_L = 1$ m^2/day, $D_T = 0.1$ m^2/day, and $\lambda = 0.01$ day^{-1}. Neglecting decay, the contaminant concentration downstream of the spill is given by

$$c^*(x, y, t) = \frac{M}{4\pi Hnt\sqrt{D_L D_T}} \exp\left[-\frac{(x - Vt)^2}{4D_L t} - \frac{y^2}{4D_T t}\right]$$

and the maximum concentration, at $x = Vt$, is given by

$$c^*_{\max}(t) = \frac{M}{4\pi Hnt\sqrt{D_L D_T}}$$

TABLE 7.18   Results for Example 7.10

t (days)	$c^*_{max}(t)$ (mg/L)	$c_{max}(t)$ (mg/L)
1	12,600	12,450
10	1,260	1,140
100	126	46.5
1,000	12.6	0.0005

which yields

$$c^*_{max}(t) = \frac{10}{4\pi(1)(0.2)t\sqrt{(1)(0.1)}} = \frac{12.6}{t}\,\text{kg/m}^3 = \frac{12,600}{t}\,\text{mg/L}$$

Accounting for first-order decay, the maximum concentration, $c_{max}(t)$, is given by

$$c_{max}(t) = c^*_{max}(t)e^{-\lambda t} = \frac{12,600}{t}e^{-0.01t}$$

Hence, the maximum concentrations at 1, 10, 100, and 1000 days are as given in Table 7.18.

(b) As time increases, the decay effect becomes more pronounced. For example, after 100 days the maximum concentration is 126 mg/L without decay compared to 46.5 mg/L with decay. The aqueous concentrations calculated should be compared with the aqueous solubility of the contaminant. If the concentration calculated exceeds the solubility of the contaminant in water, not all of the spilled contaminant will dissolve and the initial concentration at the spill location can be taken as equal to the solubility.

7.6.3   Combined Processes

In some cases, both sorption and first-order decay processes occur simultaneously. Assuming that sorption is described by the linear sorption isotherm, Equation 7.55, and that the sorbed mass decays as a first-order process, described by

$$\frac{\partial c_s}{\partial t} = -\lambda c_s \tag{7.69}$$

where c_s is the sorbed mass per unit volume of the porous medium, the mass flux per unit volume of ground water into the aqueous phase due to desorption, S^1_m, is given by

$$S^1_m = -\frac{\beta}{n}\frac{\partial c_{aq}}{\partial t} - \frac{\lambda c_s}{n} = -\frac{\beta}{n}\frac{\partial c_{aq}}{\partial t} - \frac{\lambda\beta c_{aq}}{n} \tag{7.70}$$

where c_{aq} is the aqueous concentration of the contaminant. In applications to flow in porous media, the contaminant concentration, c, used in the advection–dispersion equation is equal to the aqueous concentration, c_{aq}, hence

$$c_{aq} = c \tag{7.71}$$

and Equation 7.70 can be written as

$$S_m^1 = -\frac{\beta}{n}\frac{\partial c}{\partial t} - \lambda\frac{\beta}{n}c \tag{7.72}$$

In addition to the mass flux into the aqueous phase due to desorption, there is the additional mass, S_m^2, being removed from the ground water due to decay of the dissolved contaminant, where

$$S_m^2 = -\lambda c \tag{7.73}$$

The total rate at which mass is added to the ground water, S_m, is equal to the mass flux into the ground water due to desorption plus the mass flux due to first-order decay of the dissolved contaminant; therefore,

$$\begin{aligned}
S_m &= S_m^1 + S_m^2 \\
&= -\frac{\beta}{n}\frac{\partial c}{\partial t} - \lambda\frac{\beta}{n}c - \lambda c \\
&= -\frac{\beta}{n}\frac{\partial c}{\partial t} - \left(1 + \frac{\beta}{n}\right)\lambda c
\end{aligned} \tag{7.74}$$

Substituting this fate model into the advection–dispersion equation, Equation 7.22, and simplifying yields

$$\frac{\partial c}{\partial t} + \sum_{i=1}^{3}\frac{V_i}{R_d}\frac{\partial c}{\partial x_i} = \sum_{i=1}^{3}\frac{D_i}{R_d}\frac{\partial^2 c}{\partial x_i^2} - \lambda c \tag{7.75}$$

where R_d is the retardation factor defined by Equation 7.60. Equation 7.75 indicates that the fate and transport of a tracer that is undergoing both sorption and first-order decay is the same as if sorption is neglected, but the mean fluid velocity and dispersion coefficients are reduced by a factor $1/R_d$. Equation 7.75 can be further simplified by changing variables from c to c^*, where

$$c = c^* e^{-\lambda t} \tag{7.76}$$

Substituting Equation 7.76 into Equation 7.75 and simplifying yields

$$\frac{\partial c^*}{\partial t} + \sum_{i=1}^{3}\frac{V_i}{R_d}\frac{\partial c^*}{\partial x_i} = \sum_{i=1}^{3}\frac{D_i}{R_d}\frac{\partial^2 c^*}{\partial x_i^2} \tag{7.77}$$

This is the dispersion equation for a conservative contaminant and demonstrates that the fate and transport of a sorbing tracer undergoing first-order decay can be modeled by (1) reducing the fluid velocity and dispersion coefficients by the factor $1/R_d$, (2) neglecting both sorption and decay, and (3) reducing the resulting concentration distribution by the factor $e^{-\lambda t}$, where t is the time since the release of the tracer mass. Equation 7.77 assumes that the decay constant, λ, is the same for both the aqueous-phase contaminant and the sorbed contaminant. In cases where these decay coefficients differ, the "lumped" decay coefficient should be taken as

$$\lambda = \lambda_{aq} + \frac{\rho_b K_d}{n} \lambda_s$$

(7.78)

where λ_{aq} and λ_s are the decay coefficients in the aqueous and sorbed phases, respectively. It is noteworthy that some organic chemicals biodegrade rapidly in water but not when sorbed onto the solid matrix.

Example 7.11   Three kilograms of a contaminant is spilled over a 1-m depth of ground water that is moving with an average seepage velocity of 0.1 m/day. The longitudinal and horizontal-transverse dispersion coefficients are 0.05 and 0.005 m²/day, respectively, vertical dispersion is negligible, and the porosity is 0.2. If the retardation factor is equal to 20 and the first-order decay factor is 2 day^{-1}, calculate the concentration at the spill location after 1 day and 1 week.

SOLUTION   The concentration distribution (accounting for sorption but prior to correction for decay) is given by

$$c^*(x, y, t) = \frac{M}{4\pi H n t \sqrt{D_L D_T}} \exp\left[-\frac{(x - Vt)^2}{4D_L t} - \frac{y^2}{4D_T t} \right]$$

where $M = 3$ kg, $H = 1$ m, $n = 0.2$, $D_L = 0.05/R_d = 0.05/20 = 0.0025$ m²/day, $D_T = 0.005/R_d = 0.005/20 = 0.00025$ m²/day, $x = 0$ m, $y = 0$ m, and $V = 0.1/R_d = 0.1/20 = 0.005$ m/day. Substituting these values into the preceding equation yields

$$c^*(0, 0, t) = \frac{3}{4\pi(1)(0.2)t\sqrt{(0.0025)(0.00025)}} \exp\left[-\frac{(0.005t)^2}{4(0.0025)t} \right]$$

$$= \frac{1510}{t} \exp(-0.0025t) \qquad \text{kg/m}^3$$

Correcting for decay requires multiplying by $e^{-\lambda t}$, where $\lambda = 2$ day^{-1}, and therefore the actual concentration as a function of time is given by

$$c(0, 0, t) = c^*(0, 0, t) e^{-2t}$$

$$= \frac{1510}{t} \exp(-0.0025t - 2t)$$

$$= \frac{1510}{t} \exp(-2.0025t) \qquad \text{kg/m}^3$$

Therefore, at $t = 1$ day, $c(0, 0, t) = 205$ kg/m³ $= 205{,}000$ mg/L, and at $t = 7$ days, $c(0, 0, t) = 1.75 \times 10^{-4}$ kg/m³ $= 0.175$ mg/L.

7.7 NONAQUEOUS-PHASE LIQUIDS

Many organic compounds are only slightly soluble in water and exist in both the dissolved and insoluble (pure) phase in ground water. Pure liquids that are not dissolved are called *nonaqueous-phase liquids* (NAPLs). NAPLs have been identified at four out of five hazardous waste sites in the United States (Plumb and Pitchford, 1985) and are typically composed of either a single chemical or a mixture of several chemicals. NAPLs are further classified as light NAPLs (LNAPLs) that are less dense than water and tend to float on the water table, and as dense NAPLs (DNAPLs) that are denser than water and tend to sink to the bottom of the aquifer. Typical LNAPL and DNAPL spills are illustrated in Figure 7.11. Compounds with solubilities less than 20,000 mg/L are likely to exist as NAPLs (Prakash, 2004). The NAPLs commonly encountered at contaminated sites can be categorized into four groups on the basis of their similar chemical structures, fluid properties, and behavior in the subsurface (Adeel et al., 2000): (1) chlorinated hydrocarbons, (2) petroleum products, (3) tars and creosote, and (4) mixtures with polychlorinated biphenyls (PCBs) and oils.

Chlorinated hydrocarbons are low-molecular-weight compounds that are sparingly soluble in water, volatile in nature, and denser than water; hence they are DNAPLs.

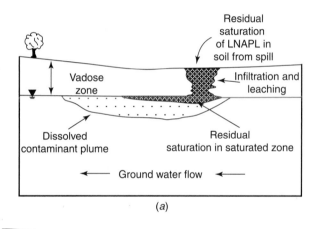

(*a*)

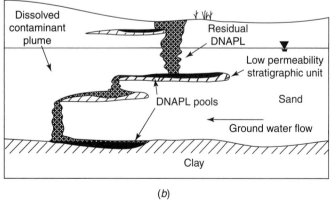

(*b*)

FIGURE 7.11 Typical (*a*) LNAPL and (*b*) DNAPL spills.

TABLE 7.19   Densities and Solubilities of NAPLs

Liquid	Density at 15°C (kg/m^3)	Solubility at 10°C (mg/L)
LNAPLs		
Medium distillates (fuel oil)	820–860	3–8
Petroleum distillates (jet fuel)	770–830	10–150
Gasoline	720–780	150–300
Crude oil	800–880	3–25
DNAPLs		
Trichloroethlene (TCE)	1,460	1,070
Tetrachloroethylene (PCE)	1,620	160
1, 1, 1-Trichloroethane (TCA)	1,320	1,700
Dichloromethane (CH_2Cl_2)	1,330	13,200
Chloroform ($CHCl_3$)	1,490	8,200
Carbon tetrachloride (CCl_4)	1,590	785
Creosote	1,110	20

Source: Schnoor (1996).

Chlorinated hydrocarbons are used predominantly as solvents and degreasers at industrial, commercial, and military facilities. Commonly encountered chlorinated hydrocarbons are tetrachloroethylene (PCE), trichloroethylene (TCE), and carbon tetrachloride. Petroleum products are typically low-molecular-weight hydrocarbons whose solubilities are similar to those of chlorinated hydrocarbons and are less dense than water; hence they are LNAPLs. Commonly encountered petroleum products are benzene, toluene, ethylbenzene, and xylene (collectively referred to as BTEX), which are among the most soluble constituents of gasoline. Tars are by-products of coke and gas production, and creosote is a widely used wood preservative. Tars and creosote are DNAPLs and are sparingly soluble in water. PCBs have been used in many industrial applications, including fire retardants in hydraulic oils and electrical transformer fluids. The production of PCBs is currently banned. PCBs are generally denser than water, and hence are DNAPLs. The densities and solubilities of several NAPLs are given in Table 7.19.

Movement of ground water past a NAPL trapped in the solid matrix of a porous medium results in the dissolution of soluble compounds and an associated downstream plume. In some cases the dissolved concentrations are sufficient to affect the density of the water significantly, inducing a vertical ground-water velocity, v_z, given by (Frind, 1982)

$$v_z = -\frac{K_z}{n_e}\left(\frac{\rho}{\rho_0} - 1\right)$$

(7.79)

where K_z is the hydraulic conductivity in the vertical direction, n_e is the effective porosity, ρ is the density of the dissolved mixture, and ρ_0 is the density of the native ground water. The relative magnitude of v_z to the horizontal seepage velocity will give an indication of the extent to which the contaminant plume moves in the same direction as the ground-water flow.

Since LNAPLs do not penetrate very deeply into the aquifer and are relatively biodegradable under natural conditions, they are generally thought to be a more manageable environmental problem than DNAPLs, which tend to be trapped deep in the aquifer (Bedient et al., 1994). Other factors that make DNAPL contamination more difficult to remediate are: (1) Chlorinated solvents do not biodegrade very rapidly and persist for long periods of time in ground water, in fact, products of microbial degradation of halogenated solvents are sometimes more toxic than the parent compounds (Parsons et al., 1984); and (2) chlorinated solvents have physical properties, such as small viscosities, that allow movement through very small fractures and downward penetration to great distances. The pattern of DNAPL penetration in aquifers is commonly referred to as *viscous fingering*. DNAPL pools on impermeable boundaries are often difficult to locate and remediate (Blatchley and Thompson, 1998). A detailed account of the fate and transport of DNAPLs in ground water can be found in Pankow and Cherry (1996).

The movement of NAPLs in ground water is governed primarily by gravity, buoyancy, and capillary forces. At low concentrations, NAPLs tend to become discontinuous and immobilized by capillary forces, and they end up trapped in the pores of aquifers, as illustrated in Figure 7.12. Under these conditions, the concentration of the NAPL is termed the *residual saturation*, which is defined as the fraction of total pore volume occupied by residual NAPL under ambient ground-water flow conditions. In the unsaturated zone, residual saturation values are typically in the range 5 to 20%, while in the saturated zone this range is typically higher and on the order of 15 to 50% (Schwille, 1988; Mercer and Cohen, 1990). Residual saturation appears to be relatively insensitive to the types of chemicals that comprise a NAPL, but is very sensitive to soil properties and heterogeneities (USEPA, 1990a). The residual saturation, S_r, of NAPLs give a good measure of how much of the contaminant will remain trapped in the soil after the free product has percolated through the soil, and the residual saturation is also a good measure of how much NAPL

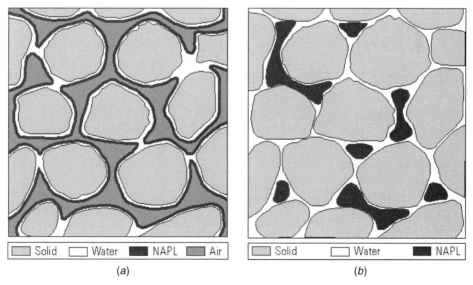

| Solid | Water | NAPL | Air |
| (a) |

| Solid | Water | NAPL |
| (b) |

FIGURE 7.12 Residual saturation in porous media: (*a*) unsaturated zone; (*b*) saturated zone. (From Abriola and Bradford, 1998.)

TABLE 7.20   Residual Saturation of Petroleum Fuels

Soil	Gasolines	Middle Distillates	Fuel Oils
Coarse gravel	0.0063	0.013	0.025
Coarse sand	0.019	0.038	0.075
Fine sand/silts	0.05	0.10	0.20

Source: American Petroleum Institute (1989).

will remain in the saturated zone after all the free product is pumped out of the aquifer. The residual saturation of various petroleum fuels in soils are given in Table 7.20, and the residual mass fraction, M_f, can be calculated using the relation

$$M_f = \frac{\rho_f n S_r}{\rho_s(1-n) + \rho_f n S_r} \tag{7.80}$$

where ρ_f is the density of the NAPL (see Appendix B.2), n is the porosity of the soil, and ρ_s is the density of the soil. In applying the residual saturations shown in Table 7.20, values of ρ_f are typically $750\,kg/m^3$ for gasoline, $800\,kg/m^3$ for middle distillates, and $900\,kg/m^3$ for fuel oils.

Example 7.12   Estimate the residual mass fraction in mg/kg when spills of gasoline in medium sand are cleaned up by pumping free product from the surface of the water table. Assume that the porosity of the aquifer is 0.23 and the density of the sand is $2600\,kg/m^3$.

SOLUTION   From the data given, $n = 0.23$, $\rho_s = 2600\,kg/m^3$, and the density of gasoline can be taken as $\rho_f = 750\,kg/m^3$. Interpolating in Table 7.20 between coarse sand and fine sand gives $S_r = 0.035$ for medium sand. Substituting the data given into Equation 7.80 gives the mass fraction, M_f, at residual saturation as

$$M_f = \frac{\rho_f n S_r}{\rho_s(1-n) + \rho_f n S_r}$$

$$= \frac{750(0.23)(0.035)}{2600(1-0.23) + 750(0.23)(0.035)} = 0.0030\,kg/kg = 3000\,mg/kg$$

Therefore, when all the free-product gasoline is removed from the contaminated soil, approximately 3000 mg/kg will remain trapped in the pores of the solid matrix. This trapped gasoline will eventually be removed by such processes as evaporation, dissolution, and biological and chemical degradation.

Even at residual saturation levels, NAPLs are capable of contaminating large volumes of water and cannot be removed easily except by dissolution in flowing ground water. The long time scales required for flowing ground water to remove residual NAPLs is illustrated in the following example.

Example 7.13   A cubic meter of aquifer has a porosity of 0.3 and contains TCE at a residual saturation of 20%. If the density of TCE is $1470\,kg/m^3$, the solubility of TCE in water is 1100 mg/L, and the mean seepage velocity of the ground water is 0.02 m/day, estimate the time it would take for the TCE to be removed by dissolution.

SOLUTION   From the data given, $n = 0.3$, and the residual saturation, S_r, is 0.20, hence the residual volume of TCE in 1 m^3 of aquifer is given by

$$\text{volume of TCE} = 0.20(0.3)(1) = 0.06 \, \text{m}^3$$

Since the density of TCE is 1470 kg/m^3, 0.06 m^3 corresponds to 0.06(1470) = 88.2 kg of TCE. With a solubility of 1100 mg/L = 1.1 kg/m^3, the volume of water required to dissolve the 88.2 kg of TCE is given by

$$\text{dissolution water required} = \frac{88.2}{1.1} = 80.2 \, \text{m}^3$$

Since the seepage velocity of the ground water is 0.02 m/day, assuming that the contaminated volume is a 1 m $\times$ 1 m $\times$ 1 m block of aquifer, the time required for 80.2 m^3 of water to flow through the 1 m^3 of contaminated aquifer is given by

$$\text{time} = \frac{80.2}{0.02n(1 \times 1)} = \frac{80.2}{0.02(0.3)(1)} = 13{,}367 \, \text{days} = 36.6 \, \text{years}$$

Hence the residual NAPL will generate a contaminant plume at saturation level (1100 mg/L) for 36.6 years! This result should be considered as somewhat approximate, since dissolution rates are highly dependent on the range and size distribution of NAPL blobs (Schnoor, 1996).

7.8 REMEDIATION OF SUBSURFACE CONTAMINATION

Subsurface contamination is commonly caused by either the careless handling of toxic substances or the leakage of toxic chemicals from storage containers (tanks) located above or below ground. Contaminants of concern include NAPLs that leave a trail of contaminant at the residual saturation level in the soil, and if a sufficient amount of contaminant is spilled, the NAPL reaches the saturated zone, where it either floats on the water table (LNAPLs) or sinks to the bottom of the aquifer (DNAPLs). Residual NAPLs within the soil and aquifer matrix act as contaminant sources, since the residual NAPL produces a dissolved plume as water either infiltrates past the NAPL in the vadose zone or flows past the NAPL in the saturated zone. In the United States, sites that are highly contaminated and pose a significant risk to the health of the public are considered candidates for classification as Superfund sites, in which case the U.S. government pays for the cleanup (from the Superfund) and seeks restitution from the entity responsible for the contamination. Superfund sites are usually clearly identified to the public, as shown in Figure 7.13 for a former oil refinery site.

The cleanup or *remediation* of contaminated soil and ground water typically requires the removal of residual NAPLs that serve as sources of contamination and/or the treatment of contaminated ground water containing contaminants dissolved from the NAPL residual. The portion of the soil or aquifer matrix containing the source NAPL is commonly called the *source zone*, and the ground water containing dissolved contaminant is called the

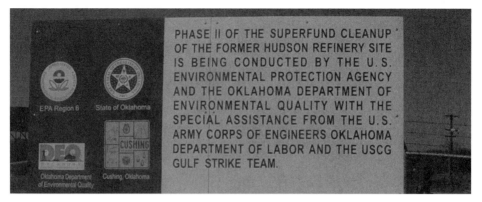

FIGURE 7.13 Sign at a Superfund site. (From USEPA, 2005a.)

contaminant plume. Since LNAPLs and DNAPLs behave much differently in soils and aquifers, remediation strategies generally reflect these differences. Most sites contaminated with LNAPLs involve the release of petroleum hydrocarbons that are aerobically degradable by indigenous microorganisms. This natural degradation typically limits the growth of dissolved LNAPL plumes. Spilled gasoline is frequently encountered in practice, and the components of gasoline that are usually of most concern are benzene, ethylbenzene, toluene, and xylene (BTEX), which collectively make up about 10% of gasoline by weight. The BTEX compounds, along with fuel additives, are the most soluble constituents of gasoline. Studies by Rice et al. (1995) and Mace et al. (1997) have demonstrated that most BTEX plumes at leaking underground (gasoline) storage tanks (LUSTs) typically do not extend more than 100 m from the downgradient edge of the source zone. Current conceptual models of biodegradation assume that significant biological transformation does not occur in the residual NAPLs but occurs in the dissolved phase. The assessment and remediation of DNAPL sites is generally much more challenging than that of LNAPL sites. Most DNAPLs of interest are chlorinated solvents that degrade slowly, if at all, under natural conditions. DNAPLs migrate vertically, have low solubilities, and can persist below the water table for decades (Johnson,1999).

7.8.1 Remediation Goals

Remediation goals are the basis for the formulation of remediation strategies. The key components of remediation goals are (1) definition of target levels, (2) delineation of a compliance zone, and (3) a statement of the time frame within which compliance will be achieved. Remediation goals are typically defined separately for contaminated soils in the vadose zone and contaminated ground water in the saturated zone.

Vadose Zone Target levels for various contaminants in soils are generally set by regulatory agencies and are based on the potential for contaminant migration to underlying ground water. Target levels for total petroleum hydrocarbons (TPHs) in soils are typically in the range 10 to 100 mg/kg (Tindall and Kunkel, 1999).

Saturated Zone Target levels for ground water remediation vary between states and are usually related to (1) maximum contaminant levels (MCLs) or maximum contaminant level goals (MCLGs) for drinking water, (2) risk based considerations, and/or (3) resource protection goals. Many regulatory agencies have adopted MCLs as target levels for ground water quality. However, in recent years some regulatory agencies have moved away from using MCLs as target levels and have required risk-based target levels (Bedient et al., 1999). Risk-based target levels take into account potential mechanisms for exposure of human or ecological receptors, the uses of the ground water, the distance between the contaminant source zone and water-supply wells, and the attenuation that occurs between the contaminant sources and water-supply wells. Other exposure pathways include inhalation of vapor and discharges to surface-water bodies. Risk-based target levels can either exceed or be less than MCLs, depending on the site-specific conditions and the acceptable risk of illness resulting from exposure to toxic compounds originating at the contaminated site. Acceptable risks of death caused by carcinogenic chemicals are typically in the range 10^{-5} to 10^{-6}. Remediation of subsurface contamination based on risk analysis is called a *risk-based corrective action* (RBCA), and guidelines for the implementation of RBCAs have been developed by the American Society for Testing and Materials (ASTM, 1995, 1998b). *Resource protection goals* apply to ground waters where no contamination is acceptable, and in these cases, target levels are either zero or background concentrations. Resource protection goals are usually the most difficult to achieve. Target cleanup levels for ground water can be applied either at the boundary of a *compliance zone* or everywhere in the ground water. Cleanup everywhere in the ground water is always preferred but is not always practical, and compliance zones are sometimes negotiated. Compliance zones are similar to mixing zones in surface-water discharges, where contaminant concentrations may exceed target levels within the compliance zone.

7.8.2 Site Investigation

The formulation of realistic remediation goals and effective cleanup strategies require a thorough investigation of the contaminated site as well as an understanding of the geochemical and microbiological processes that might affect the fate of contaminants in the subsurface. Objectives of site investigations are usually to (1) identify the contaminants of concern, (2) locate the sources of contamination, (3) determine the existing concentrations of contaminants in the subsurface, (4) characterize the subsurface hydrogeology, and (5) identify potential receptors along with any existing adverse impacts. Site investigations can be separated into vadose- and saturated-zone studies.

Vadose Zone In the unsaturated (vadose) zone, residual contaminants are typically identified using soil-gas surveys and/or soil-sample analyses. In soil-gas surveys, a probe mounted on a hollow steel pipe is pushed into the soil to a desired depth, as shown in Figure 7.14. Vapors are extracted from the probe using a vacuum pump and analyzed by a detection instrument such as an organic-vapor analyzer or portable gas chromatograph (GC). Contaminant sources are located by looking for areas where the soil-gas concentration is close to the maximum vapor concentration at the ambient temperature. In cases where mixtures of contaminants are spilled, the maximum vapor pressure, p_i, of the ith component is related to the saturation vapor pressure, p_i^o, of the pure component by *Raoult's law*, which states that

$$p_i = x_i \gamma_i p_i^o \tag{7.81}$$

FIGURE 7.14   Soil-gas probe. (From USACE, 2005c.)

where x_i is the mole fraction of the ith component and γ_i is the activity coefficient of ith component. The activity coefficient, γ_i, in Equation 7.81 accounts for nonidealities in gas behavior, and in many cases γ_i can be taken as approximately equal to unity. The concentration, c_i, of a contaminant in the vapor phase is related to the vapor pressure, p_i, by the *ideal gas law*

$$c_i = \frac{p_i m_i}{RT} \tag{7.82}$$

where m_i is the molar mass of the contaminant (g/mol), R is the universal gas constant ($=8.31$ J/K·mol), and T is the absolute temperature (K). Combining Equations 7.81 and 7.82 gives the following relation for calculating the maximum soil-gas concentration, c_i, resulting from a spill:

$$\boxed{c_i = \frac{x_i \gamma_i p_i^\circ m_i}{RT}} \tag{7.83}$$

Values of the saturation vapor pressure, p_i° and the molar mass, m_i, for several contaminants commonly found in ground water are given in Appendix B.2. In cases where the composition of the mixture is unknown, the mole fraction, x_i, in Equation 7.83 can be approximated by the mass fraction.

Example 7.14   A spill of pure benzene seeps into the ground where the ambient temperature is 18°C. (a) Calculate the maximum concentration of benzene in a soil-gas sample. (b) Contrast this with the maximum soil-gas concentration of benzene resulting from a spill of gasoline that contains a mole fraction of 1% benzene.

SOLUTION   (a) For a spill of pure benzene, the maximum vapor concentration of benzene is given by Equation 7.83, where $x_i = 1$, $\gamma_i = 1$ (for ideal case), $p_i^\circ = 7.0$ kPa $= 7000$ Pa

(Appendix B.2), $m_i = 78.1$ g/mol (Appendix B.2), $R = 8.31$ J/K·mol, and $T = 273.15 + 18 = 291.15$ K. Substituting into Equation 7.83 gives

$$c_i = \frac{x_i \gamma_i p_i^o m_i}{RT} = \frac{1(1)(7000)(78.1)}{8.31(291.15)} = 226 \text{ g/m}^3 = 226 \text{ mg/L}$$

(b) In the case of spilled gasoline, all parameters are the same as for pure benzene except $x_i = 0.01$ instead of $x_i = 1$. Therefore, Equation 7.83 gives

$$c_i = 0.01(226) = 2.26 \text{ mg/L}$$

These results demonstrate that the soil-gas concentration resulting from a spill of pure benzene is considerably higher than for a spill of gasoline, which contains only 1% benzene.

Organic vapors can migrate in the vadose zone much faster than dissolved organics in the liquid phase. Furthermore, the vapor flow velocities and direction are not controlled by hydraulic gradients as in the ground water. Spreading vapors contaminate the soil moisture and soil matrix, and vapors less dense than air can rise to the ground surface, where they can be inhaled or cause explosions if they get into basement structures. *Vapor intrusion* is defined as vapor-phase migration of volatile organic and/or inorganic compounds into occupied buildings from underlying contaminated ground water and/or soil. The presence of volatile organic compounds (VOCs) in the subsurface indicates that volatile organics may be present in either the vadose zone or the saturated zone. Good detection of organic vapors in the subsurface is possible in cases where the vadose zone is relatively thin, and organic compounds that are more suitable for detection are those with boiling points of less than 150°C, low aqueous solubility (less than 100 mmol/L), and saturation vapor pressures higher than 1.3 kPa at 20°C (Bedient et al., 1999).

As an alternative to soil-gas surveys, the presence of volatile organic compounds in the vadose zone can be detected using soil-sample headspace vapor analyses. Soil samples are collected using hand augers for depths of less than 1.5 m, backhoes are effective to depths of 3 to 4.5 m, and direct-push soil probing devices or conventional drilling rigs are usually most cost-effective for collecting soil samples at depths greater than 3 m. For the commonly used hand auger shown in Figure 7.15, the auger head [Figure 7.15(a)] is twisted by hand, using a pole attachment, into the ground to extract the soil [Figure 7.15(b)]. Soil samples are typically placed in clean glass jars with tight-fitting lids. Sample bottles, which are partially filled with soil, are either agitated or allowed to sit for awhile to permit the contaminants to volatilize. An organic-vapor analyzer or gas chromatograph is then used to measure the concentration of organic vapors in the headspace above the soil sample. However, since the vapor concentration depends on the soil texture, volatility of the contaminants, temperature of the sample, soil volume, and headspace volume, headspace vapor analyses are viewed as qualitative measures of whether contaminants are present in the soil.

Another method of detecting the presence of contaminants in a soil sample is by liquid extraction, where a liquid extract is tested for the presence of contaminants. Such tests include colorimetric and immunoassay tests (Bedient et al., 1999). Use of a black light helps to identify many NAPLs containing aromatic and other florescent hydrocarbons. Ultraviolet fluorescence tests (for fluorescent NAPLs) and hydrophobic dye methods are among the most effective ways to detect NAPLs in soil samples.

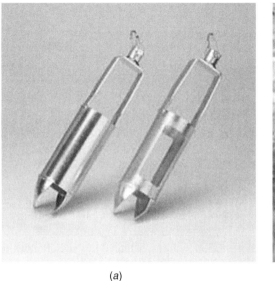

(a)   (b)

FIGURE 7.15   Hand auger. [(*a*) From Durham Geo Slope Indicator, 2005; (*b*) from MetExperts, 2005.]

Other methods, such as the installation of passive soil-gas collectors and borehole and surface geophysics, are also used to locate aqueous and nonaqueous hydrocarbons in the subsurface. Passive soil-gas collectors consist principally of activated carbon modules that are placed within the ground for a period of days to weeks, after which the collector is removed and analysed for the amount of organics sorbed from the soil gas. Ground penetrating radar, complex resistivity, and electromagnetic induction are geophysical techniques that have been applied with some success. Geophysical techniques are typically better suited to detecting subtle changes in subsurface composition, such as moving free-phase NAPL mass, than in assessing static conditions (Bedient et al., 1999).

The presence of residual NAPL in a soil or aquifer sample can be estimated by calculating whether the amount of contaminant in the sample exceeds the amount required for the contaminant to be at the solubility level in the aqueous phase and adsorbed into the soil by equilibrium partitioning. The steps to be followed in this analysis are:

1. Determine the effective solubility, c_0, and the organic-carbon sorption coefficient, K_{oc}, of the contaminant.

2. Conduct a soil-sample analysis to determine the fraction, f_{oc}, of organic carbon in the soil, the bulk density, ρ_b, of the soil, the water-filled porosity, θ_w, and the concentration, c_s, of organic compound in the saturated soil (typically in mg/kg).

3. Calculate the theoretical pore water concentration, c_w, assuming no NAPL, where

$$c_w = \frac{c_s \rho_b}{\rho_b f_{oc} K_{oc} + \theta_w} \tag{7.84}$$

4. Compare c_w with the effective solubility, c_0, such that if $c_w > c_0$, the presence of residual NAPL is indicated.

Example 7.15   Analysis of a soil sample indicates 150 mg/kg benzene, 2% organic material in the soil matrix, a bulk density of 1800 kg/m^3, and a water-filled porosity of 0.29. If the benzene in the soil is a result of a gasoline spill, where the gasoline contains a mole fraction of 1.5% benzene, determine whether gasoline is present as a NAPL in the soil. In accordance with Raoult's law, the maximum concentration of benzene in ground water can be taken as the mole fraction of benzene in gasoline multiplied by the solubility of pure benzene in water.

SOLUTION   From the data given, $c_s = 150$ mg/kg $= 1.5 \times 10^{-4}$ kg/kg, $f_{oc} = 2\% = 0.02$, $\rho_b = 1800$ kg/m^3, and $\theta_w = 0.29$. Appendix B.2 gives $K_{oc} = 10^{1.92} = 83.2$ cm^3/g $= 0.0832$ m^3/kg, and the solubility of pure benzene is 1780 mg/L. Since the spilled gasoline contains 1.5% benzene, the effective solubility of benzene derived from gasoline is 0.015(1780) = 26.7 mg/L. According to Equation 7.84, the theoretical pore water concentration, c_w, assuming no NAPL, is given by

$$c_w = \frac{c_s \rho_b}{\rho_b f_{oc} K_{oc} + \theta_w} = \frac{(1.5 \times 10^{-4})(1800)}{1800(0.02)(0.0832) + 0.29} = 82 \text{ mg/L}$$

Since c_w exceeds the solubility of benzene in water (=26.7 mg/L), the presence of of gasoline NAPL is strongly indicated.

As a rule of thumb, if the hydrocarbon concentration in the soil matrix exceeds 10,000 mg/kg (1% of soil mass), the sample probably contains some NAPL (Feenstra et al., 1991; Cohen and Mercer, 1993).

Saturated Zone   Raoult's law describes the partitioning of organics between the NAPL and aqueous phases in the saturated zone. Raoult's law states that, under ideal conditions, the aqueous-phase concentration in equilibrium with the pure constituent phase is equal to the aqueous-phase solubility multiplied by the mole fraction of the constituent in the NAPL phase. Raoult's law may be written for species i as

$$\boxed{c_i = x_i \gamma_i c_{s_i}} \tag{7.85}$$

where x_i is the mole fraction, γ_i is the activity coefficient, and c_{s_i} is the solubility of the ith component of the NAPL. The activity coefficient, γ_i is numerically set to unity for the case of an ideal NAPL mixture. Mixtures are termed ideal if the fraction of total volume and total surface area occupied by any one constituent is equal to the mole fraction of that constituent in the NAPL. Typically, NAPLs composed of similar constituents behave ideally or nearly ideally, and NAPLs composed of disparate compounds are expected to exhibit nonideal behavior. For several NAPLs of environmental concern, such as gasoline, diesel fuel, and coal tar, experiments have shown that the assumption of an ideal NAPL results in errors on the order of 10% in estimating equilibrium aqueous solubility (MacKay et al., 1991; Lee et al., 1992; Ramaswami and Luthy, 1997). Errors of this magnitude are often acceptable in comparison to measurement errors or variability in transport parameters that can result in uncertainties of one or more orders of magnitude. Thus, the assumption of an ideal NAPL yields a reasonable first estimate of the equilibrium aqueous concentration of a contaminant released from a multicomponent NAPL.

Raoult's law, Equation 7.85, is particularly useful in the case of petroleum hydrocarbons such as gasoline, which consists of more than 100 chemical constituents. A simplified mixture that has been used to represent gasoline is given in Table 7.21.

TABLE 7.21   Simplified Gasoline Mixture

Constituent	Concentration (g/L)
Benzene	8.2
Toluene	43.6
Xylene	71.8
1-Hexene	15.9
Cyclohexane	2.1
n-Hexane	20.4
Other aromatics	74.0
Other paraffins $(C_4–C_8)$	336.7
Heavy ends $(>C_8)$	145.1
Total	717.8

Source: Baehr and Corapcioglu (1987).

Example 7.16   (a) What is the maximum concentration of benzene in ground water resulting from the dissolution of a lens of pure benzene floating on the water table? (b) Contrast this with the maximum aqueous-phase concentration of benzene resulting from the dissolution of a lens of nonaqueous gasoline containing a mole fraction of 1% benzene.

SOLUTION   (a) According to Raoult's law, Equation 7.85, the maximum concentration of benzene in ground water is given by

$$c_i = x_i \gamma_i c_{s_i}$$

For pure benzene, $x_i = 1$, $\gamma_i = 1$, $c_{s_i} = 1745$ mg/L (Appendix B.2), and hence Raoult's law gives

$$c_i = 1(1)(1745) = 1745 \text{ mg/L}$$

This result verifies that for a spill of pure benzene the maximum concentration of benzene in the ground water is equal to the solubility of benzene in water.

(b) For benzene in gasoline, all of the parameters in Raoult's are the same except that $x_i = 0.01$ instead of $x_i = 1$. Hence, the maximum concentration of benzene in the ground water is given by

$$c_i = 0.01(1745) = 17.45 \text{ mg/L}$$

In practice, the presence of benzene as a NAPL would be indicated when the concentration of benzene in ground-water samples exceed 1% of the maximum concentrations calculated here. It is important to note that as the NAPL is flushed with water, the more soluble constituents are depleted, thereby altering the mole fraction.

In the saturated zone, contaminated areas are typically found by looking for locations where the contaminant concentration in the ground water exceeds 1% of the effective contaminant solubility (USEPA, 1992a; Pankow and Cherry, 1996). For example, if a mixture containing TCE was spilled in ground water and the effective solubility of TCE were 110 mg/L, the presence of the spilled mixture in the non-aqueous phase would be suggested by measuring dissolved TCE concentrations greater than 1.1 mg/L. Contaminants are seldom found at their solubility levels because of the dilution induced by monitoring wells with screen lengths much longer than than thickness of the zone where the concentration is at the solubility level.

The length of the contaminant plume in the saturated zone is usually determined by sampling the ground water along the axis of the plume, and the width of the plume is determined by sampling the ground water along one or two lines transverse to the plume axis. The depth of the plume is determined from ground water samples at several depths within the saturated zone. If contaminant extends through the full thickness of the uppermost water-bearing unit, sampling and analysis of ground water from the next underlying water-bearing stratum may be necessary to establish the vertical limit of contamination.

Monitoring Wells   Monitoring wells are essential components of all site investigations involving the saturated zone, and the proper design of monitoring wells is a prerequisite for obtaining representative measurements of ground-water quality. A typical monitoring well is illustrated in Figure 7.16. Monitoring wells basically consist of a perforated pipe

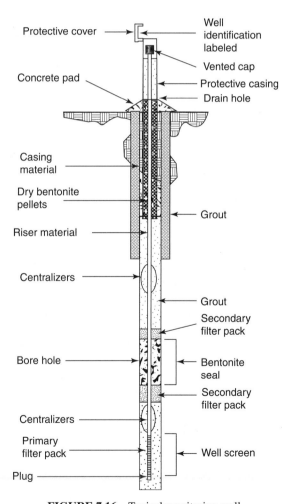

FIGURE 7.16   Typical monitoring well.

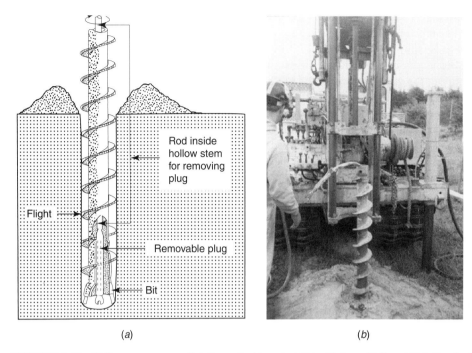

(a)   (b)

FIGURE 7.17 Hollow-stem auger. [(a) From Scalf et al., 1981; (b) from Kelley, 2005. For Scalf et al. (1981); Reprinted from Manual of *Ground-Water Quality Sampling Procedures* with permission of the National Water Well Association. Copyright 1981.]

(called the *screen* or *intake*) attached to a solid pipe (called the *casing* or *riser*) installed in a drilled borehole. The annular region between the well and the borehole is filled with sand or gravel (called the *gravel pack*) up to a level just above the screen, capped with a sealant, and backfilled with grout to the ground surface. General specifications for monitoring wells include the following (Prakash, 2004).

1. *Drilling*. Monitoring wells in unconsolidated materials are frequently installed using *hollow-stem augers*, which are rotated into the ground to bring up soil and create the boring. A typical hollow-stem auger is illustrated in Figure 7.17(a). The hollow-stem auger bores into soft soils, carrying the cuttings upward along the flights. When the desired depth is reached, the plug is removed from the drill bit and withdrawn from the hollow stem. A 50-mm-diameter monitoring well can then be inserted to the bottom of the hollow stem and the auger pulled out, leaving the small-diameter monitoring well in place. Typical internal diameters of hollow-stem augers are 11 cm (4.25 in.) for single and 30.5 cm (12 in.) for nested wells. A drill rig using a hollow-stem auger to insert water-sample tubing into the ground is shown in Figure 7.17(b). Soil samples are collected at 1.5-m (5-ft)-depth intervals using 5-cm (2-in.)-diameter split-spoon samplers. The sizes and depths may vary depending on requirements at specific locations. Physical characteristics of soil samples are recorded on the boring logs. These physical characteristics include color, visual size classifications (e.g., fine, medium, coarse), wetness (e.g., dry, moist, very wet), photoionization detector readings for VOCs, and smell. A portion of each soil sample is preserved and sent to a laboratory to analyze for the contaminants of concern and grain-size distribution.

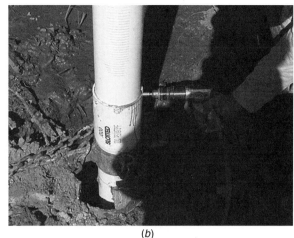

(a)   (b)

FIGURE 7.18   Installation of a well casing. (From Browns Drilling, 2005.)

2. *Casing*. Well diameters of 50 and 100 mm installed in 150- and 250-mm-diameter boreholes are most common, and most well sampling equipment can fit inside a 50-mm-diameter well. Some regulatory agencies specify the casing diameters that are required within their jurisdiction; for example, the New Jersey Department of Environmental Protection requires 100-mm-diameter monitoring-well casings under all conditions. Typically, well casings consist of Schedule 40 polyvinyl chloride (PVC) pipe that extends upward from the bottom of the borehole to about 1 m above the ground surface. The installation of a well casing is shown in Figure 7.18(*a*), and a close-up view of casing segments being screwed together is shown in Figure 7.18(*b*). In most monitoring wells the screen has the same diameter and is made of the same material as the well casing.

The diameter and material of the casing (and screen) may vary depending on site conditions, economics, and purpose. Casing and screen diameters are both influenced by the total depth of the well, since the deeper the well, the stronger the casing and screen needed to resist both lateral earth pressures and the weight of the casing itself. Pipes (casings) of a given outside diameter are made at different *schedules*, which have different wall thicknesses and inside diameters (see Appendix F.1). Heavier-schedule casings, such as Schedule 40 or Schedule 80, are stronger since they have thicker walls. Most monitoring wells are made of either PVC or stainless steel, with PTFE[2] being less common. In well screens constructed of PVC, threaded joints are generally specified since the use of glues that contain organic solvents is discouraged. Information on the compatibility of various screen and casing materials with common contaminants have been reported by Driscoll (1986), Parker et al. (1990), and Ranney and Parker (1997). These data indicate that for monitoring organics, stainless steel is the material of choice and PTFE should be avoided. For monitoring inorganics, PVC is the preferred material, and as a compromise material for monitoring both organics and inorganics, PVC appears to be the best and also is the lowest cost. In cases where PVC is the material of choice, PVC manufactured specifically for well casing should be used, and it should carry the designation *NSF wc*, which

---

[2]Polytetrafluoroethylene (PTFE) is commonly known by the brand name Teflon.

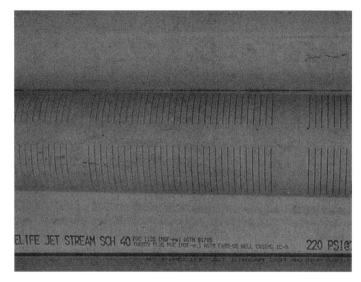

FIGURE 7.19   Well screen. (From Browns Drilling, 2005.)

indicates that the casing conforms to National Sanitation Foundation Standard 14 for potable water supply (National Sanitation Foundation, 1988). PVC should not be used if organic compounds are present as NAPLs.

3. *Screen.* The openings in well screens are generally in the form of rectangular openings called *slots*, and a typical PVC well screen is shown in Figure 7.19. Well screens are available in a variety of opening sizes, generally ranging from 0.2 to 6.4 mm (0.008 to 0.25 in.). A screen with an opening of 0.25 mm (0.010 in.) is referred to as a *10-slot screen*, and screens are commonly specified by their slot size; for example, 20-slot screen has an opening size of 0.50 mm (0.020 in.). The most commonly used screen sizes are 10- and 20-slot screens. The screen opening should be selected to retain 90% of the gravel pack.

To detect the presence of non-aqueous phase contaminants in the saturated zone, well screens should be positioned to intersect either the top (for LNAPLs) or bottom (for DNAPLs) of the water-bearing stratum. In monitoring the water table for the presence of LNAPLs, the screen must be long enough to intersect the water table over the range of annual fluctuations. In most applications, the minimum length of screen for a water-table monitoring well is 3 m, with 1.5 m above and 1.5 m below the average water table elevation. If the water table has more than a 1.5-m fluctuation, a longer screen is needed. To accurately measure the concentration of dissolved contaminants in the ground water, shorter screens with lengths less than 3 m are generally preferred, since contaminants present over a limited depth interval may be overdiluted by longer screens. If the purpose of a monitoring well is to measure the potentiometric head within the aquifer, a relatively short screen of length 1 to 2 m is usually sufficient.

4. *Gravel pack.* The annular space between the borehole and well screen is typically filled with clean sand and fine gravel, which is called the *gravel pack* or *filter pack*. A typical gravel pack in an unconsolidated (sand) formation is shown in Figure 7.20. The

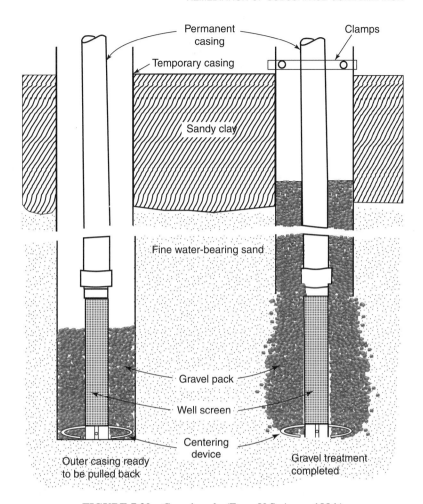

FIGURE 7.20 Gravel pack. (From U.S. Army, 1994.)

placement of such gravel packs requires the insertion of a temporary outer casing to keep the borehole open while the gravel is poured down the borehole (left) and then the withdrawal of the temporary casing to complete the gravel-pack installation (right). Gravel packs normally extend 60 to 90 cm above the top of the well screen to allow for settlement of the gravel-pack material during well development, and the gravel pack should be 5 to 8 cm thick. In cases where at least 90% of the aquifer matrix is retained by a 10-slot screen, an artificial gravel pack may not be necessary and a natural gravel pack will develop when the well is pumped (Aller et al., 1989). Gravel packs placed between the well screen and aquifer formation are typically composed of graded silica sand. Blasting sands and other general-use sands should not be used as gravel packs since they may contain materials that sorb dissolved metals, potentially compromising the integrity of the ground water sample (Bedient et al., 1999). The gravel-pack material should have an average grain size that is twice the average grain size of the aquifer matrix and a uniformity coefficient (d_{60}/d_{10}) between 2 and 3 (Driscoll, 1986). Gravel-pack gradation is designed to retain about 70%

FIGURE 7.21   Bentonite pellets. (From Dean Bennett Supply, 2005.)

of the aquifer matrix (recall that the screen slots are sized to retain 90% of the gravel pack). The annular space in the borehole above the gravel pack is generally sealed to prevent the movement of water down the gravel pack from the ground surface and other parts of the aquifer.

5. *Seal.* Annular seals typically consist of bentonite pellets or cement grout, and seals are typically 30 cm to 1 m thick. Bentonite pellets, as shown in Figure 7.21, are poured down the well casing and expand to form the seal; the pellets are typically available in 10- and 13-mm diameters and swell to 10 to 15 times their dry size when immersed in water. For nested wells in the same borehole, the bentonite pellet seal is provided in the entire length between the top of the filter pack of the deep well and the bottom of the filter pack of the shallow well and in a length of about 1 m above the filter pack of the shallow well. The annular space above the seal is typically backfilled with native material extracted from the aquifer during drilling. Monitoring wells are completed at the surface with locking caps and/or casing to prevent tampering and a concrete surface pad to protect the annular seal. In some cases, the annular space from ground surface to the top of the bentonite pellet seal is filled with a cement–bentonite mix with a cement-to-bentonite ratio of 19:1 by weight. This mix may change, depending on the site conditions (Prakash, 2004).

6. *Well development, purging, and sampling.* Monitoring wells are developed using submersible pumps until the pumped water is free of sediment. Water level, pH, specific conductivity, and temperature are recorded during well development. The well should be purged before each sampling event. If the well has been installed in a low-permeability aquifer using a dry drilling method, bailing out 3 to 10 casing volumes may be sufficient to permit collection of representative ground-water samples. If fluids have been introduced during drilling, larger volumes of water must be removed. During routine collection of ground-water samples, removal of three to five saturated casing volumes is a generally accepted minimum purge (Bedient et al., 1999). Alternatively, wells are purged until measurements of water-quality parameters such as temperature, specific conductance, and pH have stabilized. After purging, the water level in the well should be allowed to recover. Water-quality samples are sometimes collected using disposable polyethylene bailers. Specific conductance, pH, dissolved oxygen, and temperature of ground water may be measured in the field. Retrieved water samples are stored, sealed, and transported to the designated laboratory for analysis.

7. *Decontamination and waste disposal.* Drilling augers, auger bits, split-spoon samplers, and other soiled equipment should be steam cleaned prior to use, double-rinsed in tap water, and finally, rinsed with distilled water. All drill cuttings and liquid wastes, including well development water, should be containerized for appropriate disposal.

8. *Well survey.* The location and elevation of the top of the casing of a monitoring well should be surveyed and connected to appropriate state or other plane coordinates and referenced to a vertical datum. The survey information should be indicated on site maps and kept with other monitoring-well records.

Example 7.17 A monitoring well is to be installed to measure the water quality in an aquifer. A grain-size analysis of the aquifer matrix indicates that $d_{10} = 0.15$ mm, $d_{30} = 0.2$ mm, $d_{60} = 0.6$ mm, and $d_{90} = 1.6$ mm. (a) Determine whether an artificial gravel pack is necessary. If a gravel pack is required, write the specifications for the gravel pack. (b) Assuming that all natural and artificial gravel packs have the property that $d_{60} \approx 1.1 d_{50}$, determine the required slot size for the well screen.

SOLUTION (a) An artificial gravel pack is required when less than 90% of the aquifer material is retained by a 10-slot screen. The size of a 10-slot screen is 0.25 mm. From the data given, 90% of the aquifer matrix has a grain size larger than $d_{10} = 0.15$ mm. Therefore, a 10-slot screen will retain less than 90% of the aquifer matrix and an artificial gravel pack is required. The average grain size of the gravel pack should be two times the average grain size of the aquifer matrix. Taking the average grain size of the aquifer matrix as d_{50}, interpolating from the given data (between d_{30} and d_{60}) yields $d_{50} \approx 0.47$ mm. Therefore, the specifications of the gravel pack are (1) average grain size $= 2 \times 0.47$ mm $= 0.94$ mm, (2) uniformity coefficient $= 2$ to 3, (3) thickness $= 5$ to 8 cm, and (4) the gravel pack should extend 60 to 90 cm above the top of the screen.

(b) The well screen is to have a slot size that retains 90% of the gravel pack. Therefore, take the screen size as d_{10} of the gravel pack. If U_c is the uniformity coefficient of the gravel pack,

$$ d_{10} = \frac{d_{60}}{U_c} $$

Taking $d_{60} = 1.1 d_{50} = 1.1(0.94) = 1.0$ mm and $U_c = 2$ yields

$$ d_{10} = \frac{1.0}{2} = 0.50 \text{ mm} $$

Therefore, a 20-slot screen (slot size $= 0.50$ mm) is recommended.

In most site investigations, water samples are required at several depths at a single location. Under such circumstances, either a cluster of wells or multilevel samplers within a single well are used. At well clusters, each well is completed in its own borehole, and a typical well cluster is shown in Figure 7.22.

Characterization of Hydrogeology The essential geologic data required in all site investigations consist of a description of the principal stratigraphic units underlying the site, including their thickness, lateral continuity, and water-bearing properties. Geologic data are obtained by examining soil or rock samples collected from core borings. These borings

FIGURE 7.22   Cluster of monitoring wells.

are usually done during installation of monitoring wells to sample the ground water. The essential hydrogeologic data required at all site investigations consists of the hydraulic gradient, hydraulic conductivity, and effective porosity. The hydraulic gradient can be determined from water-table elevations measured at a minimum of three monitoring-well locations. Water levels are typically measured to the nearest 3 mm (0.01 ft) using a water-sensitive probe on a graduated tape. The elevation of the water table is typically obtained by subtracting the depth to water from the top-of-casing elevation. Water-table elevations should be determined at locations that are free of NAPLs floating on the water table. However, if NAPLs are floating on the surface of the water table, the thickness of the NAPL layer is multiplied by the specific gravity of the NAPL and added to the measured water-table elevation. The hydraulic conductivity in the portion of the aquifer immediately surrounding a monitoring well location can be measured using a slug test (Chin, 2006). Slug tests should generally be performed at several monitoring wells to determine the variability in the hydraulic conductivity and to provide data for calculating an average hydraulic conductivity. Pump tests can be used to characterize conditions over a larger portion of the aquifer; however, in contrast to slug tests, pump tests at contaminated sites usually require the treatment and proper disposal of all the water produced. Pump tests are expensive, and if they are used at all, are typically conducted only during the design phase of a ground-water remediation program.

Example 7.18   The piezometric heads are measured at three locations in an aquifer. Point A is located at (0 km, 0 km), point B is located at (1 km, -0.5 km), and point C is located at (0.5 km, -1.2 km), and the piezometric heads at A, B, and C are 2.157, 1.752, and 1.629 m. Determine the hydraulic gradient in the aquifer.

SOLUTION   The piezometric head, h, in the triangular region ABC can be assumed to be planar and given by

$$h(x, y) = ax + by + c$$

where a, b, and c are constants and (x, y) are the coordinate locations. Applying this equation to points A, B, and C (with all linear dimensions in meters) yields

$$2.157 = a(0) + b(0) + c$$

$$1.752 = a(1000) + b(-500) + c$$

$$1.629 = a(500) + b(-1200) + c$$

The solution of these equations is $a = -0.0002337$, $b = 0.0003426$, and $c = 2.157$. From the planar head distribution, it is clear that the components of the head gradient are given by

$$\frac{\partial h}{\partial x} = a \quad \text{and} \quad \frac{\partial h}{\partial y} = b$$

Therefore, in this case the components of the head gradient are

$$\frac{\partial h}{\partial x} = -0.0002337 \quad \text{and} \quad \frac{\partial h}{\partial y} = 0.0003426$$

which can be written in vector notation as

$$\nabla h = -0.0002337i + 0.0003426j$$

where i and j are unit vectors in the coordinate directions.

7.8.3 Remediation Strategies

Remediation strategies are designed to accomplish well-defined remediation goals. There are a wide variety of remediation strategies, and the appropriate strategy for any situation depends on several factors, including the hydrogeology of the site, the nature of the contaminants, the distribution of contaminants in the subsurface, the target levels, the time frame for cleanup, exposure of cleanup workers and the public, technical feasibiliy, and economic considerations.

Remediation strategies can be grouped as those that (1) remove residual NAPL contaminants in the soil and ground water that serve as sources of contamination, referred to as *source-zone treatment*, (2) target the treatment of ground water containing dissolved contaminants to restore the ground water to target levels, referred to as *aquifer restoration*, or (3) prevent further migration of the contaminant referred to as *migration prevention*. Some remediation technologies belong to more than one group and several commonly used remediation technologies are listed in Table 7.22. It is important to note that remediation strategies are generally designed under conditions of great uncertainty and the system design relies heavily on conceptual models, screening-level calculations, empiricism, heuristics, experience, monitoring, and refinement (Johnson, 1999). Following sound design practice does not guarantee success but simply provides a higher probability of success.

Free-Product Recovery Most regulatory programs require removal of any mobile and pumpable imiscible free-product liquid that can be collected in monitoring wells (Johnson, 1999). In the case of LNAPLs, free product is usually found within the capillary

TABLE 7.22   Groundwater Remediation Technologies

| | Objective | | |
Technology	Source-Zone Treatment	Aquifer Restoration	Migration Prevention
Free-product recovery	•		
Excavation and disposal	•		
Soil-vapor extraction	•		
Bioventing	•		
Air sparging	•	•	
Air sparging cutoff trenches			•
Pump and treat	•	•	•
Bioremediation		•	
In situ reaction walls			•
In situ containment			•
Natural attenuation		•	•

zone just above the water table; in the case of DNAPLs, free product is usually found at the bottom of the aquifer. Product recovery at DNAPL sites is not well understood. To recover LNAPL free product, a *recovery well* is typically screened in the zone containing the free product and the NAPL is pumped out of the well. However, it is important to note that free-product thickness in the recovery well does not give a true indication of the free-product thickness in the aquifer. The reason for this is that the NAPLs are distributed at varying saturation levels within the capillary fringe, which contains mostly ground water. Installation of a well screen allows the NAPL to seep into the well from the entire capillary fringe and above, filling the well up to the level where there is significant amounts of NAPL in the surrounding porous medium. Since the pores in the aquifer material surrounding the well are not fully saturated with NAPL, the depth of NAPL in the well gives a false indication of the depth of NAPL in the surrounding aquifer. The relationship between the free-product thickness in the recovery well, H_w, and the equivalent free-product thickness in the aquifer, H_a, can be estimated by the relation (Kemblowski and Chiang, 1990)

$$H_a = H_w - 3h_s \tag{7.86}$$

where h_s is the capillary rise of water in the soil, which can be estimated using Table 7.23. Since the capillary rise increases for finer aquifer materials, in fine-grained aquifers the free-product thickness measured in a monitoring well can overestimate the actual free-product thickness in the aquifer by tens of centimeters. Equation 7.86 also demonstrates that for the same amount of NAPL in the ground water, wells surrounded by fine-grained material will show a greater free-product thickness than wells surrounded by coarse-grained material. An alternative relation proposed by Hampton and Miller (1989) is

$$H_a = \frac{\rho_w - \rho_f}{\rho_w} H_w \tag{7.87}$$

TABLE 7.23 Capillary Rise in Unconsolidated Materials

Material	Grain Size (mm)	Capillary Rise (cm)
Fine gravel	2–5	2.5
Very coarse sand	1–2	6.5
Coarse sand	0.5–1	13.5
Medium sand	0.2–0.5	24.6
Fine sand	0.1–0.2	42.8
Silt	0.05–0.1	105.5

Source: Lohman (1972).

where ρ_w is the density of water and ρ_f is the density of the LNAPL. Charbeneau (2000) considered the role of soil texture on the relationship between aquifer free-product thickness, H_a, and the monitoring well thickness, H_w, and presented the relation

$$H_a = \beta(H_w - \alpha)$$

(7.88)

where α and β are constants given in Table 7.24 for LNAPLs of various densities. The parameter α in Equation 7.88 represents the minimum thickness of LNAPL in the well for the LNAPL to flow freely between the well and the aquifer. Studies by USEPA (1995) indicate that equations relating the free-product thickness in the aquifer to the free-product thickness in a monitoring well are not particularly reliable and should be used for estimation purposes only. Typically, the free-product thickness in a recovery well is 2 to 10 times thicker than the free-product thickness in the aquifer. It is important to note that a monitoring well may not contain any NAPL if the screened interval contains only residual NAPL.

TABLE 7.24 Parameters $(\alpha, \beta)^a$

Soil Texture	LNAPL Density (kg/m³)		
	700	775	850
Sand	(0.10,0.397)	(0.20,0.391)	(0.30,0.384)
Loamy sand	(0.175,0.363)	(0.25,0.352)	(0.40,0.344)
Sandy loam	(0.325,0.340)	(0.44,0.324)	(0.65,0.310)
Loam	(0.65,0.303)	(0.85,0.278)	(1.10,0.247)
Sandy clay loam	(0.55,0.252)	(0.69,0.232)	(0.90,0.211)
Silt loam	(1.00,0.273)	(1.25,0.237)	(1.60,0.195)
Silt	(1.12,0.273)	(1.45,0.234)	(1.80,0.183)
Clay loam	(1.07,0.195)	(1.35,0.166)	(1.75,0.134)
Sandy clay	(1.07,0.159)	(1.35,0.134)	(1.75,0.110)
Silty clay loam	(1.47,0.150)	(1.85,0.116)	(2.50,0.083)
Clay	(1.52,0.071)	(2.02,0.052)	(2.90,0.036)
Silty clay	(1.90,0.056)	(2.65,0.038)	(4.20,0.024)

Source: Charbeneau (2000).

[a]α is in meters and β is dimensionless.

Example 7.19 Several monitoring wells surrounding a gas station show a 30-cm-thick layer of gasoline in wells covering an area of 2500 m². The gasoline has a density of 750 kg/m³, the temperature of the ground water is 15°C, the aquifer material consists of medium sand with a (water) capillary rise of 8 cm, and the porosity of the aquifer is 0.23. Estimate the volume of gasoline floating on the water table.

SOLUTION   From the data given, $H_w = 30$ cm $= 0.30$ m, $A_{spill} = 2500$ m², $h_s = 8$ cm $= 0.08$ m, $n = 0.23$, $\rho_f = 750$ kg/m³, and at 15°C $\rho_w = 999.1$ kg/m³. Interpolating from Table 7.24 for sand grains gives $\alpha = 0.167$ m and $\beta = 0.393$. According to Kemblowski and Chiang (1990), Equation 7.86 gives the thickness H_a of gasoline in the aquifer as

$$H_a = H_w - 3h_s = 0.30 - 3(0.08) = 0.06 \text{ m} = 6 \text{ cm}$$

Using the relation proposed by Hampton and Miller (1989), Equation 7.87 gives

$$H_a = \frac{\rho_w - \rho_f}{\rho_w} H_w = \frac{999.1 - 750}{999.1}(0.30) = 0.075 \text{ m} = 7.5 \text{ cm}$$

Using the relation proposed by Charbeneau (2000), Equation 7.88 gives

$$H_a = \beta(H_w - \alpha) = 0.393(0.30 - 0.167) = 0.052 \text{ m} = 5.2 \text{ cm}$$

The volume of gasoline, V, floating on the water table is given by

$$V = nA_{spill}H_a = 0.23(2500)H_a = 575H_a$$

Therefore, the Kemblowski and Chiang equation gives $V = 575(0.06) = 35$ m³, the Hampton and Miller equation gives $V = 575(0.075) = 43$ m³, and the Charbeneau equation gives $V = 575(0.052) = 30$ m³. A conservative estimate of the volume of gasoline is 43 m³.

Pumping-well recovery schemes include (1) skimming systems for NAPLs only; (2) single, total-fluids pumps that pump water and free-product mixtures; and (3) dual recovery pump systems in which a ground-water pump is used to depress the water table, while a second pump skims the LNAPL free product that flows into the well. Skimming systems are used when there is not a significant quantity of NAPL trapped below the water table, water-table depression is not needed to accelerate flow to the well, the aquifer is composed of very low permeability materials, and it is cost-prohibitive or not possible to treat produced ground water aboveground. Single- and dual-pump recovery systems are used when a significant quantity of NAPL is trapped below the water table, a water-table depression is needed to accelerate the flow to the well, and the formation is composed of relatively permeable material. A typical dual recovery pump system is illustrated in Figure 7.23. Single-pump systems tend to emulsify the water and oil, requiring the use of an oil–water separation system, while dual-pump systems eliminate the need for an aboveground oil–water separator and minimize the load on aboveground treatment systems. In dual-pump systems, both the water pump and the product-recovery pump are installed in a single well, which must have a large enough diameter to accommodate both pumps and some product-detection probes that are used to turn the pumps on and off. Dual-pump systems have greater costs and maintenance requirements than single-pump systems. Regardless of

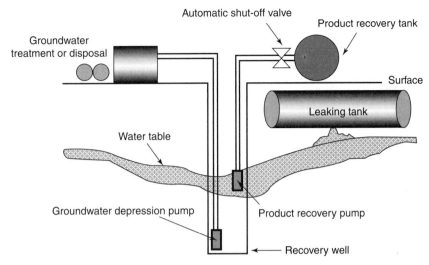

FIGURE 7.23 Dual recovery pump system. (From Federal Remediation Technologies Roundtable, 2005a.)

whether a single- or dual-pump system is used, the drawdown distribution in the NAPL layer is created by the same distribution of drawdown in the ground-water layer, and the NAPL recovery rate, Q_0 [L³/T], can be estimated by (Charbeneau, 2000)

$$Q_0 = \frac{\rho_r \beta^2 Q_w}{\mu_r n^2 H} \frac{(H_w - \alpha)^2}{H_w}$$

(7.89)

where ρ_r and μ_r are the NAPL water density and viscosity ratios, respectively (dimensionless), α [L] and β (dimensionless) are parameters given in Table 7.24, Q_w [L³/T] is the water production rate, n is the porosity, and H is the saturated thickness of the ground water [L]. It is noteworthy that for a given water production rate, the NAPL recovery rate is smaller in fine soils than in coarse soils, because β is smaller and α is larger. For NAPL thicknesses, H_w, much greater than α, Equation 7.89 indicates a linear relationship between oil recovery and and oil layer thickness. The time, t, required to reduce the monitoring-well thickness of the free product from H_{w_0} to H_{w_1} can be estimated using the relation (Charbeneau, 2000)

$$t = \frac{\pi R_c^2 n^2 \mu_r H}{\rho_r Q_w \beta} \left[F\left(\frac{H_{w_1}}{\alpha}\right) - F\left(\frac{H_{w_0}}{\alpha}\right) \right]$$

(7.90)

where R_c is the *radius of recovery* and $F(x)$ is a function defined by

$$F(x) = \frac{1}{x - 1} - \ln(x - 1)$$

(7.91)

The radius of recovery, R_c, is defined as the radial distance from the well within which the NAPL is recovered by the well and depends on the number of wells used in the recovery system, the locations of the recovery wells, and the gradients induced by the pumped water. For multiple-well systems, R_c is on the order of half the recovery well spacing, while in single-well systems R_c is bounded by the radius of influence, R, of the pumping well, and is typically less than 30 m.

Example 7.20 A spilled organic contaminant has a density of 850 kg/m³ and a dynamic viscosity of 0.002 Pa·s. Monitoring wells show a free-product thickness of 120 cm. The saturated thickness of the aquifer is 15 m, the temperature of the ground water is 20°C, the aquifer material is classified as sand, and the porosity of the aquifer is 0.39. Several recovery wells are placed within the area where the contaminant exists as a NAPL, and each well has a radius of recovery of 30 m. Allowable drawdown criteria limit the ground-water withdrawal rate at each recovery well to 12 L/min. If a single-pump system is used to recover the NAPL, calculate the fraction of NAPL by volume in the pumped water and how long it will take to remove 90% of the spilled NAPL.

SOLUTION From the data given, $\rho_f = 850$ kg/m³, $\mu_f = 0.002$ Pa·s = 2 mPa·s, $H_w =$ 120 cm = 1.20 m, $H = 15$ m, $n = 0.39$, $R_c = 30$ m, and $Q_w = 12$ L/min = 0.0002 m³/s. For sand and a NAPL density of 850 kg/m³, Table 7.24 gives $\alpha = 0.30$ m and $\beta = 0.384$. At 20°C, the dynamic viscosity of water, μ_w, is 1.002 mPa·s, and the density of water is 998.2 kg/m³. Therefore, the viscosity ratio, μ_r, and the density ratio, ρ_r, are given by

$$\mu_r = \frac{\mu_f}{\mu_w} = \frac{2}{1.002} = 1.996$$

$$\rho_r = \frac{\rho_f}{\rho_w} = \frac{850}{998.2} = 0.852$$

The NAPL recovery rate is given by Equation 7.89 as

$$Q_0 = \frac{\rho_r \beta^2 Q_w}{\mu_r n^2 H} \frac{(H_w - \alpha)^2}{H_w}$$

$$= \frac{0.852(0.384)^2(0.0002)}{1.996(0.39)^2(15)} \frac{(1.20 - 0.30)^2}{1.20}$$

$$= 3.72 \times 10^{-6} \, \text{m}^3/\text{s} = 0.22 \, \text{L/min}$$

Therefore, the total liquid production rate per well is 12 L/min + 0.22 L/min = 12.22 L/min, and hence the pumped water is 1.8% NAPL by volume. An oil–water separation system will be required to extract the NAPL from the pumped water.

The initial thickness of the NAPL layer in the aquifer, H_{a_0}, can be estimated using Equation 7.88 as

$$H_{a_0} = \beta(H_{w_0} - \alpha) = 0.384(1.20 - 0.30) = 0.346 \, \text{m}$$

Removal of 90% of the NAPL leaves a free-product thickness, H_{a_1}, of

$$H_{a_1} = 0.1 H_{a_0} = 0.1(0.346) = 0.0346 \, \text{m}$$

The thickness of the NAPL layer in a monitoring well, H_{w_1}, corresponding to the thickness of the NAPL layer in the aquifer, H_{a_1}, is given by Equation 7.88 as

$$H_{a_1} = \beta(H_{w_1} - \alpha)$$
$$0.0346 = 0.384(H_{w_1} - 0.30)$$

which gives

$$H_{w_1} = 0.390 \, \text{m}$$

Hence,

$$\frac{H_{w_0}}{\alpha} = \frac{1.20}{0.30} = 4.00$$

$$\frac{H_{w_1}}{\alpha} = \frac{0.390}{0.30} = 1.30$$

and using Equation 7.91 gives

$$F\left(\frac{H_{w_0}}{\alpha}\right) = \frac{1}{4.00-1} - \ln(4.00 - 1) = -0.765$$

$$F\left(\frac{H_{w_1}}{\alpha}\right) = \frac{1}{1.30-1} - \ln(1.30 - 1) = 4.537$$

Substituting these results into Equation 7.90 gives the time, t_{90}, required to recover 90% of the NAPL from the water table as

$$t_{90} = \frac{\pi R_c^2 n^2 \mu_r H}{\rho_r Q_w \beta} \left[F\left(\frac{H_{w_1}}{\alpha}\right) - F\left(\frac{H_{w_0}}{\alpha}\right) \right]$$

$$= \frac{\pi (30)^2 (0.39)^2 (1.996)(15)}{0.852(0.0002)0.384} [4.537 - (-0.765)]$$

$$= 1.04 \times 10^9 \, \text{s} = 33.1 \, \text{years}$$

Therefore, at the given pumping rate, it will take approximately 33 years to remove 90% of the spilled NAPL.

Free-product recovery systems frequently include interceptor trenches, which are used in cases where the water table is shallow, or some shallow, confining layer limits the maximum depth of liquid penetration to less than 6 m below the ground surface. Trenches are excavated to a depth below the liquid product, trench widths are typically 1 to 3 m depending on the stability of the soil, and the length of the trench extends beyond the limits of the free-product lens. Trench depths typically extend 1 to 2 m below the water table. One or more vertical wells are placed in the trench, which is then backfilled with very permeable gravel, and a surface seal is typically placed on top of the trench to minimize vapor losses. Drawdown in the trench must be sufficient to reverse the ground-water gradient on the down-gradient side of the trench so that the floating product cannot flow out. Liquid product

FIGURE 7.24   Soil excavation at a Superfund site. (From USACE 2005d.)

recovery rates rarely exceed 20 L/min. Vertical recovery wells are typically used where interceptor trenches are not feasible.

Excavation and Disposal   In cases where a limited amount of soil is contaminated with hazardous materials, excavation and disposal of the contaminated soil is an option to be considered. The excavated soil is transported to a hazardous-waste incinerator for complete thermal destruction of organic contaminants. This approach has been proposed for highly refractory organic compounds such as PCBs and some pesticides (Fetter, 1999), and an example of soil excavation at a Superfund site (in Virginia) is shown in Figure 7.24. The costs associated with excavation and removal of large volumes of soil to another site are usually high, and on-site treatment of soils, also called *ex situ treatment*, should always be considered as an option. On-site (ex situ) treatment usually involves processes to remove the contaminant(s) from the soil and replace the treated soil at the site. Ex situ treatment techniques include soil washing, biomounding, low-temperature thermal desorption, and high-temperature incineration. For large quantities of contaminated soil, a common reclamation method involves soil excavation and deposition in a bioreactor, where biodegradation of the contaminant is achieved in a comparatively short period of time. This requires controlling the appropriate supply of moisture, oxygen, and nutrients for the enhancement of microorganism development and growth. Total excavation of contaminated soils is sometimes not practical, such as when contamination penetrates deep into the subsurface, contaminants occur beneath a building, or NAPLs are present. A summary of ex situ remediation methods for contaminated soils is given in Table 7.25. These remediation methods are separated into physical–chemical, biological, and thermal processes.

Soil-Vapor Extraction   Soil-vapor extraction systems are designed to maximize contaminant removal by volatization. These systems extract air from the vadose zone with blowers or vacuum pumps, and treatment systems above ground remove toxic vapors from the extracted air. Soil-vapor extraction systems are frequently used to remove volatile contaminants from permeable soils and are analogous to pump-and-treat systems for ground water. A schematic diagram of a soil-vapor extraction system is shown in Figure 7.25(*a*),

TABLE 7.25   Ex Situ Remediation Methods for Contaminated Soils

Technology	Description	Application
	Physical–Chemical Processes	
Soil washing	Excavated soils are washed with water that may be amended with surfactants, chelating agents, or leaching agents to remove sorbed contaminants; resultant liquids must be treated prior to disposal.	Semivolatile organics, solvents, pesticides, metals, inorganics
Soil-vapor extraction	Soils are piled onto perforated pipes that are used to apply a vacuum, withdrawing soil gases containing volatile contaminants; resultant gases must be treated prior to emission.	Volatile organics, solvents
Solidification –stabilization	Contaminants are physically or chemically bound or enclosed within a stabilized material to reduce their mobility.	Pesticides, metals, inorganics
Soil oxidation –reduction	Oxidizing agents such as ozone, hydrogen peroxide, and chlorine or reducing agents such as alkaline polyethylene glycolate are mixed with soils in reaction vessels that may be heated; results in contaminant destruction.	Semivolatile organics, solvents, pesticides, metals, inorganics
Solvent extraction	Excavated soils are mixed with solvents, allowing the contaminants to partition into the solvent phase; treated soils are then separated from the solvent; resulting solvent must be treated prior to recovery or disposal.	Semivolatile organics, solvents, pesticides
	Biological Processes	
Slurry-phase biological treatment	Excavated soils are mixed with water, nutrients, and microbial inocula in reactors or lined lagoons to promote contaminant biodegradation.	Volatile and semivolatile organics, solvents, pesticides
Composting	Soils are mixed with amendments and piled on perforated pipes that are used to apply a vacuum, withdrawing soil gases containing volatile contaminants and drawing air into the soil pores to promote biodegradation; amendments may include bulking materials, nutrients, and microbial inocula.	Volatile and semivolatile organics, solvents, pesticides
	Thermal Processes	
Thermal desorption	Soils are placed in reactors and heated to temperatures of 95 to 300°C (low T desorption) or 300 to 500°C (high T desorption) to volatilize water and contaminants and to promote limiting oxidation; gases are treated prior to release.	Volatile and semivolatile organics, solvents

(*continued*)

TABLE 7.25   (*continued*)

Technology	Description	Application
Incineration	Soils are heated to temperatures of 850 to 1200°C in reactors in the presence of oxygen to volatilize water and contaminants and to promote oxidation; gases are treated prior to release.	Volatile and semivolatile organics, solvents, pesticides
Pyrolysis	Soils are heated to temperatures of 400 to 750°C in reactors in the absence of oxygen to volatilize water and contaminants and to promote reductive decomposition; gases are treated prior to release.	Volatile and semivolatile organics, solvents, pesticides
Vitrification	Excavated soils are heated to extreme temperatures (1600 to 2000°C) to promote melting of the solids and contaminants into a crystalline structure.	Volatile and semivolatile organics, solvents, pesticides, metals, inorganics

Source: USEPA (1993).

and a typical aboveground component of a soil-vapor extraction system is shown in Figure 7.25(*b*). In spills of complex solvent mixtures such as gasoline, soil-vapor extraction removes the components with higher vapor pressures first. The advantages of using a soil-vapor extraction system are that it creates minimal disturbance of the contaminated soil, it can be constructed from standard equipment, it is cost-effective, and it is flexible in that several variables can be adjusted during design or operation. Extraction wells are typically designed to fully penetrate the vadose zone and to extend below the water table. Penetration of the water table allows for seasonal declines in the water-table elevation and declines caused by pumping. Extraction wells typically consist of an upper 1.5 m of solid plastic casing connected to slotted plastic pipe placed in permeable packing. The borehole is typically augured and sealed around the casing to the top of the borehole with cement grout

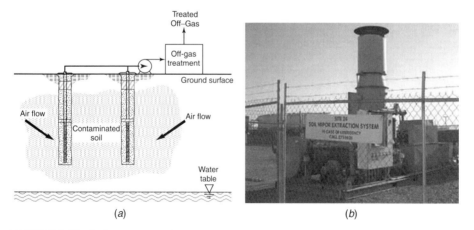

FIGURE 7.25   Soil-vapor extraction system. [(*a*) From U.S. Air Force, 2005; (*b*) from U.S. Navy, 2005.]

to prevent the direct inflow of air from the surface into the screened well intake. The typical radius of a soil-vapor extraction well ranges from 1.3 to 5 cm (Prakash, 2004). If the vadose zone is relatively thin, less than 3 m in depth, or if the contaminated soil is near the ground surface, extraction wells can be placed horizontally in vapor-extraction trenches which are excavated through the vadose zone to just above the seasonal high water table. Airflow in the soil can be enhanced at strategic locations by the installation of air-inlet or injection wells. Inlet wells allow air to be drawn into the ground at specific locations, whereas injection wells force air into the ground and can be used in closed-loop systems. Both inlet and injection wells are constructed similar to extraction wells. Vapor treatment of the extracted air may not be required if the emission rates of chemicals are low or if they are easily degraded in the atmosphere. Soil-vapor extraction is the leading method for cleaning up fuel spills and industrial solvents from unsaturated settings, and it provides one of the few alternatives to soil removal in the vadose zone where pure organic liquids often occur at residual saturation. Compounds that have been removed successfully by vapor extraction include trichloroethene, trichloroethane, tetrachloroethene, and most gasoline constituents; compounds that are less amenable to removal include trichlorobenzene, acetone, and heavier petroleum fluids. Typical treatment systems for extracted soil vapor include liquid–vapor condensation, granular activated carbon adsorption, catalytic or thermal oxidation, and biofilters.

Darcy's law can be assumed valid for the flow of air in coarse-grained soils composed of sands and gravels (Massmann, 1989). However, in applying Darcy's law to the flow of air, the density of air is not constant and the same definition of the hydraulic head as for incompressible fluids cannot be used. The more general *fluid potential*, Φ^*, must be used instead, where Φ^* is defined as (Hubbert, 1940)[3]

$$\Phi^* = gz + \int_{p_0}^{p} \frac{dp}{\rho} \tag{7.92}$$

where the density, ρ, is considered to be a function only of the fluid pressure, p, and p_0 is a reference pressure. Using Φ^*/g instead of the piezometric head $\phi\,(=p/\gamma + z)$ in the Darcy equation gives

$$\mathbf{q} = -\frac{k_a}{\mu}(\nabla p + \rho g \mathbf{k}) \tag{7.93}$$

where $\mathbf{q}$ is the bulk air-flow velocity, k_a is the intrinsic permeability of the porous matrix for airflow, and μ is the dynamic viscosity of air. Intrinsic permeabilities of porous media are typically given for cases in which the entire pore space is available for fluid flow, and in the case of airflow must generally be adjusted to account for soil moisture and NAPL residual saturation. The intrinsic permeability, k_a, for airflow can be derived from the intrinsic permeability, k, for ground water flow in saturated porous media using the relation (Stylianou and DeVantier, 1995)

$$k_a = k(1 - S_{NAPL} - S_{water})^3 \tag{7.94}$$

where S_{NAPL} is the residual saturation of the NAPL and S_{water} is soil-moisture saturation. For most airflows, the density, ρ, is small and the force associated with the pressure

---

[3]Hubbert (1940) included an additional kinetic energy term, $v^2/2$, where v is the seepage velocity. The kinetic energy term is relatively small and is usually neglected, leading to Equation 7.92.

gradient is much larger than the gravity force, hence Equation 7.93 can be closely approximated by

$$\mathbf{q} = -\frac{k_a}{\mu}\nabla p \qquad (7.95)$$

and Equation 7.95 is exact for horizontal airflows. The continuity equation for air in porous media can be written as

$$\theta_a \frac{\partial \rho}{\partial t} + \nabla \cdot (\rho \mathbf{q}) = 0 \qquad (7.96)$$

where θ_a is the volumetric air content and sources and sinks of air have been neglected. Assuming that the transport process is isothermal, the ideal gas law holds and

$$\rho = \frac{\rho_0}{p_0} p \qquad (7.97)$$

where p_0 and ρ_0 are the pressure and density at some reference state. Combining Equations 7.95 to 7.97 gives

$$\theta_a \frac{\partial p}{\partial t} + \nabla \cdot \left(p \frac{k_a}{\mu} \nabla p \right) = 0 \qquad (7.98)$$

Equation 7.98 is nonlinear in p and difficult to solve. However, in applications associated with soil-vapor extraction systems, Equation 7.98 can be linearized by assuming that the pressure, p, is equal to atmospheric pressure, p_{atm}, plus a small perturbation, p^*, where

$$p = p_{atm} + p^* \qquad (7.99)$$

and $p^* \ll p_{atm}$. Under these conditions, the viscosity of a vapor is also nearly constant and Equations 7.98 and 7.99 combine to give

$$\frac{\theta_a \mu}{k_a p_{atm}} \frac{\partial p^*}{\partial t} = \nabla^2 p^* \qquad (7.100)$$

where the intrinsic permeability, k_a, for airflow has been assumed constant. In the common case where air is extracted from a well, the pressure distribution surrounding the well is radially symmetric, and Equation 7.100 can conveniently be written in radial coordinates as

$$\frac{\theta_a \mu}{k_a p_{atm}} \frac{\partial p^*}{\partial t} = \frac{1}{r} \frac{\partial}{\partial r} \left(r \frac{\partial p^*}{\partial r} \right) \qquad (7.101)$$

This equation is to be solved subject to the following initial and boundary conditions:

$$p^*(r, 0) = 0 \qquad (7.102)$$

$$p^*(\infty, t) = 0 \qquad (7.103)$$

$$\lim_{r \to 0} r \frac{\partial p^*}{\partial r} = -\frac{Q}{2\pi(k_a/\mu)b} \qquad (7.104)$$

where b is the thickness of the airflow zone and Q is air pumpage rate. The solution of Equation 7.101 subject to Equations 7.102 to 7.104 is (Johnson et al., 1989, 1990)

$$p^* = \frac{Q}{4\pi b(k_a/\mu)}W(u) \qquad (7.105)$$

where $W(u)$ is the well function, and

$$u = \frac{r^2\theta_a\mu}{4k_a p_{atm}t} \qquad (7.106)$$

Typically, for sandy soils the pressure distribution approximates a steady state in 1 to 7 days (Bedient et al., 1999), and the steady-state pressure distribution is given by (Johnson et al., 1989)

$$p(r)^2 - p_w^2 = (p_{atm}^2 - p_w^2)\frac{\ln(r/r_w)}{\ln(R/r_w)} \qquad (7.107)$$

where p_w is the pressure at the well with radius r_w, and $p = p_{atm}$ at the radius of influence R. Using the pressure distribution given by Equation 7.107, Darcy's law can be applied to give the airflow, Q, extracted by the well as

$$Q = L\pi\frac{k_a}{\mu}p_w\frac{(p_{atm}/p_w)^2 - 1}{\ln(R/r_w)} \qquad (7.108)$$

where L is the length of the well intake (screen) and the pressures are absolute pressures. The pressure, p_w, at the well, is generally determined by the blower characteristics, and based on p_w, Equation 7.108 gives the corresponding airflow. Typically, p_w is 5 to 10 kPa below atmospheric pressure, and the radius of influence, R, is usually estimated in the field from a plot of p versus r. Typical values of R are in the range 10 to 30 m (Charbeneau, 2000), depending on soil conditions, being smaller for sandy soils and larger for silty and clayey formations (Prakash, 2004). Fortunately, Equation 7.108 is not very sensitive to R and, if no data are available, a value of 12 m can be used without a significant loss of accuracy (Johnson et al., 1990). Typically, if the intrinsic air permeability, k_a, is less than 1 darcy (10^{-12} m²), flow rates may be too low to achieve successful remediation in a reasonable time frame (Cohen and Mercer, 1993). Air permeability tests are utilized in predesign studies. In air-permeability tests, air is removed from an extraction well, measurements are made of the subsurface pressure distribution, and the measured distribution compared with Equation 7.105 to determine the air permeability, k_a. This approach is almost identical to the procedure used with the Theis equation to determine the transmissivity and storage coefficient in the saturated zone (Chin, 2006).

Equation 7.108 assumes that the airflow is steady and horizontal. However, the source of air is usually directly from the atmosphere at the ground surface, so vertical flow components might be significant. Since the ideal conditions associated with analytical solutions seldom exist in reality, analytical relations such as Equation 7.108 are most useful for screening purposes and for exploring the relationships between variables, and their practical applicability is limited to simple problems (Bedient et al., 1999). For more complex problems, numerical models are generally required (e.g., Rathfelder et al., 1991; Poulsen et al., 1996). Removal rates, $\dot{M}$, for contaminants with concentration c in the vapor phase can be approximated by

$$\dot{M} = fQc \tag{7.109}$$

where f is the fraction of pumped air that flows through contaminated soil. The contaminant concentration, c, in the pumped air can be estimated using a combination of Raoult's law, Dalton's law of partial pressures, and the ideal gas law, which give

$$c = \sum_i \frac{x_i p_i m_i}{RT} \tag{7.110}$$

where x_i is the mole fraction of component i in the liquid-phase residual, p_i is the pure-component vapor pressure at temperature T, m_i is the molar mass of component i, R is the universal gas constant, and T is the absolute temperature. Removal rates, $\dot{M}$, greater than 1 kg/day are usually required for soil-vapor extraction systems to be considered effective (Charbeneau, 2000). The minimum number of extraction wells required in a soil-vapor extraction system, N_{wells}, can be estimated by

$$N_{\text{wells}} = \frac{M}{fQct} \tag{7.111}$$

where M is the mass of contaminant spilled and t is the desired cleanup time. Uncertainty in the distribution of the intrinsic permeability can lead to substantial uncertainty in the number and location of extraction wells required for a given cleanup time, as well as uncertainty in the cleanup time corresponding to a given arrangement of extraction wells (Massmann et al., 2000). However, the flexibility of soil-vapor extraction systems allows for several adjustments during operation, including the addition or removal of wells from the system and the adjustment and redistribution of air withdrawal rates. For soil-vapor extraction to be an effective remediation strategy, the bulk concentration of hydrocarbons must typically exceed 500 mg/kg (Charbeneau, 2000), in which case the hydrocarbons will probably be present as a non-aqueous phase.

Soil-vapor extraction systems are widely used for remediating soils contaminated with volatile and semivolatile organic compounds. This popularity is due, in part, to the low cost of soil-vapor extraction relative to other available technologies, especially when contamination occurs relatively deep below the ground surface (Massmann et al., 2000).

Example 7.21 A NAPL containing 30% benzene, 60% toluene, and 10% o-xylene is to be cleaned up using soil-vapor extraction. Soil samples indicate that a mass of 10^6 kg has leaked from a large storage tank on an industrial site. The extraction wells installed at the site are to have a diameter of 150 mm, intake (screen) lengths of 5 m, and the pressure at the intake is to be held at 10 kPa below atmospheric pressure. The air temperature in the soil is measured as 20°C, the intrinsic permeability of the soil for airflow is estimated as

TABLE 7.26 Data for Example 7.21

Component	Mass Fraction	Molar Mass, m_i (g)	Mole Fraction, x_i	Saturation Vapor Pressure, p_i (kPa)
Benzene	0.30	78.1	0.34	6.9
Toluene	0.60	92.1	0.58	3.7
o-Xylene	0.10	106.2	0.08	0.9

150 darcys, and the radius of influence of each extraction well is derived from field measurements as 25 m. If 30% of the air extracted from each well passes through contaminated soil, estimate the number of wells required to clean up the spill in one year.

SOLUTION   The NAPL spilled is a mixture of benzene, toluene, and o-xylene with the properties shown in Table 7.26 (see Appendix B.2). From the data given, $T = 273.15 + 20 = 293.15$ K, and $R = 8.31$ J/K·mol. The concentration, c, of contaminant in air flowing through the contaminated soil is given by Equation 7.110 as

$$c = \sum_{i=1}^{3} \frac{x_i p_i m_i}{RT}$$

$$= \frac{1}{8.31(293.15)} [0.34(6900)(78.1) + 0.58(3700)(92.1) + 0.08(900)(106.2)]$$

$$= 159 \text{ g/m}^3$$

The flow rate, Q, of air into each extraction well is given by Equation 7.108. If the pressure at each well is maintained at 10 kPa below atmospheric pressure, and atmospheric pressure, p_{atm}, is taken as 101 kPa, $p_w = 101 - 10 = 91$ kPa. From the data given, $L = 5$ m, $k_a = 150$ darcys $= 1.48 \times 10^{-10}$ m^2, $\mu = 0.0182$ mPa·s $= 1.82 \times 10^{-5}$ Pa·s (Appendix B.3, air at 20°C), $r_w = 150/2 = 75$ mm $= 0.075$ m, and $R_0 = 25$ m. Substituting these parameters into Equation 7.108 gives

$$Q = L\pi \frac{k_a}{\mu} p_w \frac{(p_{atm}/p_w)^2 - 1}{\ln(R_0/r_w)}$$

$$= 5\pi \frac{1.48 \times 10^{-10}}{1.82 \times 10^{-5}} (91 \times 10^3) \frac{(101/91)^2 - 1}{\ln(25/0.075)}$$

$$= 0.464 \text{ m}^3/\text{s}$$

This analysis assumes that the intrinsic permeability remains constant at 1.48×10^{-10} m^2, while in reality the intrinsic permeability will increase somewhat due to the removal of NAPL. Since 30% of the extracted air flows through contaminated soil, $f = 0.30$, and Equation 7.109 gives the rate at which contaminant mass is withdrawn from each well, $\dot{M}$, as

$$\dot{M} = fQc = 0.30(0.464)(159) = 22.1 \text{ g/s} = 1912 \text{ kg/day}$$

For a 10^6-kg spill, the number of wells required to clean up the soil in one year ($=365$ days) is given by Equation 7.111 as

$$N_{\text{wells}} = \frac{M}{fQct}$$

$$= \frac{10^6}{1912(365)} = 1.43 \text{ wells}$$

Therefore, two extraction wells should clean up the spill in 1 year. This assumes that the contaminated soil is contained within the radii of influence of the two wells.

Bioventing In bioventing systems air is injected into the vadose zone with the intent to maximize in situ biodegradation of contaminants while minimizing vapor emissions. A schematic diagram of a bioventing system is shown in Figure 7.26. Bioventing systems are the reverse of soil-venting systems and have the advantage that aboveground treatment systems are not necessary. However, in bioventing systems the primary treatment process is biodegradation, compared with volatization in soil-venting processes. Soil-vapor extraction systems can do double duty as bioventing systems. Most aliphatic and monoaromatic petroleum hydrocarbons are aerobically biodegradable, with the ease of degradation decreasing with increasing molecular weight, number of aromatic rings, and increasing branched structure (Johnson, 1999). Many chlorinated solvents, such as TCE and PCE, are not readily aerobically biodegradable. The advantage of bioventing compared to enhanced bioremediation is that oxygen can be transported more easily in air (280 mg/L O_2 in air) than in water (10 mg/L O_2 in water).

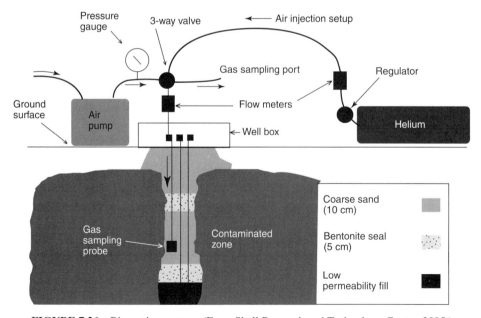

FIGURE 7.26 Bioventing system. (From Shell Research and Technology Center, 2005.)

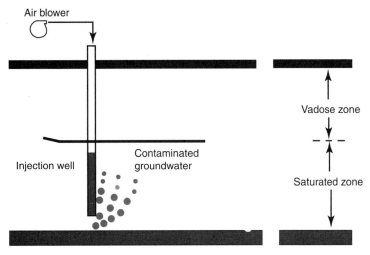

FIGURE 7.27   Air-sparging system. (From Federal Remediation Technologies Roundtable, 2005b.)

Air Sparging   *Air sparging* is a commonly used remediation technology that targets volatile organic compounds within the saturated zone. Air sparging is sometimes referred to as *in situ air stripping* or *in situ volatization* (Liban, 2000). In this process, contaminant-free air is injected below the contaminant plume through *sparge wells* or well points to volatize dissolved and adsorbed-phase VOCs and to deliver oxygen to stimulate aerobic degradation. A schematic diagram of an air-sparging system is shown in Figure 7.27. Air sparging is more effective in treating dissolved hydrocarbon plumes than for treating source areas (Bass and Brown, 1996) and is usually a feasible alternative to pump-and-treat systems. Air-sparging systems are most often combined with soil-vapor extraction (SVE) systems that remove vapors entering the vadose zone. Air sparging is also used to improve the air flow distribution near the capillary zone for bioventing purposes, to deliver hydrocarbon vapors (e.g., methane, ethane, propane) to promote the cometabolic degradation of chlorinated compounds, and to establish large circulation cells to move contaminated water to extraction wells. The primary advantages offered by air sparging are the elimination of surface water treatment equipment and disposal and the acceleration of remediation of sorbed contaminants in the capillary fringe.

A typical system has one or more subsurface points through which air is injected, the amount of air injected is usually small, and well points less than 2.5 cm in diameter are usually sufficient. Well points can be driven into many aquifers using flush-jointed steel pipe, and the most common configurations use either a vertical or a horizontal well as a sparge point. Whether remediation is sought by volatization of dissolved contaminants or through enhanced biodegradation, the volume and shape of the region of airflow are fundamental determinants of the effectiveness and efficiency of the air sparging system (Zumwalt et al., 1997). Experiments by Ji et al. (1993) using glass beads as the porous medium demonstrated that the type of airflow regime depends primarily on grain size. For grain diameters of about 4 mm and larger, bubbly flow was observed. For grain sizes of 0.75 mm or less, channel flow dominates. The transition between these two regimes occurred for grain sizes between 0.75 and 4 mm. The study by Ji et al. (1993) also showed that the channel flow regime is most likely to occur for air sparging under natural subsurface conditions. The air occupying individual air channels is continuous, and in no sense does airflow occur as a sequence of rising bubbles. It is almost impossible to predict the flow path that air channels will take between the injection point and the vadose zone. A symmetric air distribution pattern will almost never occur in a natural porous medium because small-scale heterogeneity is present even in media that appear, at large scale, to be homogeneous.

A conceptual model of injected airflow in the air-sparging process has been provided by Johnson et al. (1993). In this conceptual model, when air is injected into a well, standing water in the well bore is displaced downward and through the well screen until the air–water interface reaches the top of the well screen. The minimum air pressure (p_m) corresponding to the water column height that is displaced is given by

$$p_m = \rho_w g (L_s - L_{gw}) \tag{7.112}$$

where ρ_w is the density of water, g is the acceleration due to gravity, L_s is the depth to the top of the well screen, and L_{gw} is the depth to the water table. For injected air to penetrate the aquifer, air pressure in excess of p_m is required. This excess air pressure is commonly known as the *air entry pressure* for the formation, and it is the minimum capillary pressure required to induce airflow into a saturated porous medium. Air entry pressures can range from a few centimeters in coarse sandy soils to several meters in low-permeability clayey soils. If special diffuser screens are used to enhance air distribution, the minimum bubbling pressure for the diffuser must be overcome for air to enter the formation.

Air sparging depends on two basic processes for contaminant removal: volatization and aerobic biodegradation (Johnson et al., 1993). Where the NAPL is in contact with an air channel, contaminants will volatilize by direct evaporation from the NAPL surface. The greater contaminant mass will likely be located beyond the air channels in water-saturated zones. Thus, removal of this mass will depend on diffusive transport to the air–water interface, which is inherently a slow process. This analysis leads to the conclusion that the effectiveness of air sparging could be limited, unless the airflow also induces some degree of mixing within the water-saturated zone (Liban, 2000). Mixing induced by pulsed airflow has been shown to increase significantly the removal efficiency of VOCs induced by sparge wells (Payne et al., 1995). The reduction of airflow rates in air sparging systems to enhance bioremediation is called *biosparging*, in which case the destruction of VOCs by microbial action is greatly enhanced by the increased oxygen concentration in the saturated and vadose zone that results from sparging. In practice, many of the designs of air sparging systems are empirical and generally based on the experience of the design engineer.

Example 7.22 An air sparging system is to be used to remediate contaminated ground water. The existing contaminant plume contains PCE at 70 mg/L, and the plume extends from the water table to a depth of 2 m below the water table. Sparge wells are to be placed such that the top of the screen at each sparge well is 2.5 m below the water table. The temperature of the ground water is 20°C. (a) What is the minimum air pressure required to inject air into the sparge wells? (b) Estimate the concentration of PCE in the injected air reaching the water table. (c) If air is injected at a rate of 10 L/min, estimate the time required to remove 1 kg of PCE from the ground water.

SOLUTION   (a) From the data given, $c = 70\,\text{mg/L} = 0.07\,\text{kg/m}^3$, and $d_a = 2.5\,\text{m}$. At 20°C, $\rho_w = 998.2\,\text{kg/m}^3$. Equation 7.112 gives the air entry pressure, p_m, as

$$p_m = \rho_w g d_a = 998.2(9.81)(2.5) = 2.44 \times 10^4\,\text{Pa} = 24.4\,\text{kPa}$$

(b) According to Henry's law,

$$p_e = H c_e$$

where p_e is the equilibrium partial pressure of PCE in the gas phase (Pa), H is Henry's law constant for PCE (Pa·m³/mol), and c_e is the concentration of PCE in the ground water (mol/m³). The molar mass of PCE is $165.8\,\text{g/mol} = 0.1658\,\text{kg/mol}$, so

$$c_e = \frac{0.07}{0.1658} = 0.422\,\text{mol/m}^3$$

and Appendix B.2 gives $H = 1545\,\text{Pa·m}^3/\text{mol}$. Therefore, in the rising air,

$$p_e = H c_e = 1545(0.422) = 652\,\text{Pa} = 0.652\,\text{kPa}$$

The concentration of PCE in the sparge air reaching the water table can be estimated using the combination of Dalton's law and the ideal gas law, which gives

$$c = \frac{p_e m}{RT} = \frac{652(0.1658)}{8.31(293)} = 0.0444\,\text{kg/m}^3$$

where $R = 8.31\,\text{J/K·mol}$ and $T = 293\,\text{K}$.
   (c) Since air is injected at $Q = 10\,\text{L/min} = 1.67 \times 10^{-4}\,\text{m}^3/\text{s}$,

$$\text{removal rate of PCE from groundwater} = Qc = (1.67 \times 10^{-4})(0.0444)$$
$$= 7.4 \times 10^{-6}\,\text{kg/s}$$

and

$$\text{time required to remove 1 kg} = \frac{1}{7.4 \times 10^{-6}} = 1.35 \times 10^5\,\text{s} = 38\,\text{h}$$

Therefore, a single sparge well is capable of removing 1 kg of PCE from the ground water every 38 h.

Air-Sparging Cutoff Trenches   Air-sparging cutoff trenches remove volatile organic compounds from ground water using an air-stripping process within the trench, and a schematic diagram of a typical air-sparging cutoff trench is shown in Figure 7.28. Assuming that the water within the trench is well mixed and in equilibrium with the sparge air leaving the system, the relation between the contaminant concentration, c_0, entering the trench and the contaminant concentration, c, leaving the trench is given by (Pankow et al., 1993)

$$\boxed{\frac{c}{c_0} = \frac{1}{1 + S}} \qquad (7.113)$$

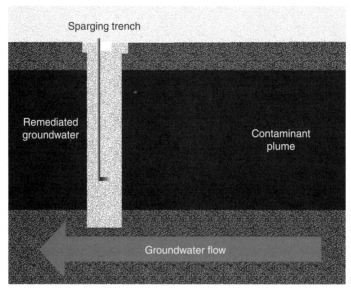

FIGURE 7.28   Air-sparging trench. (From Hydro Geo Chem, 1997.)

where

$$S = \frac{K_H Q_{in}}{Q_{air}}$$ (7.114)

and K_H is the dimensionless Henry's law constant [(mg/L air)/(mg/L water)], Q_{in} is the volumetric inflow rate of ground water into the trench, and Q_{air} is the volumetric inflow rate of air.

Example 7.23   An air-sparging cutoff trench is to be constructed to remove TCE from ground water. The specific discharge of the ground water is 0.72 m/day, the trench is 35 m long and penetrates 4 m into the saturated zone of the aquifer. Henry's law constant for TCE is 0.35 (mg/L air)/(mg/L water), the concentration of TCE in the ground water entering the trench is 100 μg/L, and the required TCE concentration leaving the trench is 5 μg/L. Estimate the required air injection rate.

SOLUTION   From the data given, $L = 35$ m, $H = 4$ m, $q = 0.72$ m/day, $c_0 = 100\,\mu$g/L, and $c = 5\,\mu$g/L. The volumetric flow rate of ground water into the trench, Q_{in}, is given by

$$Q_{in} = qLH = 0.72(35)(4) = 100.8 \text{ m}^3/\text{day}$$

Since $K_H = 0.35$ (mg/L air)/(mg/L water), Equation 7.114 gives

$$S = \frac{K_H Q_{in}}{Q_{air}} = \frac{0.35(100.8)}{Q_{air}} = \frac{35.3}{Q_{air}}$$ (7.115)

and Equation 7.113 gives

$$\frac{c}{c_0} = \frac{1}{1+S}$$

$$\frac{5}{100} = \frac{1}{1+S}$$

which yields

$$S = 19 \tag{7.116}$$

Combining Equations 7.115 and 7.116 gives

$$19 = \frac{35.3}{Q_{\text{air}}}$$

which yields

$$Q_{\text{air}} = 1.86 \, \text{m}^3/\text{day}$$

Therefore, air must be injected into the trench at $1.86\,\text{m}^3/\text{day}$ to achieve the required removal of TCE from the ground water.

Pump-and-Treat Systems   Pump-and-treat is the most commonly used remediation technology for saturated-zone ground-water cleanup in the United States (Illangasekare and Reible, 2000), and a typical pump-and-treat system is illustrated in Figure 7.29. Contaminated water can be pumped from wells, drains, buried perforated pipe, and/or trenches excavated and backfilled with gravel. On-site treatment is usually required before water can be reinjected to the aquifer, discharged to a receiving water body, used for irrigation purposes, or released to surface sanitary systems. The majority of inorganic contaminants are metals that can be removed by precipitation (through lime addition or aeration), volatile organic contaminants can be removed by air stripping, less-volatile organics can be removed by sorption onto activated carbon or biological treatment, and special methods such as ultraviolet light and electron beams can be used for the most refractory organic compounds. A typical aboveground view of a pump-and-treat system is shown in Figure 7.30. The pump, piping, and treatment systems are shown in Figure 7.30(*a*) and the discharge of treated water to a nearby lagoon is shown in Figure 7.30(*b*).

Pumping wells that extract contaminated ground water for treatment are generally located down-gradient of contaminant plumes. Pump-and-treat systems extract the ground water within the *capture zone* of the pumping well. In many cases, the capture zone is described in a time context, such as the 1- or 5-year capture zone, where the water within the volume enclosed by the time-related capture zone will be extracted by the well over that specified time. In most cases, the extent of the capture zone will ultimately stabilize, and this asymptotic capture zone is frequently used in the design of pump-and-treat systems. For a single well located in an aquifer with a regional average specific discharge, q_0, the boundary of the asymptotic capture zone is illustrated in Figure 7.31 and is given by

$$\boxed{y + \frac{Q_w}{2\pi H q_0} \tan^{-1} \frac{y}{x} = \pm \frac{Q_w}{2H q_0}} \tag{7.117}$$

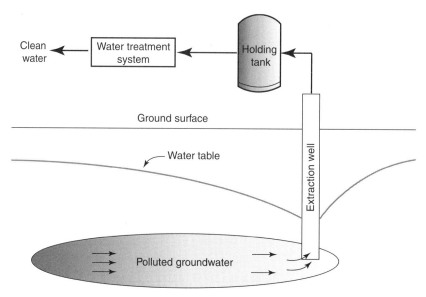

FIGURE 7.29 Pump-and-treat system. (From USEPA, 2001.)

where Q_w is the withdrawal rate at the well and H is the saturated thickness of the aquifer. Notable features of the asymptotic capture zone are that it has a maximum width of Q_w/Hq_0 upstream of the pumping well, and a stagnation point is located downstream of the pumping well, where the seepage velocity induced by the well is equal to the regional seepage velocity. Down-gradient of the stagnation point, none of the ambient ground water is captured by the pumping well. The width of the asymptotic capture zone upstream of the pumping well increases with increasing withdrawal rate, and for pump-and-treat systems to be effective the capture zone must encompass the entire contaminant plume. In many

(*a*)   (*b*)

FIGURE 7.30 (*a*) Pump-and-treat system; (*b*) disposal. (From USEPA, 2005j.)

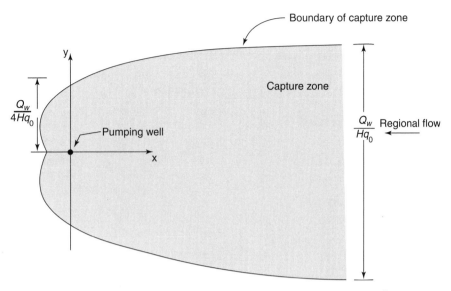

FIGURE 7.31   Asymptotic capture zone for a single well in regional flow.

cases, the pumping rate from a single well is limited by the allowable drawdown in the aquifer, and hence the size of the capture zone is also limited. Therefore, whenever the size of the contaminant plume exceeds the maximum size of a single-well capture zone, more than one pumping well must be used. In using two pumping wells, it is necessary that the capture zones of the extraction wells overlap, otherwise, a portion of the contaminant plume will pass between the two wells. It can be shown that if the distance between the extraction wells is less than or equal to $Q_w/\pi H q_0$, the capture zones will overlap (Fetter, 1999). It is frequently considered optimal to space the pumping wells such as to have the largest capture zone for a given number of wells, and in this sense $Q_w/\pi H q_0$ is considered the optimal spacing of two wells. In general, if two wells are placed a distance $2d$ apart, as illustrated in Figure 7.32, the boundary of the capture zone is defined by

$$y + \frac{Q_w}{2\pi H q_0}\left(\tan^{-1}\frac{y-d}{x} + \tan^{-1}\frac{y+d}{x}\right) = \pm\frac{Q_w}{H q_0} \tag{7.118}$$

where the maximum width of the capture zone is $2Q_w/Hq_0$. If three wells are used in a pump-and-treat system, and these wells are located at distance d apart as shown in Figure 7.33, the optimal well spacing is $1.26Q_w/\pi H q_0$, and the capture-zone boundary is defined by

$$y + \frac{Q_w}{2\pi H q_0}\left(\tan^{-1}\frac{y}{x} + \tan^{-1}\frac{y-d}{x} - \tan^{-1}\frac{y+d}{x}\right) = \pm\frac{3Q_w}{2H q_0} \tag{7.119}$$

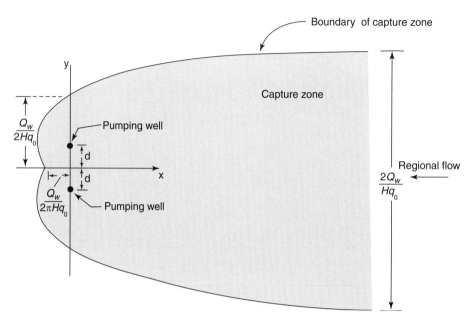

FIGURE 7.32 Asymptotic capture zone for two wells in regional flow.

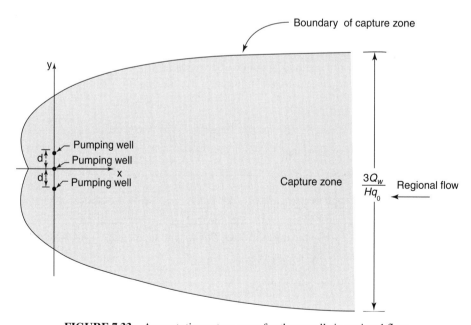

FIGURE 7.33 Asymptotic capture zone for three wells in regional flow.

TABLE 7.27   Date for Example 7.24

Number of Wells	Maximum Width of Capture Zone (m)	
1	$\dfrac{Q_w}{Hq_0}$	24
2	$\dfrac{2Q_w}{Hq_0}$	48
3	$\dfrac{3Q_w}{Hq_0}$	72

where the maximum width of the capture zone is $3Q_w/Hq_0$. In using three pumping wells, it is again necessary that the capture zones of the extraction wells overlap so that a portion of the contaminant plume does not pass between the three wells. It can be shown that if the distance between the extraction wells is less than or equal to $1.26Q_w/\pi Hq_0$ the capture zones will overlap (Domenico and Schwartz, 1998).

Example 7.24   A contaminant plume of radius 17 m is discovered at the center of an industrial site, and a pump-and-treat system is to be designed to remove the contaminant dissolved in the ground water. The aquifer has a saturated thickness of 30 m and a regional specific discharge of 1 m/day. Space limitations confine the placement of recovery wells to a maximum of 400 m downstream of the center of the plume, and drawdown limitations permit a maximum pumping rate of 500 L/min from each well. Determine the number, location, and discharge rate of the pumping wells that could be used in the pump-and-treat system.

SOLUTION   From the data given, $Q_w = 500$ L/min $= 720$ m^3/day, $q_0 = 1$ m/day, $H = 30$ m, and the maximum widths of the capture zones corresponding to one, two, and three pumping wells are given in Table 7.27. Since the width of the contaminant plume is 2×17 m $= 34$ m, a minimum of two pumping wells must be used. The next step is to see whether two wells placed 400 m downstream of the center of the plume will have a capture zone wide enough to encompass the plume. The maximum spacing of two wells for a contiguous capture zone is $2d$, where

$$2d = \frac{Q_w}{\pi Hq_0} = \frac{Q_w}{\pi(30)(1)} = 0.01061Q_w$$

and hence $d = 0.01061Q_w/2 = 0.005305Q_w$. If the plume is contained within the capture zone of a two-well system, the point $x = 400$ m, $y = 17$ m lies on the boundary of the capture zone, and Equation 7.118 gives

$$y + \frac{Q_w}{2\pi Hq_0}\left(\tan^{-1}\frac{y-d}{x} + \tan^{-1}\frac{y+d}{x}\right) = \pm\frac{Q_w}{Hq_0}$$

$$17 + \frac{Q_w}{2\pi(30)(1)}\left(\tan^{-1}\frac{17-0.005305Q_w}{400} + \tan^{-1}\frac{17+0.005305Q_w}{400}\right) = \pm\frac{Q_w}{30(1)}$$

which yields

$$Q_w = 517 \text{ m}^3/\text{day} = 359 \text{ L/min}$$

with a well spacing of

$$2d = 0.01061 Q_w = 0.01061(517) = 5.49 \, \text{m}$$

Therefore, two wells spaced 5.49 m apart, located symmetrically about a line extending 400 m downstream of the center of the plume, and pumping at least 359 L/min could be used to capture the contaminant plume.

Treated ground water is frequently pumped back into the aquifer using injection wells. This practice usually requires a reglatory permit, and there are typically water-quality standards that must be met by the injected water. The tendency of injection wells to become clogged and the need for periodic maintenance must also be considered (Fetter, 1994). Injection wells in combination with pumping wells can significantly decrease the cleanup time by creating steeper hydraulic gradients and hence ground water flowing at a faster rate to pumping wells. Several arrangements of injection and pumping wells were investigated by Satkin and Bedient (1988), who concluded that the best arrangement of wells is highly site specific. In cases where the hydraulic gradient is low, the *doublet* arrangement is particularly effective. A doublet consists of a pumping well and an injection well operating as part of a pump-and-treat system. Doublets form a circulation cell within the aquifer and are frequently used as a means of injecting and removing aquifer mitigation solutes, such as cosolvents, surfactants, or both (Kemblowski and Urroz, 1999). Typically, the recharge well is positioned directly up-gradient from the discharge well, and the magnitudes of the pumping and injection rates are the same. This arrangement isolates a circulation cell within the ground water and is sometimes referred to as *hydrodynamic isolation*. The upstream and downstream boundaries of the circulation cell are defined by the points $(-x_s, 0)$ and $(x_s, 0)$, respectively, where the x-axis is parallel to the direction of the regional ground water flow, and the origin of the coordinate system is located at the midpoint between the two wells. The stagnation points $(-x_s, 0)$ and $(x_s, 0)$ are the roots of the equation

$$\frac{q_0 H d}{Q_w} + \frac{1}{2\pi} \left[\frac{x/d}{(x/d + 1)^2} - \frac{x/d}{(x/d - 1)^2} \right] = 0 \tag{7.120}$$

where q_0 is the specific discharge of the regional flow, $2d$ is the distance between the wells, and Q_w is the pumping–injection rate. The boundary of the circulation cell is defined by

$$\frac{q_0 H y}{Q_w} + \frac{1}{2\pi} \left(\tan^{-1} \frac{y/d}{x/d + 1} - \tan^{-1} \frac{y/d}{x/d - 1} \right) = \frac{1}{2} \tag{7.121}$$

The circulation cell is symmetric with respect to the y-axis, and the circulation cell is delineated by first estimating the stagnation points, varying x between zero and x_s, and using Equation 7.121 to solve for y.

Example 7.25 A well doublet is to be used as part of a pump-and-treat system to remediate contaminated ground water. The contaminant plume is approximately 80 m long and 25 m wide. The regional hydraulic gradient is 0.25%, the average hydraulic conductivity of the aquifer is 8 m/day, and the saturated thickness of the aquifer is 30 m. If the on-site treatment system is capable of treating the pumped water at 60 L/min, what is the required spacing of the wells so that the circulation cell within the aquifer encompasses the entire plume?

SOLUTION   From the data given, $L = 80$ m, $W = 25$ m, $J = 0.25\% = 0.0025$, $K = 8$ m/day, $H = 30$ m, and $Q_w = 60$ L/min $= 86.4$ m³/day. The regional-flow specific discharge, q_0, is given by the Darcy equation as

$$q_0 = KJ = 8(0.0025) = 0.02 \, \text{m/day}$$

Requiring the upstream and downstream stagnation points $(-x_s, 0)$ and $(x_s, 0)$ to coincide with the upstream and downstream edges of the contaminant plume requires that

$$2x_s = 80 \, \text{m}$$

which yields

$$x_s = 40 \, \text{m}$$

Substituting given data into Equation 7.120 gives

$$\frac{q_0 H d}{Q_w} + \frac{1}{2\pi}\left[\frac{x/d}{(x/d+1)^2} - \frac{x/d}{(x/d-1)^2}\right] = 0$$

$$\frac{0.02(30)d}{86.4} + \frac{1}{2\pi}\left[\frac{40/d}{(40/d+1)^2} - \frac{40/d}{(40/d-1)^2}\right] = 0$$

which can be written as

$$\frac{0.278}{\chi} + 0.159\chi\left[\frac{1}{(\chi+1)^2} - \frac{1}{(\chi-1)^2}\right] = 0 \tag{7.122}$$

where

$$\chi = \frac{40}{d}$$

The only feasible solution of Equation 7.122 is

$$\chi = 2.93$$

which gives

$$d = 13.7 \, \text{m}$$

Therefore, for the upstream and downstream stagnation points in the doublet to coincide with the upstream and downstream edges of the contaminant plume, the pumping and injection wells must be $2(13.7) = 27.4$ m apart and placed symmetrically within the contaminant plume.

It is necessary to verify that the circulation cell encompasses the entire width of the contaminant plume. The maximum width of the circulation cell, $2y$, occurs when $x = 0$ and Equation 7.121 gives

$$\frac{q_0 H y}{Q_w} + \frac{1}{2\pi}\left(\tan^{-1}\frac{y/d}{x/d+1} - \tan^{-1}\frac{y/d}{x/d-1}\right) = \frac{1}{2}$$

$$\frac{0.02(30)y}{86.4} + \frac{1}{2\pi}\left(\tan^{-1}\frac{y/13.7}{0/13.7+1} - \tan^{-1}\frac{y/13.7}{0/13.7-1}\right) = \frac{1}{2}$$

which can be written as

$$0.0951\chi + 0.159[\tan^{-1}(\chi) - \tan^{-1}(-\chi)] = \frac{1}{2} \tag{7.123}$$

where

$$\chi = \frac{y}{13.7}$$

The only feasible solution of Equation 7.123 is $\chi = 1.75$ which gives $y = 24.0\,\text{m}$. The width of the circulation cell is $2(24.0) = 48.0\,\text{m}$, which is greater than the plume width of 25 m. Therefore, the plume is entirely contained within the doublet when the injection and recovery wells are spaced 27.4 m apart.

Many pump-and-treat systems have performed poorly in practice, even though their design was based on sound principles. Reasons commonly cited for the poor performance of pump-and-treat systems are the presence of hydrogeologic controls, such as zones of low hydraulic conductivity, and the presence of residual NAPLs in the water-bearing units. Contaminant-removal rates are usually high in the initial stage of operation; however, this is usually followed by a second stage in which contaminant removal rates are limited by the diffusion rate of the contaminant trapped in zones of low hydraulic conductivity. This second-stage process can cause the contaminant concentrations in the pumped water to remain marginally contaminated and approximately constant for years or even decades, possibly resulting in a failure of the pump-and-treat system to meet the target levels for ground-water remediation within a reasonable time period. In fact, contaminant levels can readily return to higher values if pumping is stopped. Pump-and-treat systems for dissolved contaminants work best when the rate of cleanup is controlled by advection. If rate-limiting processes become controlling, such as DNAPL dissolution rates or diffusion rates, pump-and-treat systems will not be efficient at mass removal. In these cases, agents to enhance the remediation such as surfactants and nutrients for microbial activity can be added to water injected into the aquifer; however, such materials should be chosen so as not to cause other types of contamination. In many cases, the cleanup of aquifers using the pump-and-treat strategy to remediate ground water to drinking water standards is simply not practical. Consequently, containment of off-site migration and risk have become the primary objectives at many pump-and-treat sites (Illangasekare and Reible, 2000).

Bioremediation Bioremediation is usually carried out in situ and involves stimulating the indigenous subsurface microorganisms by the addition of both nutrients and an electron

FIGURE 7.34   Bioremediation system. (From Pacific Northwest National Laboratory, 2005.)

acceptor. A typical bioremediation system is illustrated in Figure 7.34, where upstream injector wells add nutrients and/or oxygen to ground water flowing toward the contaminated zone, and a downstream extractor well withdraws water that has passed through the contaminated zone. Typical nutrients are nitrogen, potassium, and phosphorus, and typical electron acceptors are oxygen, nitrate, sulfate, and carbon dioxide. Oxygen is the most popular electron acceptor in aerobic biodegradation and is usually added to the ground water by sparging air, pure oxygen, hydrogen peroxide, or ozone. Care should be taken in using hydrogen peroxide, which is toxic to some microbes. In many instances the shallow subsurface has sufficient nutrient material so that the only ingredient lacking for successful bioremediation is oxygen. A wide variety of contaminants are amenable to bioremediation, including gasoline hydrocarbons, jet fuels, oils, aromatics, phenols, creosote, chlorinated phenols, nitrotoluenes, and PCBs. Bioremediation of chlorinated organic compounds such as tetrachloroethene (PCE) and trichloroethene (TCE) is most effective under anaerobic conditions, where nitrate, sulfate, and carbon dioxide can be used as electron acceptors.

For bioremediation to be a feasible remediation option, the hydraulic conductivity must be sufficiently high to allow the transport of the electron acceptor and nutrients through the aquifer, and microorganisms must be present in sufficient numbers and types to degrade the contaminants of interest. Hydraulic conductivities exceeding 1 m/day are considered sufficient for transporting nutrients and oxygen in the subsurface (Bedient et al., 1999), however, microbial growth in aquifer material can cause permeability to decrease by a factor of 1000 (Taylor et al, 1990). Bioremediation projects are generally preceded by laboratory experiments of microbial stimulation and modeling studies of nutrient delivery and transport to ensure efficient performance of the system. In *enhanced bioremediation*, a system for injection of nutrients and circulation through the contaminated portion of the aquifer is used.

When contaminants are organic compounds that can be used by microorganisms as growth substrates, it may be possible for the indigenous microbial population to biodegrade the contaminants at significant rates in the absence of engineered intervention. This is called *intrinsic bioremediation*, *natural attenuation*, or *bioattenuation*. Intrinsic bioremediation is considered a suitable remediation technology only when natural contaminant biodegradation occurs faster than migration, resulting in a stable or shrinking contaminant plume.

The major advantage associated with in situ bioremediation is that the contaminants are destroyed in place, with minimal transport to the surface. An important factor that can limit the feasibility of in situ bioremediation is the availability of the contaminant for microbial attack. That is, contaminants that have extremely low solubilities, sorb strongly to solids, or are otherwise physically inaccessible.

In Situ Reaction Walls   In situ reaction walls are excavated trenches containing material that reacts with the contaminant in the ground water. Water flows in one side of the trench and out the other side of the trench, hence the name *reaction wall*. In some cases, low-permeability cutoff walls are used to direct the ground-water flow through the reaction wall. These systems have been referred to a *funnel-and-gate systems*, where the impervious barrier is the "funnel" and the reaction wall is the "gate" (Bedient et al., 1999). Chemical, physical, and biological treatment barriers have been used in practice. As an example, granular elemental iron placed in a reaction wall induces the dechlorination of some chlorinated contaminants (e.g., TCE and PCE) and the removal of dissolved metals such as chromium, Cr(VI), through reduction and precipitation. Reaction rates are typically slow by aboveground treatment standards but are sufficiently fast for ground-water systems (Gillham and O'Hannesin, 1992; Blowes et al., 1999).

In Situ Containment   The migration of contaminants can be restricted by using various containment options. In cases where residual contaminants are in the vadose zone, the site can be capped to minimize rainfall contact and subsequent percolation of the contaminant into the ground water. Surface caps are usually constructed using either natural soils, commercially designed materials, or waste materials, and are typically sloped for rainfall to run off rather than infiltrate. Examples include clay, concrete, asphalt, lime, fly ash, and synthetic liners. *Liners* are typically used to protect ground water from leachate resulting from landfills containing hazardous materials. The type of liner depends on the type of soil and contaminant, and liner materials include polyethylene, polyvinyl chloride (PVC), many asphalt-based materials, and soil bentonite or cement mixtures.

Solidification and stabilization techniques involve treating contaminated soil to alter the physical characteristics of the soil and reduce the leachability and mobility of contaminants within the soil. Soils are usually treated in place with mixing augers or digging tools adapted to mix the additives with the contaminated soil. Common additives include cement, lime spikes with fly ash or sodium silicate, asphalt or bitumen, and various organic polymers. In some applications, combinations of additives are used together. By adjusting the quantity of additives, permanent solidification of the soil can be achieved (called *monolithic treatment*), or the soil can be made more amenable for transport after treatment. Soil solidification and stabilization techniques appear to be most effective in the stabilization of metals and organic compounds like PCBs.

Physical barriers to prevent ground water flow are commonly called *ground-water cutoff walls* and include slurry walls, grout curtains, sheet piling, and compacted liners or geomembranes. Construction of a slurry wall is illustrated in Figure 7.35. Physical barriers are most effective in shallow aquifers that are bounded below by a solid confining layer of bedrock or clay. A *slurry wall*, the most popular type of cutoff wall, is constructed by

FIGURE 7.35 Construction of a slurry wall. (From Permeable Reactive Barrier Network, 2005.)

excavating a narrow trench 0.5 to 2 m wide around a contaminated zone and filling the trench with a slurry material. The slurry acts to maintain the trench during excavation and usually consists of a mixture of soil or cement, bentonite clay, and water. Soil–bentonite cutoff walls typically allow some contaminant transport across the wall, depending on the variability in hydraulic conductivity in the bentonite (Britton et al., 2005). Trenches are usually dug using an excavator or backhoe, and sandy aquifers less than 18 m thick and underlain by an impermeable layer or bedrock are most amenable to slurry wall construction. Slurry wall construction is limited to the depths that the trench can be constructed. *Grout curtains* are constructed by injecting grout (liquid, slurry, or emulsion) into the ground through well points. Grout penetration varies from site to site but can be relatively small, requiring closely spaced injection holes: for example, every 1.5 m (Domenico and Schwartz, 1998). Ground water flow is impeded by the grout that solidifies in the interstitial pore space, and the curtain is made contiguous by injecting grout in two or three staggered rows to ensure a more or less continuous barrier. Most grouts can be injected only in materials with sand-sized and larger grain sizes. The expense of grouting and the potential for contamination-related problems in the grout limit its usefulness. *Sheet piling* involves driving interlocking sections of steel sheets into the ground. The piles are typically driven through the aquifer and down into the consolidated zone using a pile driver. Sheet piling may be less effective in coarse, dense material because the interlocking web may be disrupted during construction.

Ground water contained within cutoff walls can either be permanently isolated by a wall around the contaminant and an impervious cap, or the contaminated ground water can be pumped from the up-gradient side of the cutoff wall, treated, and injected into the ground water on the down-gradient side of the cutoff wall. The most effective approach is usually determined by economic considerations.

Natural Attenuation Natural attenuation includes physical, chemical, and biological processes that act without human intervention to reduce the mass, toxicity, mobility,

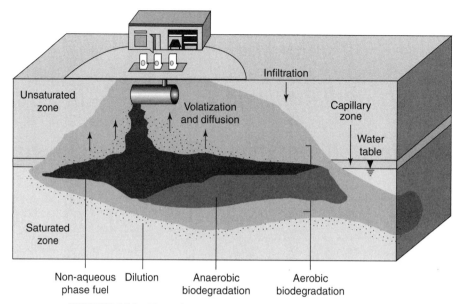

FIGURE 7.36 Natural-attenuation processes. (From USGS, 2005l.)

volume, or concentration of contaminants in soil or ground water. Conceptual illustration of the important natural attenuation processes that affect the fate of petroleum hydrocarbons in aquifers is shown in Figure 7.36. In recent years, natural attenuation has become accepted as an alternative for the management of organic compounds dissolved in ground water (Borden, 2000). Natural attenuation includes processes such as dispersion, sorption, degradation (either biodegradation or abiotic processes such as hydrolysis), volatization, and other natural processes that affect the concentration of dissolved contaminants in ground water. Volatization is relatively unimportant in most aquifers (Wiedemeier et al., 1999), except in situations where the water table is less than 5 m below the ground surface and the vadose zone consists of highly transmissive soils (Borden, 2000). Abiotic chemical transformations, including hydrolysis and elimination reactions, are not important processes for petroleum hydrocarbons but may be significant for certain chlorinated organic compounds. Biodegradation is usually the most important process in the natural removal of petroleum hydrocarbons and chlorinated solvents from ground water. In many cases, the feasibility of a natural attenuation strategy in the saturated zone depends on whether the regulatory goal is to clean up the plume to drinking water standards, or whether less stringent risk-based goals are applicable, such as preventing the plume from expanding. Since 1995, the use of natural attenuation as a remedial solution for BTEX compounds has increased dramatically (National Research Council, 2000).

Petroleum Hydrocarbons The rate and extent of hydrocarbon degradation in the subsurface depends on several factors, including the type, number, and metabolic capability of the microorganisms; temperature, pH, nutrient concentrations; and electron acceptor concentration. In many aquifers, the temperature, pH, and nutrient concentrations are within acceptable ranges for microbial growth, and an adapted microbial population is present. In these cases, the environmental factor that has the greatest influence on the rate and extent of biodegradation is the availability of suitable electron acceptors. Under aerobic conditions, bacteria utilize molecular oxygen as the electron acceptor, while anaerobic bacteria use compounds such as NO_3^-, $Fe(OH)_3$, and SO_4^{2-}.

Almost all petroleum hydrocarbons are biodegradable under aerobic conditions (Borden, 2000). Moderate- to lower-molecular-weight hydrocarbons (C_{10} to C_{24} alkanes, single-ring aromatics) appear to be the most easily degradable hydrocarbons (Atlas, 1988). In many cases the major limitation on aerobic biodegradation in the subsurface is the low solubility of oxygen in water. For example, the aerobic biodegradation of toluene (C_7H_8) can be represented by the reaction

$$C_7H_8 + 9O_2 \rightarrow 7CO_2 + 4H_2O \tag{7.124}$$

Water saturated with air contains 6 to 12 mg/L of dissolved oxygen. Complete conversion of toluene (and many other hydrocarbons) to carbon dioxide and water requires approximately 3 mg/L of oxygen for each 1 mg/L of toluene. Using this ratio, the oxygen present in water could result in the biodegradation of 2 to 4 mg/L of dissolved toluene by strictly aerobic processes. If the toluene concentration is higher, insufficient oxygen is available to support biodegradation.

When the oxygen supply is depleted and nitrate is present (or other oxidized forms of nitrogen), some facultative microorganisms will use the NO_3^- as a terminal electron acceptor instead of oxygen. For toluene, this process can be approximated by the reaction (Borden, 2000)

$$C_7H_8 + 7.2H^+ + 7.2NO_3^- \rightarrow 7CO_2 + 7.6H_2O + 3.6N_2 \tag{7.125}$$

After the available oxygen and nitrate are depleted, subsurface microorganisms may use oxidized ferric iron [Fe(III)] as an electron acceptor. Large amounts of Fe(III) are present in the sediments of most aquifers and can provide a large reservoir of electron acceptors for hydrocarbon biodegradation (Borden, 2000). A possible reaction coupling the oxidation of toluene and the reduction of Fe(III) in ferric hydroxide $[Fe(OH)_3]$ is

$$C_7H_8 + 36Fe(OH)_3 \rightarrow 7CO_2 + 36Fe^{2+} + 72OH^- + 22H_2O \tag{7.126}$$

The reduction of Fe(III) results in high concentrations of dissolved Fe(II) in contaminated aquifers. Although the exact mechanism of microbial ferric iron reduction is poorly understood, available evidence suggests that iron reduction is an important mechanism in the subsurface biodegradation of dissolved hydrocarbons.

A wide variety of petroleum organic compounds can be biodegraded by sulfate-reducing and/or methanogenic (methane-generating) microorganisms. These compounds include creosol isomers, homocyclic and heterocyclic aromatics, and unsaturated hydrocarbons. Sulfate reducers could biodegrade toluene by using toluene in the following theoretical reaction (Beller et al., 1991):

$$C_7H_8 + 4.5SO_4^{2-} + 3H_2O \rightarrow 2.25H_2S + 2.25HS^- + 7HCO_3^- + 0.25H^+ \tag{7.127}$$

and methanogenic consortia (groups of microorganisms that generate methane) could biodegrade toluene using water as an electron acceptor in the following theoretical reaction (Borden, 2000):

$$C_7H_8 + 5H_2O \rightarrow 4.5CH_4 + 2.5CO_2 \tag{7.128}$$

As water containing oxygen, nitrate, and sulfate mixes with hydrocarbon-contaminated water, bacteria attached to the aquifer matrix will consume both hydrocarbon and dissolved electron acceptors. A significant decrease in the concentration of dissolved O_2 and a significant increase in the concentration of dissolved CO_2 due to oxidation of organic matter is expected in the initial contact stages. In areas of the contaminant plume where oxygen is depleted, a significant decrease in nitrate concentration, a significant increase in dissolved iron concentration and pH due to the reduction of insoluble iron oxides in the aquifer matrix, and a significant reduction in sulfate concentration are expected.

Chlorinated Hydrocarbons   The major chlorinated solvents are carbon tetrachloride (CT), tetrachloroethene (PCE), trichloroethene (TCE), and 1,1,1-trichloroethane (TCA). These compounds are all chlorinated aliphatic hydrocarbons (CAHs) and can be transformed through biotic and abiotic processes to form a variety of chlorinated degradation products, including chloroform, methylene chloride, *cis*- and *trans*-1,2-dichloroethene (*cis*-DCE and *trans*-DCE), 1,1-dichloroethene (1,1-DCE), vinyl chloride (VC), 1,1-dichloroethane (DCA), and chloroethane (McCarty, 1996).

Many chlorinated organic compounds, including most CAHs, are resistant to aerobic biodegradation but are degradable under anaerobic conditions. Anaerobic degradation of CAHs occurs through a process called *reductive dechlorination*, where a CAH molecule serves as an electron acceptor and the chloride moiety is removed and replaced by a hydrogen, forming a less chlorinated, more reduced intermediate. Reductive dechlorination has been demonstrated under nitrate- and sulfate-reducing conditions, but the most rapid degradation rates occur under methanogenic conditions (Bouwer, 1994). The susceptibility to reductive dechlorination varies with the extent of chlorination. PCE is most easily reduced because it is the most oxidized, whereas VC is the least susceptible to reductive dechlorination because it is the most reduced. The rate of reductive dechlorination also decreases as the degree of chlorination decreases.

The pattern of degradation depends on the amount of contaminant (type 1); the amount of biologically available organic carbon in the aquifer (type 2); and the distribution, concentration, and utilization of naturally occurring electron acceptors (type 3). Depending on the site conditions, different portions of the same plume may exhibit different patterns of degradation.

Type 1 behavior occurs when the primary substrate is anthropogenic organic carbon (e.g., BTEX, landfill leachate), and this carbon drives reductive dechlorination. When evaluating natural attenuation of a plume exhibiting type 1 behavior, the investigator must determine whether the CAHs are reduced completely before the organic substrate is depleted. The substrate may be depleted during biotransformation of the CAHs and by other competing electron acceptors [oxygen, nitrate, Fe(III), and sulfate]. This behavior often results in the rapid and extensive degradation of the highly chlorinated solvents such as PCE, TCE, and DCE. If the entire contaminant plume is under type 1 behavior, PCE may be completely reduced to the nontoxic by-products ethene and ethane by the following sequence

$$PCE \rightarrow TCE \rightarrow DCE \rightarrow VC \rightarrow ethene \rightarrow ethane \qquad (7.129)$$

In many cases, VC degrades more slowly than TCE and thus tends to accumulate. Through each step of this reaction sequence, chloride ions are released, causing the dissolved chloride concentration of the ground water to increase. It is not unusual for spills of PCE and TCE to result in large concentrations of DCE and chloride as biotransformation takes place.

Type 2 behavior predominates in areas where there are relatively high levels of naturally occurring organic carbon, the organic carbon is biologically available, and this carbon serves as a substrate for reductive dechlorination. This behavior generally results in slower biodegradation of the highly chlorinated contaminants than type 1 behavior, but under the right conditions (e.g., areas with high natural organic carbon contents), this type of behavior can also result in rapid degradation of the more chlorinated compounds.

Type 3 behavior predominates in areas with lower levels of organic carbon and dissolved oxygen concentrations greater than 1 mg/L. Under these aerobic conditions, reductive dechlorination will not occur, and PCE, TCE, and DCE will not biodegrade. Under type 3 conditions, the primary mechanisms reducing contaminant concentrations are advection, dispersion, and sorption. However, VC can be rapidly oxidized.

A single chlorinated solvent plume can exhibit all three types of behavior in different portions of the plume. In some cases, PCE, TCE, and DCE may be reductively dechlorinated in the near-source area under type 1 or 2 behavior and the more reduced degradation product, VC, oxidized in the downgradient aquifer under type 3 behavior. Under these conditions, VC will be oxidized to CO_2 without the accumulation of ethene or ethane.

Assessment of Natural Attenuation   In many cases, once the contaminant source is removed, natural attenuation is sufficient to dissipate contaminant plumes within a few years (Johnson, 1999). The only design activities involved in natural attenuation is designing an appropriate monitoring protocol. The American Society for Testing and Materials (ASTM) has developed a standard for evaluating the effectiveness of remediation by natural attenuation (ASTM, 1998a). According to ASTM, primary evidence for the effectiveness of natural attenuation consists of ground-water monitoring data that delineate the extent of the contaminant plume over time. These data must be sufficient to establish whether the plume is shrinking, stable, or expanding. Shrinking and stable plumes are consistent with effective natural remediation. A secondary line of evidence is derived from geochemical indicators of biodegradation and estimates of attenuation rates. Predictions of calibrated fate and transport models that quantify the natural attenuation processes are regarded by ASTM (1998a) as optional lines of evidence as to whether natural attenuation is effective at a particular site. However, calibrated fate and transport models have proven to be useful in relating the uncertainty in hydrogeologic parameters to the uncertainty in the predicted fate and transport of a contaminant (Lu et al., 2005). Demonstration of the efficacy of natural attenuation generally requires data collected by monitoring wells, typically at least 3 years of quarterly water-quality data collected at 5 to 10 monitoring wells distributed along the dissolved plume (Johnson, 1999). Multiple years of data are needed to filter out seasonal effects and fluctuations in the elevation of the water table. The parameters measured in water-quality samples are closely tied to an understanding of the biotransformation processes and decay products described previously, and the parameters of interest are listed in Table 7.28. To protect water-supply sources, a *sentinel well* is typically placed at least 1 year's travel time up-gradient from the water supply well and along the projected centerline of the plume.

Two general approaches have been used for simulating natural attenuation processes. In the first approach, degradation of contaminants is approximated as a first-order process, and geochemical factors limiting biodegradation are not considered explicitly. In the second, more complex approach, the microbiological and geochemical factors believed to limit contaminant biodegradation are explicitly simulated. Use of the first-order decay function to simulate biodegradation can be very attractive because of the large number of

**TABLE 7.28   Recommended Analyses in Water Samples at Sites Contaminated with
Petroleum Hydrocarbons or Chlorinated Solvents**

Analysis	Petroleum Hydrocarbons	Chlorinated Solvents
Aromatic hydrocarbons (BTEX, trimethylbenzene isomers)	•	•
Chlorinated solvents		•
Total hydrocarbons (volatile and extractable)	•	
Oxygen	•	•
Nitrate	•	•
Iron (Fe^{2+})	•	•
Sulfate (SO_4^{2-})	•	•
Methane, ethane, ethene	•	•
Alkalinity	•	•
Carbon dioxide	•	
Oxidation–reduction potential	•	•
pH	•	•
Temperature	•	•
Conductivity		•
Chloride	•	•
Major cations		•
Biologically available Fe (in aquifer matrix)		•
Hydrogen (H_2)		•
Total organic carbon		•

Source: Wiedemeier et al. (1995, 1996).

available analytical models that assume this process. Furthermore, it has been shown that this assumption can provide a close match to observed hydrocarbon behavior in aquifers (Borden, 2000). Various commercially available numerical models continue to be developed for simulating oxygen-limited biodegradation.

Other Remediation Methods   There are a variety of innovative remediation methods that are not in widespread use but have been shown to be effective. Such methods include steam flooding (Falta, 2000), which involves the injection of steam into one or more injection wells and the extraction of water, NAPL, and vapors from one or more extraction wells. The practice of using steam injection for NAPL recovery is well established in the petroleum industry and has been used for enhanced oil recovery since the 1930s.

Summary of Remediation Alternatives   A summary of several of the more commonly used remediation alternatives is given in Table 7.29. The most effective method for any particular case will depend on site-specific conditions, and all methods covered in this section and listed in Table 7.29 should be given consideration.

7.9   COMPUTER MODELS

Several computer models are available for simulating water quality in ground water. These models typically provide numerical solutions to the advection–dispersion equation, or some other form of the law of conservation of mass, at discrete locations and times for

TABLE 7.29 Summary of Groundwater Remediation Alternatives

Technology	Operating Mechanism	Effectiveness
Source removal	Eliminates the source of groundwater contamination, such as leaking storage tank, surface impoundment, contaminated soil, etc.	Effective and simple.
Soil flushing	A flushing agent (water, solvent, or surfactant) is passed through the contaminant source in unsaturated soils to remove the source physically and/or chemically.	Few, if any, successful applications; may cause contaminant to spread further and thus worsen the problem.
In situ steam stripping	Similar to soil flushing except that steam is used to enhance volatilization of organic compounds.	Same as for soil flushing.
Soil vapor extraction	A vacuum is applied to unsaturated soils to withdraw volatile organic compounds.	Effective and frequently used for volatile organics in sandy soils; not effective below the water table.
Air sparging	Introduces air into aquifer via an injection well; volatile organic compounds are removed by volatilization onto injected air.	Suitable for sites with uniform sandy soils; the methodology has been effective at some sites.
Hydraulic containment (pump and treat)	A pumping well is used to capture contaminated groundwater and prevent its further migration.	Most frequently used groundwater remediation alternative.
Physical containment	The source area is enclosed by low-permeability barriers (slurry walls or sheet piles).	Requires presence of relatively shallow underlying low-permeability layer; used successfully under these conditions.
Natural in situ bioremediation	Relies on naturally occurring microorganisms to degrade organic compounds.	Used frequently and successfully for petroleum hydrocarbons in aerobic (shallow) groundwater after source removal.
Active in situ bioremediation	Introduces nutrients and/or microorganisms into the subsurface to stimulate and enhance biodegradation of organic compounds.	Successful applications are few; requires favorable geologic media and contaminants amenable to biodegradation.
Natural attenuation	Contamination is monitored, while plume dissipates by bidegradation, chemical reactions, volatilization, and dilution.	Unlikely to be successful if a continuous source is present.

Source: Shanahan (1995).

multiple interacting constituents, complex boundary conditions, spatially and temporally distributed contaminant sources and sinks, multiple fate processes, and variable flow and dispersion conditions. In engineering practice, the use of computer models to apply the fundamental principles covered in this chapter is sometimes essential. In choosing a model for a particular application, there is usually a variety of models to choose from. However, in doing work that is to be reviewed by regulatory agencies, or where professional liability is a concern, models developed and maintained by the U.S. government have the greatest credibility and, perhaps more important, are almost universally acceptable in developing permit applications and defending design protocols on liability issues. The most widely used model that has been developed and endorsed by U.S. government agencies is described briefly here.

MOC (the method of characteristics) is the most popular model used for contaminant transport in ground water. This model was originally developed by the U.S. Geological Survey (Konikow and Bredehoeft, 1978) to simulate contaminant fate and transport in two-dimensional single-layer heterogeneous aquifers for both steady and unsteady flow. The model has evolved through several updates and improvements (Goode and Konikow, 1989; Konikow et al., 1994). The processes simulated by the model include advection, hydrodynamic dispersion, sorption, and a variety of chemical reactions (Konikow and Reilly, 1998). MOC has many limitations that may preclude its use in certain applications, such as in the case of complex three-dimensional geologies. A three-dimensional version of the model (MOC3D) has been developed using MODFLOW to simulate the ground-water flow (Konikow et al., 1996).

Only one of the more widely used computer models have been cited here. Certainly, there are many other good models that are capable of performing the same tasks.

SUMMARY

Ground water is a major source of drinking water in the United States, supplying approximately 50% of the population. Contamination of ground water occurs from a variety of sources, including septic tanks, leaking underground storage tanks, irrigation return flow, leachate from solid-waste disposal sites, injection wells, and agricultural operations. Dispersion models that simulate the fate and transport of instantaneous spills, continuous point sources, and continuous plane sources are useful in preliminary analyses of exposures of human populations to contaminated ground water. Aside from the application of fate and transport models in ground water, the determination of model parameters is usually quite challenging. The dispersion coefficient of large-scale plumes in ground water is determined fundamentally by the statistics of the spatial variations in hydraulic conductivity, which are difficult to measure in the field. Limitations in available data frequently require that dispersion coefficients be estimated using empirical equations that relate the dispersivity to the length scale of the plume. Fate processes commonly accounted for in ground-water dispersion models include sorption and decay. Sorption is usually parameterized by the distribution coefficient, decay by a first-order decay factor, and these processes can be modeled by appropriate modification of conservative-contaminant models. Nonaqueous phase liquids are of special concern in ground waters, since they tend to have low solubilities, high residual concentrations, and can persist in the pores of aquifers for many years, providing a continuous source of contamination. A variety of technologies are available to remediate contaminated soil and ground water. In cases of soil contamination, the remediation objective is usually to remove the contaminant sorbed onto the soil. In cases of ground water contamination, the objective is usually to treat the ground water, either in situ or ex situ, to meet water-quality standards and to remove the subsurface source of contamination. The appropriate remediation strategy for any particular situation depends on the nature of the contaminant, the extent of contamination, potential impact, and economic considerations. Guidelines are provided to aid in the selection of the appropriate remediation technology.

PROBLEMS

7.1. Orange trees can tolerate water in the root zone with a total dissolved solids (TDS) concentration of up to 500 mg/L, and orange trees require 15 cm of water to support growth during the spring planting season. Available irrigation water and effective rainfall combined has a TDS content of 90 mg/L, soil evaporation during the spring planting season is 35 cm, and the effective rainfall is 20 cm.

  (**a**) Estimate the amount of irrigation water required and the expected TDS concentration in the root zone.

  (**b**) Determine the leaching requirement, leaching ratio, and maximum requirement for irrigation plus rainfall to avoid excessive TDS in the root zone.

7.2. Five kilograms of a conservative contaminant is spilled into the ground water and is well mixed over the top 1 m. The longitudinal and (horizontal) transverse dispersion coefficients are 0.5 and 0.05 m^2/day, respectively; vertical mixing is negligible; the mean seepage velocity is 0.3 m/day and the porosity of the aquifer is 0.2. Determine the concentrations at the spill location for the first 7 days after the spill. How are your calculated concentrations affected by the assumed depth of the spill?

7.3. Determine the maximum contaminant concentrations in Problem 7.2 for the first 7 days after the spill.

7.4. A contaminant is injected continuously over a depth of 3 m into an aquifer with a mean seepage velocity of 0.45 m/day and longitudinal and transverse dispersion coefficients of 1 and 0.1 m^2/day, respectively. The injection rate is 0.4 m^3/day with a concentration of 130 mg/L. Estimate the steady-state contaminant concentration 30 m downstream of the injection location.

7.5. For the case described in Problem 7.4, determine the distance from the injection location to the point where the contaminant concentration is 1% of the injection concentration.

7.6. A contaminant source is 5 m wide $\times$ 2 m deep and continuously releases a conservative contaminant at a concentration of 70 mg/L. The mean seepage velocity in the aquifer is 0.1 m/day, the aquifer is 10 m deep, and the longitudinal, horizontal transverse, and vertical dispersivities are 1, 0.1, and 0.01 m, respectively.

  (**a**) Determine the downstream location at which the plume will be fully mixed over the depth of the aquifer.

  (**b**) Estimate the contaminant concentrations at the water table 200 m downstream of the source after 5 years of operation.

7.7. Repeat Problem 7.6 for the case in which the contaminant undergoes biodegradation with a decay rate of 0.01 day^{-1}. Assess whether biodegradation has a significant effect on the downstream concentration.

7.8. Pure TCE exists at the residual saturation level in a portion of an aquifer that is 10 m long, 5 m wide, and 2 m deep. This volume is called the *source zone* of TCE in the aquifer. As ground water flows through the source zone, TCE is dissolved in the ground water, and the ground water exits the source zone with TCE at the solubility concentration. The ground-water flow is normal to the width of the source zone, the first-order decay coefficient of TCE in ground water is 0.02 day^{-1}, the seepage velocity is 0.3 m/day, the longitudinal dispersivity is 10 m, the horizontal-transverse dispersivity is 1 m, and the vertical-transverse dispersivity is 0.1 m. A small water-supply well is located 200 m downstream of the source zone. Determine whether the water-quality criterion of 5 μg/L for TCE will be exceeded at the well intake.

7.9. Hydraulic conductivity measurements (in m/day) on 50-cm core samples taken from an isotropic aquifer indicate a variable log-hydraulic conductivity that can be described by a mean of 3.2, variance of 1.6, and a correlation length scale of 1.3 m.

   (a) Estimate the effective hydraulic conductivity and macrodispersivity of the aquifer.

   (b) If the mean hydraulic gradient in the aquifer is 0.005 and the effective porosity is 0.15, estimate the components of the macrodispersion coefficient.

7.10. How would your dispersivity estimates change in Problem 7.9. if the aquifer were anisotropic with a horizontal correlation length scale of 1.3 m and an anisotropy ratio of 0.1?

7.11. The mean seepage velocity in an aquifer is 1 m/day, the mean pore size is 3 mm, and the molecular diffusion coefficient of a toxic contaminant in the ground water is 2×10^{-9} m^2/s. Should molecular diffusion be considered in a contaminant transport model?

7.12. Seepage velocities surrounding a municipal well field are on the order of 5 m/day, the well field is approximately circular with a radius of 70 m, and the longitudinal dispersion coefficient is on the order of 50 m^2/day. Determine whether contaminant transport from the boundary of the well field is advection or dispersion dominated.

7.13. A contaminant spill in an aquifer has resulted in a pollutant cloud that is 11 m long, 5 m wide, and 2 m deep. The pore sizes in the aquifer are on the order of 2 mm, the molecular diffusion coefficient is 10^{-9} m^2/s, the tortuosity is 1.3, and the mean seepage velocity is 0.1 m/day. Estimate the components of the dispersion coefficient that should be used in modeling plume transport.

7.14. The seepage velocity, v, surrounding a well is described by the relation

$$v = \frac{Q}{2\pi rbn}$$

where Q is the pumping rate at the well, r is the radial distance from the well, b is the aquifer thickness, and n is the effective porosity of the aquifer. At a particular well, the pumping rate is 20,000 L/min, the thickness of the saturated zone is 15 m, and the effective porosity is 0.15. Estimate the extent of the region surrounding the well where advection transport dominates macrodispersion. (*Hint:* Use the Peclet number vr/D_L as a basis for your analysis.)

7.15. Sorption onto a soil matrix can be described by the Langmuir isotherm given by

$$F = \frac{K_l \bar{S} c_{aq}}{1 + K_l c_{aq}}$$

where F is the mass of tracer sorbed per unit mass of the solid phase, c_{aq} is the aqueous concentration, K_l is the Langmuir constant, and $\bar{S}$ is the maximum sorption capacity of the soil matrix.

(**a**) If the sorption capacity of the soil matrix is 5000 mg/kg and the linear-isotherm distribution coefficient at low aqueous concentrations is 60 L/kg, estimate the Langmuir constant and express the Langmuir isotherm as a relationship between F and c_{aq} only.

(**b**) Estimate the range of aqueous concentrations for which the linear isotherm deviates by less than 10% from the Langmuir isotherm. Given that the Langmuir isotherm provides a more realistic representation of the sorption process, what is the advantage to using the linear isotherm in practice?

7.16. Determine whether the dispersivities given by Equation 7.47 are consistent with the values given in Table 7.10.

7.17. The concentration of trichloroethylene (TCE) in a ground water is measured as 100 mg/L, the fraction of organic carbon in the solid matrix is estimated as 1.5%, and the bulk density of the aquifer is approximately 1800 kg/m^3. Estimate the mass of TCE sorbed per unit volume of the aquifer.

7.18. Three kilograms of tetrachloroethylene is spilled over a 1.2-m depth of ground water and spreads laterally as the ground water moves with an average velocity of 0.2 m/day. The longitudinal and transverse dispersion coefficients are 0.05 and 0.005 m^2/day, respectively; the porosity is 0.15; the density of the soil matrix is 2.65 g/cm^3; log K_{oc} is 2.42 (K_{oc} in g/cm^3); and the organic fraction in the soil is 8%. Calculate the concentration at the spill location after 1 hour, 1 day, and 1 week. Compare these values with the concentration obtained by neglecting sorption.

7.19. Use all the empirical relationships in Table 7.12 to estimate the organic-carbon sorption coefficient of TCE. Assume that log K_{ow} of TCE is 2.29 (as shown in Table 7.13). Verify the claim that the actual value of log K_{oc}, given in Appendix B, is within one standard deviation of the of the mean of the predictions given by the empirical equations listed in Table 7.12.

7.20. An aquifer contains a contaminant spread over a 500-m^3 volume of the aquifer. The porosity of the aquifer is 0.15 and the retardation factor of the contaminant is 10. Estimate the volume of pore water that must be removed to reduce the mass of contaminant in the aquifer by 99%.

7.21. A buried drum containing 10 kg of a contaminant suddenly ruptures and spills all of its contents into the ground water over a 1-m depth. The mean seepage velocity in the aquifer is 0.5 m/day, the porosity is 0.2, the longitudinal dispersion coefficient is 1 m^2/day, the horizontal-transverse dispersion coefficient is 0.1 m^2/day, vertical mixing is negligible, and the first-order decay constant of the contaminant is 0.02 day^{-1}. Determine the maximum concentration in the ground water after 100 days. Compare this concentration to the maximum concentration without decay.

7.22. If the decay factor in Problem 7.21 could be increased by adding nutrients to the ground water, determine the required decay rate for the calculated maximum concentration to be reduced by 90%.

7.23. Five kilograms of TCE is spilled over a 0.8-m depth of ground water that moves with an average seepage velocity of 0.2 m/day. The porosity of the aquifer is 0.2, vertical dispersion is negligible, and the longitudinal and transverse dispersion coefficients are 0.1 and 0.01 m^2/day, respectively. If the retardation factor is equal to 15 and the first-order decay factor is 1 day^{-1}, calculate the concentration at the spill location after 1 day and 1 week.

7.24. What residual mass fraction can be expected when spills of fuel oil in fine sands are cleaned up by pumping free product from the ground water? How does this residual mass fraction compare with fuel-oil spills in coarse sands? Assume that the porosity is 0.20 for coarse sand, 0.30 for fine sand, and the density of sand grains is 2650 kg/m^3.

7.25. Six pumping tests have been conducted in an isotropic aquifer and the hydraulic conductivities calculated from these tests are 512, 253, 487, 619, 320, and 402 m/day. It is estimated that each of these hydraulic conductivities is characteristic of a cylinder of radius 100 m, and the correlation length scale of the hydraulic conductivities is on the order of 50 m. The temperature of the ground water is 25° C, and the characteristic pore size, d, of the aquifer can be related to the effective hydraulic conductivity, K_{eff}, of the aquifer by the Hazen formula

$$K_{eff} = 1.02 \times 10^{-3}\frac{\gamma}{\mu}d^2$$

where γ is the specific weight of the ground water, and μ is the dynamic viscosity.

(**a**) If the mean seepage velocity is 1 m/day, the porosity is 0.2, and the bulk density of the aquifer matrix is 1.8 g/cm^3, estimate the dispersion coefficient that should be used to model contaminant dispersion in the aquifer. Neglect pore-scale molecular diffusion.

(**b**) A degreasing shop spills 50 kg of trichloroethylene (C$_2$HCL$_3$) over the top 2 m of the aquifer. If the aquifer matrix contains 2% organic carbon, and the first-order decay coefficient is 0.02 day^{-1}, estimate the maximum concentration of TCE in the aquifer after 100 days. Assuming that the spill occurs over 10 min, use the solubility of TCE to determine whether TCE will exist as a NAPL in the vicinity of the spill.

7.26. A 2 m $\times$ 2 m $\times$ 3 m (deep) portion of an aquifer contains chlorobenzene at a residual saturation of 15%. If the porosity of the contaminated portion of the aquifer is 0.17, the density of chlorobenzene is 1110 kg/m^3, the solubility of chlorobenzene in water is 500 mg/L, and the mean seepage velocity of the ambient ground water is 0.05 m/day, estimate the time it would take for the chlorobenzene to be removed by dissolution.

7.27. A portion of a 500 m $\times$ 500 m industrial site has been contaminated by spillage of a 50:50 mixture (by volume) of ethylbenzene and benzene. The contaminated area is at the center of the site, has dimensions 20 m $\times$ 20 m, and the contamination extends to a depth of 2 m into the saturated zone. The thickness of the saturated zone is 25 m, the estimated hydraulic conductivity of the aquifer is 15 m/day, the effective porosity is 0.19, and the hydraulic gradient is estimated as 0.5%.

(**a**) If the contaminated area within the saturated zone contains ethylbenzene and benzene at a 25% residual saturation, estimate the steady-state concentrations of ethlybenzene and benzene on the site boundary.

(**b**) Is decay a significant factor in calculating the concentrations on the boundary?

(**c**) Estimate the time required to remove benzene from the NAPL residual by dissolution.

7.28. (a) What is the maximum concentration of trichloroethene (TCE) that can be found in a soil-gas sample at $20°C$?

(b) If a liquid consisting of 35% TCE is spilled, what is the maximum concentration of TCE expected in the soil gas?

7.29. What is the maximum concentration of o-xylene in a plume of contaminated ground water? Contrast this with the maximum concentration of o-xylene in ground water resulting from a spill of gasoline containing a mole fraction of 9% o-xylene.

7.30. If the mass of an organic compound in a soil sample per unit mass the solid matrix is c_s [M/M], the bulk density of the soil is ρ_b, the water-filled porosity is θ_w, the fraction of organic carbon in the soil is f_{oc}, and the organic-carbon sorption coefficient is K_{oc}, show that the equilibrium concentration, c_w, of the organic compound in the interstitial ground water is given by

$$c_w = \frac{c_s \rho_b}{\rho_b f_{oc} K_{oc} + \theta_w}$$

Explain the implications if c_w is calculated from measurements of c_s, ρ_b, θ_w, f_{oc}, and K_{oc}, and the calculated value of c_w exceeds the saturation concentration of the compound.

7.31. A grain-size analysis of an aquifer matrix indicates that $d_{10} = 0.10\,\text{mm}$, $d_{30} = 0.15\,\text{mm}$, $d_{50} = 0.50\,\text{mm}$, and $d_{90} = 1.2\,\text{mm}$.

(a) Determine whether an artificial gravel pack will be necessary for a monitoring well that is to be installed in this aquifer.

(b) If a gravel pack is required, write the specifications for the gravel pack and determine the required slot size for the well screen. Assume that $d_{60} \simeq 1.1\,d_{50}$ for the gravel pack.

7.32. The piezometric heads measured at three locations in an aquifer are 3.62, 2.20, and 2.10 m. The coordinates of the measurement locations are (1 km, 2.5 km), (2.3 km, 1.4 km), and (1.7 km, 1.3 km). Determine the hydraulic gradient in the aquifer.

7.33. A major spill of chloroethane at an industrial site results in a lens of pure chloroethane in the ground water. Monitoring wells indicate a thickness of 62 cm extending over an area of 5000 m². A site investigation shows that the temperature of the ground water is $18°C$, the aquifer material consists of coarse sand with a capillary rise of 11 cm, and the porosity of the aquifer is 0.20. Estimate the volume of pure chloroethane.

7.34. Monitoring wells installed near a gas station with leaking underground storage tanks (LUSTs) show a 2.3-m-thick layer of pure gasoline floating on top of the ground water over a circular area of radius 30 m. The density and viscosity of the gasoline are $750\,\text{kg/m}^3$ and $0.31\,\text{mPa}\cdot\text{s}$, respectively.

(a) If the aquifer material consists of sandy loam with a porosity of 0.23, a saturated thickness of 27.3 m, and a hydraulic conductivity of 5 m/day, estimate how much free-product gasoline exists in the saturated zone of the aquifer.

(b) Economic considerations indicate that a dual-pump system would be most effective in removing the free-product gasoline from the top of the water table.

The Cooper–Jacob equation can be used to relate the drawdown, s_w, at the recovery well to the pumping rate, Q_w, such that

$$s_w = \frac{Q_w}{4\pi T}\left(-0.5772 - \ln\frac{r_w^2 S}{4Tt}\right)$$

where T is the transmissivity (=hydraulic conductivity $\times$ saturated thickness of aquifer), r_w is the radius of the well, S is the specific yield ($\approx$ porosity), and t is the time since pumping began. Local regulations permit a maximum drawdown of 2 m after 1 year, and a 150-mm-diameter recovery well is expected to have a radius of influence of 50 m. Determine the maximum allowable pumping rate at the recovery well and the corresponding gasoline recovery rate.

7.35. Monitoring wells at a contaminated site show a NAPL floating on the water table with a free-product thickness of 150 cm. Laboratory tests indicate that the NAPL has a density of 750 kg/m^3 and a dynamic viscosity of 0.0012 Pa·s. The aquifer has an average hydraulic conductivity of 4 m/day, the saturated thickness of the aquifer is 20 m, and the temperature of the ground water is 17°C. The aquifer material is classified as loamy sand and has a porosity of 0.30. Each recovery well installed at the site has a radius of recovery of 25 m, and the discharge of each recovery well is 15 L/min. Calculate the fraction of NAPL by volume in the pumped water and how long it will take to remove 90% of the spilled NAPL.

7.36. Derive Equation 7.108.

7.37. An organic compound containing 60% chloroethane and 40% ethylbenzene has been spilled and is to be cleaned up using soil vapor extraction. Soil samples indicate that a mass of 1000 kg has been spilled. The extraction wells to be used have a diameter of 125 mm, screen lengths of 3 m, and the intake pressure is to be held at 5 kPa below atmospheric pressure. The air temperature in the soil is 18°C, the intrinsic permeability of the soil for airflow is 110 darcys, and the radius of influence of a vapor extraction well is 20 m. If 25% of the air extracted from each well passes through contaminated soil, estimate the number of wells required to clean up the spill in 2 weeks.

7.38. A significant amount of PCE is spilled into the vadose zone over 100 m^2 of a site. The vadose zone is 2.5 m thick, the porosity is 0.3, the density of the soil matrix is 2600 kg/m^3, and the organic-carbon content of the soil is 2%. A soil-sample analysis indicates a PCE concentration of 3500 mg/kg, and it is estimated that 50% of the pore space is filled with water.

(a) Verify that the PCE exists as a NAPL, and estimate the residual saturation of PCE in the soil.

(b) A soil vapor extraction (SVE) system is to be used to remediate the contaminated soil. The SVE system is to use 100-mm-diameter extraction wells that penetrate the entire vadose zone, the radius of influence of each well can be taken as 25 m, and the pressure at the intake will be held at 15 kPa below atmospheric pressure. If the soil temperature is 20°C, and the intrinsic permeability is 300 darcys, estimate how many extraction wells will be required to remediate the soil in 1 year.

7.39. A soil-vapor extraction well is to be designed to remediate a soil contaminated by a spill of 100 kg of a mixture of 40% tetrachloroethene and 60% trichloroethene. The vadose zone is 3.5 m thick and has an intrinsic permeability of 100 darcys for the airflow. The radius of influence of the extraction well is 30 m, and the soil vapor is to be pumped out with a blower having a performance curve

$$p_{atm} - p_w = 53 - 3.25Q^2$$

where p_{atm} is the atmospheric pressure (kPa), p_w is the pressure at the well intake (kPa), and Q is the vapor extraction rate (m³/s).

(a) If the extraction well has a diameter of 100 mm and the temperature of the air in the soil is 20° C, calculate the rate at which the blower will remove air from the vadose zone.

(b) If the extraction well is placed such that 50% of the pumped air flows through the contaminated zone, approximately how long will it take to remediate the soil?

7.40. A contaminant plume 50 m long and 10 m wide is to be removed using a pump-and-treat system. The aquifer has a saturated thickness of 25 m and a regional specific discharge of 0.5 m/day. If the pumping wells can be placed a maximum of 200 m downstream of the center of the plume, and the maximum allowable pumping rate is 650 L/min from each well, determine the required number, location, and discharge rate of the pumping wells.

7.41. A contaminant plume 50 m long and 60 m wide is to be remediated using a pump-and-treat system that can handle 100 L/min. The pumping and injection wells are to be placed in a doublet arrangement to prevent further migration of the plume in the aquifer. The regional hydraulic gradient is 0.15%, the average hydraulic conductivity is 15 m/day, and the saturated thickness of the aquifer is 25 m. What well spacing between the pumping and injection wells would you recommend?

7.42. Ethylbenzene is to be treated in situ using an air sparging trench. The trench is to be designed to reduce the concentration of ethylbenzene from 20 mg/L to 1 mg/L. The hydraulic conductivity of the aquifer is 10 m/day, the regional hydraulic gradient is 0.30%, the temperature of the ground water is 20° C, and the trench is to be 50 m long and penetrate 5 m into the saturated zone of the aquifer. The Henry's law constant for ethylbenzene may be found in Table 4.5. Estimate the required air injection rate.

CHAPTER 8

OCEANS AND ESTUARIES

8.1 INTRODUCTION

Ocean shorelines, coastal waters, and coral reefs are used for a variety of commercial and recreational activities, and the quality of waters in the vicinity of ocean shorelines is generally of concern to government regulators. Pollutants frequently found in coastal waters are bacteria, turbidity, and excess nutrients, with the primary sources of pollution being urban runoff and disposal of domestic wastewater (USEPA, 2000a). Coral reefs are fragile, unique, and vibrant ecosystems that support a variety of organisms, algae, plants, and animals. In addition, coral reefs typically support thousands of jobs and billions of dollars in annual revenues from tourism, recreation, and fishing. In a 1990 survey of the health of the world's oceans, the United Nations Group of Experts on the Scientific Aspects of Marine Pollution (GESAMP) placed domestic wastewater (sewage) discharges to the sea at the top of the list of concerns, with heavy-metal and oil pollution given a much lower priority. Pathogen contamination by sewage discharges can result in the closure of shellfishing areas and bathing beaches, and high nutrient levels from sewage discharges can result in algae blooms that consume oxygen and attract predators such as sea urchins and crown-of-thorn sea stars, which destroy living coral.

An *estuary* is defined as a semiclosed coastal body of water having a free connection with the open ocean and containing a measurable quantity of seawater (USEPA, 1984a). Estuaries vary greatly in size and shape and are more commonly known as *bays, lagoons, harbors, inlets*, or *sounds*, although not all water bodies by those names are estuaries. Estuaries commonly occur at the lower reaches of rivers, where ocean tides and river flows interact. A typical estuary is illustrated in Figure 8.1, where the view is toward the open ocean. Some familiar estuaries in the United States include San Francisco Bay, Puget Sound, Chesapeake Bay, and Tampa Bay. Estuaries are typically the sink for all pollution

FIGURE 8.1 Mawddach estuary, UK. (From Newbould, 2005.)

activities that take place in their contributing watershed, and runoff from urban areas is the leading cause of impairments to surveyed estuaries in the United States (National Water Research Institute, 2004). River deltas, such as the Mississippi River delta and the Nile River delta, are not normally considered as estuaries since saltwater intrusion in the deltas of large rivers is minimal and the freshwater impact extends out into the sea. Whether the confluence of a river with a sea is called an estuary or a delta depends on the flow and amount of sediment carried by the river.

8.2 OCEAN-OUTFALL DISCHARGES

Ocean outfalls are used by many coastal communities to discharge treated domestic wastewater into open-ocean waters. In some cases, treated domestic wastewater is blended with wastewater from other sources prior to discharge (Fergen et al., 1999). The wastewater discharged by an ocean outfall experiences rapid mixing in the immediate vicinity of the outfall, with dilution resulting from the entrainment of ambient seawater as the buoyant freshwater plume rises in the denser saltwater environment. As ambient seawater is entrained, the effluent plume becomes denser, rising until the plume density equals the density of the ambient seawater. If the ocean is (density) stratified, there is the possibility that the plume will be "trapped" below the surface. Conversely, if the ocean is unstratified, the density of the freshwater plume can never equal the density of the surrounding ocean water (no matter how much of the ocean water is entrained) and the plume reaches the ocean surface, possibly forming a noticeable *boil*. Trapping of wastewater plumes beneath the surface is desirable because contaminants that make it to the surface are more easily transported to the shore. An illustration of a discharge port and a plume rising (and being trapped) in a stratified environment is given in Figure 8.2. Density stratification in coastal waters may be caused by temperature differences (due to heating of the upper layers) and/or by salinity differences (due to river inflows). An internal hydraulic jump sometimes occurs near the boil, causing dilutions that are three to five times higher than at the center of the boil (Wright et al., 1991). Ocean currents advect the plume away from the outfall,

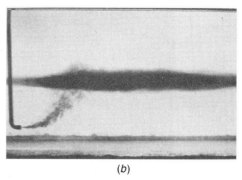

(a)   (b)

FIGURE 8.2   (*a*) Discharge port; (*b*) plume rising in a stratified environment. (From Fan and Brooks, 1969.)

and spatial variations in these currents result in further mixing of the effluent plume. The region in the immediate vicinity of the outfall, where mixing is dominated by buoyancy effects, is called the *near field*, and the region farther away from the outfall, where mixing is dominated by spatial variations in ocean currents, is called the *far field*.

Ocean outfalls must generally be designed to minimize the adverse impacts of waste-water discharges on both benthic and pelagic marine communities. *Benthic* communities live on or within the bottom substrate of the ocean, while *pelagic* communities live within the water column with little or no association with the bottom. Benthic organisms are sometimes called *benthos*. Pelagic organisms are further divided into organisms that can swim and move freely of currents, collectively called *nekton*, and organisms that drift with ocean currents, called *plankton*. Examples of plankton include single-celled plants and fish larvae. Categories of plankton are *phytoplankton* and *zooplankton*. Phytoplankton are microscopic plants that are generally considered to be at the base of the food web and include fungi, bacteria, and various algae. *Zooplankton* are microscopic animals and include crustaceans such as copepods and krill. Spiny lobsters, mollusks, fish, and jellyfish spend part of their life cycle as plankton. Marine ecosystems are generally sustained by the transport of plankton by currents, and the protection of these ecosystems depends on min-imizing the adverse impacts of contaminant discharges on the ocean environment. Thresholds of adverse impacts are usually reflected in aquatic-life water-quality criteria for ocean waters.

The U.S. Environmental Protection Agency (USEPA) aquatic-life water-quality criteria for ocean waters are stated in terms of the magnitude, duration, and frequency of exposure of marine organisms to contaminants. The EPA recommends criteria for maximum con-centration (CMC) to protect marine organisms from acute (short-duration) effects, and cri-teria for continuous concentration (CCC) to protect marine organisms from the chronic (long-duration) effects of toxic substances. CMC limits are typically based on 4-day (96-h) bioassays, and CCC limits are typically based on 7-day bioassays. Both CMC and CCC limits must not be exceeded more than once every three years. If the effluent discharged from an ocean outfall does not meet the ambient water-quality criteria, which is usually the case, regulatory agencies usually allow the delineation of a *mixing zone* surrounding the outfall, within which there is sufficient dilution that the ambient water-quality criteria are met on and beyond the boundary of the mixing zone. However, some quality standards, typically CMCs, must typically be met everywhere within the water body, including within the mixing zone. There are usually statutory limits for the maximum size of a mixing zone surrounding an ocean outfall. In the state of Florida, mixing zones in open-ocean waters are required to have areas less than or equal to 502,655 m^2 (Florida DEP, 1995), which is equal to the area enclosed by a circle of radius 400 m, whereas New Jersey defines a mixing zone as the area within 100 m of the effluent discharge. Rapid dilution of discharged effluent within Florida mixing zones must be ensured by the use of multiport diffusers or single-port outfalls designed to achieve at least a 20:1 dilution of the effluent prior to reaching the surface. Contaminants of concern in domestic wastewater discharges typically include residual chlorine and/or chlorine-produced oxidants. Human-health water-quality criteria must be considered in addition to aquatic-life water-quality criteria, with the most restrictive criteria taking precedence. The primary human-health criteria associated with the discharge of treated domestic wastewater into the oceans are associated with imbibing of contaminated seawater at sea-bathing beaches and the ingestion of contaminated seafood, with the latter risk typically being much greater than the first.

The design of ocean outfalls and the analysis of plume dilution have been the subject of research for many years and are now fairly mature fields (Fischer et al., 1979; Wood et al., 1993; Carvalho et al., 2002). However, the analysis of far-field mixing is still an evolving area of research, and no systematic protocol has yet emerged to predict far-field mixing processes accurately. Far-field processes determine the impact of ocean discharges on beaches and are important in selecting acceptable outfall locations, since the wastewater discharged must continuously be removed from where the wastewater is discharged.

A typical ocean-outfall diffuser is illustrated in Figure 8.3. Diffusers contain multiple ports, with each port designed to discharge effluent at the same rate. Outfall pipelines that transport wastewater from onshore treatment plants to the diffusers are generally buried to the point where the water is deep enough to protect them from wave action, usually about 10 m. Beyond the buried portion, the outfall pipe rests on the bottom of the ocean, with a flanking of rock to prevent currents from undercutting it where the bottom is soft. The construction of an ocean outfall and the lowering of a diffuser to the seafloor are illustrated in Figure 8.4. In cases where the outfall pipe and diffuser are entirely buried under the ocean bottom, riser-nozzle assemblies are used to discharge the effluent into the ocean. Most ocean outfalls are installed at depths ranging from 30 to 70 m and at distances from shore

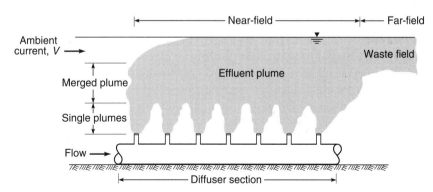

FIGURE 8.3 Wastewater discharge from an ocean-outfall diffuser. (From Chin, David A., *Water-Resources Engineering.* Copyright © 2000. Reprinted by permission of Pearson Education, Inc., Upper Saddle River, NJ.)

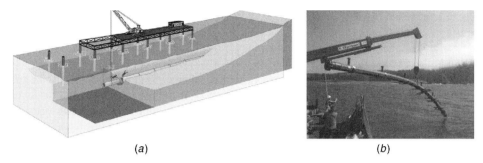

(a)   (b)

FIGURE 8.4   Outfall construction: (*a*) laying an outfall pipe; (*b*) lowering a diffuser to the seafloor. (From Christchurch City Council, 2005.)

of 1 to 8 km, and the characteristics of several ocean outfalls are shown in Table 8.1. It is interesting to note that the discharge per unit length of most outfalls is on the order of 0.01 m²/s.

In shallow waters, plumes originating from individual ports in multiport diffusers tend to reach the ocean surface prior to merging, in which case dilution of the effluent is determined by the dynamics of the individual plumes discharged from each of the ports. The effects of surface waves on near-field dilution at shallow-water outfalls are usually neglected; in certain cases, however, these effects may be significant (Chin, 1987, 1988). In deepwater outfalls, plumes originating from individual ports typically merge together well in advance of reaching the ocean surface or being trapped due to density stratification.

8.2.1 Near-Field Mixing

The dynamics of near-field mixing or *initial dilution* depends on whether the effluent plumes originating at the diffuser ports merge prior to either being trapped or reaching the ocean surface. In cases where adjacent plumes do not merge, each plume behaves (approximately) independently and identically, and near-field mixing can be inferred from the analysis of a single plume. In cases where adjacent plumes merge well in advance of either being trapped or reaching the ocean surface, the (merged) plume behaves as if the effluent were discharged from a long slot or line. Such plumes are called *line plumes*.

Single Plumes   Consider the case of a single effluent plume as illustrated in Figure 8.5. The effluent is discharged through a port of diameter D at velocity u_e, where the effluent has a density ρ_e and contains a contaminant at concentration c_e. Consider further that the ambient ocean water has a density ρ_a (assuming unstratified conditions), a depth-averaged velocity u_a, and we are interested in calculating the contaminant concentration c at a distance y above the discharge port. The relationship between the contaminant concentration, c, and the parameters controlling the dilution of the effluent plume can be written in the following functional form:

$$c = f_1\,(c_e,\, u_e,\, D,\, \rho_e,\, \rho_a,\, g,\, u_a,\, y) \tag{8.1}$$

Assuming that the density difference between the effluent and seawater is small compared to their absolute densities, the kinematics of the effluent plume does not depend explicitly

TABLE 8.1 Characteristics of Several Ocean Outfalls

Name	Location	Design Discharge[a] (m³/s)	Depth (m)	Distance Offshore (m)	Diffuser Length (m)	Ports	Port Spacing (m)	Port Diameter (m)	Port Orientation	Year Operation Began
Hyperion	Los Angeles	18.4	59.4	8,389	2,114	165	14.6	0.17–0.21	—	1960
Deer Island	Boston	13.9	30	13,000	1,600	50	32	—	8-port heads	2000
Orange County	California	12.7	56.4	6,522	1,829	500	3.7	0.075–0.105	—	1971
San Diego	San Diego	10.3	62.5	3,505	819	56	1.8	0.203–0.229	—	1963
White Point No. 4	Los Angeles	9.66	54.1	2,268	1,352	740	1.8	0.051–0.091	—	1965
Central District	Miami	7.89	28.2	5,730	39	5	9.8	1.22	Vertical	—
South Bay	San Diego	7.71	29.0	5,800	1,204	—	—	—	—	1998
White Point No. 3	Los Angeles	6.57	62.5	2,408	731	100	7.3	0.165–0.191	—	1956
North District	Miami	5.52	29.0	3,350	110	12	12.2	0.61	Horizontal	—
West Point	Seattle	5.49	68.6	930	183	200	0.91	0.114–0.146	—	1965
Sand Island	Honolulu	4.64	67–72	2,780	1,031	285	7.3	0.076–0.090	—	1975
Broward County	—	3.51	32.5	2,120	—	1	—	1.37	Horizontal	—
Hollywood, FL	Hollywood	2.37	28.5	3,060	—	1	—	1.52	Horizontal	—
Delray Beach	Delray Beach	0.72	29	1,590	—	1	—	0.76	Horizontal	—
Boca Raton	Boca Raton	0.60	27.3	1,510	—	1	—	0.91	45°	—

[a]Discharges are average design discharges.

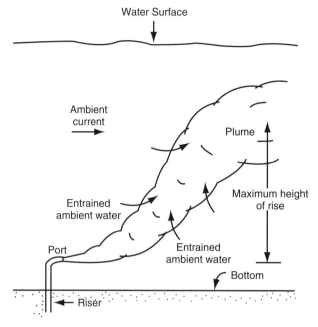

FIGURE 8.5   Single plume. (From Metcalf and Eddy, 1989.)

on the absolute density of the effluent plume and ambient seawater but on the difference in densities and the resultant buoyancy effect. This approximation is called the *Boussinesq approximation*. The buoyancy effect is measured by the effective gravity, g', defined by the relation

$$g' = \frac{\rho_a - \rho_e}{\rho_e} g \qquad (8.2)$$

and the functional expression for the contaminant concentration in the effluent plume at a distance y above the discharge port, Equation 8.1, can be written as

$$c = f_2 (c_e, u_e, D, g', u_a, y) \qquad (8.3)$$

This relationship can be simplified by defining a *volume flux*, Q_0, *specific momentum flux*, M_0, and *specific buoyancy flux*, B_0, by the relations

$$Q_0 = u_e \pi \left(\frac{D}{2} \right)^2 \qquad (8.4)$$

$$M_0 = Q_0 u_e = u_e^2 \pi \left(\frac{D}{2} \right)^2 \qquad (8.5)$$

$$B_0 = Q_0 g' = u_e \pi \left(\frac{D}{2} \right)^2 g' \qquad (8.6)$$

The variables Q_0, M_0, and B_0 involve only u_e, D, and g' and can therefore be used instead of u_e, D, and g' in Equation 8.3 to yield the following functional expression for the contaminant concentration in the effluent plume:

$$c = f_3(c_e, Q_0, M_0, B_0, u_a, y) \tag{8.7}$$

Based on the Buckingham pi theorem, Equation 8.7 can be expressed as a relationship between four dimensionless groups, and the following groupings are particularly convenient:

$$\frac{Q_0 c_e}{u_a L_b^2 c} = f_4\left(\frac{L_M}{L_Q}, \frac{L_M}{L_b}, \frac{y}{L_b}\right) \tag{8.8}$$

where L_M, L_Q, and L_b are length scales defined by the following relations:

$$L_Q = \frac{Q_0}{M_0^{1/2}} = \frac{\sqrt{\pi}}{2} D \tag{8.9}$$

$$L_M = \frac{M_0^{3/4}}{B_0^{1/2}} \tag{8.10}$$

$$L_b = \frac{B_0}{u_a^3} \tag{8.11}$$

The length scale L_Q measures the distance over which the port geometry influences the motion of the plume, the length scale L_M measures the distance over which the initial momentum effect is important, and the length scale L_b measures the distance to where the ambient current begins to become more important than the plume buoyancy in controlling the motion of the plume. The length scale L_M is sometimes referred to as the *jet/plume transition length scale*, and L_b is sometimes referred to as the *plume/crossflow length scale* (Méndez-Díaz and Jirka, 1996). In practical terms, if x is the distance along the plume centerline, then the port geometry influences plume dilution where $x < L_Q$, specific momentum flux controls plume dilution where $L_Q < x < L_M$, specific buoyancy flux controls plume dilution where $L_M < x < L_b$, and ambient currents control plume dilution where $x > L_b$.

Defining the plume dilution, S, by the relation

$$S = \frac{c_e}{c} \tag{8.12}$$

the functional relationship given by Equation 8.8 can be written in the form

$$\frac{S Q_0}{u_a L_b^2} = f_4\left(\frac{L_M}{L_Q}, \frac{L_M}{L_b}, \frac{y}{L_b}\right) \tag{8.13}$$

In most sewage outfalls, the port geometry has a relatively minor influence on the dilution of the effluent plume. Under these circumstances, the dilution becomes insensitive to the value of L_Q, and the functional expression for the plume dilution, Equation 8.13, can be written as

$$\frac{S Q_0}{u_a L_b^2} = f_5\left(\frac{L_M}{L_b}, \frac{y}{L_b}\right) \tag{8.14}$$

This relationship can be further reduced by considering the physical meaning of the length-scale ratio L_M/L_b. When $L_M/L_b \gg 1$, the ambient currents overwhelm the plume buoyancy before the buoyancy overwhelms the effluent momentum, and the plume motion can be expected to be dominated by ambient currents rather than buoyancy. On the other hand, when $L_M/L_b \ll 1$, the plume buoyancy becomes an important factor in plume dilution, in advance of the ambient currents dominating plume motion. According to Lee and Neville-Jones (1987), for many ocean outfalls, L_M/L_b is sufficiently small that these plumes are either buoyancy dominated over the entire depth, where $y/L_b < 5$, or influenced by both buoyancy and ambient currents, where $y/L_b > 5$. Lee and Neville-Jones (1987) termed these regimes the *buoyancy-dominated near field* (BDNF) and *buoyancy-dominated far field* (BDFF), respectively, and suggested the following formulas for horizontal plume discharges in unstratified ambient seawater:

$$\text{BDNF:}\quad \frac{SQ_0}{u_a L_b^2} = 0.31 \left(\frac{y}{L_b}\right)^{5/3} \qquad \text{when} \quad \frac{y}{L_b} < 5 \tag{8.15}$$

$$\text{BDFF:}\quad \frac{SQ_0}{u_a L_b^2} = 0.32 \left(\frac{y}{L_b}\right)^{2} \qquad \text{when} \quad \frac{y}{L_b} \geq 5 \tag{8.16}$$

where S is the minimum dilution in the surface boil generated by the discharge and y is the depth of the discharge below the ocean surface. A laboratory snapshot of a plume with significant ambient flow is shown in Figure 8.6. It is relevant to note that neither Equation 8.15 nor 8.16 includes the momentum length scale, L_M, because in both cases the initial discharge momentum does not influence the plume dilution significantly. Equations 8.15 and 8.16 include the blocking effect of the established wastefield at the water surface and can also be used to describe the dilution of vertical discharges since, for buoyancy-dominated discharges, vertical and horizontal discharges behave similarly (Huang et al., 1998). Rearranging Equations 8.15 and 8.16 reveal that the ambient current speed, u_a, is absent from the BDNF equation and the effluent buoyancy, $Q_0 g'$, is absent from the BDFF equation.

FIGURE 8.6   Single plume with ambient flow. (From Wright, 1977.)

Huang and colleagues (1998) have noted that there is a discontinuity in the predictions of Equations 8.15 and 8.16 at $y/L_b = 5$ and that Equation 8.15 does not accurately quantify the plume dilution in stagnant environments, where y/L_b approaches zero. Huang and colleagues (1998) suggested using the following relationship to describe the dilution in the BDNF, transition, and BDFF regimes:

$$\frac{SQ_0}{u_a y^2} = 0.08 \left(\frac{y}{L_b}\right)^{-1/3} + \frac{0.32}{1 + 0.2 \, (y/L_b)^{-0.5}} \qquad (8.17)$$

In cases where the ambient current is equal to zero (i.e., a stagnant environment), the minimum dilution is more conveniently described by

$$S = 0.08 \frac{B_0^{1/3}}{Q_0} y^{5/3} \qquad (8.18)$$

Equation 8.18 is consistent with results derived originally by Rouse et al. (1952), assuming a wastefield thickness of 10 to 15% of the water depth, and this assumed wastefield thickness for individual plumes has been validated experimentally by Lee and Jirka (1981). The application of the near-field dilution formulas is illustrated by the following example.

Example 8.1   The Central District outfall in Miami (Florida) discharges treated domestic wastewater at a depth of 28.2 m from a diffuser containing five 1.22-m-diameter ports spaced 9.8 m apart. The average effluent flow rate is 5.73 m³/s, the 10-percentile ambient current is 11 cm/s, and the density of the ambient seawater is 1.024 g/cm³. The density of the effluent can be assumed to be 0.998 g/cm³. Determine the length scales of the effluent plumes and calculate the minimum dilution. Neglect merging of adjacent plumes.

SOLUTION   Calculate the basic characteristics of the effluent plume.

$$\text{Effective gravity, } g' = \frac{\Delta\rho}{\rho} g = \frac{1.024 - 0.998}{0.998}(9.81) = 0.256 \text{ m/s}^2$$

$$\text{Port discharge, } Q_0 = \frac{5.73}{5} = 1.15 \text{ m}^3/\text{s}$$

$$\text{Port area, } A_p = \frac{\pi}{4}(1.22)^2 = 1.169 \text{ m}^2$$

$$\text{Port velocity, } u_e = \frac{Q_0}{A_p} = \frac{1.15}{1.169} = 0.984 \text{ m/s}$$

$$\text{Momentum flux, } M_0 = Q_0 u_e = 1.15(0.984) = 1.13 \text{ m}^4/\text{s}^2$$

$$\text{Buoyancy flux, } B_0 = Q_0 g' = 1.15(0.256) = 0.294 \text{ m}^4/\text{s}^3$$

The length scales are derived from these plume characteristics as follows:

$$L_Q = \frac{Q_0}{M_0^{1/2}} = \frac{1.15}{(1.13)^{1/2}} = 1.08 \text{ m}$$

$$L_M = \frac{M_0^{3/4}}{B_0^{1/2}} = \frac{(1.13)^{3/4}}{(0.294)^{1/2}} = 2.02 \text{ m}$$

$$L_b = \frac{B_0}{u_a^3} = \frac{0.294}{(0.11)^3} = 221 \text{ m}$$

Based on these length scales, it is to be expected that port geometry will only be important within 1.08 m of the discharge port, buoyancy will be the dominant factor in plume motion after 2.02 m, and the ambient current will not dominate the plume motion before the plume surfaces ($y \ll L_b$). Using Equation 8.15 to calculate the dilution yields

$$\frac{S(1.15)}{0.11(221)^2} = 0.31 \left(\frac{28.2}{221} \right)^{5/3} \tag{8.19}$$

which gives $S = 47$.

Using the equation proposed by Huang et al. (1998), Equation 8.17, yields

$$\frac{S(1.15)}{0.11(28.2)^2} = 0.08 \left(\frac{28.2}{221} \right)^{-1/3} + \frac{0.32}{1 + 0.2(28.2/221)^{-0.5}}$$

which gives $S = 28$. Since the Huang et al. (1998) equation is supported by a wide range of field data, accounts for transitional effects, and is asymptotically correct as the currents approach zero, this equation is given more weight and dilution at the Central District outfall is estimated as 28.

In cases where the ocean is density stratified, there is no guarantee that the plume will reach the ocean surface, and the previous equations for unstratified receiving waters, Equations 8.15 to 8.18, are not applicable. For a linear density gradient in a stagnant environment ($u_a = 0$), the maximum height of rise, z_{max}, can be estimated using the relation

$$\boxed{z_{\text{max}} = 3.98 B_0 \, \epsilon^{-3/8}} \tag{8.20}$$

where ϵ is a stratification parameter defined by

$$\epsilon = -\frac{g}{\rho_0} \frac{d\rho_a}{dz} \tag{8.21}$$

where ρ_0 is the ambient density at the discharge location ($z = 0$) and $\rho_a(z)$ is the ambient density at a distance z above the discharge location. The stratification parameter, ϵ, given by Equation 8.21, is related to the *Brunt–Vaisala* or *buoyancy frequency, N*, by the relation

$$N^2 = \epsilon = -\frac{g}{\rho_0} \frac{d\rho_a}{dz} \tag{8.22}$$

and many engineers prefer to work in terms of the buoyancy frequency, N, rather than the stratification parameter, ϵ (e.g., Rubin and Atkinson, 2001). Typical values of the buoyancy frequency, N, are on the order of 10^{-3} Hz. The centerline dilution, S, at the terminal level, z_{max}, can be estimated using the relation (Morton et al, 1956; Jirka and Lee, 1994)

$$
S = 0.071 \frac{B_0^{1/3}}{Q_0} z_{max}^{5/3}
\tag{8.23}
$$

In cases where there is an ambient current ($u_a \neq 0$), the maximum rise height can be estimated using the relation (Wright, 1977)

$$
z_{max} = 1.8 \left(\frac{u_a}{\epsilon^{1/2}} \right)^{2/3} \left(\frac{B_0}{u_a^3} \right)^{1/3}
\tag{8.24}
$$

Wright (1977) showed that in the case of a nonzero current, mixing up to the maximum rise height, z_{max}, is insensitive to stratification, and the dilution can be estimated using Equation 8.16 taking y equal to z_{max}.

Line Plumes   Plumes discharged from multiport diffusers generally increase in diameter as they rise in the water column, and in some cases adjacent plumes merge together prior to reaching their terminal rise height. In cases where adjacent plumes merge, the dilution equations for single plumes cannot be used to predict the near-field dilution. Laboratory experiments by Papanicolaou (1984) on vertical axisymmetric plumes indicate that the plume radius, R, at a distance y above the discharge port can be estimated by

$$
R = R_0 + 0.105y
\tag{8.25}
$$

where R_0 is the initial radius of the plume. Therefore, if the spacing between adjacent ports is s_p, merging occurs whenever

$$
R - R_0 = \frac{s_p}{2}
\tag{8.26}
$$

and combining Equations 8.25 and 8.26 gives

$$
s_p = 0.21y
\tag{8.27}
$$

This important relation states that whenever the port spacing exceeds 21% of the diffuser depth, adjacent plumes are not expected to merge, and single-plume equations can be used to estimate the near-field dilution. For standard diffusers in shallow waters, maximum dilution is achieved using the closest port spacing such that adjacent plumes do not merge (Wood et al., 1993). Shallow-water diffusers are typically used in depths less than 30 m and are common off the east coast of the United States and around countries such as New Zealand that have extensive continental shelves. In deeper waters it is seldom practical to have the large port spacings necessary to prevent adjacent plumes from merging, and

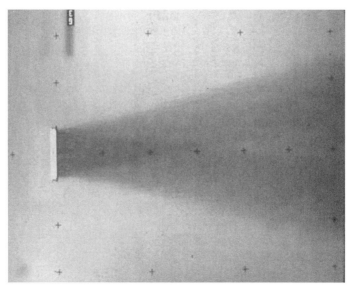

FIGURE 8.7   Plan view of a line plume. (Courtesy of Philip J. W. Roberts.)

plumes from individual ports typically merge together well below the rise height. In cases where adjacent plumes merge together well below the rise height, and the diffuser length is much greater than the water depth, the effluent plumes behave very much as if they were discharged from a slot rather than from separate ports. A plan view of a (laboratory-scale) line plume with an ambient current perpendicular to the diffuser is shown in Figure 8.7. Line plumes are characteristic of deepwater outfalls, such as those typically found off the west coast of the United States.

Consider the case of a buoyant jet that is discharged through a slot of width B at velocity u_e; the effluent has a density ρ_e and contains a contaminant at concentration c_e. Consider further that the ambient ocean has a density ρ_a (assuming unstratified conditions), a depth-averaged velocity u_a, and we are interested in calculating the concentration c at a distance y above the discharge slot. The relationship between the contaminant concentration, c, and the parameters controlling the dilution of the effluent plume can be written in the functional form

$$c = f_1'(c_e, u_e, B, \rho_e, \rho_a, g, u_a, y) \tag{8.28}$$

As in the case of a single round plume, it can be assumed that the density differences are small compared with the absolute densities (the Boussinesq assumption), and therefore the kinematics of the plume does not depend explicitly on the absolute densities of the effluent plume and ambient seawater but on the difference in densities and the resultant buoyancy effect, which is parameterized by the effective gravity, g', defined by Equation 8.2. The functional expression for the contaminant concentration in the effluent plume at distance y above the discharge port, Equation 8.28, can therefore be written as

$$c = f_2'(c_e, u_e, B, g', u_a, y) \tag{8.29}$$

This relationship can be simplified by defining the volume flux, q_0, specific momentum flux, m_0, and specific buoyancy flux, b_0, by the relations

$$q_0 = u_e B \tag{8.30}$$

$$m_0 = q_0 u_e = u_e^2 B \tag{8.31}$$

$$b_0 = q_0 g' = u_e g' B \tag{8.32}$$

The variables q_0, m_0, and b_0 involve only u_e, B, and g', and they can therefore be used instead of u_e, B, and g' in Equation 8.29 to yield the following functional expression for the contaminant concentration in the effluent plume:

$$c = f_3'(c_e, q_0, m_0, b_0, u_a, y) \tag{8.33}$$

On the basis of the Buckingham pi theorem, Equation 8.33 can be expressed as a relationship between four dimensionless groups, and the following groupings are particularly convenient:

$$\frac{q_0 c_e}{u_a y c} = f_4'\left(\frac{l_M}{l_Q}, \frac{l_M}{l_m}, \frac{y}{l_m}\right) \tag{8.34}$$

where l_M, l_Q, and l_m are length scales defined by the following relations:

$$l_Q = \frac{q_0}{b_0^{1/3}} = \left(\frac{u_e^2 B^2}{g'}\right)^{2/3} \tag{8.35}$$

$$l_M = \frac{m_0}{b_0^{2/3}} = \left(\frac{u_e^4 B}{g'^2}\right)^{1/3} \tag{8.36}$$

$$l_m = \frac{m_0}{u_a^2} = \frac{u_e^2 B}{u_a^2} \tag{8.37}$$

The length scale l_Q measures the distance over which the port geometry influences the motion of the plume, the length scale l_M measures the distance to where the plume buoyancy begins to become more important than the discharge momentum in controlling the motion of the plume, and the length scale l_m measures the distance to where the ambient current begins to become more important than the jet momentum in controlling the motion of the plume. The length scale l_Q is sometimes referred to as the *discharge/buoyancy length scale*, the length scale l_M is sometimes referred to as the *jet/plume transition length scale*, and l_m is sometimes referred to as the *jet/crossflow length scale* (Méndez-Díaz and Jirka, 1996). Defining the plume dilution, S, by the relation

$$S = \frac{c_e}{c} \tag{8.38}$$

the functional relationship given by Equation 8.34 can be written in the form

$$\frac{S q_0}{u_a y} = f_5'\left(\frac{l_M}{l_Q}, \frac{l_M}{l_m}, \frac{y}{l_m}\right) \tag{8.39}$$

In most sewage outfalls, the port geometry has a relatively minor influence on the dilution of the effluent plume. Under these circumstances, the dilution becomes insensitive to the value of l_Q, and the functional expression for the plume dilution, Equation 8.39, can be written as

$$\frac{Sq_0}{u_a y} = f_6'\left(\frac{l_M}{l_m}, \frac{y}{l_m}\right) \tag{8.40}$$

This relationship can be further reduced by considering the physical meaning of the length-scale ratio l_M/l_m. Using the definitions of l_M and l_m given by Equations 8.36 and 8.37,

$$\frac{l_M}{l_m} = \left(\frac{u_a^3}{b_0}\right)^{2/3} \tag{8.41}$$

which measures the relative importance of the ambient flow and buoyancy on the dynamics of the plume motion. An ambient/discharge Froude number, F_a, is commonly used in practice (Roberts, 1977; Méndez-Díaz and Jirka, 1996) and is defined by

$$F_a = \frac{u_a}{b_0^{1/3}} \tag{8.42}$$

The ratio l_M/l_m can be expressed in terms of F_a by the relation

$$\frac{l_M}{l_m} = F_a^2 \tag{8.43}$$

and therefore the functional expression for the plume dilution becomes

$$\frac{Sq_0}{u_a y} = f_7'\left(F_a, \frac{y}{l_m}\right) \tag{8.44}$$

Experiments to determine an empirical equation for the relationship given by Equation 8.44 were conducted by Roberts (1977) for cases where the initial momentum of the jet is negligible. Under these circumstances, the effluent momentum flux is negligible, which means that m_0 is not included in the dimensional analysis and Equation 8.44 becomes

$$\frac{Sq_0}{u_a y} = f_8'(F_a) \tag{8.45}$$

This functional relationship, as derived experimentally by Roberts (1977), is shown in Figure 8.8, where θ indicates the direction of the current relative to the diffuser, and Roberts' Froude number, F, is related to the ambient/discharge Froude number, F_a, by the relation

$$F = F_a^3 \tag{8.46}$$

The results shown in Figure 8.8 clearly indicate that currents perpendicular to the diffuser produce the greatest dilutions. For example, when $F = F_a^3 \approx 100$, the perpendicular alignment

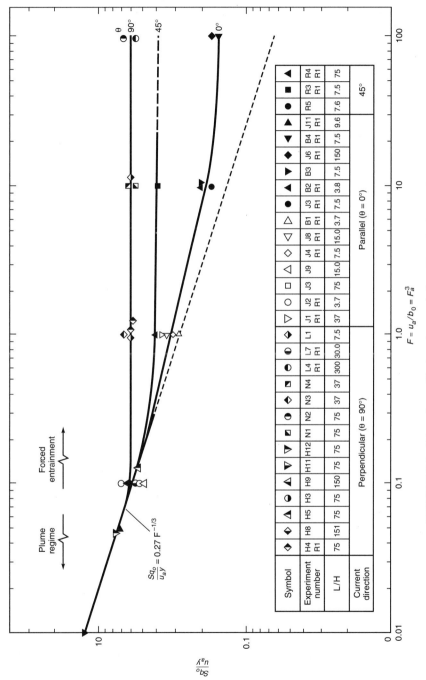

FIGURE 8.8 Line-plume dilution characteristics. (From Roberts, 1977.)

results in a dilution that is four times greater than that for a parallel alignment. In cases where F_a is small, the ambient flow has a minimal effect compared to the effect of buoyancy on the plume dynamics. Under these circumstances, the plume behavior is approximately the same as for a stagnant ambient, which according to Roberts (1977) is given by

$$\boxed{\frac{Sq_0}{u_a y} = 0.27 F_a^{-1}} \tag{8.47}$$

Fischer et al. (1979) have noted that the laboratory experiments of Roberts (1977) were performed at low Reynolds numbers, and application of the results to field-scale diffusers was left unresolved. Subsequent analyses by Wood et al. (1993) indicate that the Roberts results are likely to yield conservative estimates of dilution at field scales. Experimental studies by Méndez-Díaz and Jirka (1996) indicate that ambient currents have a significant effect on the plume whenever $F_a > 0.6$ and that the Roberts results may be limited to cases where the effluent jets are located very close to the bottom of the water column.

Example 8.2 The Central District outfall in Miami (Florida) discharges treated domestic wastewater at a depth of 28.2 m from a diffuser containing five 1.22-m-diameter ports spaced 9.8 m apart. The average effluent flow rate is 5.73 m³/s, the 10-percentile ambient current is 11 cm/s, and the density of the ambient seawater is 1.024 g/cm³. The density of the effluent can be assumed to be 0.998 g/cm³. Assuming that the diffuser can be treated as a line source, calculate the expected minimum dilution for currents perpendicular and parallel to the diffuser. Assess whether it is reasonable to treat the diffuser as a line source.

SOLUTION   For five ports spaced 9.8 m apart, the length of the diffuser, L, can be taken as

$$L = 4(9.8) = 39.2 \text{ m}$$

and the effective gravity, g', is

$$g' = \frac{\rho_a - \rho_e}{\rho_e} g = \frac{1.024 - 0.998}{0.998}(9.81) = 0.256 \text{ m/s}^2$$

The volume flux, q_0 is given by

$$q_0 = \frac{Q}{L} = \frac{5.73}{39.2} = 0.146 \text{ m}^2/\text{s}$$

and the buoyancy flux, b_0, is given by

$$b_0 = q_0 g' = 0.146(0.256) = 0.0374 \text{ m}^3/\text{s}^3$$

From the data given, $u_a = 11$ cm/s $= 0.11$ m/s, and hence the Froude number, F, in Figure 8.8 is

$$F = \frac{u_a^3}{b_0} = \frac{0.11^3}{0.0374} = 0.0356$$

Figure 8.8 indicates that for $F = 0.0356$, the minimum dilution is independent of the direction of the current, and that

$$\frac{Sq_0}{u_a y} = 0.27 F_a^{-1} = 0.27 F^{-1/3}$$

$$= 0.27(0.0356)^{-1/3} = 0.82$$

and therefore the minimum dilution, S, for currents at any angle to the diffuser is given by

$$S = 0.82 \frac{u_a y}{q_0} = 0.82 \left[\frac{0.11(28.2)}{0.146} \right] = 17.4$$

The minimum dilution in the waste field above the diffuser is 17.4.

Since the diffuser depth, y, is 28.2 m, the minimum port spacing, s_p, such that adjacent plumes do not merge is given by Equation 8.27 as

$$s_p = 0.21 y = 0.21(28.2) = 5.92\,\text{m}$$

Since the actual port spacing of 9.8 m exceeds the minimum port spacing of 5.92 m, adjacent plumes do not merge and it would be unreasonable to treat the diffuser as a line source. Comparing the calculated line-plume dilution ($= 17.4$) with the calculated single-plume dilution ($= 28$) in Example 8.1 confirms the assertion that multiport diffusers with noninteracting plumes achieve a higher dilution than line plumes for a given diffuser length.

In the case of a stratified ambient where $F < 0.1$ and the stratification profile is linear, the maximum height of rise and minimum dilution in the waste field are found to be independent of the angle of the current relative to the diffuser (Jirka and Lee, 1994). In this case ($F < 0.1$), the maximum rise height, z_{max}, can be estimated using the relation

$$z_{max} = 2.6 g b_0^{1/3} \epsilon^{-1/2} \tag{8.48}$$

where b_0 is the buoyancy flux defined by Equation 8.32, and ϵ is the stratification parameter defined by Equation 8.21. The corresponding plume centerline (minimum) dilution, S, can be estimated by

$$\boxed{S = 0.37 \frac{b_0^{1/3}}{q_0} z_{max}} \tag{8.49}$$

where q_0 discharge per unit length of diffuser. Rubin and Atkinson (2001) have suggested that the coefficient in Equation 8.48 should be 3.6 instead of 2.6, and that the coefficient in Equation 8.49 should be 0.24 instead of 0.37. It is recommended that the most conservative value of the dilution be assumed in design applications. For $0.1 < F < 100$, the rise height and minimum dilution for a perpendicular alignment can be estimated by

$$z_{max} = 2.5 \frac{b_0^{1/2} \epsilon^{-1/2}}{u_a^{1/2}} \tag{8.50}$$

$$S = 0.4 \frac{b_0^{1/3}}{q_0} z_{max} F^{1/6}(2.19 F^{1/6} - 0.52) \tag{8.51}$$

Equations 8.50 and 8.51 are only applicable where z_{max} is less than the water depth and the current is perpendicular to the diffuser. Similar equations for other current alignments can be found in Roberts et al. (1989).

Design Considerations   The primary purpose of a multiport diffuser is to distribute the effluent between discharge ports such that the required initial dilution is achieved. The hydraulic design of diffusers determines the shape and dimensions of the both the diffuser and discharge ports such that (1) the flows are distributed evenly through the discharge ports, (2) the velocities through the discharge ports are sufficient to prevent seawater intrusion into the diffuser, (3) the port diameters are large enough to prevent clogging, and (4) the velocities in the diffuser pipe are sufficient to prevent deposition of suspended material within the pipe.

The discharge from each port, Q_0, can be estimated by (Fischer et al., 1979)

$$Q_0 = C_D A_p \sqrt{2g\,\Delta h} \qquad (8.52)$$

where C_D is a discharge coefficient, A_p is the port area, g is the acceleration due to gravity, and Δh is the difference in total head across the port. For ports with rounded entrances cast directly into the wall of the diffuser, C_D can be estimated by (Fischer et al., 1979)

$$C_D = 0.975\left(1 - \frac{V_d^2}{2g\,\Delta h}\right) \qquad (8.53)$$

and for sharp-edged ports

$$C_D = 0.63 - 0.58\,\frac{V_d^2}{2g\,\Delta h} \qquad (8.54)$$

where V_d is the velocity in the diffuser pipe. The head outside a diffuser remains constant along the diffuser and equal to the elevation of the ocean surface, and frictional head losses within the diffuser cause the head difference, Δh, across the ports to decrease with distance along the diffuser. According to Equation 8.52, to maintain the same discharge through all ports, the port area, A_p, must be increased along the diffuser to compensate for the decrease in the head difference, Δh. Head losses along the diffuser can be calculated using the Darcy–Weisbach equation, and a useful algorithm for calculating head losses and port discharges in diffusers can be found in Fischer et al. (1979). For horizontal diffusers, the ports can be sized such that the port discharges are equal regardless of the total discharge. However, in cases where the diffuser is laid on a slope, the distribution of port discharges depends on the total effluent discharge, and is not the same for all flows.

For a diffuser to flow full, the ratio of the total port area downstream of any pipe section to the area of the pipe section should not exceed 0.7 and should ideally be between 0.3 and 0.7 (Fischer et al., 1979). This criterion usually requires that the diameter of the diffuser pipe be reduced at discrete points along the diffuser.

Seawater intrusion into the diffuser can be prevented by keeping the port Froude number, F_p, much greater than unity (Wood et al., 1993), where F_p is defined by

$$F_p = \frac{u_e}{\sqrt{g'D}} \qquad (8.55)$$

FIGURE 8.9   Diffuser ports. (Courtesy of Philip J. W. Roberts.)

where u_e is the port discharge velocity, g' is the effective gravity, and D is the port diameter. Ports must be smooth, bellmouthed, large enough to prevent clogging, and made of material resistant to mussel and weed growth. There is much debate concerning minimum port sizes, with typical recommended sizes of 200 mm for unscreened effluent (Brown, 1988), 65 mm for secondary-treated effluent, and 50 mm for tertiary-treated effluent (Wilkinson, 1990). In some cases, the diffuser port is covered by a rubber valve that collapses to prevent seawater flow into the diffuser. Such a rubber port valve is shown in Figure 8.9, where a diffuser port without the valve is also shown.

The diffuser diameter must be such that the velocity in the diffuser exceeds the critical velocity required to prevent the deposition of suspended solids. The critical velocity generally depends on the level of treatment of the effluent, and guidelines for estimating the critical velocity can be found in Ackers (1991). As effluent is discharged through ports along the pipe, the flow rate in the diffuser pipe decreases gradually, and the diffuser diameter has to be reduced to maintain the pipe velocity above the critical value.

Designing the length and port spacing in shallow-water diffusers is usually a fairly straightforward process in which the required initial dilution, S, and the diffuser depth, y are both given, and the required port discharge, Q_0, can be calculated using Equation 8.17 or 8.18 for nonmerging plumes. Dividing the total effluent discharge by Q_0 gives the number of ports, N_p, and the port spacing, s_p, must be at least equal to 21% of the depth, as given by Equation 8.27. Practical considerations usually require a port spacing that is a multiple or fraction of the length of a pipe section. The required diffuser length, L, can be calculated using the relation

$$L = (N_p - 1)s_p \tag{8.56}$$

There are a wide variety of designs used in practice, with port diameters typically between 10 and 30 cm and port spacings between 1 and 10 m (Rubin and Atkinson, 2001). The ports may be simple holes cut into the distribution pipe or they may be nozzles at the tips of risers that direct flow from the pipe. Designing deepwater outfalls is more complex than designing shallow-water outfalls, since density gradients in the water column usually cause the effluent plumes to be trapped below the water surface, and merging of adjacent plumes are more commonplace. Detailed guidance on the design of deepwater outfalls may be found in Fischer et al. (1979) and Wood et al. (1993).

Example 8.3   A diffuser is to be located in 30 m of water, discharge 5.73 m³/s of secondary-treated domestic wastewater, and provide an initial dilution of 20:1. The critical velocity required to prevent deposition in the diffuser is 60 cm/s, and the total head at the upstream port will be maintained at 32 m by an onshore pumping station. The density of the wastewater can be taken as 998 kg/m³, the seawater density is 1024 kg/m³, and the design ambient current is 11 cm/s. Design the length of the diffuser, port spacing, number of ports, and the diameter of the most upstream port in the diffuser.

SOLUTION   From the data given, $S = 20$, $y = 30$ m, $Q = 5.73$ m³/s, $\rho_e = 998$ kg/m³, $\rho_a = 1024$ kg/m³, and $u_a = 11$ cm/s $= 0.11$ m/s. The effective gravity, g', can be derived from the given data as

$$g' = \frac{\rho_a - \rho_e}{\rho_e} g = \frac{1024 - 998}{998}(9.81) = 0.256 \text{ m/s}^2$$

and the plume/crossflow length scale, L_b, can be derived as

$$L_b = \frac{B_0}{u_a^3} = \frac{Q_0 g'}{u_a^3} = \frac{Q_0(0.256)}{0.11^3} = 192 Q_0$$

The diffuser will be designed as a shallow-water diffuser with nonmerging plumes. The dilution of each individual plume is given by Equation 8.17 as

$$\frac{SQ_0}{u_a y^2} = 0.08 \left(\frac{y}{L_b} \right)^{-1/3} + \frac{0.32}{1 + 0.2(y/L_b)^{-0.5}}$$

$$\frac{20 Q_0}{0.11(30)^2} = 0.08 \left(\frac{30}{192 Q_0} \right)^{-1/3} + \frac{0.32}{1 + 0.2(30/192 Q_0)^{-0.5}}$$

which gives

$$Q_0 = 1.84 \text{ m}^3/\text{s}$$

Since the total discharge, Q, is 5.73 m³/s, the required number of ports, N_p, is given by

$$N_p = \frac{Q}{Q_0} = \frac{5.73}{1.84} = 3.1 \text{ ports}$$

Therefore, four ports should be used in the diffuser. Using four ports gives a discharge of $5.73/4 = 1.43$ m³/s per port, and Equation 8.17 gives a corresponding dilution of 25.

Since the diffuser depth, y, is 30 m, the minimum port spacing is given by Equation 8.27 as

$$s_p = 0.21y = 0.21(30) = 6.3 \, \text{m}$$

Using this port spacing, the required diffuser length, L, is given by

$$L = (N_p - 1)s_p = (4 - 1)6.3 = 18.9 \, \text{m}$$

For a critical velocity, V_c, of 60 cm/s, the maximum allowable diameter, D_d, of the diffuser is given by

$$\frac{\pi}{4}D_d^2 = \frac{Q}{V_c} = \frac{5.73}{0.60}$$

which yields

$$D_d = 3.48 \, \text{m}$$

Therefore, any diffuser with diameter less than or equal to 3.48 m will be sufficient to maintain a self-cleansing velocity in the diffuser.

At the most upstream port in the diffuser, the head difference, Δh, across the port is given by

$$\Delta h = 32 \, \text{m} - 30 \, \text{m} = 2 \, \text{m}$$

Using ports with rounded entrances cast directly into the wall of the diffuser, Equation 8.53 gives the coefficient of discharge, C_D, as

$$C_D = 0.975 \left(1 - \frac{V_d^2}{2g \, \Delta h} \right)$$

$$= 0.975 \left(1 - \frac{0.60^2}{2(9.81)(2)} \right) = 0.966$$

The port discharge is then given by Equation 8.52 as

$$Q_0 = C_D A_p \sqrt{2g \, \Delta h}$$

$$1.43 = 0.966 \left(\frac{\pi}{4}D_p^2 \right) \sqrt{2(9.81)(2)}$$

which yields $D_p = 0.549$ m. Therefore, a port diameter of 0.549 m is required to discharge the effluent at the required discharge rate under the available head difference. The port area, A_p, is given by

$$A_p = \frac{\pi}{4}D_p^2 = \frac{\pi}{4}(0.549)^2 = 0.237 \, \text{m}^2$$

the effluent velocity, u_e, is given by

$$u_e = \frac{Q_0}{A_p} = \frac{1.43}{0.237} = 6.03 \, \text{m/s}$$

and the port Froude number, F_p, is given by Equation 8.55 as

$$F_p = \frac{u_e}{\sqrt{g'D_p}} = \frac{6.03}{\sqrt{0.256(0.549)}} = 16.1$$

Since $F_p \gg 1$, saltwater intrusion into the diffuser is not expected to be a problem. Also, since the minimum port diameter recommended for discharging secondary-treated effluent is 65 mm, clogging of the port is not expected to be a problem.

In summary, the diffuser should be 18.9 m long, 3.48 m in diameter (maximum), and have four ports spaced 6.3 m apart. The first port along the diffuser should have a diameter of 0.549 m.

8.2.2 Far-Field Mixing

Far-field mixing occurs after initial plume dilution (near-field mixing). Far-field mixing is dominated by spatial and temporal variations in ocean currents, and far-field models are generally applicable when the momentum and buoyancy fluxes of the effluent plume are overwhelmed by the ocean currents. The transition from near field to far field is illustrated in Figure 8.10, where the concentrated boil is caused by the surfacing plume, and advection of the plume downstream is indicative of the far-field transport mechanism. The dispersion coefficient in the ocean increases with the size of the contaminant plume, a fact that is observed in practically all tracer experiments in the ocean (Okubo, 1971) and is attributed to the fact that as a tracer cloud grows, the cloud experiences a wider range of velocities, which leads to increased growth rates and larger diffusion coefficients. Okubo

FIGURE 8.10   Transition from near- to far-field mixing. (From Wood, 2005a.)

(1971) analyzed the results of several field-scale dye experiments and derived an empirical expression for the oceanic diffusion coefficients as a function of the size of the tracer cloud. Within observed tracer clouds, Okubo (1971) used the area enclosed by each concentration contour to define a circle of radius r_e enclosing the same area as the irregular concentration contour, and then calculated the variance, σ_{rc}^2, of the entire tracer cloud using the relation

$$\sigma_{rc}^2 = \frac{\int_0^\infty r_e^2 c 2\pi r_e \, dr_e}{\int_0^\infty c 2\pi r_e \, dr_e} \tag{8.57}$$

where $c(r_e, t)$ is the concentration with an equivalent circular contour of radius r_e at time t. The characterization of the variance of a tracer distribution by σ_{rc}^2 can be compared with the variance characterization of bivariate Gaussian distributions, which requires specification of the variances along both the major and minor principal axes. If the tracer distribution is Gaussian and the variances along the major and minor axes are σ_x^2 and σ_y^2, respectively, it can be shown that

$$\sigma_{rc}^2 = 2\sigma_x \sigma_y \tag{8.58}$$

Okubo (1971) plotted σ_{rc}^2 versus time for several instantaneous dye releases in the surface layers of coastal waters and showed that the following empirical relationship provided a reasonably good fit to the data observed:

$$\boxed{\sigma_{rc}^2 = 0.0108 t^{2.34}} \tag{8.59}$$

where σ_{rc}^2 is measured in cm^2 and t is the time since release, measured in seconds. Okubo (1971) fitted data for times ranging from 2 hours to nearly 1 month. The variance of a tracer cloud as a function of time can be used to calculate an apparent diffusion coefficient, K_a, using the relation

$$K_a = \frac{\sigma_{rc}^2}{4t} \tag{8.60}$$

Defining the length scale, L, of the tracer cloud by

$$L = 3\sigma_{rc} \tag{8.61}$$

Okubo (1971) used the results of field-scale dye studies to plot the apparent diffusion coefficient, K_a, versus the length scale, L, of the cloud. These results are shown in Figure 8.11. These data show a good fit to the empirical relation

$$\boxed{K_a = 0.0103 L^{1.15}} \tag{8.62}$$

where K_a is in cm^2/s and L is in centimeters. Equation 8.62 is widely used in practice to estimate the apparent diffusion coefficient as a function of length scale for contaminants

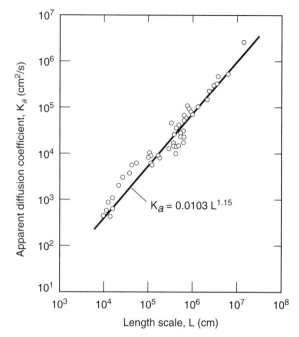

FIGURE 8.11   Apparent diffusion coefficient versus length scale in coastal waters. (From Chin, David A., *Water-Resources Engineering.* Copyright © 2000. Reprinted by permission of Pearson Education, Inc., Upper Saddle River, NJ.)

released into the ocean. More recent studies reported by Zimmerman (1986) and Roberts (1999) have shown that for travel times of a few hours, typical apparent diffusion coefficients are on the order of 0.1 to 1 m^2/s; for time scales up to a day or so, vertical velocity shear is usually most important, resulting in apparent diffusion coefficients on the order or 10 m^2/s; and for longer times, horizontal shear dominates and the apparent diffusion coefficient may grow to 100 to 1000 m^2/s.

Example 8.4   (a) Compare the apparent diffusion coefficient of an oil spill having a characteristic size of 50 m with an oil spill whose size is 100 m. (b) Estimate how long it would take for a small oil spill to grow to a size of 50 m.

SOLUTION   (a) The apparent diffusion coefficient, K_a, is related to the length scale, L, of the oil spill by

$$K_a = 0.0103 L^{1.15} \qquad \text{cm}^2/\text{s}$$

When $L = 50\,\text{m} = 5000\,\text{cm}$,

$$K_a = 0.0103(5000)^{1.15} = 185\,\text{cm}^2/\text{s}$$

and when $L = 100\,\text{m} = 10{,}000\,\text{cm}$,

$$K_a = 0.0103(10{,}000)^{1.15} = 410\,\text{cm}^2/\text{s}$$

Therefore, when the oil spill doubles its size from 50 m to 100 m, the apparent diffusion coefficient more than doubles, going from 185 cm²/s to 410 cm²/s.

(b) The variance of the oil spill as a function of time can be estimated by

$$\sigma_{rc}^2 = 0.0108t^{2.34} \quad cm^2$$

or

$$\sigma_{rc} = 0.104t^{1.17} \quad cm$$

Defining the length scale, L, by $L = 3\sigma_{rc}$,

$$L = 0.312t^{1.17} \quad cm$$

When $L = 50\,m = 5000\,cm$,

$$5000 = 0.312t^{1.17} \quad cm$$

which leads to

$$t = 3930\,s = 65.5\,min = 1.09\,h$$

Therefore, a small oil spill will take approximately 1.09 h to grow to a size of 50 m.

For multiport diffusers, far-field models typically assume that the contaminant source is a rectangular plane area perpendicular to the mean current at the trapping level of the plume, as shown in Figure 8.12, where the width of the plane area (source) is equal to the

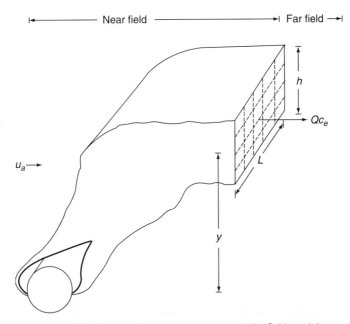

FIGURE 8.12   Interface between near- and far-field models.

length of the diffuser and the depth of the plane area is equal to the plume thickness at the trapping height (e.g., Chin, 1985; Huang et al., 1996). For surfacing plumes in unstratified environments, the initial plume thickness can be taken as 30% of the total depth (Koh, 1983). Assuming that the contaminant mass flux leaving the diffuser is equal to the contaminant mass flux across the source plane at the trapping level,

$$Qc_e = u_a Lhc_0 \tag{8.63}$$

where Q is the effluent volume flux, c_e is the contaminant concentration in the effluent, u_a is the ambient velocity, L is the diffuser length, h is the height of the waste field, and c_0 is the contaminant concentration crossing the source plane. Equation 8.63 can be put in the more convenient form

$$\boxed{c_0 = \frac{Qc_e}{u_a Lh}} \tag{8.64}$$

Brooks (1960) proposed a simple far-field model that assumes a constant ambient velocity, u_a, and neglects both vertical diffusion and diffusion in the flow direction. These assumptions are justified by the observations that the diffusive flux in the flow direction is usually much smaller than the advective flux, vertical diffusion is damped significantly by buoyancy, and the vertical diffusion coefficient is much smaller than the transverse diffusion coefficient. The governing steady-state advection–diffusion equation is therefore given by

$$\boxed{u_a \frac{\partial c}{\partial x} = \frac{\partial}{\partial y}\left(\varepsilon_y \frac{\partial c}{\partial y}\right)} \tag{8.65}$$

with boundary conditions

$$c(0, y) = \begin{cases} c_0, & |y| < \dfrac{L}{2} \\ 0, & |y| > \dfrac{L}{2} \end{cases} \tag{8.66}$$

$$c(x, \pm\infty) = 0 \tag{8.67}$$

where x is in the direction of flow, y is in the transverse (horizontal) direction, and ε_y is the transverse diffusion coefficient. Brooks (1960) assumed that the transverse diffusion coefficient, ε_y, increases with the size of the plume in accordance with the *four-thirds law* originally proposed by Richardson (1926) and supported by the field results of Okubo (1971). According to the four-thirds law, the transverse diffusion coefficient, ε_y, is given by

$$\varepsilon_y = k\sigma_y^{4/3} \tag{8.68}$$

where σ_y is the standard deviation of the cross-sectional concentration at a distance x from the source and k is a constant. An analytic expression for the resulting concentration distribution is not available, but the maximum concentration (along the plume centerline) is given by

$$c(x,0) = c_0 \text{ erf} \left[\sqrt{\frac{3}{2\left[1 + \frac{2}{3}\left(\beta x/L\right)\right]^3 - 1}} \right]$$

(8.69)

where β is defined by

$$\beta = \frac{12\varepsilon_0}{u_a L}$$

(8.70)

and ε_0 is the transverse diffusion coefficient at $x = 0$. The Brooks (1960) formulation given by Equation 8.69 has been widely used to predict the far-field mixing of ocean-outfall discharges, and its popularity is no doubt due to its simplicity and the fact that the maximum concentration is expressed in terms of measurable quantities.

Models of far-field mixing are still in their infancy, and it is still common practice to neglect spatial variations in the ambient velocity and parameterize the mixing process by an empirical diffusion coefficient that is unrelated to the spatial and temporal characteristics of the currents at the outfall site (e.g., the Brooks model). This pragmatic approach is usually a result of economic constraints, which typically allow for the deployment of only a single current meter in the vicinity of an ocean outfall, thereby precluding any measurement of the spatial characteristics of the velocity field. Ideally, if the ambient currents could be measured at several locations within the far field, far-field mixing of effluent plumes could be simulated much more accurately than is possible at present using conventional methods. More comprehensive models of far-field mixing have been proposed by Chin and Roberts (1985) and Chin and colleagues (1997).

Example 8.5 A 39.2-m-long multiport diffuser discharges treated domestic wastewater at a rate of 5.73 m³/s and at a depth of 28.2 m. If the ambient current is 11 cm/s, estimate the distance downstream of the diffuser to where the dilution is equal to 100.

SOLUTION   Equation 8.64 estimates the initial dilution in the far-field model as

$$\frac{c_e}{c_0} = \frac{u_a L h}{Q}$$

where $u_a = 11 \text{ cm/s} = 0.11 \text{ m/s}$, $L = 39.2 \text{ m}$, $h = 0.3(28.2) = 8.46 \text{ m}$, $Q = 5.73 \text{ m}^3/\text{s}$, and hence

$$\frac{c_e}{c_0} = \frac{u_a L h}{Q} = \frac{0.11(39.2)(8.46)}{5.73} = 6.37$$

The initial plume dilution for the far-field model is therefore estimated as 6.37. This estimate of initial dilution is certainly not as accurate as using a near-field model to estimate the initial dilution, but this approximation is acceptable when the far-field dilution is much greater than the near-field dilution. If a near-field model is used to estimate the initial dilution, Equation 8.64 is still valid, but the initial height of the waste field, h, is no longer taken as 30% of the depth. Equation 8.69 gives the far-field dilution as

$$\frac{c_{max}}{c_0} = \mathrm{erf}\left[\sqrt{\frac{3}{2\left[1 + \frac{2}{3}(\beta x/L)\right]^3 - 1}}\right]$$

and the total dilution, c_e/c_{max}, is given by

$$\frac{c_e}{c_{max}} = \frac{c_e}{c_0}\frac{c_0}{c_{max}} = 6.37\left\{\mathrm{erf}\left[\sqrt{\frac{3}{2\left[1 + \frac{2}{3}(\beta x/L)\right]^3 - 1}}\right]\right\}^{-1} \tag{8.71}$$

where

$$\beta = \frac{12\varepsilon_0}{u_a L} \tag{8.72}$$

The diffusion coefficient at the diffuser, ε_0, can be estimated using the diffuser length, $L = 39.2\,\mathrm{m} = 3920\,\mathrm{cm}$ and the Okubo relation (Equation 8.62) as

$$\varepsilon_0 = 0.0103L^{1.15} = 0.0103(3920)^{1.15} = 140\,\mathrm{cm^2/s} = 0.014\,\mathrm{m^2/s}$$

The parameter, β, is therefore given by Equation 8.72 as

$$\beta = \frac{12\varepsilon_0}{u_a L} = \frac{12(0.014)}{0.11(39.2)} = 0.0390$$

Substituting into Equation 8.71 to find the distance, x, from the diffuser where the dilution is 100 gives

$$100 = 6.37\left\{\mathrm{erf}\left[\sqrt{\frac{3}{2\left[1 + \frac{2}{3}(0.0390x/39.2)\right]^3 - 1}}\right]\right\}^{-1}$$

which simplifies to

$$0.0637 = \mathrm{erf}\left[\sqrt{\frac{3}{2(1 + 0.000663x)^3 - 1}}\right]$$

Using the error function tabulated in Appendix E.1 gives

$$x = 10{,}200\,\mathrm{m} = 10.2\,\mathrm{km}$$

Hence, the dilution reaches 100 at a distance of about 10.2 km downstream of the diffuser.

The Brooks model of far-field dispersion assumes that the contaminant in the discharged wastewater is conservative. In cases where the contaminant is not conservative and the decay process can be approximated by a first-order reaction with decay parameter k, the concentration distribution along the plume centerline given by Equation 8.156 can be multiplied by e^{-kx/u_a} to give

$$c(x, 0) = c_0 e^{-kx/u_a} \, \text{erf}\left[\sqrt{\frac{3}{2\left[1 + \frac{2}{3}\left(\beta x/L\right)\right]^3 - 1}} \right] \tag{8.73}$$

The decay coefficient, k, is frequently expressed in terms of the time, T_{90}, required for 90% reduction in mass due to decay, where

$$T_{90} = \frac{\ln 10}{k} \approx \frac{2.30}{k} \tag{8.74}$$

Fecal coliforms are nonconservative tracers that are frequently used as indicator organisms in far-field dispersion models, particularly with regard to assessing the potential contamination of beaches by pathogenic microorganisms. Inactivation of fecal coliforms in the ocean is caused primarily by solar radiation, with secondary causes of inactivation being nutrient deficiency, temperature, and predation by natural microbiota. Field studies from around the world have reported T_{90} values for fecal coliform in the range 0.6 to 24 hours in daylight and 60 to 100 hours at night (Wood et al., 1993). A recent study in Mamala Bay, Hawaii, showed that during the daytime hours, a conservative estimate of T_{90} for *E. coli* was 6.9 h, and during the night negligible decay could be assumed (Roberts, 1999). Decay is seldom considered in near-field models, since the time scale of near-field mixing is typically on the order of minutes, which is much shorter than the time scale of decay processes. In far-field models, it is reasonable to neglect decay if the travel time is less than the nighttime duration. Worst-case scenarios for assessing the impact of ocean outfalls on beaches correspond to assuming a persistent onshore (wind-induced) current during the nighttime hours. Such worst-case scenarios have been observed in ocean outfalls (e.g., Stewart, 1973).

Example 8.6 An ocean outfall discharges domestic wastewater 6 km offshore, and under worst-case conditions an onshore wind causes surface currents of 22 cm/s to move the contaminant plume directly toward recreational beaches. A near-field model indicates an initial dilution of 20, and application of the Brooks model for a conservative contaminant indicates a dilution of 110 between the beaches and the zone of initial dilution (ZID). If the fecal coliform concentration in the wastewater effluent is 40,000 cfu per 100 mL and the time for 90% decay in seawater is 6 h, determine the fecal coliform concentration on the beaches. How does this compare with the coliform concentration if decay is neglected?

SOLUTION From the data given, $x = 6\,\text{km} = 6000\,\text{m}$, $u_a = 22\,\text{cm/s} = 0.22\,\text{m/s}$, $c_e = 40,000\,\text{cfu}$ per 100 mL, and the near- and far-field dilutions (neglecting decay) are given by

$$\frac{c_e}{c_0} = 20 \quad \text{and} \quad \frac{c_0}{c_b} = 110$$

where c_e is the coliform concentration in the effluent, c_0 is the coliform concentration after initial dilution, and c_b is the coliform concentration on the beaches. Neglecting decay, c_b is given by

$$c_b = c_e \frac{c_0}{c_e} \frac{c_b}{c_0} = 40{,}000 \left(\frac{1}{20}\right)\left(\frac{1}{110}\right) = 18/100\,\text{mL}$$

If decay is considered, $T_{90} = 6\,\text{h}$ and the decay coefficient, k, is given by Equation 8.74 as

$$k = \frac{\ln 10}{T_{90}} = \frac{2.30}{6} = 0.383\,\text{h}^{-1} = 0.000106\,\text{s}^{-1}$$

and

$$\frac{c_b}{c_0} = \frac{1}{110} e^{-kx/u_a} = \frac{1}{110} e^{-(0.000106)(6000)/0.22} = 0.000505$$

Therefore, the far-field dilution is given by

$$\frac{c_0}{c_b} = \frac{1}{0.000505} = 1980$$

and the fecal coliform concentration on the beaches, c_b, is given by

$$c_b = c_e \frac{c_0}{c_e} \frac{c_b}{c_0} = 40{,}000 \left(\frac{1}{20}\right)\left(\frac{1}{1980}\right) = 1/100\,\text{mL}$$

Therefore, consideration of coliform decay yields an order-of-magnitude reduction in the expected level of fecal coliforms on the beaches.

8.3 WATER-QUALITY CONTROL IN ESTUARIES

The high dissolved-solids concentrations and salinities of estuarine waters typically make them impractical for use as water-supply sources. However, estuarine waters, along with adjacent wetlands, are extremely productive habitats that support large populations of commercial fish, shellfish, and wildlife. Estuaries typically contain many different types of habitats, including shallow open waters, freshwater and salt marshes, sandy beaches, mud and sand flats, rocky shores, oyster reefs, mangrove forests, tidal pools, sea grass, kelp beds, and wooded swamps. Estuaries are visited by a number of migratory species that breed either in freshwater (e.g., salmon) or in salt water (e.g., American eel) and contamination of estuaries can have much more serious consequences for aquatic systems than would be apparent from a casual examination of the kinds of organisms present in the estuary at any one time. Tens of thousands of birds, mammals, fish, and other wildlife depend on estuarine habitats as places to live, feed, and reproduce. For example, pelicans are commonly found in estuaries, and a cluster of pelicans in the San Jose estuary (in Mexico) is shown in Figure 8.13. Estuaries provide ideal spots for migratory birds to rest and refuel during their journeys, and many species of fish and shellfish rely on the sheltered waters of estuaries as protected places to spawn, giving them the nickname "nurseries of the sea."

FIGURE 8.13 Pelicans in San Jose estuary, Mexico. (From Graciela Tiburcio Pintos, 2005.)

Hundreds of marine organisms, including most commercially valuable fish species, depend on estuaries at some point during their development. From the human perspective, tourism, fisheries, and other commercial activities thrive on the wealth of natural resources estuaries supply. Estuaries provide habitat for more than 75% of commercial fish catch in the United States and for 80 to 90% of the recreational fish catch.

Pathogenic bacteria and viruses do not generally pose a threat to estuarine aquatic life; however, shellfish can accumulate pathogens, causing disease when harvested and consumed by humans. Therefore, the harvest and sale of shellfish from estuaries and ocean waters with excess bacteria are strictly regulated.

A very important water-quality issue in estuaries is excess nutrients such as nitrogen and phosphorus. As in lakes, nutrient discharges from human activities can promote enhanced growth of algae, resulting in eutrophic conditions that can stifle fish growth and damage shellfish habitats. It has been estimated that 53% of estuaries in the United States experience hypoxia (reduced oxygen levels) or anoxia for at least part of the year (Novotny, 2003). Harmful marine algal blooms caused primarily by increased nutrient discharges into estuaries have been responsible for an estimated $1 billion in economic losses during the 1990s.

Many large population and industrial centers have developed adjacent to estuaries because of the easy access to both the ocean and inland river systems for water transportation. The wastes from these large population/industrial centers have often been discharged into estuarine waters, with little or no awareness of the biological importance of the receiving estuary or of the tendency of pollutants to be recycled within the estuary. In the United States, the National Estuary Program (NEP), established by Congress in 1987 in amendments to the Clean Water Act, has the primary objective of protecting estuaries of national significance that are threatened by degradation caused by human activity. The program is administered by the U.S. Environmental Protection Agency (USEPA), and a sample of estuaries participating in the NEP are listed in Table 8.2. It is apparent from the table

Table 8.2   Sample of Major Estuaries in the United States

Estuary	Area (km²)	Watershed Size (km²)	Features	Major Threats
Galveston Bay	1550	10,980	Average depth is 2.1 m.	Variety of contaminants originating from sewage discharges and urban runoff.
Long Island Sound	3420	43,560	Average depth is 19.2 m.	Low dissolved oxygen, toxics, pathogens, floatable debris, living resources and habitat. Domestic and industrial wastewater are significant sources.
San Francisco Bay	4140	155,400	Two major inflow rivers: Sacramento and San Joaquin. Depth of central bay averages 13.1 m, southern and northern areas 4.6–5.2 m, deepest point 110 m below sea level, under Golden Gate Bridge.	Variety of contaminants originating from sewage discharges, agricultural and urban runoff.
Tampa Bay	1040	5700	Four major inflow rivers: Hillsborough, Alafia, Manatee, and Little Manatee. On average, only 3.7 m deep.	Excess nitrogen from stormwater and wastewater discharges, habitat loss.

that sewage discharges and runoff from urban and agricultural areas are the major contributors to water-quality degradation in estuaries.

8.3.1   Classification of Estuaries

Analysis of estuaries can be simplified by using a classification system to compare similar types of estuaries. The most common classifications are based on geomorphology and stratification.

Geomorphological Classification of Estuaries   Geomorphological classification of estuaries categorizes estuaries into one of the following four classes: (1) drowned river valleys (coastal plain estuaries), (2) fjords, (3) bar-built estuaries, and (4) other estuaries produced by tectonic activity, faulting, landslides, and volcanic eruptions (USEPA, 1984). These types of estuaries are described as follows:

1. *Coastal plain estuaries* are gently sloping toward the mouth and are sometimes stratified; examples are Chesapeake Bay, Delaware River estuary, and New York Bight.

2. *Fjords* are formed by glaciation and are typically stratified deep and narrow gorges, found mostly in Norway and Alaska, although Puget Sound is an example of an estuary formed by glacial scouring.

3. *Bar-built estuaries* are coastal bodies of water enclosed by the deposition of a sand-bar off the coast. They are typically well mixed and are found off the Gulf coast or South Atlantic regions of the United States; an example is Pamlico Sound in North Carolina.

4. *Other estuaries* that do not fit into the preceding three categories include San Francisco Bay, which was formed by tectonic forces.

Shallow partially mixed estuaries are very productive and sensitive to the input of nutrients; elongated deep estuaries are typically migration routes for anadromous and catadromous fish and are less sensitive to nutrient inputs.

A satellite view of San Francisco Bay, an example of a tectonic estuary, is shown in Figure 8.14. San Francisco Bay was formed when the block between two active faults was placed under tension and dropped down slowly; the faults are visible on the satellite image. The San Andreas Fault is indicated by arrows on the left, while the trace of the Hayward Fault is indicated by the two arrows on the right side of the image. During the last ice age, sea level was more than 100 m lower than it is today, and this basin was isolated from the Pacific Ocean. With the melting of the glaciers in North America and Europe, sea level rose, and seawater flooded in through the Golden Gate into the estuary. Freshwater is supplied to the estuary by the Sacramento River, shown in the upper right of Figure 8.14.

Stratification Classification of Estuaries   Stratification is most often used for classifying estuaries influenced by tides and freshwater inflows. Three stratification classes of estuaries are:

1. *Highly stratified (salt wedge) estuaries* have large river discharges flowing into them.
2. *Partially mixed estuaries* have medium river discharges.
3. *Vertically homogeneous estuaries* have small river discharges.

The primary parameter than can be used to classify the stratification potential of estuaries is the *Richardson number*, Ri, for estuaries, defined as (Fischer et al., 1979)

$$\mathrm{Ri} = \frac{\Delta\rho}{\rho}\,\frac{gQ_f}{WU_t^3} \tag{8.75}$$

where $\Delta\rho$ is the difference between freshwater and seawater density, typically $25\,\mathrm{kg/m^3}$; ρ is the reference density, typically $1000\,\mathrm{kg/m^3}$; g is gravity ($\mathrm{m/s^2}$), Q_f is the freshwater inflow ($\mathrm{m^3/s}$); W is the width of the estuary (m); and U_t is the mean tidal velocity (m/s). If Ri is large (>0.8), the estuary is expected to be strongly stratified and dominated by density currents, and if Ri is small (<0.08), the estuary is expected to be well mixed. Transition from a well-mixed to a strongly mixed estuary occurs in the range

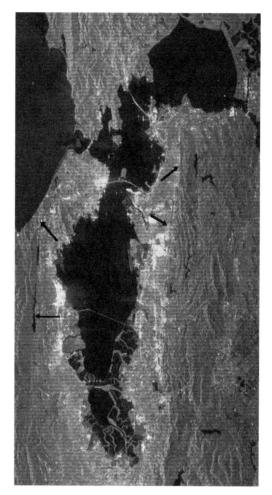

FIGURE 8.14   Satellite view of San Francisco Bay, California. (From NASA, 2005b.)

$0.08 <$ Ri < 0.8. The stratification classifications of several estuaries in the United States are shown in Table 8.3.

8.3.2 Physical Conditions

Flow and Circulation   Regardless of their mode of formation, estuaries tend to be highly productive ecosystems because of the nature of the estuarine circulation pattern that characterizes these systems. Since freshwater is less dense than salt water of comparable temperature, there is a natural tendency in estuarine systems for the freshwater from land runoff to flow from the head (entrance) to the mouth (exit) of the estuary along the surface, whereas seawater moves in and out with the tides along the bottom. This typical pattern of water movement in an estuary is illustrated in Figure 8.15. As indicated by the curved arrows in Figure 8.15, there is typically some upward mixing of seawater into the

Table 8.3 Stratification Classification of Estuaries

Type	River Discharge	Example
Highly stratified	Large	Mississippi River (Louisiana)
		Mobile River (Alabama)
Partially mixed	Medium	Chesapeake Bay (Maryland, Virginia)
		James River estuary (Virginia)
		Potomac River (Maryland, Virginia)
Vertically homogeneous	Small	Delaware River estuary (Delaware, Pennsylvania, New Jersey)
		Biscayne Bay (Florida)
		Tampa Bay (Florida)
		San Francisco Bay (California)
		San Diego Bay (California)

Source: USEPA (1984a.)

freshwater, so that some of the seawater that enters the estuary near the bottom flows back out near the surface. As a result, there is a net outflow of water (freshwater mixed with some salt water) at the mouth of the estuary in the upper part of the water column and a net inflow of seawater in the lower part of the water column. If the flux of freshwater into the estuary at the head is large compared to the tidal in-and-out flux of seawater, there is generally a sharp demarcation between the freshwater at the top of the water column and the salt water below. Due to the mixing of salt water and freshwater, this sharp transition region gradually blurs as one approaches the mouth of the estuary. Such an estuary is commonly referred to as a *salt-wedge estuary* because of the shape of the saltwater "wedge" in the lower part of the water column when the estuary is viewed in longitudinal profile.

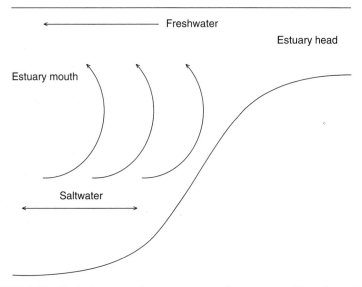

FIGURE 8.15 Typical pattern of water movement in an estuary. (From Laws, 2000.)

If the in-and-out flux of the tides is large compared to the flux of freshwater, then fresh-water and seawater tend to be thoroughly mixed together throughout the estuary except in the immediate vicinity of the head. Such an estuary is commonly referred to as a *well-mixed estuary*. Undoubtedly, many estuaries are best classified as having circulation patterns intermediate between those of typical salt-wedge and well-mixed estuaries. The important point to keep in mind is that there is a net outflow of water near the surface and a net inflow near the bottom at the mouth of all estuaries, regardless of whether the details of the circulation pattern correspond most closely to those of a salt-wedge, well-mixed, or intermediate-type situation (Laws, 2000). Since there is a net inflow and upward mixing of seawater at the bottom of an estuary, detritus that sinks out of the mixed layer at the surface and regenerated nutrients from the deeper portions of the estuary are constantly being carried back into the estuary and mixed up into the surface waters. Suspended organic matter that drifts out of the estuary on the surface current, and that subsequently sinks offshore or is eaten and then excreted offshore, tends to be swept back into the estuary by the net influx of bottom water. Figure 8.16 depicts the cycling of nutrients and organic matter in an estuary as influenced by the estuarine circulation pattern. The physical circulation pattern in estuaries provides a natural mechanism for recycling food and inorganic nutrients and thereby maintains a high level of production in the estuarine system. In a similar manner, pollutants introduced into an estuary tend to be recycled in the same manner as nutrients. Consequently, estuaries are undesirable receiving waters for pollutants, since the pollutants tend not to be directly washed out to sea and dispersed. If an effluent is discharged into the bottom water layer of a stratified estuary and has neutral buoyancy, it may move upstream until it reaches the end of the saltwater wedge and comes under the influence of the seaward-moving upper layers of water. If it has negative buoyancy, it may never travel seaward at all. As a general rule, estuarine circulation patterns can be expected to magnify the impact of pollutants discharged into them.

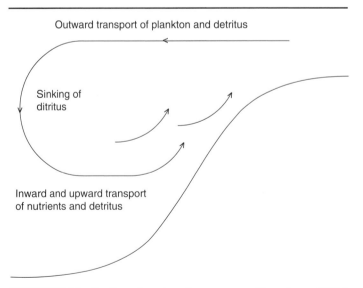

FIGURE 8.16 Cycling of nutrients in an estuary. (From Laws, 2000.)

Estuaries are particularly susceptible to problems associated with cultural eutrophication because the estuarine circulation pattern tends to recycle discharged nutrients and hence magnify their impact on production and biomass. Estuaries are naturally eutrophic systems because of the high efficiency with which the estuarine circulation pattern recycles nutrients and, because of their high productivity, estuaries such as those found along the coastline of the United States account for some important coastal fisheries and serve as breeding and/or nursery grounds for many marine organisms that one usually associates with the open ocean, such as sharks and whales.

An important parameter used to measure flow in estuaries is the *flushing time*, which can be defined as the average time needed to remove a particle from a point in the estuary into the sea (Novotny, 2003). Long flushing times are usually associated with poor water-quality conditions, and factors influencing flushing times are tidal ranges, freshwater inflows, and wind (USEPA, 1984a). The *replacement time* is the time required for all the water contained in the estuary to be replaced by tidal flows, and the replacement time, t_R, can be estimated using the relation

$$t_R = 0.4 \frac{L^2}{E_L} \qquad (8.76)$$

where L is the length of the estuary [L] and E_L is the longitudinal dispersion coefficient [L^2/T]. The magnitudes of longitudinal dispersion coefficients in several river estuaries in the United States and the UK are given in Table 8.4. Estuaries with flushing times less than 1 to 2 weeks do not provide enough residence time for the development of algal blooms and generally exhibit better water quality.

Salinity Distribution The difference in density between outflow and inflow produces secondary currents that ultimately affect the salinity distribution across an estuary. The

Table 8.4 Longitudinal Dispersion Coefficients in Select Estuaries

Estuary	River Flow (m^3/s)	Dispersion Coefficient (m^2/s)
Delaware River (DE, NJ)	70.8	150
Hudson River (NY)	141.6	600
East River (NY)	0	300
Cooper River (SC)	283.2	900
Savannah River (GA, SC)	198.2	300–600
Lower Raritan River (NJ)	4.25	150
Hudson ship channel (TX)	25.5	800
Cape Fear River (NC)	28.3	60–300
Potomac River (MD, VA)	15.5	30–300
Compton Creek (NJ)	0.3	30
Wappinger and Fishkill Creeks (NY)	0.05	15–30
San Francisco Bay		
Southern bay		18–180
Northern bay		46–1800
Thames River (UK)	Low river flow	53–84
	High river flow	338

Source: USEPA (1984a) and Fischer et al. (1979).

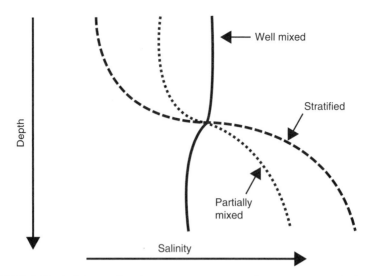

FIGURE 8.17   Salinity gradient in an estuary. (From National Oceanography Centre, 2005.)

salinity distribution is important because it affects the fauna and flora within the estuary, and the salinity distribution is indicative of the mixing characteristics of the estuary as they may affect the dispersion of pollutants within the estuary. Although the salinity distribution has only a modest influence on mixing in open oceans, it has a dominant effect on mixing in estuaries. Highly stratified (stable) estuaries offer significant resistance to mixing. The salinity of an estuary ranges from effectively zero at the head of an estuary to the value of undiluted ocean water, approximately 35 g/kg, at the mouth of the estuary. The corresponding range of water densities, from 1000 to 1025 kg/m³, is roughly five times larger than the range caused by typical environmental temperature differences. Consequently, salinity profiles act more prominently than temperature profiles in estuarine stratification. Typical (vertical) salinity profiles in estuaries are illustrated in Figure 8.17. In stratified estuaries, there is a significant vertical variation in salinity, with a relatively uniform top layer over a relatively uniform bottom layer, with a sharp salinity gradient in between. In well-mixed estuaries, there is a relatively small vertical variation in salinity, and partially mixed estuaries have vertical salinity distributions somewhere in between those of stratified and well-mixed estuaries.

The two most important sources of freshwater to an estuary are stream flow and precipitation, with stream flow generally representing the greatest contribution to the estuary. The location of the salinity gradient in a river-controlled estuary is to a large extent a function of stream flow, and the location of the iso-concentration lines may change considerably depending on whether stream flow is high or low. This in turn may affect the biology of the estuary, resulting in population shifts as biological species adjust to changes in salinity. Most estuarine species are adapted to survive temporary changes in salinity either by migration or some other mechanism (e.g., mussels can close their shells). However, many estuarine organisms cannot withstand these changes indefinitely. Of particular concern are flood-control dams that produce controlled discharges to an estuary rather than relatively short but massive discharge during high-flow periods. Dams operated to impound water for water supplies during low-flow periods may drastically alter the pattern of freshwater flow to an estuary, and although the annual discharge may remain the same, seasonal changes may have significant impact on the estuary salinity distribution and its biota.

The response of an estuary to rainfall events depends on the intensity of rainfall, the drainage area, and the size of the estuary. Movement of the salt front in an estuary is dependent on both tidal influences and freshwater flow to the estuary. Variations in salinity generally follow seasonal patterns such that the salt front will occur farther down-estuary during the rainy season than during the dry season. The salinity profile may also vary from day to day, reflecting the effect of individual rainfall events, and may undergo major changes due to extreme meteorological events.

Substrate   The bottom sediments or *substrate* of most estuaries is a mix of sand, silt, and mud that has been transported and deposited by ocean currents or by freshwater sources. Typically, sand is the predominant substrate in areas of significant tidal current, whereas silt and mud are predominant in slower tidal zones. Tidal velocity in some Norwegian fjords during high- and low-tide peaks may exceed 2 to 3 m/s, and much of the estuarine substrate is mobile and in flux. Rocky areas may also be present, particularly in fjord-type estuaries.

It is often difficult for plants to colonize estuaries because of the lack of suitable anchorage points and because of the turbidity of the water, which restricts light penetration. Submerged aquatic vegetation (SAV) (macrophytes) develops in sheltered areas where silt and mud accumulate. These plants, which include sea grass (Figure 8.18), help to slow currents, leading to further deposition of silt. The growth of plants often keeps pace with rising sediment levels so that over a long period of time substantial deposits of sediment and plant material may be seen. SAV serves very important roles as habitat and as a food source for much of the biota of an estuary, and major estuary studies have shown that the health of SAV communities serves as an important indicator of estuary health (USEPA, 1994).

FIGURE 8.18   Submerged aquatic vegetation (sea grass). (From State of Texas, 2005.)

Segmentation   Segmentation of an estuary can provide a useful framework for evaluating the influence of estuarine physical characteristics such as circulation, mixing, salinity, and geomorphology on the attainability of designated uses that are associated with water-quality standards. *Segmentation* is the compartmentalization of an estuary into subunits with homogeneous physical characteristics. In the absence of water pollution, physical characteristics of different regions of the estuary tend to govern the suitability for major water uses (USEPA, 1994). Once the segment network is established, each segment can be subjected to a use-attainability analysis. In most cases, the segmentation process offers a useful management structure for monitoring compliance with water-quality goals. An example of a segmented estuary is Tampa Bay, which is shown in Figure 8.19. The major bay segments are Hillsborough Bay (HB), old Tampa Bay (OTB), middle Tampa Bay subsegments (MTB*xx*), and lower Tampa Bay (LTB). The solid dots in Figure 8.19(*a*) are the locations of sea grass elevation survey sites.

The segmentation approach is an evaluation tool that recognizes that an estuary is an interrelated ecosystem composed of chemically, physically, and biologically diverse areas. It assumes that an ecosystem as diverse as an estuary cannot be managed effectively as a single unit because different uses and associated water-quality goals will be appropriate and feasible for different regions of the estuary. However, after developing a network based on physical characteristics, boundaries can be refined with available chemical and biological data to maximize the homogeneity of each segment.

A potential source of concern about the construction and utility of segmentation, particularly for use-attainability analyses, is that an estuary is a fluid system with only a few

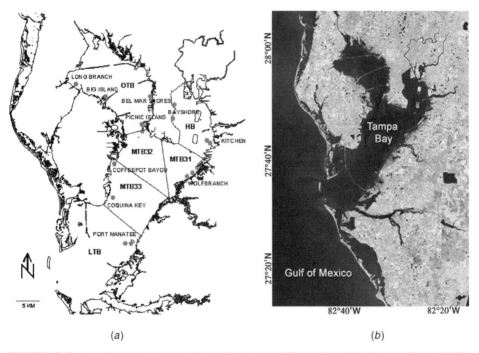

(*a*)   (*b*)

FIGURE 8.19   (*a*) Segmentation and (*b*) satellite view of Tampa Bay, Florida. [(*a*) From USGS, 2005h; (*b*) from USGS, 2005i.]

obvious boundaries, such as the sea surface and the sediment–water interface. Fixed boundaries may seem unnatural to scientists, managers, and users, who are more likely to view the estuary as a continuum than as a system composed of separable parts (USEPA, 1994).

To account for the dynamic nature of an estuary, estuarine circulation patterns should be a prominent factor in delineating the segment network. Circulation patterns control the transport of and residence times for heat, salinity, phytoplankton, nutrients, sediment, and other pollutants throughout an estuary. The variations in salinity concentrations from the head to the mouth of an estuary typically produce a separation of biological communities based on salinity tolerances or preferences.

The use of simplified geometry, such as uniform depth and cross-sectional area for simulating estuarine systems is highly judgemental, and the consequences of such simplifications should be considered. Mathematical models such as the WASP 7 model are available that can represent these systems with a far higher degree of sophistication.

8.3.3 Chemical Conditions

The most critical water-quality indicators for designated-use attainment in an estuary are usually dissolved oxygen, chlorophyll a, nutrients, and toxicants. Dissolved oxygen (DO) is an important water-quality indictor for all fisheries uses. In evaluating use attainability, assessments of DO impacts should consider the relative contributions of three different sources of oxygen demand: photosynthesis/respiration demand from phytoplankton, water-column demand, and benthic oxygen demand. If use impairment is occurring, assessments of the significance of each oxygen sink can be used to evaluate the feasibility of achieving sufficient pollution control to attain the designated use.

Chlorophyll a is the most popular indicator of algal concentrations and nutrient overenrichment, which in turn can be related to diurnal DO depressions due to algal respiration. The nutrients of greatest concern in an estuary are nitrogen and phosphorus. Typically, the control of phosphorus levels can limit algal growth near the head of the estuary, while the control of nitrogen levels can limit algal growth near the mouth of the estuary; however, these relationships depend on factors such as nitrogen–phosphorus (N/P) ratios and light-penetration potential. Excessive phytoplankton concentrations, as indicated by chlorophyll a levels, can cause wide diurnal variations in surface DO, due to daytime photosynthetic oxygen production and nighttime oxygen depletion by respiration and depletion of bottom DO associated with the decomposition of dead algae. Excessive chlorophyll a levels also result in shading, which reduces light penetration for submerged aquatic vegetation (SAV). The prevention of nutrient overenrichment is probably the most important water-quality requirement for a healthy SAV community (USEPA, 1994).

Sewage-treatment plants are typically the major source of nutrients, particularly phosphorus, to estuaries in urban areas. Agricultural land uses and urban land uses represent significant nonpoint sources of nutrients, particularly nitrogen. It is important to base control strategies on an understanding of the sources of each type of nutrient, in both the estuary and its feeder streams. Point sources of nutrients are typically much more amenable to control than are nonpoint sources. Because phosphorus removal from municipal wastewater discharges is typically less expensive than nitrogen removal, the control of phosphorus discharges is often the method of choice for the prevention or reversal of use impairment in the upper estuary (i.e., tidal fresh zone). However, nutrient control in the upper reaches of the estuary my cause algal blooms in the lower reaches. This occurs

FIGURE 8.20   Alum spill into Penobscot Bay, Maine. (From Penobscot Bay Watch, 2005.)

because phosphorus control in the upper reaches may reduce algal blooms there, but in so doing increases the amount of nitrogen transported to the lower reaches, where nitrogen is the limiting nutrient causing a bloom there.

Potential impacts from toxic substances such as pesticides, herbicides, heavy metals, and chlorinated effluents must generally be considered in use-attainability studies. The presence of certain toxicants in excessive concentrations within bottom sediments of the water column may prevent the attainment of designated uses (particularly, fisheries propagation and harvesting and sea-grass habitat uses) in estuaries that satisfy water-quality criteria for DO and chlorophyll a. An example of industrial operations that can affect the designated use of an estuary segment is shown in Figure 8.20, which illustrates the impact of alum spilling into Penobscot Bay (Maine) from an industrial operation. The area affected is discolored (white) and is at risk for further impact.

8.3.4 Biological Conditions

Salinity, light penetration, and substrate composition are the most critical factors that affect the distribution and survival of plant and animal communities in an estuary. Colonizing plants and animals must be able to withstand periodic submersion and desiccation as well as fluctuating salinity, temperature, and dissolved oxygen.

The depth to which attached plants may become established is limited by turbidity because plants require light for photosynthesis and estuaries are typically turbid because of large quantities of detritus and silt contributed by surrounding marshes and rivers. Algal growth may also hinder light penetration. If not enough light penetrates to the lower depths, animals cannot rely on visual cues for habitat selection, feeding, or finding a mate.

Several generalizations concerning the responses of estuarine organisms to salinity have been noted (USEPA, 1994):

- Organisms living in estuaries subjected to wide salinity fluctuations can withstand a wider range of salinities than species that occur in high-salinity estuaries.
- Intertidal-zone animals tend to tolerate wider ranges of salinities than do subtidal and open-ocean organisms.
- Low intertidal species are less tolerant of low salinities than are high intertidal species.
- More sessile animals are likely to be more tolerant of fluctuating salinities than are organisms that are highly mobile and capable of migrating during times of salinity stress.

Estuaries are generally characterized by low diversity of species but high productivity. Examples of animal species found in estuaries are shown in Figure 8.21.

8.3.5 Use-Attainability Evaluations

Assessments of designated uses relative to aquatic-life protection is generally evaluated in terms of biological measurements and indexes. Physical and chemical factors are also considered in determining the highest-level use an estuary can achieve. Physical and chemical evaluations can proceed on several levels, depending on the level of detail required and amount of knowledge available about the estuary. As a first step, the estuary is classified in terms of physical processes so that it can be compared with reference estuaries in terms of differences in water quality and biological communities, which can be related to anthropogenic influences such as pollution discharges.

The second step is to perform desktop or simple computer-model calculations to improve the understanding of spatial and temporal water-quality conditions in the present system. These calculations include continuous point source and simple box model calculations.

A third step is to perform detailed analyses through the use of more sophisticated computer models. These tools can be used to evaluate a system's response to removing individual point- and nonpoint-source discharges, so as to assist with assessments of the causes of any use impairment.

8.4 COMPUTER MODELS

Several computer models are available for simulating the water quality in oceans and estuaries. These models typically provide numerical solutions to the advection–dispersion equation, or some other form of the law of conservation of mass, at discrete locations and times for multiple interacting constituents, complex boundary conditions, spatially and temporally distributed contaminant sources and sinks, multiple fate processes, and variable flow and dispersion conditions. In engineering practice, the use of computer models to apply the fundamental principles covered in this chapter is sometimes essential. In choosing a model for a particular application, there is usually a variety of models to choose from.

(a)

(b)

(c)

FIGURE 8.21 Estuarine organisms: (*a*) sea star; (*b*) crab; (*c*) seaweed. (From CSIRO Marine and Atmospheric Research, 2005.)

However, in doing work that is to be reviewed by regulatory agencies, or where professional liability is a concern, models developed and maintained by the U.S. government have the greatest credibility and, perhaps more important, are almost universally acceptable in developing permit applications and defending design protocols on liability issues. Several of the more widely used models that have been developed and endorsed by U.S. government agencies are described briefly here.

PLUMES (Baumgartner et al, 1994) is a computer model used for simulating the dilution of freshwater discharges into ocean waters. The PLUMES model contains four well-established submodels for near- and far-field dilution. The near-field models are applicable to both single- and multiport discharges, and the far-field models are implementations of the Brooks far-field dispersion equation (Equation 8.69). The PLUMES model is useful in cases of complex stratification, variable currents, and transient outfall discharges. The PLUMES model is currently maintained by the EPA Pacific Ecosystems Branch in Newport, Oregon.

CORMIX (Cornell mixing zone expert system) is a water-quality modeling and decision support system designed for environmental impact assessment of mixing zones resulting from wastewater discharges from point sources. CORMIX characterizes the near-field mixing of water and wastewater plumes discharged to surface waters using solutions for submerged single and multiport discharges and buoyant surface discharges. The system emphasizes the role of boundary interaction to predict plume geometry and dilution in relation to regulatory mixing zone requirements. As an expert system, CORMIX is a user-friendly application that guides water-quality analysts in simulating a site-specific discharge configuration. To facilitate its use, ample instructions are provided, suggestions for improving dilution characteristics are included, and warning messages are displayed when undesirable or uncommon flow conditions occur.

CE-QUAL-W2 (Cole and Buchak, 1995) is a two-dimensional water-quality and hydrodynamic code supported U.S. Army Waterways Experiment Station. The model has been widely applied to stratified surface-water systems such as lakes, reservoirs, and estuaries and computes water levels, horizontal and vertical velocities, temperature, and 21 other water-quality parameters (such as dissolved oxygen, nutrients, organic matter, algae, pH, the carbonate cycle, bacteria, and dissolved and suspended solids).

Only a few of the more widely used computer models have been cited here. Certainly, there are many other good models that are capable of performing the same tasks.

SUMMARY

Oceans and estuaries are water bodies that are unique to coastal areas. Coastal communities commonly discharge treated domestic wastewater into the ocean using submerged outfalls. Discharged effluent generally undergoes two distinct phases of mixing: near- and far-field mixing. Near-field mixing occurs in the immediate vicinity of the outfall and is influenced primarily by the buoyancy of the freshwater discharge; far-field mixing occurs when mixing is dominated by the ambient ocean currents. Near-field mixing models are fairly mature, and available formulations yield fairly accurate predictions of initial dilution. Far-field models are still in the development stage; however, Brooks' far-field model can be used to approximate far-field mixing. Estuaries have a free connection with the open ocean and contain measurable quantities of fresh and salt water. Typically, there is a net outflow of water (freshwater mixed with some salt water) at the mouth of an estuary in the upper part of the water column, and a net inflow of seawater in the lower part of the water column. Estuaries are particularly susceptible to problems associated with cultural eutrophication because the estuarine circulation pattern tends to recycle discharged nutrients and hence magnify their impact on production and biomass. The salinity distribution in an estuary is influenced significantly by the freshwater inflow rate, and the resulting density distribution has a dominant effect on mixing in estuaries. The most critical water-quality indicators for designated-use attainment in an estuary are dissolved oxygen, chlorophyll _a_, nutrients, and toxicants. Respiration demand from phytoplankton, water-column demand, and benthic oxygen demand are the major oxygen sinks, and sewage-treatment plants, agricultural, and urban land uses are typically the major source of nutrients. Salinity, light penetration, and substrate composition are the most critical factors that affect the distribution and survival of plant and animal communities in a estuary.

PROBLEMS

8.1. A single-port outfall discharges treated municipal wastewater into an unstratified stagnant ocean at a depth of 15 m. The diameter of the outfall port is 0.7 m, the discharge velocity is 3 m/s, the density of the wastewater is 998 kg/m³, and the density of the ambient seawater is 1024 kg/m³. Estimate the plume dilution.

8.2. Repeat Problem 8.1 for the case in which the ambient current is 15 cm/s. Determine the approximate current speed at which the plume changes from the BDNF to the BDFF regime.

8.3. An ocean outfall discharges at a depth of 31 m from a diffuser containing seven 1-m-diameter ports spaced 10 m apart. The average effluent flow rate is 7.5 m³/s, the ambient current is 8 cm/s, the density of the ambient seawater is 1.024 g/cm³, and the density of the effluent is 0.998 g/cm³. Determine the length scales of the effluent plumes, and calculate the minimum dilution. Neglect merging of adjacent plumes.

8.4. Consider an outfall that is required to discharge treated domestic wastewater at 3 m³/s in 20 m of water, where the 10-percentile current is 5 cm/s.

   (a) What would the wastewater dilution be if the wastewater is simply discharged out of the end of the 900-mm-diameter outfall pipe?

   (b) What would the wastewater dilution be if the wastewater were discharged through closely spaced ports along a 12-m-long diffuser? Assume typical values for the wastewater and seawater densities.

8.5. Repeat Problem 8.3, treating the diffuser as a line source.

   (a) Calculate the dilution for currents perpendicular and parallel to the diffuser.

   (b) Determine the effluent discharge rate at which the plume dilution becomes independent of the current direction.

8.6. Use Equation 8.47 to derive an expression for the dilution of a slot plume in a stagnant environment. Express the dilution in terms of y, g', L, and Q.

8.7. An ocean-outfall diffuser is to be designed to discharge 1.75 m³/s of sewage effluent at a depth of 20 m and achieve a near-field dilution of 25:1. The effluent density is 998 kg/m³ and the seawater density is 1024 kg/m³.

   (a) If the dilution effect of the ambient currents is neglected, determine the required length of the diffuser. State your assumption(s) regarding the port spacing.

   (b) If the effluent is simply discharged out of the end of the outfall pipe, what dilution would be achieved? Based on your analysis, would a diffuser or end-of-pipe discharge be preferable in this case?

8.8. A diffuser is to be located in 28 m of water, discharge 3.90 m³/s of primary-treated domestic wastewater, and provide an initial dilution of 30:1. The critical velocity required to prevent deposition in the diffuser is 60 cm/s, and the total head at the upstream port will be maintained at 31 m. The density of the wastewater can be taken as 998 kg/m³, the seawater density is 1024 kg/m³, and the outfall is to be designed for stagnant ambient conditions. Design the diffuser length, port spacing, number of ports, and the diameter of the most upstream port in the diffuser.

8.9. A 15-kg slug of Rhodamine WT dye is released instantaneously at one point into the ocean, and the concentration distribution of the dye is measured every 3 h for the 12-h duration of daylight when the dye can be seen. The horizontal variance of the dye cloud as a function of time is given in Table 8.5. Estimate the apparent diffusion coefficient as a function of time. Compare your results to the Okubo relation (Equation 8.62).

Table 8.5   Data for Problem 8.9

Time, t (h)	$\sigma_{x'}^2$ (cm^2)	$\sigma_{y'}^2$ (cm^2)
0	1.1×10^4	1.2×10^4
3	3.3×10^7	3.0×10^7
6	1.5×10^8	1.6×10^8
9	4.1×10^8	3.9×10^8
12	7.9×10^8	7.7×10^8

8.10. (a) Compare the apparent diffusion coefficient of a contaminant cloud with a characteristic size of 100 m with the apparent diffusion coefficient of a contaminant cloud whose size is 200 m.

(b) Estimate how long it would take for a small contaminant spill to grow to a size of 100 m.

8.11. A 50-m-long multiport diffuser discharges effluent at a rate of 6.5 m³/s and at a depth of 32 m. If the ambient current is 20 cm/s, estimate the distance downstream of the diffuser to where the dilution is equal to 150. Assume that the initial wastefield thickness is 30% of the depth.

8.12. If the near-field mixing at the outfall described in Problem 8.11 is analyzed more closely using the Roberts line-plume model, compare the assumed wastefield thickness (30%) with that derived from using the near-field dilution model. Assume that the current is perpendicular to the outfall, the density of the discharge is 988 kg/m³, the density of the seawater is 1025 kg/m³, and the average dilution in the plume is $\sqrt{2}$ times the minimum dilution.

8.13. A regulatory mixing zone has a boundary that is 500 m away from an ocean outfall when the ambient current is 15 cm/s. Within the mixing zone, near- and far-field models indicate dilutions of 18 and 40, respectively, for conservative contaminants contained in the wastewater effluent.

(a) What is the conservative-contaminant dilution on the boundary of the mixing zone?

(b) If a contaminant has a 90% decay time of 2 h in seawater, what is the dilution of this contaminant on the boundary of the mixing zone?

8.14. Why is decay not considered in near-field dilution models?

8.15. A near-field ocean-outfall dilution model indicates that the length, L (m), of a diffuser required to achieve a minimum initial dilution S is given by

$$L = 0.277S^{3/2}$$

The outfall is to be located 20 km from a popular beach, and winds are expected to induce a maximum current of 30 cm/s toward the beach. The fecal coliform concentration in the effluent is 80,000 cfu per 100 mL, and the maximum allowable fecal coliform count on the beach is 200 cfu per 100 mL. Estimate the minimum outfall length required. How would you justify neglecting fecal coliform decay in your analysis? What other considerations would govern selection of the diffuser length?

8.16. (a) The city of Hollywood, Florida, discharges treated domestic wastewater through a single-port outfall located 3.05 km offshore. This outfall is simply an open-ended pipe of diameter 1.52 m at a depth of 28.5 m. The 10th percentile current at the outfall location is 8 cm/s, and the discharge from the outfall varies in the range 0.88 to 1.93 m³/s, with an average value of 1.40 m³/s. The average density of the seawater is 1.024 g/cm³, and the average density of the effluent is 0.998 g/cm³. Use a length-scale analysis to assess whether the ambient current has a significant effect on the near-field plume dilution. Estimate the dilution expected at the outfall. Compare the calculated dilution to the dilution when the currents are totally neglected. Does this result confirm your length-scale analysis? The Florida Administrative Code 62-4.244(3)(c) has the following requirements for all outfalls in the state of Florida: "Rapid dilution shall be ensured by the use of multiport diffusers, or a single port outfall designed to achieve a minimum of 20:1 dilution of the effluent prior to reaching the surface." Is the Hollywood outfall in compliance with these regulations?

(b) An engineer working for the city of Hollywood proposes replacing the single-port outfall with a three-port diffuser having 1-m-diameter ports spaced 5 m apart. This astute engineer asserts that it would be best to space the ports such that the plumes do not merge. Compare the outfall dilution when the plumes do not merge with the dilution for merging plumes.

(c) Assuming a near-field dilution of 30, a three-port diffuser with merging plumes, and an effluent coliform concentration of 100,000 cfu per 100 mL, estimate the coliform count at a beach directly inland from the outfall. If the coliform criteria for the beach is 50 cfu per 100 mL, should the beach be closed to the public?

CHAPTER 9

WATERSHEDS

9.1 INTRODUCTION

A *watershed* is an area that captures atmospheric precipitation (rain, snow) and drains the resulting surface runoff to a surface-water body such as a river, lake, or estuary. In the case of rivers and streams, a watershed is typically associated with the segment of the stream between its upstream end and a particular location along the stream, usually the terminal outlet or confluence with another stream. A typical stream watershed is illustrated in Figure 9.1. Large watersheds are sometimes called *basins* (e.g., the Colorado River Basin or the Susquehanna River Basin). Conceptually, all land on the surface of the Earth can be divided into distinct watersheds, with a water body or river segment associated with each watershed. Since contaminant input to a water body is mostly generated from activities within its watershed, the control of polluting activities and processes within a watershed is a fundamental component of water-quality control. This approach is commonly referred to as *water-quality-based watershed management*. This approach works best when there is a single governmental entity with jurisdiction over the entire watershed, which occurs mostly in watersheds on local and regional scales. On a global scale this approach is much more difficult to implement; in 1990, nearly 40% of the world population lived in international watersheds, with 23 of these watersheds being shared by 4 to 12 countries (Jørgensen et al., 2005)

The primary goal of water-quality-based watershed management is to control activities within a watershed and thereby limit the input of contaminants into the receiving water body. For receiving water bodies that serve as sources of drinking water, source-water protection plans (SWPPs) must be developed that delineate the watershed, inventory potential contamination sources, and determine the susceptibility the water supply to identified contamination sources. Other goals of watershed management include flood protection, erosion control, and habitat preservation. Some management goals are unique to particular

Water-Quality Engineering in Natural Systems, by David A. Chin
Copyright © 2006 John Wiley & Sons, Inc.

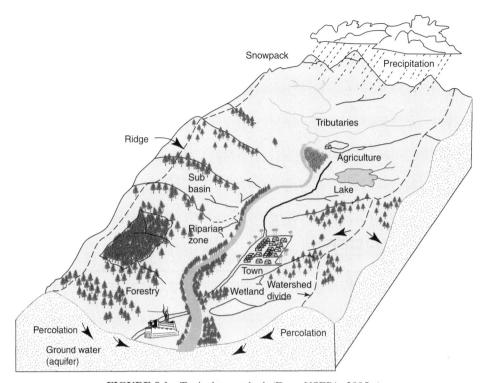

FIGURE 9.1   Typical watershed. (From USEPA, 2005a.)

areas, for example metals loading to streams and impairment of water quality from hardrock metal mines are common problems in mountainous watersheds in the western United States (Caruso, 2005).

Effective watershed management involves many scientific disciplines, including geology, biology, chemistry, hydrology, and hydraulics; however, socioeconomic and political considerations are frequently deciding factors. Public awareness of the relationship between polluting activities within a watershed and the quality of the receiving water body can contribute significantly to water-quality protection. To this end, signs such as that shown in Figure 9.2 can be quite effective.

Key regulatory requirements that affect activities within watersheds relate to source-water protection and restoration of impaired waters. Source-water protection plans include the identification, assessment, and control of all contaminant sources within the watershed that could significantly affect the quality of the water supply. Impaired waters are water bodies that do not meet their water-quality standards. If impaired waters exist within a watershed, the contaminant sources responsible for the pollution must be identified and pollutant loadings compatible with meeting the water-quality standards must be determined and enforced. Such pollutant loadings are commonly called total maximum daily loads (TMDLs). Source-water protection plans and the calculation of TMDLs are both essential components of watershed management and are covered in this chapter.

The two primary land uses that affect the quality of ground and surface waters are urban and agricultural land uses. This chapter covers pollution sources commonly associated with urban and agricultural lands, the fate and transport of pollutants originating on these

FIGURE 9.2   Watershed sign. (From City Austin, 2005.)

lands, and best management practices that can be used to limit pollution originating from these lands.

9.2 SOURCE-WATER PROTECTION

In the United States, all states are required to develop source-water protection plans (SWPPs) to protect public water supplies from contaminant sources within the catchment area of drinking-water intakes. The catchment area of a drinking-water intake is typically the same as the watershed associated with the water body containing the intake, and this catchment area is called the *source-water protection area* (SWPA) of the intake. Key components of SWPPs are as follows:

1. *Delineation of the source-water protection area.* Water-supply intakes can be either surface-water or ground-water intakes, and the delineation of the corresponding protection area depends on the type of intake.

For ground-water systems, available information about ground-water flow and recharge is used to determine the source-water protection area. Ground-water intakes typically consist of wells (vertical or horizontal) or infiltration galleries, and SWPAs are typically delineated using either distance, ground-water flow boundaries, time of travel, drawdown, or assimilative capacity. Water-supply wells very close to surface-water bodies can be classified as Ground Water Under the Direct Influence of Surface Water (GWUDI) and

(a)   (b)

FIGURE 9.3 Wellhead protection: (*a*) sign; (*b*) wellheads. [(*a*) From South Dakota DNR, 2005; (*b*) from USGS, 2005k.]

treated as surface-water sources for delineation purposes. For ground-water intakes, source-water protection areas are commonly referred to as *wellhead protection areas*, and an example of a sign identifying a wellhead protection area and wellheads are shown in Figure 9.3. Wellhead protection areas are typically divided into three zones, with an inner zone given the most protection, an intermediate zone where detailed contaminant inventories are conducted and best management practices required, and an outer zone where the potential impact of contaminants on drinking water intakes is regarded as minimal. The inner zone is usually delineated using a distance criterion, the intermediate zone (also known as the *inventory region*) delineated by time of travel, and the outer zone delineated by a hydrogeologic boundary.

Surface-water systems draw water from a stream, river, lake, or reservoir; and the land area in the watershed upstream of the intake must generally be delineated. It is common to divide the watershed area into the following zones: (1) an inner zone closest to the intake where most types of contamination sources may be found to be significant; and (2) an outer zone that includes more distant areas. The inner zone is typically delineated by an upstream distance or travel time (based on the average annual high flow) plus a minimum buffer width, while the outer zone is typically delineated by the watershed boundary.

2. *Inventory of the actual and potential sources of contamination.* Sources of pollutants that could potentially contaminate the water supply must be identified. This inventory usually results in a list and a map of facilities and activities within the delineated area that could release contaminants into the watershed of the river or lake being used for water supply. Some examples of potential pollutant sources include urban runoff from streets and lawns, farms and other entities that apply pesticides and fertilizers, and sludge-disposal sites.

3. *Determination of the susceptibility of the water source to contamination.* Inventory results are combined with other relevant information to estimate the likelihood of contamination of the water supply by potential sources of contamination. *Susceptibility* is defined by USEPA as "the potential for a public water supply to draw water contaminated by inventories sources at concentrations that would pose concern" (USEPA, 1997c). Susceptibility determinations are typically based on the fate and transport of contaminants within the SWPA. The use of prescribed buffer zones and travel times are the most common approaches to controlling the susceptibility of drinking-water intakes to contamination (USEPA, 1997a, b). Buffer zones can be created through utility ownership or usage

restrictions, thereby providing for filtration of runoff through natural vegetation. No standards for the creation of buffer zones around surface-water sources exist (AWWA, 1990), but the required buffer areas take into account exposure to contamination and the degree of treatment provided by the buffer. Wetlands are frequently used as buffers for water-supply intakes, and preservation of wetlands maintains the natural filtration and cleansing provided by these critical areas. In Florida, potential contaminant sources located within a source water protection area are evaluated and ranked as posing a low, medium, or high threat of contaminant release to the source water protection area. This analysis includes an evaluation of the likelihood of a release affecting the public water system.

4. *Releasing the results of the assessments to the public.* The information contained in the source-water protection plan must be is summarized for the public. Such summaries help communities better understand the potential threats to their water supplies and identify priority needs for protecting the water from contamination. Local communities, working in cooperation with local, regional, and state agencies, can use the information in the source-water protection plan to create a broader source-water protection program to address current and future threats to the quality of their drinking-water supplies.

Communities typically use a wide array of source-water protection methods to prevent contamination of drinking-water supplies. One management option involves regulations such as prohibiting or restricting land uses that may release contaminants in critical source-water protection areas. Along with regulations, many communities hold local events and distribute information to educate and encourage citizens and businesses to recycle used oil, limit their use of pesticides, participate in watershed cleanup activities, and a variety of other prevention activities. Another approach to source water protection is the purchase of land or the creation of conservation easements to serve as a protection zone for the drinking-water source.

Statutory drinking-water standards apply to public water systems, which provide water to at least 15 connections or 25 persons at least 60 days out of the year (most cities, towns, schools, businesses, campgrounds, and shopping malls are served by public water systems). The 10% of Americans whose water comes from private wells (individual wells serving fewer than 25 persons) are not protected by these statutory drinking-water standards. People with private wells are responsible for making sure that their own drinking water is safe. Bottled water is regulated by the U.S. Food and Drug Administration as a food product and is required to meet standards equivalent to those set by the U.S. Environmental Protection Agency (USEPA) for drinking water.

9.3 WATERSHED-GENERATED POLLUTANT LOADS

Quantification of the relationship between the contaminant loads discharged into a water body and the quality of the receiving water is essential in determining the maximum contaminant loading that is consistent with meeting water-quality standards in the receiving water body. Such relationships between contaminant loading and water quality in the receiving body have been described in previous chapters for rivers, lakes, wetlands, estuaries, oceans, and ground water. Contaminants of most concern in such analyses are typically sediments, nutrients, metals, and pathogens (Caruso, 2005).

Contaminant loads on water bodies come from three main terrestrial sources: (1) point sources, (2) nonpoint sources, and (3) background (natural) sources. Point sources are

FIGURE 9.4   Outfall pipe. (From University of Arizona, 2005 Photo by Barbara Tellman.)

primarily localized sources of contamination that enter water bodies via outfall pipes, and such an outfall pipe is shown in Figure 9.4. Nonpoint sources typically consist of areas over which contaminants are distributed and are subsequently transported to receiving water bodies. Urban roadways and parking lots are common examples of nonpoint sources, where various automobile-related contaminants are deposited and the drainage system collects the contaminant-laden runoff and transports it to a receiving water. A typical example of a nonpoint source is shown in Figure 9.5. Contaminant loads resulting from nonpoint sources is collectively referred to as *diffuse pollution*.

The *loading capacity* of a water body is defined as the maximum contaminant load, including the background load (BL), that a water body can accept and still meet its

FIGURE 9.5   Typical nonpoint source of water pollution. (From University of Arizona, 2005.)

water-quality standards. The loading capacity typically includes a margin of safety. The *total maximum daily load* (TMDL) is the component of the loading capacity allocated to anthropogenic effects and loads and is given by

$$\boxed{TMDL = LC - BL - MOS} \tag{9.1}$$

where LC is the loading capacity, BL is the background load, and MOS is the margin of safety. The margin of safety takes into account uncertainties in the relationship between effluent quality and receiving-water quality, including inaccuracies in water-quality models. The *waste load allocation* (WLA) is the portion of the surface-water loading capacity allocated to existing and future point-source discharges. The WLA serves as a basis for determining the water-quality-based effluent limitations on point-source discharges required by National Pollutant Discharge Elimination System (NPDES) permits and monitoring requirements. The *load allocation* (LA) is the portion of the surface-water loading capacity allocated to existing and future nonpoint sources, and this allocation is the basis for development of nonpoint-source restoration plans. Since the loading capacity (LC) is equal to the sum of the waste load allocation (WLA) and the load allocation (LA), the TMDL given by Equation 9.1 can be expressed in the form

$$TMDL = (WLA + LA) - BL - MOS \tag{9.2}$$

In cases where the receiving water is impaired, the contaminant load entering the water body exceeds the TMDL and contaminant-load reductions are required. These load reductions must come from the point sources and/or the nonpoint sources. Reduction of contaminant loading from point sources is fairly straightforward and can be made by adjusting (downward) the allowable loads in the (NPDES) discharge permit. On the other hand, required reductions in contaminant loading from nonpoint sources must be made by changing land-use practices in the contributing watershed. Understanding the relationship between the contaminant loading on a water body and land uses in the contributing watershed is the basis for environmentally sound land management. Typically, *best management practices* are instituted for the land uses, and fate and transport models are used to relate the land-use practices to the nonpoint contaminant loads. These relationships are a primary focus of this chapter. Water-body restoration is not possible if the watershed sources of pollution are not controlled.

Models used to simulate pollution generation at the source and its movement from the source area to the receiving water body are called *loading models*, and such models are an integral part of the TMDL process. In most cases, the loading models provide flow (hydrograph) and concentration (pollutograph) distributions at various points in the watershed and entry points into the receiving water body. Models used to simulate the fate and transport of contaminants within water bodies are called *receiving-water models*. Clearly, both loading models and receiving-water models must be used in determining effective strategies to control the quality of water in natural systems.

For simulating overland pollutant transport, hydrology must be calibrated and determined first, followed by sediment, and finally, pollutant transport (Novotny, 2003). Any errors that appear in the hydrology or erosion component will be transferred and magnified in all subsequent components. *Calibration* of a model involves varying the parameters of the model within acceptable ranges until a satisfactory agreement between the measured and computed output values is achieved. Model calibration is a subjective process requiring experience, and an evaluation of what constitutes a calibrated model is purely judgmental (USEPA, 1992b). *Validation* means demonstrating that the model output is realistic and in agreement with observations; however, data used in calibration must not be used in validation. Calibration and validation of models are essential components in TMDL calculations.

Reviews of existing contaminant-transport models applicable to diffuse-pollution modeling of urban and agricultural watersheds can be found in Donigan et al. (1995) and USEPA (1997e), and new models are constantly being developed (e.g., Donohue et al., 2005). There are two commonly used approaches to modeling diffuse pollution: lumped-parameter modeling and distributed-parameter modeling. *Lumped-parameter models* treat the watershed, or a significant portion of it, as one unit, while *distributed-parameter models* divide the watershed into smaller homogeneous units with uniform characteristics. In distributed-parameter models, each subarea is described by a mass-balance equation, and mass-balance equations for the entire system are solved simultaneously at each time step. Both lumped- and distributed-parameter models can be used inside or attached to a geographical information system (GIS) shell. Watershed models may range from modeling small uniform segments of less than 1 ha to entire watersheds with areas of several hundred square kilometers. Models can be designed to run on either an event or continuous basis. *Single-event models* simulate the response of a watershed to a single (rainfall) event, while *continuous models* simulate the response of the watershed to several consecutive events. These models typically use time steps ranging from a fraction of an hour to a day, with water and pollutant mass balanced at each time step.

The formulation of toxic standards in water bodies does not allow a simple application of steady-state models, since toxic criteria are related to the probability (frequency) of excursions of concentrations. For example, the 1-hour average concentration of a priority pollutant can exceed the maximum concentration criterion (CMC) no more than once in 3 years, and for chronic toxicity the 4-day average concentration may exceed the continuous concentration criterion (CCC) no more than once in 3 years. Consequently, modeling must include these statistical considerations.

Methods used to control contaminant loads on water bodies can be divided into: source-control, hydrologic modification of source areas, reduction of transport of pollutants from the source area to the receiving water, and end-of-pipe removal of pollutants. These approaches are described briefly as follows:

1. *Source-control* reduces or eliminates sources of pollution and includes such measures as restricting the use of polluting and harmful chemicals, reducing discharge rates (such as requiring wastewater reuse), litter-control programs in urban areas, and soil conservation in agricultural areas.

2. *Hydrologic modification of the source area* is used primarily to control pollution that is carried by surface runoff. Reduction of imperviousness, detention and retention of surface runoff in urban areas, and soil-conservation best management practices in agricultural areas are common examples of hydrologic modification.

3. *Reduction of contaminant transport* is determined primarily by the type of drainage system in the watershed. Typical urban sewerage and storm-sewer collection systems do not provide opportunities for attenuation of pollutants, while natural drainage systems and overland flow provide many opportunities and places for sedimentation and filtration of

pollutants in surface runoff. Pollutant attenuation can be enhanced by installation of buffer strips along drainageways and the use of catch basins before urban stormwater enters a sewer system.

4. *End-of-pipe removal of pollutants* is typically accomplished by some form of treatment. Sedimentation ponds and wetlands at the drainage outlet point are common examples.

The approaches described above are applicable to diffuse pollutant loads from nonpoint sources. For point sources of pollution, abatement generally consists of end-of-pipe control.

Example 9.1 The analysis of a water-quality-impaired lake indicates that the maximum phosphorus load consistent with the lake meeting the phosphorus standard is 800 kg/yr. Background atmospheric loading of phosphorus is estimated to be 50 kg/yr, and the current NPDES permits for discharges into the lake allow an annual loading of 700 kg/yr. (a) Determine the TMDL of the lake and the acceptable loading from nonpoint sources in the watershed. (b) If the current diffuse loading from nonpoint sources is estimated to be 300 kg/yr, recommend an equitable reduction in the load allocations. Assume a 10% factor of safety in the estimated loading capacity of the lake.

SOLUTION   (a) From the data given, LC = 800 kg/yr, BL = 50 kg/yr, and MOS = 0.10 (800) = 80 kg/yr (corresponding to a 10% factor of safety). According to Equation 9.1, the total maximum daily load (TMDL) is given by

$$\text{TMDL} = \text{LC} - \text{BL} - \text{MOS} = 800 - 50 - 80 = 670 \text{ kg/yr}$$

The loading capacity (LC) of the lake is the sum of the waste-load allocation (WLA) from point sources and the load allocation (LA) from nonpoint sources according to the relation

$$\text{LC} = \text{WLA} + \text{LA}$$

Taking LC = 800 kg/yr and WLA = 700 kg/yr yields

$$800 = 700 + \text{LA}$$

which yields LA = 100 kg/yr. Therefore, the acceptable phosphorus loading from nonpoint sources is 100 kg/yr.

(b) Since the current diffuse loading is 300 kg/yr, reduction in phosphorus loading is needed. If r is the percent reduction in the WLA and LA that will yield an acceptable total phosphorus load of 800 kg/yr,

$$800 = \left(1 - \frac{r}{100}\right) 700 + \left(1 - \frac{r}{100}\right) 300$$

which yields $r = 20\%$. Therefore, the NPDES permits should be revised downward to a waste-load allocation of $(1 - 0.2)700 = 560$ kg/yr, and the load allocation from nonpoint sources should be reduced to $(1 - 0.2)300 = 240$ kg/yr. The reduced loading from nonpoint sources could be accomplished by instituting best management practices.

9.4 URBAN WATERSHEDS

Urban development is associated with population increases, development of transportation networks and other urban infrastructure, and the increased presence of pollution sources associated with traffic density and commercial and industrial development. Urbanization invariably increases the percentage of impervious surfaces in a watershed, and imperviousness and drainage are by far the most important parameters to which pollution loads can be related. Impervious surfaces include pavement on highways, sidewalks, parking lots, driveways, and rooftops. As population density increases, there are more emissions from cars, more pet waste and litter deposited on impervious surfaces, more pesticides and oils from parked vehicle drippings and maintenance, and more of other wastes that are washed directly into storm sewers or dumped into storm drains. Flow over an impervious area resulting from stormwater runoff is shown in Figure 9.6. As a result of increased imperviousness, less rainfall percolates into the ground and more flows overland, taking with it the pollutants that accumulate on the impervious surface.

Urban drainage systems deal with three types of flows: wastewater (sewage) flow, stormwater runoff, and infiltration inflow. Wastewater flows are typically routed to a wastewater treatment plant and treated prior to discharge through a pipe or diffuser into a receiving body. Stormwater runoff occurs during wet-weather conditions and originates from streets, roads, and other impervious areas. Flows in drainage systems also occur under dry-weather conditions from sources such as irrigation and infiltration inflow from ground water. The term *urban runoff* collectively refers to water from rain, irrigation, or other sources that flows over the land surface in an urban setting, carrying chemicals, microorganisms, and excessive sediments directly into receiving waters and causing pollution (National Water Research Institute, 2004). In some cases, particularly in older cities, wastewater flows and stormwater runoff are combined into a single pipe that runs to a wastewater treatment plant, and combined flows in excess of the plant capacity are frequently discharged without treatment into a receiving water body. Infiltration inflow into

FIGURE 9.6 Stormwater runoff over an impervious surface. (From Lake Superior Streams, 2005.)

sewers originates from ground-water infiltration into sewer pipes, leaking street manhole covers, cross-connections with stormwater conduits, and other sources. Although infiltration inflow tends to be clean water, it increases the volume of flow in the conduits and increases the frequency of overflows of untreated wastewater from sewer systems that bypass the treatment plant, thereby increasing the volume of pollutant entering receiving water bodies. Other deleterious effects of infiltration inflow are that the wastewater arrives at the treatment plant diluted, which decreases the efficiency of the treatment process. In areas that are less pervious (less than 40% imperviousness), storm sewers may not be needed and storm drainage can be accomplished by roadside swales, grassed waterways, small creeks, and canals. This type of natural drainage is typical of suburban communities. Areas with imperviousness less than 10% typically have minimal impact on water quality.

In areas with storm sewers, the two primary modes of drainage are without pretreatment and with pretreatment. In drainage systems without pretreatment, sometimes called *positive-drainage systems*, stormwater runoff is collected and conveyed directly to the nearest water body. In drainage systems with pretreatment, stormwater is collected and detained or partially retained on site before it enters stormwater conduits. This type of sedimentation pretreatment or infiltration removes a significant amount of pollution prior to the collected runoff being discharged into off-site receiving waters.

Streams in urban environments are usually stressed physically, chemically, and biologically. The increased magnitude and frequency of runoff increases the frequency of a stream reaching its critical erosive velocity, at which point the stream begins to erode. This causes deepening, widening, straightening, and sedimentation problems. Most urban stream corridors have been straightened, enclosed, or channelized. Such practices increase channel slopes, which tends to transport the problems downstream (DeBarry, 2004), and removes habitat for essential aquatic species, thus degrading the biodiversity.

9.4.1 Sources of Pollution

The major sources of water pollution in urban areas are as follows:

1. *Atmospheric deposition.* Atmospheric deposition of pollutants is divided into wet and dry surface loading. *Wet loading* originates from contaminants absorbed from the air by rain and snow, whereas *dry loading* results from atmospheric fallout. Atmospheric loading originates from both local and distant sources, and industrial (urban and transportation) and agricultural activities are the most frequent contributors to the pollution content of atmospheric deposits. In most larger cities, the deposition rate of atmospheric particulates in wet and dry fallout ranges from 7 to more than $30\,g/m^2 \cdot$ month. Higher deposition rates occur in congested downtown and industrial zones, and lower rates are typical of residential and other low-density suburban zones. There have been several location-specific studies of atmospheric deposition rates (e.g., Ng, 1987), but these results cannot be extended to other areas.

2. *Street refuse.* Particles with sizes larger than dust ($>60\,\mu$m) are considered as street refuse or street dirt. These deposits can be further divided into median-sized deposits ($60\,\mu$m to 2 mm) and litter (>2 mm). Sources of street dirt are numerous and very often difficult to control; however, it is reasonable to assume that a part of the dirt originates from mechanical breakdown of larger litter particles.

3. *Vegetation.* In residential areas, fallen leaves and vegetation residues, including grass clippings, typically dominate street refuse composition during the fall season. During

FIGURE 9.7   Motor-vehicle source of water pollution.

defoliage a mature tree can produce 15 to 25 kg of organic leaf residue which contains significant amounts of nutrients. Fallen leaves are about 90% organic and contain 0.05 to 0.28% phosphorus. Leaf fallout in urban areas and its wash-off into storm sewers is a significant source of biodegradable organics.

4. *Urban animals and birds.* Fecal matter of urban animals, including pets and birds, is a significant source of bacterial contamination in urban runoff. Fecal coliforms in urban runoff have been found at levels greater than 10^6 cfu per 100 mL (Bannerman et al., 1993), however, more typical values are on the order of 20,000 cfu per 100 mL; these levels tend to be higher in the warm months and lower in the cold months (USEPA, 1983a). Although these fecal coliform levels are high, this indicator may not be useful in identifying health risks in urban runoff.

5. *Traffic.* Motor-vehicle traffic, such as shown in Figure 9.7, is directly responsible for deposition of substantial amounts of pollutants, including toxic hydrocarbons, metals, asbestos, and oils. In addition to exhaust emissions, tire wear, solids carried on tires and vehicle bodies, wear and breakdown of parts, and loss of lubrication fluids add to the pollutant inputs contributed by traffic. Vehicular loss of oil on roads and parking lots has been found to be a major source of polyaromatic hydrocarbons (PAHs). Regulatory actions to control exhaust emissions and vehicle wear, mandatory vehicle emission testing, and improved fuel additives all contribute to per-vehicle reduction in traffic-related pollution. However, increased traffic volume causes increased pollution, particulary of heavy metals. It has been noted that streets paved entirely with asphalt have loadings about 80% higher than those of all-concrete streets (Sartor and Boyd, 1972). Of all types of urban and highway pollution, oil and hydrocarbons may be the most troublesome (Ellis and Chatfield, 2000, 2001). It has been reported that of the 50,000 tons of oil sold each year in the UK, 20,000 to 24,000 tons end up in highway and road runoff, and another 40% is unaccounted for and probably also enters the environment. Ellis and Chatfield (2000, 2001) concluded that the majority of highly toxic high-molecular-weight PAH fractions found in diffuse urban runoff can be related to traffic emissions and leakages that are washed from the impermeable surface during precipitation events.

6. *Other sources of toxic chemicals.* Urban runoff is the major source of toxic pollutants such as toxic metals and organic toxic compounds such as PAHs and pesticides. The primary sources of these compounds are summarized in Table 9.1.

TABLE 9.1 Sources of Toxic and Hazardous Substances in Urban Runoff

Pollutant	Source		
	Automobile Use	Pesticide Use	Industrial/Other Use
Heavy metals			
Copper	Metal corrosion	Algicide	Paint, wood preservative, electroplating
Lead	Gasoline, batteries	—	Paint, lead pipe
Zinc	Metal corrosion, tires, road salt	Wood preservative	Paint, metal corrosion
Chromium	Metal corrosion	—	Paint, metal corrosion, electroplating
Halogenated aliphatics			
Methylene chloride	—	Fumigant	Plastics, paint remover, solvent
Methyl chloride	Gasoline	Fumigant	Refrigerant, solvent
Phthalate esters			
Bis(2-ethylexy)phthalate	—	—	Plasticizer
Butylbenzyl phthalate	—	—	Plasticizer
D-N-butyl phthalate	—	Insecticide	Plasticizer, printing inks, paper, stain, adhesive
Polycyclic aromatic hydrocarbons			
Chrysene	Gasoline, oil, grease	—	—
Phenanthrene	Gasoline	—	Wood/coal combustion
Pyrene	Gasoline, oil, asphalt	Wood preservative	Wood/coal combustion
Other volatiles			
Benzene	Gasoline	—	Solvent
Chloroform	Formed from salt	Insecticide	Solvent, formed from chlorination
Toluene	Gasoline, asphalt	—	Solvent
Pesticides and phenols			
Lindane (γ-BHC)	—	Mosquito control, seed pretreatment	—
Chlordane	—	Termite control	—
Dieldrin	—	Insecticide	Wood processing
Pentachrolophenol	—	—	Wood preservative, paint
PCBs	—	—	Electrical, insulation, several other industrial applications
Asbestos	Brake and clutch lining, tire additives	—	Insulations

Source: USEPA (1990d).

7. *Nutrients from fertilizers.* Excessive use of lawn fertilizers can be a significant source of phosphorus and nitrogen in runoff from landscaped urban surfaces. Research has shown that runoff from areas with a large percentage of land cover in lawns can contribute significantly to phosphorus levels in receiving waters (Bannerman et al., 1993).

8. *Deicing chemicals.* The application of deicing chemicals and abrasives to provide safe driving conditions during winter is practiced in the snowbelt areas of the United States, all of Canada, and in many European countries. In the United States, road deicing salts are applied at rates of 75 to 330 kg/km of highway (street) lane. The leading states in highway salt use in the United States are Minnesota, Michigan, New York, Ohio, and Pennsylvania (Novotny, 2003). Typical road salt is 96 to 98% sodium chloride. About half of states use calcium chloride ($CaCl_2$) at lower temperatures, either as a liquid or as a dry mixture with salt (NaCl). Calcium chloride has a lower freezing point than salt and is often applied when temperatures range from -25 to $0°C$. Approximately 100 million tons of $CaCl_2$ are applied to U.S. roads compared to 10 million tons of road salt. Mixtures of sand and other abrasives with road salt in various proportions have been used in many communities and by some state highway departments. Abrasives can clog urban storm drains and roadside swales and generate significant cleanup costs in urban areas.

9. *Airport deicing.* All aircraft anti-icing and deicing chemicals used in North America are based on formulations of either ethylene or propylene glycol. Glycols themselves are not acutely toxic; however, deicing and anti-icing mixtures have been found to have significant chronic toxicity (Novotny, 2003). Airport runoff containing high concentrations of glycols can be toxic to animals if they drink it (pets and animals might like the taste). Glycols are biodegradable in soils and aquatic environments and have very high BOD. Consequently, runoff containing deicing chemicals poses a great hazard to the oxygen levels in receiving waters and must generally be treated before discharge. Five-day BOD in snowmelt water from airports may be as high as 22,000 mg/L, and at an application rate for large aircraft of 1000 L per aircraft, the BOD load from a single application is equivalent to the sewage BOD daily load from about 10,000 people (Novotny et al., 1999).

10. *Erosion.* Urban erosion can be divided into surface erosion of pervious surfaces and channel erosion. Surface erosion is driven by the energy of rainfall and overland flow, while channel erosion is related to the channel flow rates. Construction-site erosion is most severe and can be responsible for the major part of the sediment load in urban and suburban streams. A typical urban construction site is shown in Figure 9.8. In the United States alone, over 80 million tons/yr of sediment is washed from construction sites into surface-water bodies (Novotny, 2003). Although this load is only a fraction of the total erosion load, it is most devastating to the local urban streams that drain small watersheds also affected by washoff of contaminated solids from impervious areas and many urban activities. Urban pervious surfaces in humid areas are usually well protected by vegetation and yield pollutant inputs only during extreme storm events.

11. *Cross-connections and illicit discharges into storm sewers.* Illicit discharges and cross-connections can contribute significantly to the pollutant loads in storm sewers (Field et al., 2000). Nonstormwater discharges in storm sewers originate primarily from sewage and industrial wastewater leaking from sanitary sewers, failing septic tanks, ground-water infiltration containing contaminated ground water, and vehicle maintenance activities. Deliberate dumping into storm sewers and catch basins of used oil or waste paint are illegal in most instances; however, such practices are especially common and troublesome.

FIGURE 9.8   Typical residential construction site.

In some (older) cities in the United States, and in many other countries, combined-sewer overflow (CSO) occurs and is a concern. Combined sewers carry both storm runoff and sewage. Typically, urban combined-sewer systems are designed to carry flow that is about four to eight times the average dry-weather flow (sewage flows), while treatment plants serving these systems are typically designed to handle mixed flows that are four to six times the average dry weather flow. Typically, without CSO control, overflows from combined systems in most urban areas occur on average between 10 and 60 times per year (Novotny, 2003). Average water-quality characteristics of CSOs are shown in Table 9.2. Organic solids accumulate and sewer slime grows during dry periods, and hence far more pollutants accumulate in combined sewers than in storm sewers, which are idle and mostly dry between rainfall events. A nationwide assessment has shown that the annual load from CSOs for BOD are about the same as effluent from wastewater treatment plants treating wastewater from the same area (USEPA, 1983a). The annual discharges of suspended solids and lead are approximately 15 times higher in CSOs than in secondary treatment plant effluents. Annual loads of total nitrogen and phosphates from wastewater treatment plant effluents are typically four and seven times higher, respectively, than from CSOs.

TABLE 9.2   Average Water Quality of Combined-Sewer Overflows

Parameter	Average Concentration
BOD_5 (mg/L)	115
Suspended solids (mg/L)	370
Total nitrogen (mg/L)	9–10
Phosphate (mg/L)	1.9
Lead (mg/L)	0.37
Total coliform (MPN/100 mL)	10^2–10^4

Source: USEPA (1978).

9.4.2 Fate and Transport Processes

A major study called the National Urban Runoff Program (NURP) (USEPA, 1983a) was conducted in the United States between 1978 and 1983 to investigate the relationship between urban land uses and pollutant loadings resulting from surface runoff from urban areas. The NURP study included analyses of thousands of storms in 28 experimental watersheds located in a wide variety of locations in the United States. A key finding from the NURP study was that uncontrolled discharges from storm-sewer systems that drain runoff from residential, commercial, and light industrial areas carried more than 10 times the annual loading of total suspended solids (TSS) than did discharges from municipal treatment plants that provide secondary treatment (USEPA, 1999a). The NURP study also showed that urban stormwater runoff carries higher annual loadings than effluent from secondary treatment plants for the following water-quality parameters: chemical oxidation demand (COD), lead, copper, oil and grease, and polyaromatic hydrocarbons (PAHs).

The NURP study indicated that there is no consistent pattern of concentration within runoff events, and the NURP study focused on the statistical evaluation of the *event mean concentration* (EMC), defined as

$$\text{EMC} = \frac{\text{mass of pollutant contained in runoff event}}{\text{total volume of flow in the event}} = \frac{\Sigma Q_i C_i}{\Sigma Q_i} \tag{9.3}$$

where Q_i are the flow coordinates on the runoff hydrograph and C_i are the corresponding concentrations on the pollutograph. The EMC is equal to the flow-weighted mean concentration. The NURP study found that geographical location, land-use category, runoff volume, and other factors are statistically unrelated to EMCs and do not explain site-to-site or event-to-event variability. A possible reason for this is that it is not the land or land use that causes pollution, but pollution is more related to inputs and polluting activities that occur on the land. In most cases of practical interest, the total mass and EMC of a pollutant are far more important than the individual concentration distributions within individual runoff events. An interesting finding of the NURP study was that the EMCs for most pollutants follow a log-normal distribution, and median values and coefficients of variation (COVs) of EMCs found by NURP for the the major urban land-use categories are shown in Table 9.3.

TABLE 9.3 Event-Mean Concentrations for Urban Land Uses

Pollutant	Units	Residential		Mixed		Commercial		Open/Nonurban	
		Median	COV	Median	COV	Median	COV	Median	COV
BOD	mg/L	10	0.41	7.8	0.52	9.3	0.31	—	—
COD	mg/L	73	0.55	65	0.58	57	0.39	40	0.78
SS	mg/L	101	0.96	67	1.14	69	0.85	70	2.92
Total Pb	μg/L	144	0.75	114	1.15	104	0.68	30	1.52
Total Cu	μg/L	33	0.99	27	1.32	29	0.81	—	—
Total Zn	μg/L	135	0.84	154	0.78	226	1.07	195	0.66
TKN	μg/L	1900	0.73	1288	0.50	1179	0.43	965	1.00
NO_{2+3}-N	μg/L	736	0.83	558	0.67	572	0.48	543	0.91
Total P	μg/L	383	0.69	263	0.75	201	0.67	121	1.66
Soluble P	μg/L	143	0.46	56	0.75	80	0.71	26	2.11

The *coefficient of variation* is defined as the standard deviation of the observations divided by the mean. In some other site-specific studies, values significantly outside the ranges shown in Table 9.3 have been reported; such as COD levels as high as 1000 mg/L, suspended solids (SS) up to 2000 mg/L, and total phosphorus as high as 15 mg/L (Abernathy, 1981). The contribution of the watershed to the SS, BOD, COD, N, and P in urban runoff reflects the results of soil erosion, the leaching of nutrients from exposed soils and *detritus*,[1] and transportation by the runoff of various accumulated wastes in the watershed during dry weather.

In the United States, sewage treatment plants are required to remove at least 85% of the BOD and SS from raw sewage, and keep the BOD and SS of treated wastewater below 30 mg/L. Given the typical characteristics of urban runoff shown in Table 9.3, this indicates that urban runoff could contain BOD and SS concentrations comparable to or much higher than the concentrations in treated sewage.

Using the NURP data shown in Table 9.3, the unit load of a pollutant in runoff from ungaged urban watersheds can be estimated using the relation

$$\boxed{\text{load (kg/ha)} = 0.01 C \times P \times \text{EMC}} \tag{9.4}$$

where C is the runoff coefficient (dimensionless) defined by

$$C = \frac{\text{volume of runoff}}{\text{volume of rainfall}} \tag{9.5}$$

and P is the rainfall depth (mm). The units of the EMC in Equation 9.4 are mg/L ($= \text{g/m}^3$), and the runoff coefficient, C, can be estimated using the NURP data plotted in Figure 9.9. According to Schueler (1987), the runoff coefficient, C, can be approximated by

$$C = 0.05 + 0.009I \tag{9.6}$$

where I is the percent imperviousness of the catchment area. Utilization of Equation 9.4 to calculate the pollutant load in runoff from urban areas was called the *simple method* by Schueler (1987) and is considered applicable to areas less than 2.6 km² (1 mi²). The rainfall depth, P, in Equation 9.4 can be expressed in the form

$$P = P_o P_j \tag{9.7}$$

where P_o is the rainfall depth over the desired time interval and P_j is a factor that corrects P_o for storms that produce no runoff. The factor P_j is typically equal to the fraction of annual or seasonal rainfall that does not produce runoff, and for individual storms, $P_j = 1$.

Example 9.2 An urban (residential) area is estimated to be 40% impervious and to have an average annual rainfall of 1320 mm, and it is estimated that 50% of the annual rainfall does not produce any runoff. Assuming that the EMC distribution of suspended solids is

---

[1]Detritus is the organic debris formed by the decomposition of plants and animals.

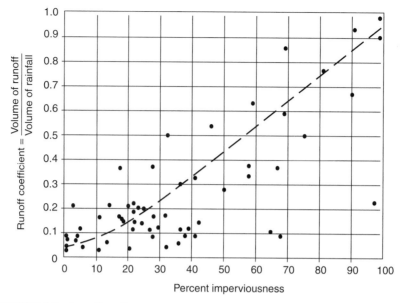

FIGURE 9.9   Runoff coefficients derived from NURP studies. (From USEPA, 1983a.)

normal, estimate the median and 90th percentile load of suspended solids contained in the runoff.

SOLUTION   From the NURP data in Table 9.3, the median suspended-solids (SS) concentration in runoff from a residential area is 101 mg/L, and the coefficient of variation is 0.96. Since the suspended-solids EMC distribution is normal, the mean, μ, and standard deviation, σ, of the EMC are given by

$$\mu = 101 \text{ mg/L}$$

$$\sigma = 0.96(101 \text{ mg/L}) = 97 \text{ mg/L}$$

The 90th percentile frequency factor for a normal distribution is obtained from Appendix D as 1.28, and hence the 90th percentile EMC for suspended solids, EMC_{90}, is given by

$$EMC_{90} = \mu + 1.28\sigma = 101 \text{ mg/L} + 1.28(97 \text{ mg/L}) = 225 \text{ mg/L}$$

For an imperviousness of 40%, the runoff coefficient, C, can be estimated from Figure 9.9 as 0.34. From the data given, $P_o = 1320$ mm, $P_j = 0.50$, and hence the effective annual precipitation, P, is given by Equation 9.7 as

$$P = P_o P_j = 1320(0.50) = 660 \text{ mm}$$

Equation 9.4 gives the median (= mean) annual load of suspended solids as

$$\text{load (kg/ha)} = 0.01 C \times P \times EMC$$

$$= 0.01(0.34)(660)(101) = 227 \text{ kg/ha}$$

TABLE 9.4   Annual Pollutant Loads from Various Sources

Land Use	SS [kg/(ha · yr)]	BOD [kg/(ha · yr)]	COD [kg/(ha · yr)]	Total N [kg/(ha · yr)]	Total Phosphate-P [kg/(ha · yr)]
Urban runoff	641	47	310	8.8	1.1
Cropland					
Cultivated	7940	5–10	40–80	15.5	1.4
Noncultivated	2040	5–10	40–80	16	0.18
Pastureland	2270	10–15	40–80	20	0.15
Rangeland	2720	2–3	10–15	3	0.5
Forest	410	2–3	10–15	0.5–3.5	0.01–0.15
Raw sewage	683	683	1194	137	35

Source: Laws (2000).

and the 90th percentile load as

$$\text{load (kg/ha)} = 0.01 C \times P \times \text{EMC}_{90}$$

$$= 0.01(0.34)(660)(225) = 505 \text{ kg/ha}$$

These estimated pollutant-load statistics reflect a typical urban watershed and could be significantly different, depending on local conditions such as the amount of ongoing construction within the area.

The EMC approach to estimating pollutant loads in urban runoff should be used with caution, since storm-drainage systems in some urban watersheds contain *dry-weather flow* that does not originate from rainfall runoff and yet accounts for a significant portion of the annual pollutant load discharged by the system. The dry-weather flow in a storm-drainage system typically originates from such sources as ground-water inflow, permitted discharges, illicit connections, excess irrigation, automobile washing, and other residential and commercial uses. A case in point is the dry-weather flow from the Ballona Creek catchment in southern California, which accounts 54, 19, 33, and 44% of the average-annual load of chromium, copper, lead, and nickel discharged through the storm drains (McPherson et al., 2005).

Many water-quality problems are caused not only by urban runoff but depend on the total pollutant load discharged annually into the water body. An indication of the annual pollutant load in urban runoff relative to pollutant loads from other sources is shown in Table 9.4. The loads shown in Table 9.4 are for a Cincinnati watershed reported by Weibel (1969). Since secondary treatment of domestic wastewater removes at least 85% of the SS and BOD from raw sewage, the loads shown in Table 9.4 indicate that urban runoff can contribute about half as much BOD and six to seven times as much SS to a receiving body of water as treated sewage, and that treated sewage is a far more important source of nutrient loading than urban runoff.

In contrast to using only Equation 9.4 to estimate pollutant loadings in urban runoff, a composite mass loading approach can be utilized to predict the contaminant load in urban runoff. The following example illustrates this approach.

Example 9.3   A watershed contains a variety of land uses and soils, and the pollutant loadings for total phosphorus are given in Table 9.5. The watershed area is predominantly

TABLE 9.5   Data for Example 9.3

| Land Use | Total P (kg/ha/yr) for Soil Group: | | | |
	A	B	C	D
Open space	0.09	0.09	0.09	0.09
Meadow	0.09	0.09	0.09	0.09
Newly graded	3.47	5.26	6.27	7.39
Forest	0.09	0.09	0.09	0.09
Commercial	1.79	1.79	1.79	1.79
Industrial	1.46	1.51	1.51	1.46
Residential				
0.05 ha or less	1.79	1.79	1.90	1.96
0.10–0.13 ha	1.01	1.23	1.23	1.23
0.2–0.4 ha	0.78	1.01	1.06	1.06
0.8–1.6 ha	0.27	0.37	0.37	0.37
Smooth surface	2.24	2.24	2.24	2.24

type B soil and consists of 24.3 ha of residential land with 0.4-ha lots, 8.9 ha of commercial area, 4.1 ha of industrial area, and 2.0 ha of open space. Estimate the annual loading of total phosphorus at the catchment outlet.

SOLUTION   Using the data given, the annual phosphorus loading from each type of area in the watershed is calculated summed. These calculations are summarized in Table 9.6. Therefore, based on this analysis, it is expected that the catchment runoff will contain a total phosphorus load of 46.8 kg/yr.

Although wet-weather urban runoff originates from both pervious and impervious surfaces, most urban runoff comes from impervious surfaces and typically only larger storms yield appreciable runoff from pervious areas. Pollutant loads from impervious areas are generally related to the accumulation of solids on roadways, and a two-step procedure is commonly used to estimate pollutant loads from these areas. In the first step, a model is used to quantitatively predict the buildup of solids as a function of time, and in the second step a model is used to quantitatively predict the wash-off of the accumulated solids. To obtain the loading of pollutants, the sediment loads are multiplied by the pollution content

TABLE 9.6   Results for Example 9.3

Land Use	Area (ha)	Percent of Total (%)	Soil Group	Unit Loading Rate (kg/ha/yr)	Total (kg/yr)
Residential	24.3	62	B	1.01	24.5
Commercial	8.9	23	B	1.79	16.0
Industrial	4.1	10	B	1.51	6.1
Open space	2.0	5	B	0.09	0.2
Total	39.3	100			46.8

of the sediment. Although exact calculation of pollutant accumulation and wash-off on street surfaces is highly uncertain, the models described below have been incorporated into urban watershed models, and the users of such models should be familiar with the basic concepts embodied in these models.

Buildup Models   Almost all street refuse can be found within 1 m of the curb, and accumulation of pollutant mass is typically expressed per meter of curb. The mass-balance accumulation function for street solids is typically expressed in the form

$$\frac{dm}{dt} = p - \xi m \tag{9.8}$$

where m is the amount of pollutants in curb storage [ML^{-1}], t is time [T], p is the sum of all pollutant inputs [ML^{-1}T^{-1}], and ξ is a removal coefficient [T^{-1}]. An expression for estimating the removal coefficient, ξ, is given by (Novotny et al., 1985)

$$\xi = 0.0116e^{-0.08H} \, (\mathrm{TS} + \mathrm{WS}) \tag{9.9}$$

where ξ is in day^{-1}, H is the curb height in cm, TS is the traffic speed in km/h, and WS is the wind speed in km/h. The coefficient of correlation between Equation 9.9 and the measured data was reported to be 0.86. Equation 9.8 can be integrated to yield the following expression for the pollutant buildup as a function of time:

$$\boxed{m(t) = \frac{p}{\xi}(1 - e^{-\xi t}) + m(0)e^{-\xi t}} \tag{9.10}$$

where $m(0)$ is the initial load of solids. In these types of processes there is always a tendency to attain an equilibrium whereby Equation 9.8 yields

$$\frac{dm}{dt} = 0 \qquad \text{when} \quad p = \xi m_{\mathrm{eq}} \tag{9.11}$$

and Equation 9.8 can be written in terms of the equilibrium solids accumulation, m_{eq}, as

$$\frac{dm}{dt} = -\xi(m - m_{\mathrm{eq}}) \tag{9.12}$$

This expression elucidates the fact that the accumulation rate of solids in street-curb storage can be either positive or negative, depending on whether the initial pollutant load, $m(0)$, is greater or smaller than the equilibrium load m_{eq}. Statistical evaluation and validation of the buildup equation, Equation 9.10, are described by Novotny et al. (1985), and these results showed that the removal coefficient, ξ, in medium-density residential areas was fairly constant, attaining values of about 0.2 to 0.4 day^{-1}, meaning that approximately 20 to 40% of the solids accumulated near the curb on the street surface is removed daily by wind and traffic.

The input, p, of solids into curb storage from the three major sources: refuse deposition (p_r), atmospheric dry deposition (p_a), and traffic (p_t) can be combined into the following loading estimate (Novotny, 2003):

$$p = p_r + p_a + p_t \tag{9.13}$$

where p_r, p_a, and p_t pollutant inputs can be estimated using the relations

$$p_r = \text{LIT} \tag{9.14}$$

$$p_a = \tfrac{1}{2}(\text{ATMFL})\,(\text{SW}) \tag{9.15}$$

$$p_t = \tfrac{1}{2}(\text{TE})\,(\text{TD})\,(\text{RCC}) \tag{9.16}$$

where LIT is the litter and street-refuse deposition rate (g/m · day), ATMFL is the dry atmospheric deposition rate (g/m² · day), SW is the street width (m), TD is the traffic emission rate (g/axle · m), TE is the traffic density (axles/day), and RCC is the road condition factor.

A model for sediment accumulation during the winter is very different from that described by Equation 9.10, since the effect of wind and traffic on sediment accumulation on streets is much different during the winter, and snow removal practices and application of deicing chemicals further complicate the process. Snow piles are effective traps of street pollutants, including large amounts of salt. Since solids and pollutants are incorporated into snow and are not removed from the snowpack by wind and traffic, the accumulation rate is almost linear and hence much higher than in nonwinter periods. For this reason, the quantity of accumulated pollutants near the curb at the end of the snow period is very high. Bannerman et al. (1983) documented that street loads of sediment and toxic metals are at their highest level from the onset of the first significant spring melt through the first spring rainfall event. The possibility of reducing the quantity of street pollutants during winter is very limited, since the use of sweeping equipment is not possible when snow piles and frozen ice are located on the sides of the streets.

Wash-off Models   The simplest and most widely used model of wash-off was introduced by Sartor et al. (1972, 1974) and is based on the following first-order removal concept:

$$\frac{dm}{dt} = -k_U i m \tag{9.17}$$

where m is the mass of solids remaining on the surface [M]; t is the time [T]; k_U is a constant called the *urban wash-off coefficient*, which depends on street-surface characteristics [L⁻¹]; and i is the rainfall intensity [LT⁻¹]. Equation 9.17 integrates to

$$\boxed{m(t) = m(0)\,\exp(-k_U i t)} \tag{9.18}$$

where $m(0)$ is the initial mass stored on the street. The constant k_U has been found almost independent of particle size within the range studied, $10\,\mu$m to 1 mm, and is commonly taken as $0.19\,\text{mm}^{-1}$ by urban runoff models that utilize this concept.

The authors of the STORM urban runoff model (Hydrologic Engineering Center, U.S. Army Corps of Engineers, 1975) modified Equation 9.18 by assuming that not all the solids are available for transport, is which case

$$m(t) = am(0) \exp(-k_U it) \tag{9.19}$$

where a is the *availability factor* (dimensionless), which accounts for the nonhomogeneous makeup of particles and the variability in travel distance of dust and dirt particles. The availability factor can be estimated using the relation

$$a = 0.057 + 0.04 i^{1.1} \tag{9.20}$$

where i is in mm/h, and the maximum value for a is 1.0. Equations 9.18 and 9.19 have both been incorporated into most widely used urban-runoff models.

Models describing the fate and transport of contaminants that are not connected to sediment transport have also been developed. For example, Massoudieh et al. (2005) developed a model to simulate the fate and transport of roadside-applied herbicides, which can have significant adverse impacts on primary productivity in receiving waters.

Example 9.4 A 100-m segment of an 8-m-wide urban roadway drains into a nearby lake where water-quality standards are being violated. Field reconnaissance indicates that the roadway segment has a solids deposition rate of 10 g/m · day, a dry-deposition rate of 1 g/m² · day, a traffic emission rate of 0.05 g/axle · m, a traffic density of 50 axles/day, and a road condition factor of 0.8. The height of the curb is 20 cm, the average traffic speed is 60 km/h, and the average wind speed is 8 km/h. It is estimated that the urban washoff coefficient is 0.1 mm⁻¹. Ten days before a recent 2-h storm produced 50 mm of runoff, the solids on the roadway segment was estimated to be 0.6 kg/m. (a) Estimate the mass of solids entering the lake from the storm runoff. (b) What is the equilibrium level of solids on the roadway?

SOLUTION (a) From the data given, LIT = 10 g/m · day, ATMFL = 1 g/m² · day, SW = 8 m, TE = 0.05 g/axle · m, TD = 50 axles/day, RCC = 0.8, H = 20 cm, TS = 60 km/h, WS = 8 km/h, $m(0)$ = 0.6 kg/m = 600 g/m, k_U = 0.1 mm⁻¹, and i = 50/2 mm/h = 25 mm/h. The mass of solids per unit curb length, $m(t)$, as a function of time, t, is given by Equation 9.10 as

$$m(t) = \frac{p}{\xi}(1 - e^{-\xi t}) + m(0)e^{-\xi t} \tag{9.21}$$

The solids input into curb storage, p, is given by Equations 9.13 to 9.16 as

$$p = p_r + p_a + p_t = \text{LIT} + \tfrac{1}{2}(\text{ATMFL})(\text{SW}) + \tfrac{1}{2}(\text{TE})(\text{TD})(\text{RCC})$$

$$= 10 + \tfrac{1}{2}(1)(8) + \tfrac{1}{2}(0.05)(50)(0.8)$$

$$= 44 \text{ g/m} \cdot \text{day}$$

The removal coefficient, ξ, is given by Equation 9.9 as

$$\xi = 0.0116e^{-0.08H} \, (TS + WS) = 0.0116e^{-0.08(20)} \, (60 + 8) = 0.16 \text{ day}^{-1}$$

Substituting $p = 44 \text{ g/m} \cdot \text{day}$, $\xi = 0.16 \text{ day}^{-1}$, $t = 10$ days, and $m(0) = 600 \text{ g/m}$ into Equation 9.21 gives the mass of solids along the roadway at the beginning of the storm as

$$m(t) = \frac{44}{0.16} \, [1 - e^{-0.16(10)}] + 600e^{-0.16(10)} = 98 \text{ g/m}$$

The mass of solids along the roadway at a time t after the start of the storm is given by Equation 9.19 as

$$m(t) = am(0) \, \exp(-k_U it) \tag{9.22}$$

where the availability factor, a, is given by Equation 9.20 as

$$a = 0.057 + 0.04i^{1.1} = 0.057 + 0.04(25)^{1.1} = 1.44$$

Since $a > 1$, take $a = 1$. Substituting $a = 1$, $m(0) = 98 \text{ g/m}$, $k_U = 0.1 \text{ mm}^{-1}$, $i = 25 \text{ mm/h}$, and $t = 2 \text{ h}$ into Equation 9.22 yields the residual mass along the curb at the end of the storm as

$$m(t) = 1(98) \, \exp[-0.1(25)(2)] = 0.67 \text{ g/m}$$

Since the mass of solids remaining along the roadway after the storm is 0.67 g/m and the prestorm mass along the roadway is 98 g/m, the wash-off is equal to 98 g/m − 0.67 g/m = 97.3 g/m. Since the roadway segment is 100 m long,

$$\text{total solids in storm runoff} = 100 \text{ m} \times 97.3 \text{ g/m} = 9730 \text{ g} = 9.73 \text{ kg}$$

Therefore, it is estimated that the storm washes approximately 10 kg of solids from the roadway pavement into the lake.

(b) The equilibrium mass of solids, m_{eq}, on the roadway pavement is given by Equation 9.11 as

$$m_{eq} = \frac{p}{\xi} = \frac{44}{0.16} = 275 \text{ g/m}$$

The accumulated mass on the roadway is expected to approach 275 g/m in the absence of any washoff.

9.4.3   Best Management Practices

There are a variety of methods that are used to attenuate urban pollution prior to discharge into water bodies. These methods fall into one of the following five categories: (1) prevention, (2) source control, (3) hydrologic modification, (4) reduction of pollutants and flows in the conveyance systems, and (5) end-of-pipe pollution controls. Prevention practices prevent the deposition of pollutants in the urban landscape; source-control practices prevent pollutants from coming into contact with precipitation and stormwater runoff, hydrologic modification minimizes the runoff formation from precipitation, reduction of pollutants and flows in the conveyance systems involve special channel features for pollutant attenuation such as swales and particle removal structures, and end-of-pipe controls include such treatment features as wetlands that are immediately prior to discharge into a receiving water body.

Source-Control Measures   Source-control measures are the most effective in controlling pollution from urban runoff. These measures commonly include reduction of pollutant accumulation on the impervious surfaces, reduction of erosion of pervious lands, and on-site runoff infiltration. Some of the most advanced suburban developments have zero-discharge stormwater management where urban runoff is dissipated by a succession of best management practices that promote storage, infiltration, and evapotranspiration (Novotny, 2003). The most effective source-control measures are described below.

1. *Removal of solids from street surfaces.* Removal of solids from street surfaces is a commonly practiced source-control measure. This includes litter-control programs and street cleaning. Litter includes paper, vegetation residues, animal feces, bottles, broken glass, and plastics. In the fall season, leaves are typically the most dominant component of street litter, and it has been shown that litter-control programs can reduce the amount of deposition of pollutants by as much as 50%. It has been estimated that the average tree drops 14.5 to 26 kg of leaves per tree per year, and the leachate from leaves and lawn clippings is a source of phosphorus in urban runoff. Pet waste can be a source of fecal bacteria, nutrients, and oxygen-demanding compounds in urban runoff when allowed to be deposited on sidewalks and urban streets. Many communities have ordinances that regulate pet waste and require proper disposal of the waste by pet owners.

Street sweeping involves the removal of dust and debris, pet fecal waste, and trash from parking lots and street surfaces. The most common street sweeper uses a truck-mounted rotating gutter broom to remove particles from the street gutter and place them in the path of a large cylindrical broom which rotates to carry the material onto a conveyor belt and into a hopper. Street sweepers come in many makes and models, such as those shown in Figure 9.10 *Street flushing* is another practice that washes streets by water jets delivered

(a)   (b)

FIGURE 9.10   Street sweepers. [(*a*) From City of Vernon, British Columbia, 2005; (*b*) from Cornell University, Environmental Compliance Office, 2005.]

from tanker trucks. Flushing cleans the entire street and not just a narrow strip near the curb. Sweeping is more common in the United States, while flushing is more practiced in Europe. The effect of street sweeping on reducing pollutant loads is related more to frequency of sweeping and less to sweeper efficiency. Infrequent sweeping (less than 1 week) has a poor effect irrespective of the pollutant-removal efficiency of the sweeper (Novotny and Olem, 1994). In many cases, street sweeping is practiced for its aesthetic rather than water-quality benefits.

2. *Erosion control.* Erosion from pervious areas is often a source of pollution. Temporary or permanent seeding of grass, sodding, and mulching are used to reduce erosion. Such measures are important and mandated in some areas for control of pollution caused by construction activities. Covering an exposed area with any of a number of available mulches generally increases surface roughness and storage, protects the surface against rainfall impact, and subsequently reduces erosion.

3. *Control of surface application of chemicals.* Measures to control surface application of chemicals include control of herbicide use on pervious grassed areas (lawns and golf courses). Control of pollution caused by chemical use by individual homeowners on their lands is difficult due to a lack of legal instruments.

4. *Control of urban and highway pollution during winter.* Deicing chemicals, such as sodium chloride (NaCl), are the predominant cause of pollution from winter runoff. To diminish the environmental threats of deicing chemicals and abrasives, the selection of deicing compounds and their application rates must be judicial and targeted. Advancements in weather forecasting by the use of road/weather information systems (RWIS) in connection with thermal mapping of roads and anti-icing technology, and the development of new, more environmentally safe deicing chemicals and chemical mixtures lead to reductions in chemical application rates (Novotny et al., 1999). The practice of prewetting road surfaces with liquid (not granular) slurry of a chemical has been found to be an effective way to reduce application rates.

Hydrologic Modifications   Hydrologic modifications of urban watersheds include measures and practices that reduce the volume and intensity of urban runoff entering the storm-sewer or combined-sewer system. Hydrologic modifications can be divided into the following three practices:

- Practices that increase permeability and enhance infiltration, such as the use of pervious pavements or vegetation filter strips
- Practices that increase on-site storage
- Practices that reduce the size of impervious areas directly connected to the sewer system

In applying practices that increase permeability and enhance infiltration, care must be taken since the longevity of these practices can be severely limited by lack of pretreatment, poor construction practices, application in unsuitable sites, lack of regular maintenance, and faulty design. Several hydrologic modifications are described in more detail below.

1. *Porous pavements.* Porous pavement is an alternative to conventional pavement whereby rainfall is allowed to percolate through the pavement into the subbase. Porous pavements are made either from asphalt or concrete in which fine filler fractions are missing or are modular. The primary benefit of porous pavements is a significant reduction or

FIGURE 9.11   Porous pavement. (Courtesy of David A. Chin.)

even complete elimination of surface runoff from an otherwise impervious area. Aquifer recharge by infiltrated water is a second benefit, and reduced need for storm drainage is a third benefit. If subsoils are very permeable, there may be no need for installing storm drains. Porous pavement is most feasible when subsoils are permeable and ground slopes are relatively flat. In areas with poorly draining subsoils or if the pavement is installed over an existing impervious base, a drainage system can be installed. Porous pavement has an excellent potential for use in parking areas and on side streets. Clogging may occur during construction and/or operation; however, this can be remedied by flushing and sweeping. An example of a porous pavement in Maimi, Florida, is shown in Figure 9.11. Guidelines for the design of porous pavements are given in Table 9.7. The porosity of porous pavements is typically on the order of 0.25, and the thickness is typically on the order of 7.5 to 10 cm. The construction cost of porous pavement is about the same or even less than for conventional pavement when savings on storm-drainage infrastructure is included.

TABLE 9.7   Guidelines for Design of Porous Pavement

Parameter	Design Criteria
Drainage area	Up to 4 ha (10 acres)
Minimum infiltration rate of subsoil	1.2 cm/h (0.5 in./h)
Minimum separation from ground-water table	0.5 to 0.9 m (2 to 3 ft)
Maximum pavement slope	5%
Maintenance	Frequent vacuum sweeping of fine sediment

Source: Schueler et al. (1991).

FIGURE 9.12   Storage of rooftop runoff. (From India Together, 2005; photo by S. Vishwanath.)

2. *Increasing surface storage.* Rooftop storage on flat roofs, temporary ponding, and restriction of stormwater inlets are used for control of combined-sewer overflows and to reduce flooding. Diverting rooftop-collected rainwater into storage tanks and subsequent reuse for irrigation and other nonpotable water supply has been practiced in many countries for centuries and is a feasible stormwater management and reuse alternative. An example of such a practice in India is shown in Figure 9.12.

3. *Decreasing directly connected impervious area.* Directly connected impervious area (DCIA) is defined as the impermeable area that drains directly into the drainage system without passing over pervious area. Roadways are the most common example of DCIA where runoff from the pavement is routed directly to a stormwater inlet, such as shown in Figure 9.13. It is widely recognized that minimization of DCIA is by far the most effective method of runoff quality control because it delays the peaks of flows into the sewers and maximizes infiltration. Practices used to minimize DCIA include disconnecting roof drains from storm sewers, permitting surface runoff to overflow onto adjacent pervious surfaces, and directing stormwater runoff to infiltration structures such as dry wells, infiltration basins, and ditches.

4. *Increasing infiltration.* Infiltration basins and infiltration trenches are the most commonly used structures for increasing the infiltration of stormwater runoff. These practices are described in more detail below.

a. *Infiltration basins.* An infiltration basin is made by constructing an embankment or by excavating an area down to relatively permeable soils. Infiltration basins store

FIGURE 9.13   Direct connection between a roadway and a stormwater inlet. (Courtesy of David
A. Chin.)

stormwater runoff temporarily until it infiltrates through the bottom and sides of the
basin. These basins are normally dry and can be incorporated into the landscape design
as open areas or even recreational areas such as sports fields. An example of an
infiltration basin draining a large parking lot is shown in Figure 9.14, where the concrete-
lined drainage channel leading from the parking lot to the infiltration basin is clearly
apparent at the far end of the basin. Infiltration basins need to drain down and dry out

FIGURE 9.14   Infiltration basin. (Courtesy of David A. Chin.)

in a reasonable period of time to prevent sealing of the bottom by a slime layer of algae, bacteria, and fungi. If water is allowed to sit in the bottom of the basin for more than 72 h (= 3 days) in most climates, the conditions to allow slime formation are high (Novotny, 2003). The following formula can be used to calculate the maximum allowable ponding depth in a basin to achieve a given design ponding time:

$$d_{max} = fT_p \qquad (9.23)$$

where d_{max} is the maximum design depth, f is the soil infiltration rate, and T_p is the design ponding time. To maintain the infiltration capacity of the basin, it is important that excessive sediment loadings be avoided. Studies in Florida have found that infiltration basins with grass bottoms tend to perform longer than basins with earthen bottoms. Guidelines for the design of infiltration basins are given in Table 9.8. Maintenance needs include annual inspections and inspections after large storms, mowing at least twice per year, debris removal, erosion control, and control of nuisance odor or mosquito problems. Deep tilling may be needed at 5- to 10-year intervals to break up a clogged surface layer.

b. *Infiltration trenches.* A conventional infiltration trench is a shallow excavated trench that has been backfilled with stone to create an underground reservoir. Infiltration trenches work similar to infiltration basins and have similar pollutant-removal capabilities. An infiltration trench under construction is shown in Figure 9.15, where the coarse aggregate used to fill the trench and the porous pipe used to deliver runoff to the trench (via a catch basin) are clearly apparent. Guidelines for the design of infiltration trenches are given in Table 9.9. Maintenance requirements for infiltration trenches include inspections annually and after large storms, buffer-strip maintenance and mowing, and rehabilitation of the trench when clogging begins to occur. Surface clogging can be remedied by replacing the top layer of the trench, but bottom clogging requires the removal of all of the filter and stone aggregate. Experience in Maryland is that about one in five conventional trenches installed in the 1970s and 1980s failed to operate as designed immediately after construction, and barely half of all conventional infiltration trenches typically operate adequately after 5 years (Schueler et al., 1991; Galli, 1992). To minimize the likelihood that the trench becomes clogged during construction, the entire area contributing to the infiltration device should be stabilized before construction of the trench. Oil and grease should be removed before they enter infiltration

TABLE 9.8 Guidelines for Design of Infiltration Basins

Parameter	Design Criteria
Drainage area	2 to 20 ha (5 to 50 acres)
Minimum infiltration rate	1.25 cm/h (0.5 in./h)
Maximum ponding time	72 h (24 to 36 h for ponds planted with grass)
Minimum ground-water table	1.2 m (4 ft) below the bottom
Minimum distance from buildings	6.6 m (22 ft)
Inlet control	Prefiltration of settleable solids

Source: Minnesota Pollution Control Agency (1989).

FIGURE 9.15   Infiltration trench. (Courtesy of David A. Chin.)

TABLE 9.9   Guidelines for Design of Infiltration Trenches

Parameter	Design Criteria
Drainage area	0.8 to 2 ha (2 to 5 acres)
Minimum infiltration rate	0.7 cm/h (0.27 in./h)
Minimum distance from buildings	6.6 m (22 ft)
Minimum separation from ground water	0.6 to 0.9 m (2 to 3 ft)
Inlet control	Prefiltration of settleable solids

Source: Minnesota Pollution Control Agency (1989) and Schueler et al. (1991).

trenches since these contaminants are difficult to remove and present a threat to ground water.

Both infiltration basins and trenches are prone to clogging by deposited solids, and to increase their life span when sediment-laden stormwater flows are anticipated, sediment traps prior to inflow are recommended. The best pretreatment device is a grassed filter or buffer strip along the periphery of the basin or trench, which should be at least 6.6 m wide (Harrington, 1989). Typical long-term pollutant-removal rates for infiltration basins and trenches are given in Table 9.10.

Attenuation of Pollutants   The most common practices used to attenuate and reduce the transport of pollutants from sources to receiving water bodies are filter strips and buffer zones. These are described in more detail below.

TABLE 9.10   Typical Long-Term Removal Rates for Infiltration Basins and Infiltration Trenches

Pollutant	Typical Removal Rate (%)
Sediment	75–90
Total phosphorus	50–70
Total nitrogen	45–60
Biological oxygen demand	70–80
Metals	75–90
Bacteria	75–90

Source: Schueler (1987).

Filter Strips   Filter strips are vegetated sections of land designed to accept runoff as overland sheet flow from upstream developments or flow from a highway or parking lot. Filter strips remove pollutants from runoff by filtering, provide some infiltration, and slow down the runoff flow to promote sedimentation. An example of a filter strip adjacent to a parking lot is shown in Figure 9.16, where runoff from the parking lot enters the filter strip through cuts in the curb (on the left) and runoff from the walkway (on right) runs directly

FIGURE 9.16   Filter strip. (From Cornell University, Environmental Compliance Office, 2005.)

TABLE 9.11   Guidelines for Design of Grass Filter Strips

Parameter	Design Criteria
Filter width	Minimum width 15 to 23 m (50 to 75 ft), plus additional 1.2 m (4 ft) for each 1% slope
Flow depth	5 to 10 cm (2 to 4 in.)
Filter slope	Maximum slope 5%
Flow velocity	Maximum flow velocity of 0.75 m/s (2.5 ft/s)
Grass height	Optimum grass height of 15 to 30 cm (6 to 12 in.)
Flow distribution	Should include a flow spreader at the upstream end to facilitate sheet flow across the filter

Source: Minnesota Pollution Control Agency (1989) and Novotny and Olem (1994).

into the filter strip. The filter strip shown in Figure 9.16 has an underdrain to promote infiltration and subsurface drainage. In areas with (NRCS) type A or type B soils, filter strips can facilitate infiltration without underdrains. Filter strips cannot treat high-velocity flows and therefore are generally used for small drainage areas. Grass filter strips provide higher pollutant removal rates than do grass swales. The difference between filter strips and swales is in the type of flow. Flow depths in filter strips are less than the heights of the grasses, thus creating laminar flows that enhance settling and filtering. Flow in swales is concentrated and flow depths are greater than the heights of the grasses, which usually results in turbulent flow. Vegetated filter strips are feasible in low-density developments with small drainage areas and areas bordering roads and parking lots. Guidelines for the design of grass filter strips are given in Table 9.11. Dense grass needs to be maintained in filter strips, and gully and channel formation should be prevented. Grass heights should remain at 15 cm (6 in.) or greater. Pesticide and fertilizer use should be limited to the minimum necessary for dense growth. Filter strips are effective in removing sediment and sediment-associated pollutants such as bacteria, particulate nutrients, pesticides, and metals. Infiltration is an important removal mechanism in filter strips, and many pollutants (including phosphorus) are dissolved or associated with very fine particles that move into the soil with infiltrating water. The overland-flow distance at which 100% of the sediment is removed by a filter strip is called the *critical distance*, and in a study of Bermuda grass, more than 99% of sand was removed over a distance of 3 m (10 ft), 99% of silt over 15 m (45 ft), and 99% of clay over 120 m (400 ft) (Wilson, 1967).

Environmental Corridors and Buffer Zones   Vegetated land adjacent to a receiving water body acts as a buffer between a polluting urban area and the receiving water body. These zones are commonly called *environmental corridors*. These corridors lose their efficiency if a storm-drainage outlet bypasses these vegetated areas and discharges directly into the receiving water body or into a channel with concentrated flows that is directly connected to the water body. Treatment processes for storm runoff, such as vegetated filters, infiltration basins, detention and retention ponds, and wetlands, are incorporated into the landscape of the corridor.

Buffer strips made of uneven shoreline vegetation may also be used to attenuate runoff pollutants that otherwise would reach the receiving water body. A typical buffer strip adjacent to a river, more commonly called a *riparian buffer*, is shown in Figure 9.17. A 30-m-wide

FIGURE 9.17 Riparian buffer strip. (From NRCS, 2005c.)

buffer strip is recommended for protection of surface-water reservoirs from which water is used for drinking water supply in states where residential development is permitted in such watersheds (Nieswand et al., 1990). In some countries, urban and agricultural land-use practices are greatly restricted, or not permitted at all, in watersheds of water-supply reservoirs. Buffer strips are ineffective on steep slopes with loose soils.

Sediment Barriers and Silt Fences   These are small temporary structures used at various points within or at the periphery of a disturbed area to detain runoff for a short period of time and trap heavier sediment particles. Sediment barriers are built with a variety of materials, such as straw bales, filter fabric attached to a wire or wood fence, filter fabric on straw bales, and gravel and earth berms. These barriers are placed in the path of sediment-laden runoff, commonly from construction sites and surface mining. A typical silt fence is shown in Figure 9.18. Such sediment barriers should not be placed across a drainageway that carries a large volume of runoff.

Collection-System Pollution Control   Collection-system pollution-control practices include methods and practices for removing pollutants from runoff after they leave the source area. Such methods include grassed waterways and channel stabilization, riprap and gabions, and catch basins. These are described below.

Grassed Waterways and Channel Stabilization   When conveyance channels are used, channel erosion can be controlled by designing hydraulically stable channels (Chin, 2006). Grassed waterways are probably the most inexpensive but effective means of conveying and treating water. In a simple case, a grassed roadside swale will perform as well or better than a more expensive buried storm sewer (Novotny, 2003). A typical grassed swale, shown in Figure 9.19, is a shallow channel with side slopes on the order of 3:1 (H:V) and is used only for conveyance of surface runoff, hence is mostly dry and contains

FIGURE 9.18   Silt fence. (From ACF Environmental, 2005.)

water only during and following rain. Similar to grass filters, swales and grassed water-ways remove pollutants by slowing down the flow, filtering by grasses, infiltration, and nutrient uptake by vegetation. However, in contrast to grass filters, the grasses in swales are submerged and flow in the waterways are turbulent and hence less effective for removal of particulates than grass filters. Generally, the lower the slope and velocity, the better the treatment performance of swales. For effective pollutant-removal control, the depth in the

FIGURE 9.19   Typical grassed roadside swale. (From Purdue Research Foundation, 2005.)

TABLE 9.12 Maximum Permissible Velocities in Vegetated Swales

Cover	Range of Channel Gradient (%)	Permissible Velocity (m/s)
Turcote, midland, and coastal Bermuda grass	0–5.0	1.8
	5.1–10.0	1.5
	>10	1.2
Reed canary grass, Kentucky, 31 tall fescue, Kentucky bluegrass	0–5.0	1.5
	5.1–10.0	1.2
	>10	0.9
Red fescue	0–5.0	0.75
Annuals—ryegrass	0–5.0	0.75

Source: Minnesota Pollution Control Agency (1989) and Novotny and Olem (1994).

swale should not be greater than 30 to 45 cm, and the velocity should not exceed 0.9 to 1.8 m/s. Higher slopes can be reduced by check dams and/or by temporary barriers made of straw bales and cloth. Grassed swales are typically designed on the basis of a maximum allowable velocity, with the velocity in the channel calculated using the Manning equation. Typical allowable velocities for grass swales are shown in Table 9.12. The Manning equation used to estimate the capacity of the grassed channel, Q (m³/s), is given by

$$Q = A\frac{1}{n}R^{2/3}S_0^{1/2} \tag{9.24}$$

where A is the cross-sectional flow area of the channel (m²), n is the Manning roughness coefficient for the channel, R is the hydraulic radius, and S_0 is the slope of the channel. The velocity of flow, V, in the channel is given by

$$V = \frac{Q}{A} \tag{9.25}$$

and the Manning roughness coefficient, n, is related to retardance of the grass species, derived from Table 9.13, and can be estimated using the relations in Table 9.14.

Riprap and Gabions Riprap is a layer of loose rock or concrete blocks placed over an erodible soil surface, and gabions are blocks made of wire mesh filled with smaller rocks or gravel. These are illustrated in Figure 9.20. Both riprap and gabions are used primarily for channel stabilization in high-erosion zones such as sharp bends, channel drops, and flow-energy dissipators (stilling basins) below the outlets of sewers, narrow bridges, and near connections with lined high-velocity channels.

Catch Basins A catch basin is a chamber or well, usually built at the curbline of a street, through which stormwater is admitted into the sewer system. A catch basin typically has as its base a sump which should be large enough to provide storage for trapped debris. A typical catch basin, and its connection to the drainage system, is shown in

TABLE 9.13   Retardance in Grass-Lined Channels

Retardance	Cover	Condition
A	Reed canary grass	Excellent stand, tall (average 90 cm)
	Yellow bluestem *Ischaemum*	Excellent stand, tall (average 90 cm)
	Weeping lovegrass	Excellent stand, tall (average 75 cm)
B	Smooth bromegrass	Good stand, mowed (average 30–40 cm)
	Bermuda grass	Good stand, tall (average 30 cm)
	Native grass mixture (little bluestem, blue grama, and other long and short Midwest grasses)	Good stand, unmowed
	Tall fescue	Good stand, unmowed (average 45 cm)
	Lespedeza sericea	Good stand, not woody, tall (average 50 cm)
	Grass–legume mixture–timothy, smooth	Good stand, uncut (average 50 cm)
	Tall fescue, with bird's foot trefoil or lodino	Good stand, uncut (average 45 cm)
	Blue grama	Good stand, uncut (average 35 cm)
	Kudzu	Very dense growth, uncut
		Dense growth, uncut
	Weeping lovegrass	Good stand, tall (average 60 cm)
		Good stand, unmowed (average 35 cm)
	Alfalfa	Good stand, uncut (average 30 cm)
C	Bahia	Good stand, uncut (15–18 cm)
	Bermuda grass	Good stand, mowed (average 15 cm)
	Redtop	Good stand, headed (40–60 cm)
	Grass–legume mixture–summer	Good stand, uncut (15–20 cm)
	Centipede grass	Very dense cover (average 15 cm)
	Kentucky bluegrass	Good stand, headed (15–30 cm)
	Crabgrass	Fair stand, uncut (average 25–120 cm)
	Common *Lespedeza*	Good stand, uncut (average 30 cm)
D	Bermuda grass	Good stand, cut to 6-cm height
	Red fescue	Good stand, headed (30–45 cm)
	Buffalo grass	Good stand, uncut (8–15 cm)
	Grass–legume mixture–fall, spring	Good stand, uncut (10–13 cm)
	Lespedeza sericea	After cutting to 5-cm height; very good stand before cutting
	Common *Lespedeza*	Excellent stand, uncut (average 10 cm)
E	Bermuda grass	Good stand, cut to 4-cm height
	Bermuda grass	Burned stubble

Source: Coyle (1975) and FHWA (1996).

Figure 9.21; the catch basin is shown on the left. To maintain the effectiveness of catch basins for pollutant removal, they must be cleaned about twice a year, depending on local conditions.

Detention–Retention Facilities   Ponds and storage basins are the backbone of urban stormwater management. The combination of a pond (storage and pretreatment), wetland

TABLE 9.14 Manning n for Vegetative Linings[a]

Retardance	Formula for n
A	$n = \dfrac{1.22R^{1/6}}{30.2 + 19.97\log(R^{1.4}S_0^{0.4})}$
B	$n = \dfrac{1.22R^{1/6}}{37.4 + 19.97\log(R^{1.4}S_0^{0.4})}$
C	$n = \dfrac{1.22R^{1/6}}{44.6 + 19.97\log(R^{1.4}S_0^{0.4})}$
D	$n = \dfrac{1.22R^{1/6}}{49.0 + 19.97\log(R^{1.4}S_0^{0.4})}$
E	$n = \dfrac{1.22R^{1/6}}{52.1 + 19.97\log(R^{1.4}S_0^{0.4})}$

Source: Kouwen et al. (1980).
[a]R in meters.

(treatment), and infiltration or irrigation infrastructure can result in zero-discharge sustainable stormwater disposal and reuse. The sizing of storage basins can best be accomplished by continuous simulation models, since it requires consideration of storage changes resulting from transient inflow and the rate at which the storage basin is emptied after a storm. The constraining parameter is either the number of overflows allowed per year or the total volume captured. The basin storage volume needed to capture a certain portion of runoff and the number of overflows are interrelated. The two types of detention basins used for quality control of urban runoff are wet detention ponds and extended or modified dry ponds.

A *wet detention pond* has a permanent pool of water, and a simple wet pond acts as a settling facility with relatively low efficiency. A typical wet detention pond is shown

(a)

(b)

FIGURE 9.20 (*a*) Riprap; (*b*) gabions. [(*a*) From USACE, 2005b; (*b*) from NRCS 2000d.]

FIGURE 9.21   Catch basin and drainage-system connection. (From Queens University, 2005.)

in Figure 9.22, where the corrugated metal outlet riser is shown on the left. Accumulated solids must be removed (dredged) to maintain the removal efficiency and the aesthetics of the pond. Improper design and maintenance can make such facilities an eyesore and a mosquito-breeding mudhole. A well-designed engineered wet detention pond consists of (1) a permanent water pool, (2) an overlying zone in which the design runoff volume temporarily increases the depth of the pool while it is stored and released at the

FIGURE 9.22   Wet detention pond. (From City of Greensboro, 2005.)

FIGURE 9.23   Dry detention basin. (From Bow River Basin Council, 2005.)

allowable peak-discharge rate, and (3) a shallow littoral zone that acts as a biological filter. The depth of wet detention ponds should be between 1 and 3 m, and it is important that the side slope of the pond be less than 5:1 (H:V) to minimize the danger of drowning.

A *dry detention pond* or *dry detention basin* is a stormwater detention facility that is normally dry and is designed to store stormwater temporarily during high-peak-flow runoff events. A typical dry detention basin is shown in Figure 9.23. A safety overflow spillway is part of the basin and is used for conveyance when the storage capacity of the basin is exhausted. The outlet of dry detention basins used for flood control is typically sized for large storms, and smaller but polluting runoff events will pass through such basins mostly without appreciable attenuation of the pollution load. Such dry detention basins are ineffective for urban runoff quality control. By combining the dry detention basin with an infiltration system located at the bottom of the basin, the pollution-control capability is enhanced and can be effective for pollution removal. Such basins are called *modified dry detention basins*. Modified dry detention basins, also known as *extended dry detention basins*, are effective pollution-control devices.

In snowbelt zones of North America and Europe, some detention ponds receive flows with extremely high concentrations of salt. High salinity increases the solubility of metals in sediments and water, and the sediment accumulated in ponds can actually become a source of metals rather than a sink. Hence, the removal efficiencies in winter are much less than in a nonwinter period.

Hartigan (1989) compared removal efficiencies of modified (extended) dry basins and wet detention ponds. He reported that removal of total phosphorus in wet ponds is two to three times greater than in modified dry basins (50 to 60% versus 20 to 30%) and 1.3 to 2 times greater for total nitrogen (30 to 40% versus 20 to 30%). For other pollutants, the average removal rates for wet detention ponds and extended dry detention basins were very similar: 80 to 90% for total dissolved solids, 70 to 80% for lead, 40 to 50% for zinc, and 20 to 40% for BOD and COD.

Storage facilities used for flood control and water-quality control have different objectives and design criteria. Facilities designed solely for flood control use statistically rare design storms and may be ineffective for water-quality control. *Dual-purpose ponds* are used for both flood control and water-quality control. Design guidelines for dual-purpose ponds are as follows (Novotny, 1995):

- Wet-detention permanent-storage volume should provide an average hydraulic residence time of at least 14 days. Use the wettest 14-day period during an average year to calculate the hydraulic residence time.
- A bleed-down volume for water-quality control should be provided for the first 25 mm of runoff from the directly connected impervious area.
- The flood-control volume is calculated for a design storm with a recurrence interval from 10 to 100 years. The flood-control volume should include the treatment volume.
- Outlets are designed such that no more than one-half of the water-quality volume is discharged in the first 60 hours following a storm event.
- Side slopes should be 6:1 or flatter to provide a littoral shelf from the shoreline out to a point 0.6 to 1 m below the permanent pool elevation. Side slopes above the littoral zone should be no steeper than 4:1.
- Inlet structures should be designed to dissipate the energy of water entering the pond to prevent short circuiting.
- Ponds that receive stormwater from contributing areas with more than 50% impervious surface or that are a potential source of oil and grease contamination must include a baffle, skimmer, and grease trap to prevent these substances from being discharged from the pond.
- Native plant species should be planted up to a depth of 1 m below the permanent pool.
- A length/width ratio of 4:1 to 7:1 is preferred.

Most urban surface runoff is treated using some form of settling basin. Since many of the pollutants in urban runoff are associated with suspended solids, and settling of the suspended solids will usually remove a significant portion of the BOD, nutrients, hydrocarbons, metals, and pesticides from the water. The effectiveness of settling basins on removing metals depends on the partitioning of metals between the adsorbed (particulate) phase and the aqueous (dissolved) phase, since only metals in the particulate phase are removed by settling basins. Studies by Dean et al. (2005) in an experimental urban watershed in Baton Rouge, Louisiana, demonstrated that Cd and Cu partitioned nearly equally between particulate and dissolved phases, Zn was primarily particulate bound, and Pb was highly particulate bound.

In contrast to surface-water settling basins, it is sometimes possible to route land runoff to ground-recharge basins where the runoff can percolate into the soil. Such basins require that sufficient land area be available and that the soil be sufficiently porous. Ground-water recharge basins are classified as either dry or wet, with the latter containing a permanent pool of water. In areas where the water table is sufficiently low, percolation of runoff may also be accomplished with the use of *dry wells*, which are pits or trenches in porous soil backfilled with rock. An important concern with the use of ground-recharge basins is the possibility that pollutants in the runoff will contaminate ground-water supplies. However, this possibility is minimal to the extent that the pollutants are associated with suspended solids, since percolation of water through the soil typically removes close to 100% of the suspended solids. Pathogens derived from animal wastes are also removed with virtually 100% efficiency by the percolation process. However, many pollutants are not effectively removed by percolation, and the possibility of ground-water contamination from urban runoff does exist.

Overflow from combined-sewer systems is sometimes diverted to tunnels or tanks for later treatment. Berlin, Germany, for example, has a system of underground tanks to handle overflow from the combined-sewer system that serves one-third of West Berlin. The largest treatment system for combined-sewer overflow is the tunnel-and-reservoir plan (TARP) used by the city of Chicago. TARP consists of a system of huge tunnels, underground reservoirs, pumping stations, and treatment facilities in which excess combined-sewer flow is stored for later treatment.

Attenuation of Priority Pollutants Priority pollutants include regulated and nonregulated pollutants that are toxic to humans and can potentially occur at harmful levels in the water environment. Priority pollutants can exist in both dissolved and adsorbed-particulate phases, with only the dissolved phase being bioavailable and toxic. For nonpolar organic chemicals, the proportion between these two phases is given by the octanol–water partition coefficient, K_{ow}, and similar partition coefficients are applied to metals and some polar organic compounds, for which the partitioning coefficient would be pH dependent. In a study of the relationship between adsorbed polyaromatic hydrocarbons (PAHs) and particle size in urban runoff, it was found that most of these contaminants can be found on silt-sized particles (2 to 63 μm), with clay-sized particles (<2 μm) less contaminated (Herrmann and Kari, 1990). It is suspected that the organic particulates to which these contaminants are most attracted are in these size fractions. It was also found that priority pollutants with high partition coefficients ($K_{ow} > 10^5$ cm^3/g) exhibited a more pronounced "first-flush" effect than did those that are more dissolved. In similar studies, it has been demonstrated that in urban runoff, zinc and cadmium exhibited a preference for the dissolved phase, whereas lead is dominantly in the suspended phase (Morrison et al., 1984). Most organics and metals are associated with the particulate fractions (Pitt and Baron, 1989). Sand particles do not adsorb most priority pollutants, and BMPs that effectively remove only coarser particles might be ineffective for removal of priority pollutants.

Particulate organic carbon is the best surrogate parameter for estimating removal efficiencies of nonionic toxic chemicals (DDT, PCBs, PAHs, and a large number of organic pesticides) with higher octanol–water partition coefficients ($K_{ow} > 10^5$ cm^3/g). In the absence of particulate organic carbon measurements, volatile suspended solids may be substituted as a measure. The efficiency of various practices for removal of nonionic priority organic pollutants may be correlated to the removal or fate of particulate organic matter, although the fate models may not always be simple. Compounds with low octanol–water partition coefficients cannot be removed with particulates; their removal efficiency is related to their biodegradability and volatization.

When toxic metals are added to water from both natural and human-made sources, they undergo complexation with ligands. *Ligands* are organic and inorganic chemical constituents that combine with metals in a chemical complex. Iron and manganese oxyhydrates provide the strongest adsorption sites, followed by particulate organics and clays. The adsorption reactions are strongly pH dependent, ranging from zero adsorption at low pH to 100%

adsorption at high pH. The change from no adsorption to full adsorption is fairly sharp, usually over 2 pH units. At pH > 7, most metals are complexed, while at pH < 5, the concentration of free metal ions increases dramatically.

The best surrogate parameter for estimating the efficiency of stormwater management systems for removal of metals would be removal of suspended sediment without its sand fractions and removal of particulate organic carbon. The efficiency of removal of metals could be correlated to the removal efficiency of clay and silt fractions of the sediment and the removal efficiency of particulate organic carbon (or volatile suspended solids) parameters. During the winter in snowbelt regions, the high salinity of urban and highway runoff dramatically reduces the partition coefficient for metals (Novotny et al., 1999). This reduces the efficiency of best management practices that accumulate the particulate metals fraction in the sediment, in some cases to the point that metals are leached from the sediments and the BMPs become a source rather than a sink of metals.

Based on the above-mentioned and other considerations, the following conclusions can be drawn regarding the efficiencies of various BMPs for the control of priority pollutants:

- Not all BMPs designed for the removal of traditional pollutants may be effective in the removal of priority pollutants.
- Removal efficiency should be evaluated by considering the octanol–water partition coefficient for polar organics, biodegradability, and volatility.
- BMPs designed to remove solids (such as settling ponds) may remove the mass of priority pollutants associated with particulates but might not be effective for removing toxicity that is attributed to dissolved components.
- The most effective BMPs are ponds enhanced with wetlands, wetlands only, water hyacinth ponds, overland flow systems, and sand–peat filters. Usually, these BMPs provide 80 to 99% removal of priority pollutants. A typical constructed wetland used to treat stormwater runoff is shown in Figure 9.24.

FIGURE 9.24 Constructed wetland for stormwater runoff. (From Bow River Basin Council, 2005.)

- Some priority pollutants, such as toxic metals and several organics, are best immobilized in an anaerobic environment by binding to sulfides and other complexing ligands, or are best biodegraded in an anoxic or anaerobic environment. Wetlands, sand–peat filters, and facultative lagoons may provide such a reducing environment.

- Priority pollutants with higher volatility can be removed in systems that provide a high degree of aeration and exposure to the atmosphere. Such systems include aerated ponds and overland-flow systems.

- BMPs with moderate removal efficiencies include most ponds, grassed swales, sand filters, and porous pavements. Since such systems remove primarily solids, their efficiency for removal of pollutants with low octanol–water partitioning and toxicity reduction may be diminished.

- BMPs using infiltration may be moderately effective for removal of pollutants with higher octanol–water partitioning coefficients and for some metals, but these methods are unsuitable for control of pollutants with low octanol–water partition coefficients and for some metals (e.g., zinc) that have a high dissolved fraction in the runoff.

- Street sweeping and similar surface control of pollutants may be ineffective but may be required for aesthetic enhancement and maintaining cleanliness of the area.

- There is no single BMP that would be effective for all priority pollutants. Typically, a combination of units are required.

9.5   AGRICULTURAL WATERSHEDS

Agricultural operations have a great potential to affect the water environment adversely, particularly from nonpoint-source runoff, hazardous-waste disposal, and habitat destruction. Agriculture is currently the greatest nonpoint-source threat to drinking-water quality in the United States (Trust for Public Land, 2005). In addition, conversion of land to agriculture has involved dramatic changes in the landscape, such as deforestation, wetland drainage, and irrigation of arid lands. Agriculture uses about 70% of all freshwater supplies in the United States, which makes it the largest user of freshwater resources (Ongley, 1996). The major agricultural pollutants responsible for water-quality impairment are salts, nutrients (nitrogen and phosphorus), and pesticides. Inadequate drainage has resulted in salinization of soil and irrigation return flows, and subsequent use of chemicals has made irrigation return flow a pollution hazard. Streams and lakes in agricultural areas typically exhibit excessive algal growths and eutrophication caused by discharges of nutrients from fields and animal operations. In some locations, the ground water is unsuitable for human consumption, due to high nitrate content and contamination by organic chemicals, many of them carcinogenic. The release of nutrients from agricultural operations into receiving waters can occur due to nutrients in runoff or seepage from croplands, rupture of manure-storage lagoons, or accidental spills of fertilizer.

Compounds such as ammonia and nitrate that contain a single nitrogen atom are commonly referred to as *fixed-nitrogen species*, and certain groups of bacteria are capable of converting gaseous nitrogen to fixed nitrogen in the form of ammonium ion (NH_4^+). Energy from the oxidation of biomass to CO_2 reduces the nitrogen in N_2 to NH_4^+ according to the reaction

$$3CH_2O + 2N_2 + 3H_2O + 4H^+ \rightarrow 3CO_2 + 4NH_4^+ \tag{9.26}$$

This process is of particular interest in agriculture because fixed nitrogen is a necessary nutrient for plant growth. Bacteria that fix nitrogen live in a symbiotic relationship in the root system of legumes (e.g., beans, peas). When these plants are grown, soils become enriched in fixed nitrogen rather than depleted, and utilizing legumes in a crop rotation system can reduce the need for chemical fertilizers.

Intensive crop production requires the addition of fertilizers to sustain crop yields. The most widely used fertilizers are lime (to maintain a proper soil pH), nitrogen (N), phosphorus (P), and potassium (K). The primary water-quality concerns related to the application of fertilizers are the eutrophication of surface waters and the contamination of ground waters with nitrates. To ensure optimal crop growth, nitrogen is usually added to the soil in the form of commercial (chemical) fertilizer or manure. The most common forms of commercial fertilizer are ammonium nitrate and anhydrous ammonia (NH_3). Table 9.15 lists typical yields and nutrient requirements for a variety of crops. Continuous cropping of land with a single crop can result in depletion of the nutrients in the soil, and crop rotation is often used as an alternative to fertilizers to improve soil fertility. Losses of nitrogen from the soil may occur as the result of surface runoff, leaching of nitrate, or the release of nitrogen gas (N_2), nitrous oxide (N_2O), and ammonia (NH_3) to the atmosphere. Although the mobility of ammonia-nitrogen and organic-nitrogen in the soil is limited, nitrate is very mobile. Consequently, leaching losses occur when nitrate is present in the soil water and there is downward movement of water in the soil. If oxygen is not present in the soil, denitrifying bacteria can convert nitrate to nitrogen and nitrous oxide, which can be lost to the atmosphere. This process, called *denitrification*, typically occurs in the anaerobic environment of waterlogged soils.

The pollution of ground waters with nitrate is widespread in agricultural areas, and it has been reported that about 12% of domestic water-supply wells in agricultural areas in the United States have nitrate concentrations in excess of the 10-mg/L drinking water standard (USGS, 1999).

Agricultural land is generally divided into the following categories for inventory purposes:

- Dryland cropland
- Irrigated cropland
- Partureland
- Rangeland
- Forestland
- Confined animal feeding operations
- Specialty areas (e.g., aquaculture, orchard crops, and wildlife land)

These divisions are useful in water-quality planning efforts and nonpoint-source pollution control since each type of land area has a distinct set of pollutants of concern associated with that land use, and because most current best management practice reference guides are divided by these categories rather than by pollutant. The branch of forestry dealing with the development and care of forests is called *silviculture*, and this area is frequently lumped with agriculture, which is the practice of cultivating the land or raising stock.

9.5.1 Sources of Pollution

The major pollutants associated with agriculture include sediment, nutrients (especially N and P), pesticides and other toxins, pathogens, and salinity. A common symptom of the high

TABLE 9.15 Approximate Yields and Nutrient Content of Selected Crops

Crop	Yield/ha	kg N/ha	kg P/ha
Alfalfa	9 tonnes	225	30
Barley, grains	100 bushels	39	7
Barley, straw	2.2 tonnes	17	2
Beans,[a] dry	75 bushels	84	11
Cabbage	45 tonnes	165	18
Clover,[a] red	4.5 tonnes	89	11
Clover, white	4.5 tonnes	145	11
Corn, grain	270 bushels	151	27
Corn, stover	10 tonnes	111	18
Corn, silage	56 tonnes	225	34
Cotton, lint and seed	2.2 tonnes	67	13
Cotton, stalks	2.2 tonnes	50	7
Lettuce	45 tonnes	100	13
Oats, grain	22 bushels	62	11
Oats, straw	4.5 tonnes	28	9
Onions	17 tonnes	50	9
Peanuts,[a] nuts	3.4 tonnes	123	7
Potatoes, tubers	990 hundredweight	106	14
Potatoes, vines	2.2 tonnes	100	9
Rice, grains	225 bushels	62	13
Rice, straw	5.6 tonnes	34	4
Rye, grains	75 bushels	39	4
Rye, straw	3.4 tonnes	17	4
Sorghum, grain	150 bushels	56	11
Sorghum, stubble	6.7 tonnes	73	9
Soybean,[a] grain	111 bushels	179	18
Soybean,[a] straw	2.2 tonnes	28	4
Sugar beets, roots	45 tonnes	95	16
Sugar beets, tops	27 tonnes	123	11
Sugarcane, stalks	67 tonnes	112	22
Sugarcane, tops	29 tonnes	56	11
Tobacco	3.4 tonnes	129	11
Tomatoes, fruits	56 tonnes	162	22
Tomatoes, vines	3.4 tonnes	78	11
Wheat, grain	123 bushels	73	16
Wheat, straw	3.4 tonnes	22	2
Bermuda grass	—	540–670	—
Fescue	—	300	—
Medium mature forest, deciduous	—	30–60	—
Medium mature forest, evergreen	—	20–30	—

Sources: Stewart et al. (1975) and Powell (1976).

[a]Legumes do not require fertilizer nitrogen and can fix atmospheric nitrogen.

FIGURE 9.25   Excessive algae growth in agricultural drainage canal. (From USGS, 2005j.)

nutrient loads in agricultural runoff is the excessive amounts of algae that are commonly found in drainage canals in and around agricultural areas, as illustrated in Figure 9.25. Different types of agricultural land uses are more likely than others to contribute certain pollutants. Table 9.16 lists typical sources and types of pollution in runoff and subsurface percolation from agricultural and silvicultural operations.

There are several poor agricultural practices that adversely affect the diffuse loading of pollutants on water bodies from agricultural and silvicultural areas, such as:

- Livestock permitted uncontrolled access to riparian areas. This causes sloughing of streambank soils and degrades stream bank vegetation.

TABLE 9.16   Agricultural Land Use Versus Type of Pollution

Land Use	Contaminants of Concern
Dryland cropland	Sediment, adsorbed nutrients, and pesticides
Irrigated cropland	Sediment, both absorbed and dissolved nutrients and pesticides, traces of certain metals, salts, and sometimes bacteria, viruses, and other microorganisms
Pastureland	Bacteria, nutrients, sediment, and sometimes pesticides
Rangeland	Sediment, bacteria, nutrients, and occasionally, metals or pesticides
Forestland	Sediment, organic materials, and adsorbed nutrients due to logging operations
Confined animal feeding operations	Bacteria, viruses, and other microbes;both dissolved and adsorbed nutrients, sediment, organic material, salts, and metals
Specialty areas	
Aquaculture	Dissolved nutrients, bacteria, and other pathogens
Orchards and nurseries	Nutrients (generally dissolved), pesticides, salts, bacteria, organic material, and trace metals
Wildlife land	Bacteria and nutrients if wildlife populations become unbalanced

- Crop-production systems that plow fields right up to the edge of the stream bank. This practice destabilizes the banks, causing them to collapse, smother riparian vegetation, and increase the sediment load in the adjacent channel.
- Pollutants contained in eroded topsoil and irrigation return flows, interflow originating from tile drainage and leaching of excess irrigation water, and ground water base flow containing high nitrate content due to overfertilization.
- Draining of low-lying areas, primarily wetlands, for crop production. This practice causes loss of riparian vegetation through clearing, loss of natural high water table from drainage, and loss of flood-carrying capacity.

Erosion and soil loss by surface runoff and nitrate leaching into ground water are the predominant sources of pollution from cropland. Other effects include leaching of agricultural chemicals and pesticides. The disturbing activity associated with tillage increases the erosion potential of croplands. With the exception of arid lands, soil loss by erosion from fields is at least an order of magnitude higher than background loads. Soil erosion is the major cause of diffuse pollution, and sediment is also the most visible pollutant (Novotny, 2003). It has been reported that of nutrient (N and P) losses from cropland, about 90% is associated with soil loss (Alberts et al., 1978; McDowell et al., 1989). Losses of nutrients from cropland represent a relatively small portion of applied fertilizer; however, the concentrations in the runoff almost always exceed the criteria for preventing accelerated eutrophication of receiving water bodies. The contribution of pesticides to diffuse pollution is also of great concern. For example, measurements by Klaine et al. (1988) have shown that just over 1% of total atrazine applied to a small agricultural watershed was lost with runoff; however, concentration in runoff reached as high as $250\,\mu g/L$.

Soils have the capacity to retain many pollutants in their particulate form, which are far less environmentally damaging than in the dissolved form. This is particularly relevant to the soil retention of phosphates, hydrophobic organic chemicals, ammonium, and metals. The capacity of soils to retain and absorb pollutants depends on its composition and redox status. The most important component is the soil organic matter, followed by pH, clay content, soil moisture, and cation-exchange capacity. Typically, at some point the soil becomes saturated by the pollutant and larger quantities are released in dissolved form into ground water and base flow of surface waters. The first indication that the soil retention capacity has been exhausted for some pollutants is the dramatic increase in nitrate pollution in ground and surface waters. Well-aerated agricultural soils have a lower retention capacity for nitrogen that is readily nitrified into mobile nitrate forms. For other pollutants, as long as the soil retention capacity is not exhausted, the result is a net accumulation of pollutants in the soil. Typically, most of the phosphorus applied is retained in the soil and saturation is typically reached within a few years of excess phosphorus application.

Pollution from animal operations can be divided into that from pastures and that from concentrated animal feeding operations (feedlots). Concentrated feedlots that can manage 1000 or more animals are statutory point sources, and a permit is required for their wastewater disposal. Livestock wastes account on average for 30% of the total phosphorus load in European inland waters, and all other sectors of agriculture account for an additional 17% (Economic Commission for Europe, 1992). In is interesting to note that phosphorus production by one dairy cow or heifer is on the order of 18 kg/yr, and of that amount, a significant portion may reach the receiving water body, depending on the proximity of the farm to the watercourse and on the degree of pollutant attenuation during overland flow. The phosphorus load by one cow is equivalent to 18 to 20 humans. The BOD concentration of

TABLE 9.17   Pollutant Concentration from Animal Operations

	BOD$_5$ (mg/L)	COD (mg/L)	Total N (mg/L)	Total P (mg/L)	Reference
Runoff from feedlot	800–11,000	3000–30,000	100–2100	10–500	Miner et al. (2000)
Runoff from grazed pasture	—	—	4.5	7	Robins (1985)
Untreated sewage	160	235	30	10	Novotny et al. (1989)

barnyard runoff exceeds that of sewage by two orders of magnitude, and therefore barnyard runoff can cause significant oxygen depletion in receiving waters. Typical concentrations of BOD$_5$, COD, total nitrogen, and total phosphorus in feedlot runoff, runoff from grazed pasture, and untreated sewage are shown in Table 9.17. Barnyard runoff also carries pathogenic microbes such as the protozoan *Cryptosporidium*, with other well-known diseases transmitted from livestock to humans, including salmonellosis, staphylococcus, tetanus, foot-and-mouth disease, mad cow disease, and tuberculosis. The majority of feedlot wastes reaching surface waters are transported by surface runoff. Biosolids disposal from domestic (urban) wastewater treatment plants on agricultural land can also be a significant source of pollution if these biosolids are not treated to reduce the levels of pathogenic microorganisms (Brewster et al., 2005). Enteric viruses are often concentrated in sewage sludge because they have a tendency to combine with solids (Sano et al., 2001). Alkaline conditions resulting from the addition of lime (CaO) to biosolids is effective in inactivating viruses.

Pastureland and rangeland account for the largest proportion of total land use in the United States and include about 40% of all nonfederal land. In *pastures*, animals roam and feed on natural vegetation; *rangeland* typically refers to large-scale but low-density animal operations in the arid and semiarid western United States. Pastureland is a significant source of diffuse pollution when proper erosion-control practices are not in place or when grazing livestock are allowed to approach or enter surface waters. It has been reported that sediment yield from pastures in Washington State was minimized when vegetative cover remained greater than 50%, regardless of animal trampling disturbances (Smolen et al., 1990). Overgrazing and allowing livestock to approach and enter watercourses are the major pollution activities on pastureland and rangeland. If these activities are controlled, pollution from such land can be minimal.

Undisturbed forests or woodland represent the best protection of lands from sediment and pollution losses. Woodlands and forests have high resistance to surface runoff due to ground mulch and terrain roughness. Even lowland forests with high water tables absorb large amounts of precipitation and actively retain water and contaminants. Uncontrolled logging operations (clearcutting) disturb the forest resistance to erosion, and observations indicate that almost all sediment reaching waterways from forestlands originate from the construction of logging roads and from clearcuts.

9.5.2   Fate and Transport Processes

The two main pathways by which agricultural pollutants travel from their sources to receiving water bodies are via overland flow into a surface water or via infiltration and percolation through the soil into the ground water. In the case of overland flow, contaminants tend to be adsorbed onto eroded sediment particles and transported with these particles. In the case of percolation through the soil into the ground water, contaminants exist in both the aqueous (dissolved) phase and the sorbed (solid) phase attached to soil particles; and only the dissolved phase of the pollutant affects the saturated zone. Fate and transport processes for both overland flow and subsurface flow are described in the following sections.

Erosion   Each year, millions of tons of soil and weathered geological material are washed from the land surface into receiving water bodies, and human activities and land use can increase the rate of erosion dramatically. For example, deforestation in areas of high precipitation and high slope can be devastating to both water quality and flooding.

Land uses or watershed modifications that produce elevated sediment yields are considered polluting activities. Eroded soil particles carry pollutants that can be harmful to the ecology of receiving water bodies and to humans. However, the strong affinity of fine sediments—primarily, clay and organic particulates—to adsorb and make the pollutants biologically unavailable is considered by some as a partial water-quality benefit of sediment discharges.

The term *denudation* refers to the geomorphological process of weathering or breakdown of parent rock materials, entrainment of the weathered debris, and transportation and deposition of the debris. The term *erosion* is often used synonymously with denudation, although erosion applies to entrainment and transportation of debris by water and wind but not to weathering. *Geomorphology* is the science dealing with the shaping of Earth surfaces, including erosion, tectonic processes, weathering, and other stresses, including those caused by humans. Erosion processes can be divided into sheet erosion, rill erosion, gully erosion, stream bank erosion, floodplain scour, and shoreline or bluff erosion. *Sheet erosion* occurs when sediment is entrained and transported by sheet flow, typically over very flat areas such as roadways, *rill erosion* occurs when sediment is entrained and transported in small eroded channels called *rills*, and *gully erosion* occurs when sediment is entrained and transported by runoff in large eroded channels called *gullies*. Sheet erosion, rill erosion, and gully erosion are sequential, as illustrated in Figure 9.26. Typically, rills are only several centimeters in depth and occur mainly in cultivated soils, while gullies are formed when rills grow to 0.5 to 30 m in depth. Gully erosion is often defined for agricultural land in terms of channels too deep to be easily remedied with ordinary farm tillage equipment. Examples of a rill and gully are shown in Figure 9.27, where the size of the rill is scaled by a person's hand and the size of the gully is scaled by a person's height.

Erosion is usually measured in tons/ha per unit time (year or season) or per storm. The ease with which surface soils give way to erosion is called *erodibility*, and watersheds can typically be divided into areas where sediment is eroded and areas where sediment is deposited. The eroding process is called *degradation*, and sediment deposition is called *aggradation*. Erosion and soil loss are not a major problem in flat watersheds with slopes in the range 0 to 2%. In such watersheds, drainage tile flow and/or irrigation return flow are the primary causes of pollution. *Erosion control* implies an action to reduce soil loss and subsequent delivery of sediment from the source area to the receiving water body. Erosion control is typically accomplished by land management, buffer strips, channel modification, sediment traps, and other structural and nonstructural practices.

Several factors affect sediment transport in overland flow, the most important of which are as follows:

1. *Rainfall impact* detaches soil particles and keeps them in suspension for the duration of a rainfall. The intensity of rainfall impact at the scale of individual soil particles is illustrated in Figure 9.28. Erosion caused by rainfall impact is called *splash erosion*.

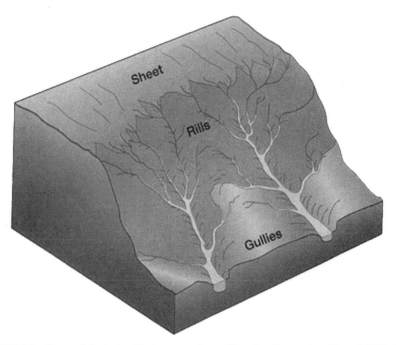

FIGURE 9.26   Sequential relationship between sheet, rill, and gully erosion. (From NRCS, 2005e.)

2. *Overland-flow energy* detaches soil particles from small rills, and together with some interrill contribution, the particles remain is suspension as long as overland flow persists.

3. *Vegetation* slows flow and filters out particles during shallow-flow conditions.

4. *Infiltration* filters out the particles from the overland flow.

5. *Small depressions and ponding* allow particles to be deposited because of reduced flow velocity.

6. *Change of slope of overland flow* associated with drainage-area concavity oftens flattens the slope near the drainage channel and steepens the slope uphill.

(a) (b)

FIGURE 9.27   (*a*) Rill; (*b*) gully. [(*a*) From NRCS, 2005e; (*b*) from USDA, 2005.]

FIGURE 9.28   Rainfall impact and splash erosion. (From NRCS, 2005e.)

The major soil properties related to erosion are soil texture and composition. Soil texture determines the permeability and erodibility of soils; and higher permeability soils are less hydrologically active. Vegetation influences sediment yields by dissipating rainfall energy, binding the soil and increasing porosity by its root system, and reducing soil moisture by evapotranspiration, thereby increasing infiltration.

The *universal soil loss equation* (USLE) is the most widely used estimator of soil loss caused by upland erosion. The USLE was originally formulated by Wischmeier and Smith (1965) and predicts primarily sheet and rill erosion. The current version of the universal soil loss equation is called the *revised universal soil loss equation* (RUSLE); however, the acronym USLE is commonly interpreted to be the latest version of the universal soil loss equation, which is currently the RUSLE. With this understanding, the USLE is given by

$$A = RK(LS)CP \qquad (9.27)$$

where A is the soil loss [metric tons (tonnes)/ha] for a given storm or period, R is the rainfall energy factor or rainfall erosivity factor, K is the soil erodibility factor, LS is the slope-length factor, C is the cropping management (vegetative cover) factor, and P is the erosion control practice factor. The USLE (Equation 9.27) expresses the rate of soil loss per unit area due to erosion by rain. The *rainfall energy factor*, R, is equal to the sum of the rainfall erosion indices for all storms during the period of prediction, usually given on a per-year basis. For any prediction period, the rainfall energy factor is given by

$$R_r = I \sum_{i=1}^{N} (2.29 + 1.15 \ln X_i) \qquad (9.28)$$

FIGURE 9.29   Annual rainfall erosivity factor, R_r (tons/acre). (From Stewart et al., 1975.)

where i is the storm index, N is the number of storms in the prediction period, I is the maximum 30-minute rainfall intensity of all storms in the prediction period (cm/h), and X_i is the average rainfall intensity during the ith storm (cm/h). Average annual rainfall energy factors, R_r, have been determined for the United States and are shown in Figure 9.29. More detailed rainfall energy factors are also available for several individual states. Both erosion by rainfall energy (interrill erosion) and detachment of soil particles by overland flow (rill erosion) contribute to soil loss. Thus, the rainfall factor, R, should also include the effect of runoff. A modification of Equation 9.28 was proposed by Foster et al. (1977) as

$$R = aR_r + bcQq^{1/3} \qquad (9.29)$$

where a and b are weighting parameters ($a + b = 1$), c is an equality coefficient, Q is the runoff volume (cm), and q is the maximum runoff rate (cm/h). It was suggested that the detachment of particles by runoff and rain energy is about evenly divided ($a = b = 0.5$), the equality coefficient in SI units is 15, and substituting the values of a, b, and c into Equation 9.29 gives the overall rainfall factor, R, in the USLE (Equation 9.27) as

$$\boxed{R = 0.5R_r + 7.5Qq^{1/3}} \qquad (9.30)$$

In applying Equation 9.30, it should be noted that the proportion between rainfall and runoff erosivity may vary greatly between regions. The *soil erodibility factor*, K, in the USLE is a measure of potential erodibility of soil relative to erosion over a 22-m-long overland flow length on a 9% slope in clean-tilled continuous fallow. This factor is a function of soil texture

TABLE 9.18 **Magnitude of Soil Erodibility Factor, K**

Soil Texture	Organic Matter Content (%)		
	0.5	2	4
Sand	0.05	0.03	0.02
Fine sand	0.16	0.14	0.10
Very fine sand	0.42	0.36	0.28
Loamy sand	0.12	0.10	0.16
Loamy fine sand	0.24	0.20	0.16
Loamy very fine sand	0.44	0.38	0.30
Sandy loam	0.27	0.24	0.19
Fine sandy loam	0.35	0.30	0.24
Very fine sandy loam	0.47	0.41	0.35
Loam	0.38	0.34	0.29
Silt loam	0.48	0.42	0.33
Silt	0.60	0.52	0.42
Sandy clay loam	0.27	0.25	0.21
Clay loam	0.28	0.25	0.21
Silty clay loam	0.37	0.32	0.26
Sandy clay	0.14	0.13	0.12
Silty clay	0.25	0.23	0.19
Clay	—	0.13–0.2	—

Source: Stewart et al. (1975).

and composition, and values of K can be estimated using Table 9.18. The slope-length factor, LS, in Equation 9.27 is a function of the overland runoff length and slope, and it is a dimensionless parameter that adjusts the soil-loss estimates for effects of length and steepness of the field slope. The LS factor can be estimated using the following relation (Stewart et al., 1975):

$$LS = \left(\frac{L}{22.1}\right)^m (0.065 + 0.04579S + 0.0065S^2) \qquad (9.31)$$

where L is the length (m) from the point of origin of the overland flow to the point where the slope decreases to the extent that deposition begins or to the point at which runoff enters a defined channel, S is the average slope (%) over the runoff length, and m is an exponent dependent on the slope steepness, as given in Table 9.19. If the average slope is used in calculating the LS factor, the average slope will underestimate the LS factor when the actual slope is convex and overestimate the erosion when the actual slope is concave.

TABLE 9.19 **Exponent Parameter in Estimating Slope-Length Factor**

Slope (%)	m
<1	0.2
1–3.5	0.3
3.5–4.5	0.4
≥4.5	0.5

TABLE 9.20   Values of *C* for Cropland, Pasture, and Woodland

Land Cover or Land Use	C
Continuous fallow tilled up and down slope	1.0
Shortly after seeding and harvesting	0.3–0.8
For crops during main part of growing season	
Corn	0.1–0.3
Wheat	0.05–0.15
Cotton	0.4
Soybean	0.2–0.3
Meadow	0.01–0.02
For permanent pasture, idle land, unmanaged woodland	
Ground cover 85–100%	
As grass	0.003
As weeds	0.01
Ground cover 80%	
As grass	0.01
As weeds	0.04
Ground cover 60%	
As grass	0.04
As weeds	0.09
For managed woodland	
Tree canopy	
75–100%	0.001
75–100% to 40–75%	0.002–0.004
75–100% to 20–40%	0.003–0.01

Source: Wischmeier and Smith (1965), Wischmeier (1972), and Stewart et al. (1975).

To minimize these errors, large areas should be broken up into areas of fairly uniform slope, and the smaller *LS* factor will control the amount eroded. The *cropping management factor*, *C*, also called the *vegetation cover factor*, in Equation 9.27 estimates the effect of ground-cover conditions, soil conditions, and general management practices on erosion rates. It is a dimensionless quantity with a value of 1 for *continuous fallow ground*, which is defined as land that has been tilled up and down the slope and maintained free of vegetation and surface crusting. The effect of vegetation on erosion rates results from canopy protection, reduction of rainfall energy, and protection of soil by plant residues, roots, and mulches. Typical magnitudes of *C* for agricultural land, permanent pasture, and idle rural land are shown in Table 9.20. Generally, *C* reflects the protection of the soil surface against the impact of rain droplets and subsequent loss of soil particles. Grassed and urban areas have *C* factors similar to those for permanent pasture. The *erosion control practice factor*, *P*, in the USLE (Equation 9.27) accounts for the erosion-control effectiveness of such land treatment as contouring, compacting, established sedimentation basins, and other control practices. Values of *P* for various agricultural practices are shown in Table 9.21. These values of *P* are highly empirical and should be used only as a first approximation. On land slopes with more than a 15% gradient, it is commonly recommended that an agricultural field operation be shifted to contour strip cropping, with proper selection of crops (Jørgensen et al., 2005).

The universal soil loss equation (USLE) should be applied with due caution since it was specifically designed for the following applications (Wischmeier, 1976):

TABLE 9.21   **Values of *P* for Agricultural Land Uses**

| Slope (%) | Contouring | Strip Cropping and Terracing | |
		Alternate Meadows	Close-Grown Crops
0–2.0	0.6	0.3	0.45
2.1–7.0	0.5	0.25	0.40
7.1–12.0	0.6	0.30	0.45
12.1–18.0	0.8	0.40	0.60
18.1–24.0	0.9	0.45	0.70
>24.0	1.0	1.0	1.0

Source: Wischmeier and Smith (1965).

- Predicting average annual soil movement from a given field slope under specified land use and management conditions
- Guiding the selection of conservation practices for specific sites
- Estimating the reduction of soil loss attainable from various changes that farmers might make in their cropping system or cultural practices
- Determining how much more intensively a given field could be cropped safely if contoured, terraced, or strip-cropped
- Determining the maximum slope length on which given cropping and management can be tolerated in the field
- Providing local soil loss data to agricultural technicians, conservation agencies, and others to use when discussing erosion plans with farmers and contractors.

The accuracy of the USLE is increased if it is combined with a hydrological rainfall-excess model. Note that the rainfall-erosivity factor, R, has a value greater than zero for every rainfall; hence, erosion and soil loss is anticipated by the soil loss equation for any precipitation. A hydrological rainfall-excess model in combination with the USLE would eliminate erosion by rainfall with no excess rain.

All of the soil that is eroded from the land surface in a watershed does not end up in the receiving water body. The amount of eroded soil that ends up in the water body is called the *sediment yield*, and the fraction of the gross erosion that ends up as sediment yield is called the *delivery ratio*; hence,

$$\text{delivery ratio} = \frac{\text{sediment yield}}{\text{gross erosion}} \tag{9.32}$$

The gross erosion in the denominator of Equation 9.32 is given by the USLE (A in Equation 9.27), and delivery ratios for agricultural lands are typically in the range 1 to 30% (Novotny et al., 1986).

Example 9.5   A 200-ha cotton farm in central Georgia consists of predominantly clay loam soil with approximately on 2% organic matter, and the ground surface has approximately a 3% slope. During the off season, the cotton fields are plowed up and down the land slope. The annual runoff volume is 30 cm and the maximum runoff rate is 5 cm/h. (a) If the typical overland-flow distance from the beginning of overland flow to a drainage stream is 1.4 km, estimate the average annual soil loss. (b) If the delivery ratio is estimated as 15%, what is the annual sediment loading on the receiving water body?

SOLUTION (a) The annual soil loss can be estimated using the universal soil loss equation (USLE), Equation 9.27, given by

$$A = RK(LS)CP$$

According to Figure 9.29, the average annual rainfall-erosivity factor, R_r, for central Georgia is

$$R_r = 300 \text{ tonnes/acre} = 2.24 \times 300 \text{ tonnes/ha} = 672 \text{ tonnes/ha}$$

From the data given, $Q = 30$ cm and $q = 5$ cm/h; hence, the rainfall factor, R, can be estimated by Equation 9.30 as

$$R = 0.5R_r + 7.5Qq^{1/3} = 0.5(672) + 7.5(30)(5)^{1/3} = 721 \text{ tonnes/ha}$$

For clay loam soil with 2% organic matter, Table 9.18 gives $K = 0.25$. From the data given, the distance from the origin of overland flow to the drainage channel, L, is 1.4 km = 1400 m, the average slope, S, is 3%; Table 9.17 gives $m = 0.3$, and Equation 9.31 gives the slope-length factor, LS, as

$$LS = \left(\frac{L}{22.1}\right)^m (0.065 + 0.04579S + 0.0065S^2)$$

$$= \left(\frac{1400}{22.1}\right)^{0.3} [0.065 + 0.04579(3) + 0.0065(3)^2]$$

$$= 1.37$$

For cotton, Table 9.20 gives the crop-management factor, C, as 0.4. During the off-season, when the land is plowed up and down the land slope, Table 9.20 gives $C = 1$; therefore, the average annual value of C can be taken as

$$C = \frac{1.0 + 0.4}{2} = 0.7$$

Since no special erosion-control measures are implemented, $P = 1$, and the USLE, Equation 9.27, gives

$$A = RK(LS)CP = 721(0.25)(1.37)(0.7)(1) = 173 \text{ tonnes/ha}$$

The annual soil erosion from the 200-ha farm is predicted to be on the order of 173 tonnes/ha $\times$ 200 ha = 34,600 tonnes.

(b) If the delivery ratio is 15%, the sediment load on the receiving water is estimated as 0.15 $\times$ 34,600 tonnes = 5190 tonnes.

Soil Pollution Chemicals used in agriculture are a major source of soil pollution, and when this pollution becomes excessive, these pollutants frequently end up affecting water

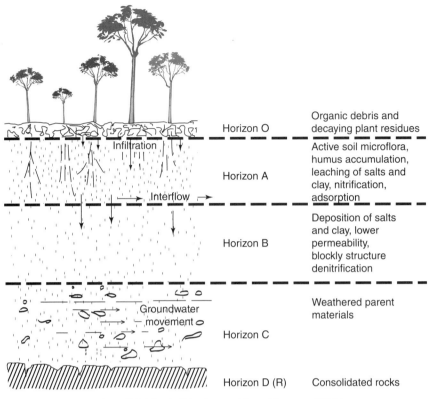

Horizon O — Organic debris and decaying plant residues

Infiltration

Horizon A — Active soil microflora, humus accumulation, leaching of salts and clay, nitrification, adsorption

Interflow →

Horizon B — Deposition of salts and clay, lower permeability, blockly structure denitrification

Groundwater movement

Horizon C — Weathered parent materials

Horizon D (R)   Consolidated rocks

FIGURE 9.30   Soil horizons. (From Novotny, 2003.)

quality. For example, the widespread occurrence of the pesticide atrazine in ground water in the United States cornbelt and in Europe (e.g., Po River in Italy) is a serious environmental threat. The buffering capacity of soil is the only barrier between surface topsoil contamination and the pollution of ground and surface waters, where the buffering capacity is either zero or very small.

The soil profile is divided into three layers or horizons, as illustrated in Figure 9.30. A *soil horizon* is defined as a layer of similar soil color, texture, structure, or porosity oriented approximately horizontally to the Earth's surface. A typical soil profile may also contain a surface layer of decaying organic debris (detritus) with little soil called the *O horizon* or *organic horizon*. The *A horizon*, which is usually several centimeters to a fraction of a meter thick, is the soil of greatest concern, since roots, soil microorganisms, and organics are found there in their greatest densities. It is also a layer of considerable leaching and is often referred to as *topsoil*. The A horizon is typically scraped and sold for lawn topsoil or stockpiled for reapplication on construction sites. The *B horizon*, underlying the A horizon, is a subsoil where most leached salts, chemicals, and clay from the A horizon are deposited. It usually has less organic matter and few plant roots, since only large plants and trees have root systems penetrating subsoils. The B horizon is commonly referred to as the *subsoil*, and the accumulation of material leached from the A horizon often results in a higher-density soil with clay accumulation and pronounced soil structure. The *C horizon* extends from the bottom of the B horizon to the top of the parent bedrock from which the

FIGURE 9.31   Soil profile. (From NRCS, 2005e.)

soil evolved by weathering. The C horizon has characteristics of the unconsolidated, weathered parent material.

The A horizon is of considerable importance to diffuse pollution studies since it is the layer where most adsorption and biochemical degradation of pollutants takes place. The microbial processes by which pollutants and nutrients (nitrogen and phosphorus) are decomposed or transformed are confined primarily to the A horizon. A typical soil profile observed in the field is illustrated in Figure 9.31, where the darker portion of the soil profile corresponds to the A horizon. The soil profile shown in Figure 9.31 is scaled by the leaf and stem of a corn plant. The soil profile shown in Figure 9.31 is from central Iowa, where the average topsoil depth has decreased from around 40 cm to around 20 cm during the 100-year period from 1900 to 1999 (NRCS, 2005e). Only soluble (mobile) pollutants can penetrate deeper soil zones and eventually pollute ground water. In most cases, soil horizon B is less permeable than the topsoil, resulting in occasional *interflow*, which is lateral flow between the A and B horizons. Saturated zones are generally located in the C horizon.

The infiltration capacity at the ground surface is related to the texture of the surface soil horizon, which is determined by the sand, silt, and clay fractions. Soils can be divided into *soil separates* according to particle-size ranges shown in Table 9.22, and the soil texture is derived from the relative proportions of sand, silt, and clay using the USDA soil-texture triangle, shown in Figure 9.32. NRCS soil maps and GIS databases are commonly used to delineate the geographical distribution of various types of soils. Slope classifications are frequently added as subscripts to a soil-series name, and the USDA slope classifications

TABLE 9.22   USDA Soil Separates

Separate	Diameter (mm)
Very coarse sand	2.00–1.00
Coarse sand	1.00–0.50
Medium sand	0.50–0.25
Fine sand	0.25–0.10
Very fine sand	0.10–0.05
Silt	0.05–0.002
Clay	<0.002

Source: Foth (1990).

used for this purpose are given in Table 9.23. As can be seen from the table, slope classes overlap due to localized soil and topographic conditions. Topographic slopes are important for assessing runoff and erosion potential in a watershed. In classifying soils, the name of the soil reported in the soil map has two parts: a local name and a slope. For example, in the soil map of southeastern Wisconsin, a code RoB signifies Rosseau loamy fine sand in the slope category B (2 to 6% slope).

Many regions have contaminated soils caused by intensive agricultural practices. Examples include phosphate-saturated soils, nitrate leaching, and leaching of certain organic pesticides such as atrazine and other chemicals. Certain agricultural practices may

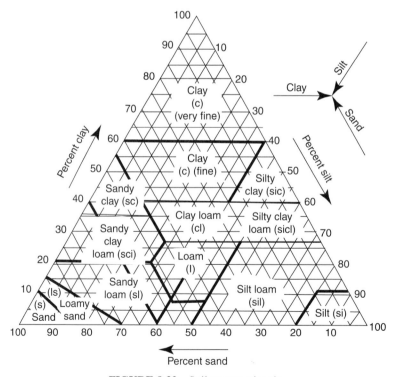

FIGURE 9.32   Soil texture triangle.

TABLE 9.23   USDA Slope Classifications

Class	Slope Groupings
A or no slope class	0–3%, 0–5%
B	0–8%, 3–8%
C	8–15%, 8–20%, 8–25%
D	15–25%, 15–30%
E	25–70%
F	45–65%

increase or decrease the mobility of some pollutants. For example, farming practices that increase the pH levels in soils decrease the mobility of metals. Also, the higher organic content of farmed soils increases the retention of some organic chemicals, keeps them in an oxidized state, and maintains an aerobic environment, which may be more favorable for biological degradation.

Organic matter is an integral part of soils and sediments. The organic content of soils varies from less than 1% to more than 40% in some organic soils, feedlot soils, wetlands, and aquatic sediments. The organic content of soils is usually reported as percent organic matter or percent organic carbon. Soils that are aerated usually have smaller organic content than soils that are saturated with water most of the time. The organic content of soils and sediments, often called *humus*, is a product of the biodegrading process by microorganisms. It is rich in nutrients and remains for long periods as an important supply for microorganisms. A significant part of the soil organic matter is not biodegradable, and almost all of the organic matter is contained in the A and O horizons of the soil profile. The organic matter in the soil plays an important role in retaining other pollutants and making them immobile. It has a high storage capacity for metals and organic chemicals, and it is concentrated at the interface between the soil and the atmosphere as well as between the soil and plants.

Since particulate pollutant transport is part of the sediment erosion and movement processes, many models use an arbitrary proportionality factor called the *potency factor* or *transmission coefficient* to relate the sediment loading to that of other contaminants. This relationship is given by

$$Y_i = p_i Y_s \tag{9.33}$$

where Y_i is the loading or concentration of contaminant i, p_i is the potency factor for the contaminant, and Y_s is the loading or concentration of sediment from the soil. Potency factors can be related to the concentration of the contaminant in the parent topsoil and the enrichment factor for the contaminant according to the relation

$$p_i = S_{si} \cdot \mathrm{ER}_i \tag{9.34}$$

where S_{si} is the concentration of the contaminant in the topsoil and ER_i is the enrichment ratio of the contaminant between the source and the point of interest or watershed outlet. The enrichment ratio is the concentration of contaminant in the eroded material, on the sediment in runoff or stream flow divided by its concentration in the parent soil material. Enrichment ratios depend on the soil and contaminant of interest and are typically in the range 1.0 to 4.0. Enrichment ratios are generally greater than 1 because it is mostly the finer soil fraction that is eroded, and this finer soil fraction has a greater sorption capacity per unit mass than the bulk soil.

Example 9.6   Hillsdale Lake in eastern Kansas is surrounded by a 369-km^2 watershed in which the topsoil has an average phosphorus concentration of 0.71 g of phosphorus per kilogram of soil. Measurements in 2002 indicated an annual sediment load into the lake of 21.9×10^6 kg and a typical enrichment ratio of 2.4 for phosphorus. Estimate the annual phosphorus loading from sediment-laden runoff into Hillsdale Lake in 2002.

SOLUTION   From the data given, $S_s = 0.71$ g/kg, ER $= 2.4$, and $Y_s = 21.9 \times 10^6$ kg/yr. The potency factor, p, for phosphorus is given by Equation 9.34 as

$$p = S_s \cdot \text{ER} = 0.7(2.4) = 1.7 \text{ g/kg}$$

and the annual phosphorus loading, Y, on Hillsdale Lake is given by Equation 9.33 as

$$Y = pY_s = 1.7(21.9 \times 10^6) = 37.2 \times 10^6 \text{ g P/yr} = 37{,}200 \text{ kg P/yr}$$

This phosphorus loading (37,200 kg/yr in 2002) can be associated with the phosphorus concentration in the lake to assess whether reductions in phosphorus loading are needed.

Pollutants exist in soils, sediments, and sediment-laden water in several phases. They can be precipitated and/or strongly adsorbed to particles (solid phase), be dissolved or dissociated in water or soil moisture (liquid phase), volatilize, or be gasified. The capacity of soils and sediments to adsorb and retain contaminants depends on their composition. For organic micropollutants, the most important component in soils is the organic particulate matter, which has the strongest binding capacity. For inorganic contaminants such as toxic metals, the adsorbing capacity of both organic and inorganic soil-sediment particulates should be considered. The adsorbing capacity is related to the surface area of the particles, and hence small particles such as clay minerals have the highest adsorbing capacity. The important capacity-controlling parameters are the soil organic matter content and the cation-exchange capacity, which is determined by the surface area of the particles. Most clays and organic matter have a net negative charge, making them effective in holding positively charged particles (cations) such as Al, Fe, H^+, and Ca. The cation-exchange capacity (CEC) in milliequivalents per 100 grams of soil can be estimated using the relation

$$\text{CEC} = 2.5(\% \text{ organic matter}) + 0.57(\% \text{ clay}) \tag{9.35}$$

Adsorption or complexation of the contaminant into the particulate form also immobilizes contaminants and makes them biologically unavailable in most cases. Generally, water-soluble (hydrophilic) compounds are weakly adsorbed onto soil particles; hence, they have higher bioavailability and leach more easily into ground water. The mobility of an organic compound in soils and sediments is related to the octanol–water partitioning coefficient, K_{ow}. Sorption equilibria are quantified by the isotherms described in Section 7.6.1. The Langmuir, Freundlich, and linear isotherms are all widely used and accepted, although the linear isotherm is perhaps the most used. In cases where a fraction of the contaminant is adsorbed, the total concentration of the contaminant in the soil, c_T, is equal to the sum of the aqueous (dissolved) concentration, c_{aq}, and adsorbed-particulate concentration, c_p; hence,

$$c_T = \theta c_{aq} + c_p = \theta c_{aq} + F\rho_b \qquad (9.36)$$

where θ is the water content of the soil (dimensionless), F is the adsorbed concentration of the contaminant [M/M], and ρ_b is the bulk density of the soil [M/L^3]. If adsorption is described by a linear isotherm, the adsorbed concentration, F, and aqueous concentration, c_{aq}, are related by

$$F = K_d c_{aq} \qquad (9.37)$$

where K_d is the distribution coefficient [L^3/M]. Combining Equations 9.36 and 9.37 gives the following relationship between the aqueous concentration, c_{aq}, and the total concentration, c_T:

$$\boxed{c_{aq} = \frac{1}{\theta + K_d \rho_b} c_T} \qquad (9.38)$$

Example 9.7 A 1-m^3 sample of soil is found to contain 53 g of atrazine (a herbicide) and to have a water content of 0.15 and a bulk (dry) density of 1610 kg/m^3. If the soil is 1% organic carbon and has an estimated distribution coefficient of 1.6 mL/g, estimate the concentration of atrazine in the pore water. The solubility of atrazine is 33 mg/L.

SOLUTION   From the data given, $c_T = 53$ g/m^3 = 53 mg/L, $\theta = 0.15$, $\rho_b = 1610$ kg/m^3, and $K_d = 1.6$ mL/g = 1.6×10^{-3} m^3/kg. Substituting these data into Equation 9.38 yields

$$c_{aq} = \frac{1}{\theta + K_d \rho_b} c_T = \frac{1}{0.15 + (1.6 \times 10^{-3})(1610)} (53) = 19 \text{ mg/L}$$

Therefore, the concentration of atrazine in the pore water is estimated as 19 mg/L. Since the solubility of atrazine is 33 mg/L, application of Equation 9.38 is validated. If the pore concentration calculated were greater than the solubility, the actual pore-water concentration of atrazine would be equal to the solubility, some pure-phase atrazine would be in the soil, and Equation 9.38 would not be valid.

It is interesting to note that the drinking-water standard for atrazine is 0.003 mg/L. Therefore, a pore-water concentration of 19 mg/L indicates a significant liklihood of ground-water contamination.

The potential for volatization of a substance is related to the saturated vapor pressure of the substance in the air above the soil interface; however, actual volatization from the soil is also affected by many other factors, such as atmospheric air movement, temperature, and soil characteristics. If a chemical exists in a soil in the vapor phase in addition to the liquid and solid phases, Equation 9.36 is expanded to include the vapor phase as

$$c_T = \theta c_{aq} + F\rho_b + ac_g \qquad (9.39)$$

where a is the volumetric air content ($a = n - \theta$, where n is the volumetric porosity) and c_g is the vapor density of the chemical [M/L^3 of soil air]. The relation between the vapor density and corresponding concentration of the chemical in (pore) water solution is given by *Henry's law* as

$$c_{aq} = K_H c_g \tag{9.40}$$

where K_H is Henry's constant for the chemical (dimensionless). Care should be taken in using K_H since it is commonplace in technical references to define the dimensionless Henry's constant as the inverse of K_H. Combining Equations 9.39 and 9.40 gives the following relationship between the aqueous concentration, c_{aq}, and the total concentration, c_T:

$$\boxed{c_{aq} = \frac{1}{\theta + K_d \rho_b + (n - \theta)/K_H} c_T} \tag{9.41}$$

Example 9.8   A 1-m^3 sample of soil is found to contain 53 g of atrazine and to have a water content of 0.15, a porosity of 0.20, and a bulk (dry) density of 1610 kg/m^3. If the soil has an estimated distribution coefficient of 1.6 mL/g, and the (dimensionless) Henry's constant of atrazine is 1.03×10^7, estimate the concentration of atrazine in the pore water and assess the effect of atrazine volatization on the aqueous concentration.

SOLUTION   From the data given, $c_T = 53$ $g/m^3 = 53$ mg/L, $\theta = 0.15$, $n = 0.20$, $K_H = 1.03 \times 10^7$, $\rho_b = 1610$ kg/m^3, and $K_d = 1.6$ mL/g $= 1.6 \times 10^{-3}$ m^3/kg. Substituting these data into Equation 9.41 yields

$$c_{aq} = \frac{1}{\theta + K_d \rho_b + (n - \theta)/K_H} c_T$$

$$= \frac{1}{0.15 + (1.6 \times 10^{-3})(1610) + (0.20 - 0.15)/(1.03 \times 10^7)} (53) = 19 \text{ mg/L}$$

Therefore, the concentration of atrazine in the pore water is estimated as 19 mg/L. Example 9.7 had exactly the same parameters, neglected vaporization, and yielded the same aqueous concentration of 19 mg/L. This result confirms that atrazine volatization has a negligible effect on the fate and transport of atrazine in the water environment.

Biological degradation of a chemical usually implies a breakdown by living microorganisms to more simple compounds, ultimately to carbon dioxide, water, methane, ammonium, and possibly to other simple by-products. Biotransformation of chemicals in soil is accomplished by microorganisms or fungi, and biodegradation may occur in both aerobic and anaerobic environments. Biodegradation is commonly represented by *Monod's equation*,

$$\frac{dc_{aq}}{dt} = -\frac{\mu_m X c_{aq}}{K_s + c_{aq}} \tag{9.42}$$

where μ_m is the maximum substrate utilization rate [$M/L^3 \cdot T$], X is the microbial biomass per unit volume of pore water [M/L^3], and K_s is the half-saturation constant for the chemical

[M/L^3]. For small concentrations of chemicals in the soil and a sufficient and constant microbial population, biodegradation is commonly described by the first-order reaction

$$\frac{dc_{aq}}{dt} = -k_b c_{aq} \qquad (9.43)$$

where k_b is the decay constant [T^{-1}]. If c_{aq} is large relative to K_s, Equation 9.42 indicates that biodegradation is described by the linear equation

$$\frac{dc_{aq}}{dt} = \text{constant} \qquad (9.44)$$

Typically, due to lack of information, it is not possible to quantify specific degradation rates. The only information usually available for many organic chemicals is their half-life and/or overall persistence of the chemical in the soil. Using the first-order reaction given by Equation 9.43, the aqueous concentration (in the soil water) as a function of time is given by

$$c_{aq}(t) = c_{aq}(0)e^{-k_d t} \qquad (9.45)$$

where $c_{aq}(t)$ and $c_{aq}(0)$ are the aqueous concentrations at time t and zero, respectively, and k_d is the overall degradation coefficient. Using Equation 9.45, the half-life, $t_{0.5}$, is related to the overall degradation coefficient, k_d, by the relation

$$t_{0.5} = -\frac{\ln(0.5)}{k_d} \qquad (9.46)$$

The half-lives of several pesticides (insecticides) in soils are given in Table 9.24.

Example 9.9   The half-life of atrazine is estimated to be 100 days. If the pore-water concentration of atrazine in a soil is found to be 19 mg/L, estimate how long it will take for the concentration of atrazine to decrease to the drinking-water standard of 0.003 mg/L.

SOLUTION   From the data given, $t_{0.5} = 100$ days, $c_{aq}(0) = 19$ mg/L, and $c_{aq}(t) = 0.003$ mg/L. The overall degradation coefficient, k_d, is given by Equation 9.46 as

$$k_d = -\frac{\ln(0.5)}{t_{0.5}} = -\frac{\ln(0.5)}{100} = 0.00693 \text{ day}^{-1}$$

TABLE 9.24   Half-Lives of Insecticides in Soil

Insecticide	Half-Life (months)
Aldrin	3–8
Chlordane	10–12
DDT	~30
Dieldrin	~27
Heptachlor	8–10
Lindane	12–20

Source: Kuhnt (1995).

The time, t, for the concentration of atrazine to decay from 19 mg/L to 0.003 mg/L is given by Equation 9.45, where

$$c_{aq}(t) = c_{aq}(0)e^{-k_d t}$$
$$0.003 = 19e^{-0.00693t}$$

which yields

$$t = 1263 \text{ days}$$

Therefore, it will take approximately 3.5 years ($= 1263$ days) for the concentration of atrazine to decay to the drinking-water standard, provided that no additional atrazine is added to the soil.

The fate and transport of several specific contaminants originating in the soil environment are described in the following sections.

Phosphorus Phosphorus originates from natural weathering of the phosphate mineral apatite, industrial fertilizers, organic fertilizers (manure), sewage, and phosphate detergents. Land application of liquid manure (an organic fertilizer) from a hog-feeding operation in central Iowa is shown in Figure 9.33. Net accumulation of phosphorus in soils has been recorded worldwide. The only mechanism for removing phosphorus from soils is by plant uptake and subsequent harvest, with removal of plants from the field. Consequently, most agricultural and suburban soils exhibit a steady accumulation of phosphorus. Measures of phosphorus concentration include *total dissolved phosphorus* (TDP), which includes both the inorganic and organic dissolved fractions, and *total phosphorus* (TP),

FIGURE 9.33 Organic fertilizer application. (From NRCS, 2005e.)

which includes the TDP and *total particulate phosphorus* (PP). Total particulate phosphorus is that contained in material retained by a 0.45-μm filter.

The concentration of phosphorus in agricultural soils is typically high, and it is commonly found that the Langmuir isotherm best describes the adsorption process. The *Langmuir isotherm* allows for a maximum sorption capacity on the solid matrix and is defined by the relation

$$F = \frac{K_l \overline{S} c_{aq}}{1 + K_l c_{aq}} \tag{9.47}$$

where F is the adsorbed concentration [M/M], K_l is the *Langmuir constant* [L^3/M], and $\overline{S}$ is the maximum sorption capacity [M/M]. Values of K_l and $\overline{S}$ in acid soils can best be estimated by

$$K_l = 0.061 + 170{,}000 \times 10^{-\text{pH}} + 0.027(\% \text{ clay}) + 0.076(\% \text{ organic C}) \tag{9.48}$$

$$\overline{S} = -3.5 + 10.7(\% \text{ clay}) + 49.5(\% \text{ organic C}) \tag{9.49}$$

where K_l is in L/mg and $\overline{S}$ is in μg/g. The coefficient of multiple correlation for Equations 9.48 and 9.49 were 0.53 and 0.83, respectively (Novotny, 2003). For calcerous soils (versus acid soils) calcium compounds control the solubility of phosphorus, and the distribution of particulate and dissolved P is governed by the solubility diagram shown in Figure 9.34.

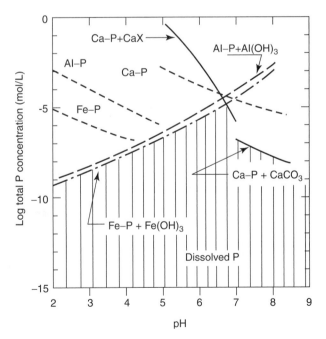

FIGURE 9.34 Solubility of calcium, aluminum, and iron phosphates in soil. (From Hsu and Jackson, 1960.)

Example 9.10   Fertilizer is applied to a field at an annual rate of 200 kg/ha and is plowed in and distributed uniformly to a depth of 0.3 m. The soil has a porosity of 0.4, a bulk density of 1600 kg/m^3, and consists of silt loam with 15% clay, 60% silt, and 25% sand. The pH of the pore water is 6.5, and the organic carbon content is 1.5%. (a) Assuming that the pores are saturated with water, estimate the concentration of P in the pore water. (b) What is the retention capacity of the soil for P? (c) If the annual plant uptake of P is 70 kg/ha, estimate the time for the soil to be saturated with P.

SOLUTION   (a) Since the soil is acidic (pH 6.5), 15% clay, and 1.5% organic C, the Langmuir constant, K_l, and the maximum sorption capacity, $\bar{S}$, can be estimated using Equations 9.48 and 9.49 as

$$K_l = 0.061 + 170{,}000 \times 10^{-\text{pH}} + 0.027(\% \text{ clay}) + 0.076(\% \text{ organic C})$$

$$= 0.061 + 170{,}000 \times 10^{-6.5} + 0.027(15) + 0.076(1.5) = 0.634 \text{ L/mg}$$

$$= 0.000634 \text{ L/}\mu\text{g}$$

$$\bar{S} = -3.5 + 10.7(\% \text{ clay}) + 49.5(\% \text{ organic C}) = -3.5 + 10.7(15) + 49.5(1.5)$$

$$= 231 \, \mu\text{g/g}$$

Using the calculated values of K_l and $\bar{S}$, the Langmuir sorption isotherm is given by Equation 9.47 as

$$F = \frac{K_l \bar{S} c_{\text{aq}}}{1 + K_l c_{\text{aq}}} = \frac{0.000634(231)c_{\text{aq}}}{1 + (0.000634)c_{\text{aq}}} = \frac{0.146 c_{\text{aq}}}{1 + 0.000634 c_{\text{aq}}}$$

where c_{aq} is in μg/L and F is in μg/g. Since P is applied at a rate of 200 kg/ha and is mixed over the top 0.3 m of soil, the total inorganic P content of the soil, c_T, is given by

$$c_T = \frac{200 \text{ kg/ha} \times 10^9 \, \mu\text{g/kg}}{0.3 \text{ m} \times 10^4 \text{ m}^2/\text{ha} \times 10^3 \text{ L/m}^3} = 66{,}700 \, \mu\text{g/L}$$

The P in the soil is distributed between the adsorbed and aqueous phases; therefore, the total concentration, c_T, is given by Equation 9.36. Taking the moisture content, θ, equal to the porosity ($= 0.4$), and given the bulk density, ρ_b, as 1600 kg/m^3 = 1600 g/L, Equation 9.36 can be combined with the (Langmuir) adsorption isotherm to give

$$c_T = \theta c_{\text{aq}} + F \rho_b$$

$$66{,}700 = 0.4 c_{\text{aq}} + \frac{0.146 c_{\text{aq}}}{1 + 0.000634 c_{\text{aq}}} (1600)$$

which yields

$$c_{\text{aq}} = 348 \, \mu\text{g/L}$$

Therefore, the pore-water concentration of P is 348 μg/L, and water infiltrating through the soil and into the ground water will have this concentration.

(b) The retention capacity of the soil for P is equal to $\overline{S} = 231\ \mu g/g = 2.31 \times 10^{-4}\ kg/kg$, and since the P is distributed over the top 0.3 m of soil and the bulk density of the soil is $1600\ kg/m^3$, the retention capacity is given by

$$\text{retention capacity} = 231 \times 10^{-6}\ kg/kg \times 1600\ kg/m^3 \times 0.3\ m$$

$$\times 10^4\ m^2/ha = 1100\ kg\ P/ha$$

(c) Since P is applied at an annual rate of 200 kg/ha and the plant uptake is 70 kg/ha · yr, the time, T, for the soil to become saturated can be approximated by

$$T = \frac{1100\ kg/ha}{200\ kg/ha \cdot yr - 70\ kg/ha \cdot yr} = 8.5\ \text{years}$$

Therefore, after approximately 8.5 years the soil will be saturated with P and any excess P applied will not be retained by the topsoil soil and will generally be passed through to the underlying layers with infiltrated water originating from either rainfall or irrigation. This could cause a serious water-quality problem, particularly for nearby surface-water bodies.

Inorganic phosphorus in soils is largely unavailable for plant or leaching; therefore, excess phosphorus has to be added to provide sufficient nutrients for crops, resulting in further phosphorus buildup. Most of the phosphorus applied is contained in the particulate phase and moves with eroded soil. Before saturation, very little phosphorus is leached into ground water or moves with soil water. When the soil becomes saturated with phosphorus, significant leaching may occur. In intensively worked agricultural areas, saturation is typically reached in decades and sometimes in less than 20 years (Salomons and Stol, 1995). The enrichment ratio for P eroded with soil particles has been found to be around 2, although not much recent data is available (Stoltenberg and White, 1953).

Numerous models have been developed for estimating the transfer of phosphorus from land to water in agricultural catchments; however, recent research has indicated that the applicability of these models depends significantly on the spatial resolution in which these models are applied (Brazier et al., 2005). Typically, models applied on a 1-km^2 scale can be expected to perform adequately.

Nitrogen In contrast to phosphorus and many organic chemicals, nitrogen does not readily accumulate in the soil and is readily transported to the ground water in the form of nitrate, NO_3^-, and, to a far lesser degree, ammonium, NH_4^+. Nitrate pollution of ground water is becoming a worldwide problem. For nitrates, subsurface flow and not erosion may be the primary transport process that carries nitrogen from the source area to receiving water bodies. Consequences of increased nitrate concentrations are "blue baby" sickness, caused by excessive nitrate in drinking water, and excessive algal growth in transitional surface waters such as estuaries (Neal and Heathwaite, 2005).

Sources of soil nitrogen include soil fertilizers and nitrogen fixation from the atmosphere by soil bacteria and legumes, plant residues, and precipitation. The application of anhydrous ammonia fertilizer in Cedar County, Iowa, is illustrated in Figure 9.35. In suburban areas, in addition to fertilizing lawns, significant amounts of nitrogen enter soils and subsoils from the seepage areas of household septic systems. Nitrogen in soil is contained mainly in soil organic matter, or in cases of ammonium ions, can be sorbed by clays and

FIGURE 9.35   Application of anhydrous ammonia fertilizer. (From NRCS, 2005e.)

organic matter. Ammonium-adsorption isotherms that are typical of sand, silt, and clay are given by the following Freundlich relations (Preul and Schoepfer, 1968):

$$\text{Sand:} \quad F = 1.4c_{aq}^{0.96} \tag{9.50}$$

$$\text{Silt:} \quad F = 7.1c_{aq}^{0.78} \tag{9.51}$$

$$\text{Clay:} \quad F = 15.6c_{aq}^{0.69} \tag{9.52}$$

where F is the adsorbed concentration in μg/g and c_{aq} is the aqueous concentration in mg/L.

Example 9.11   A sandy soil contains 1.5% organic matter and 0.10% N measured as total Kjeldahl nitrogen ($=$ organic N $+$ NH$_4^+$), of which 70% is organic N. (a) If the moisture content of the soil is 0.25, the porosity of the soil is 0.40, the wilting point of the soil is 15%, and the bulk density of the soil is 1600 kg/m^3, determine the amount of mobile N (available) and immobile N. (b) If a 4-cm storm results in 1.5 cm of runoff and 1.5 tonnes/ha of soil loss, and the enrichment ratio for organic nitrogen is 1.5, estimate the N loss in the runoff. Assume that the runoff mixes with the soil moisture over a topsoil depth of 0.4 cm.

SOLUTION   (a) Organic nitrogen is contained only in the soil particulates and there is no equilibrium with the aqueous phase. Therefore, the organic-N concentration in the aqueous phase can be taken as zero. Equilibrium between the sorbed and aqueous ammonia-N is given by Equation 9.50, which can be rewritten here as the following isotherm:

$$F = 1.4c_{aq}^{0.96}$$

Since ammonia-N in the soil is $(1 - 0.70)0.10\% = 0.03\% = 300\,\mu g/g$ and the bulk density, ρ_b, is $1600\,kg/m^3 = 1600\,g/L$, the total ammonia-N content of the soil, c_T, is given by

$$c_T = 300\,\mu g/g \times 1600\,g/L \times 10^{-3}\,mg/\mu g = 480\,mg/L$$

The ammonia-N in the soil is distributed between the adsorbed and aqueous phase, therefore the total concentration, c_T, is given by Equation 9.36. The moisture content, θ, is 0.25, and Equation 9.36 can be combined with the adsorption isotherm to give

$$c_T = \theta c_{aq} + F\rho_b$$
$$480 = (0.25)c_{aq} + 1.4c_{aq}^{0.96}(1600)(10^{-3}\,mg/\mu g)$$

which yields $c_{aq} = 234\,mg/L$. The corresponding adsorbed concentration of ammonia-N is given by the sorption isotherm as

$$F = 1.4c_{aq}^{0.96} = 1.4(234)^{0.96} = 263\,\mu g/g$$

The mobile N is contained in the pore water as ammonium-N at a concentration of $234\,mg/L$. The immobile N is contained in the particulate soil as organic-N at $70\,\mu g/g$ $(= 0.07\%)$ and ammonia-N at $263\,\mu g/g$, for a total of $70\,\mu g/g + 263\,\mu g/g = 333\,\mu g/g$.

(b) Since the total sorbed N is $333\,\mu g/g = 333\,g/tonne$, the soil loss is 1.5 tonnes/ha, and the enrichment ratio, ER, is 1.5, the fixed (sorbed) N loss, Y_{SN}, in the runoff is given by

$$Y_{SN} = (\text{sorbed N}) \times (\text{soil loss}) \times ER = 333\,g/tonne \times 1.5\,tonnes/ha \times 1.5 = 749\,g/ha$$

Assuming that the runoff mixes with the soil moisture over a topsoil depth of 0.4 cm, a mass balance gives (for a storm depth of 4 cm),

soil content of dissolved N in top 0.4 cm of soil = N in runoff mixed with soil water

$$(\theta - \theta_{wilt}) \times 0.4\,cm \times c_{aq} = (n \times 0.4\,cm + 4\,cm)c$$

where θ is the moisture content of the soil $(= 0.25)$, θ_{wilt} is the wilting point $(= 0.15)$, c_{aq} is the aqueous N concentration $(= 234\,mg/L)$, n is the porosity $(= 0.40)$, and c is the concentration of dissolved N in the runoff. Substituting these parameters into the mass balance equation gives

$$(0.25 - 0.15) \times 0.4\,cm \times 234\,mg/L = (0.40 \times 0.4\,cm + 4\,cm)c$$

which yields $c = 2.25\,mg/L = 2.25\,g/m^3$. Since the depth of runoff is 1.5 cm, the dissolved N in the surface runoff, Y_{DN}, is given by

$$Y_{DN} = 2.25\,g/m^3 \times 1.5\,cm \times 0.01\,m/cm \times 10^4\,m^2/ha = 338\,g/ha$$

The total N loss in the runoff is therefore given by

$$\text{total N loss} = Y_{SN} + Y_{DN} = 749\,g/ha + 338\,g/ha = 1087\,g/ha$$

Nitrogen is lost from the soil primarily by erosion, crop harvesting, and nitrate leaching. In higher-pH calcareous soils, ammonium ion is converted to gaseous ammonia that can volatilize from soils or be dissolved in soil water as nonionized ammonia.

The nitrification and denitrification processes (described in Chapter 2) commonly occur in soils. Over 90% of the fertilizer in the United States is in the form of ammonium salts. If the ammonium is applied to an aerated microorganism-rich soil such as farmland, nitrification occurs, resulting in the conversion of NH_4^+ to NO_3^-. The optimum temperature for nitrification is 22°C and the rate of nitrification decreases rapidly on both sides of the temperature curve. The rate of nitrification essentially ceases below 10°C and above 45°C (Stanford et al., 1973). Also, since nitrifying bacteria depend on water as their living environment, the nitrification rate decreases with decreasing moisture content. The process of denitrification usually occurs when soils become saturated with water, which is typical of aquatic sediments and wetland substrate. Denitrification is minor or nonexistent when soils are aerated. Wetlands exhibit more than 90% removal of nitrate N, less if N arrives in the wetland in ammoniacal or organic form. The primary removal mechanisms are by denitrification, by ammonification of the organic nitrogen, and by the uptake of the ammonia by plants and heterotrophic microorganisms.

A simple steady-state model for estimating nitrate loading on ground water was proposed by Mills et al. (1982). This model assumes a constant annual load of nitrogen fertilizer or wastewater in the form of organic nitrogen, and is given by

$$L = 1000XN[1 - v(1 - O) - U] \tag{9.53}$$

where L is the annual steady-state nitrate-N load to ground water (kg/ha), X is the average annual solids waste–organic fertilizer application rate (tonnes/ha), N is the nitrogen fraction of the applied compound, O is the organic fraction of the applied compound, v is the fraction of applied nitrogen that volatilizes (a soil-pH dependent variable), and U is the average crop uptake of nitrogen.

Example 9.12   A potato farmer spreads 3700 tonnes/ha of cow manure annually over his land each year. If cow manure contains 2% nitrogen, is 85% organic, 5% of the applied nitrogen volatilizes, and 90% of the applied nitrogen is taken up by the potato crop, estimate the nitrogen loading on the ground water.

SOLUTION   From the data given, $X = 3700$ tonnes/ha, $N = 0.02$, $v = 0.05$, $O = 0.85$, and $U = 0.90$. Substituting into Equation 9.53 yields

$$L = 1000\,XN\,[1 - v(1 - O) - U]$$
$$= 1000(3700)(0.02)[1 - 0.05(1 - 0.85) - 0.90]$$
$$= 6845\,\text{kg/ha}$$

This ground-water loading (6845 kg N/ha) should be assessed in terms of the associated ground-water concentration of nitrate-N relative to the drinking-water standard of 10 mg/L (of nitrate-N).

Toxic Metals   The toxic metals of most concern in soils are arsenic (As), cadmium (Cd), chromium (Cr), copper (Cu), lead (Pb), and zinc (Zn). For agricultural soils, inputs from the atmosphere, fertilizers, certain pesticides, sewage sludge disposal, and manure are the most significant. An example of sewage sludge disposal on agricultural land is shown in Figure 9.36. When metals are added to water or soil from both natural and human-made sources, they undergo complexation with ligands, which can be both inorganic and organic. Because the metals exist in aqueous solution as positively charged cations, ligands are mostly negatively charged anions that bound the metal ion. Examples of inorganic ligands include CO_3^{2-}, Cl^-, S^{2-}, PO_4^{3-}, NO_3^-, and others. Organic ligands are humic substances that form from the decomposition of vegetation. Complexation is important because the free metallic ions (e.g., Cd^{2+}, Cu^{2+}, Pb^{2+}, Zn^{2+}) or methyl-metal complexes are far more toxic than other less soluble complexes. Many metal complexes and metal forms themselves are not biologically available and hence are not toxic.

Inorganic arsenic compounds have been used in agriculture as pesticides and defoliants for many years. On a worldwide basis, 80 to 90% of the produced arsenic is used in agriculture and accumulates in the soil (Nriagu and Pacyna, 1988). Due to the serious environmental problems caused by agricultural and lake applications, use of arsenic compounds was banned in the United States in 1967. Other more common sources of arsenic include atmospheric deposition of emissions from burning coal and smelting of

FIGURE 9.36   Land application of sewage sludge. (From South West Water, 2005.)

ores, sewage sludge disposal into soil, animal manure application, and fertilizer application. Once incorporated into soils and sediments, arsenic reverts to arsenate, which is strongly held by the clay fraction. Chemically, arsenic is very close to phosphorus and sometimes interferes with phosphate determinations, and vice versa.

Organic Chemicals   The use of organic chemicals is credited with substantially increasing yields of agricultural crops and assisting in controlling pests and diseases such as malaria and other diseases transmitted by insects. Organic chemicals deposited on soil surfaces, such as by pesticide application shown in Figure 9.37, can (1) be transported into the atmosphere by volatization, (2) be transported into ground water by leaching from the soil, (3) be transported into surface runoff by leaching and attaching onto eroded sediment; (4) degrade (photochemically and microbially), and (5) be taken up by plants. *Drift* is that portion of an application that does not reach the target area. The extent of drift losses and the subsequent dispersion and transport of chemicals in air is governed primarily by meteorological conditions and method of application. Field measurements indicate that significant volatization losses may occur if pesticides are not incorporated into the soil. Under field conditions, the volatization process is continuous, although its highest rate occurs immediately following pesticide application. Photodecomposition occurs only to pesticides located on the soil surface, and in most cases, photodegradation reactions are comparatively slow and their rate depends on the physical state of the pesticides (Schnoor et al., 1987). Microbiological degradation is considered to be the major pathway of degradation of many pesticides in soils and sediments. The efficiency of this mechanism depends on such environmental factors as temperature, moisture, organic matter content, aeration, pH, and pesticide concentration. In general, organochlorine pesticides are the most resistant to degradation. Dissipation of atrazine from soils through uptake by corn, sorghum, and other crops has been indicated by experiments, but the significance of plant uptake as a fate process for pesticides is typically small.

FIGURE 9.37   Pesticide application. (From NRCS, 2005e.)

Contamination of ground water can occur through leaching. Downward movement of agriculturally applied pesticides is controlled by soil type and chemistry, pesticide composition, and climate factors. The leachability of a compound from soil depends primarily on the degree of adsorption of the chemical on the soil particles. Sorption of nonionic (nonpolar) pesticides—organochlorine and organophosphorus insecticides—in soils and sediments is correlated primarily with organic particulate content and to a much lesser degree with clay. Soil and sediment pH controls the overall charge of the molecule and its absorptivity in soil organic matter and clay. Organochlorine insecticides, which have low solubility, are least mobile in soils and sediments, followed by organophosphorus insecticides. The water-soluble acidic herbicides are most mobile. Other pesticides, including triazines, atrazine, phenyl urea, and carbamates, have an intermediate degree of mobility.

Pesticides in soils are continuously subjected to dissipation processes and their concentration decreases gradually after application. The mechanisms of dissipation include adsorption on soil particles and subsequent loss by erosion. A methodology to assess the vulnerability of ground water to pesticide contamination based on observations in monitoring wells was reported by Worrall and Besien (2005). Such assessments are an integral component of ensuring drinking-water sources are adequately protected from pesticide contamination.

9.5.3 Best Management Practices

Best management practices (BMPs) are methods and practices for preventing or reducing nonpoint source pollution to a level compatible with water-quality goals. When selecting BMPs, it is important that the pollutants and the forms in which they are transported be known. Many BMPs listed in the technical handbooks of land-resource agencies were designed to protect soil or grass resources or to provide economic stability for farmers by improving productivity. For example, while conservation tillage is extremely effective for erosion control, it is not very effective in removing or controlling soluble pollutants from runoff or leached water.

BMPs can be selected in two ways: (1) to control a known or suspected type of pollution (e.g., phosphorus or bacteria) from a particular source (e.g., runoff from a cornfield or a dairy feedlot); or (2) to prevent pollution from a category of land-use activity (such as agriculture row-crop farming or containerized nursery irrigation return flow). The recommended procedure for selecting BMPs to solve water-quality problems is as follows (USDA, 1980, 1988; Gordon and Hansen, 1989; Brach, 1990):

1. Identify the water-quality problem: for example, annual summer algal blooms in a lake.

2. Identify the pollutants contributing to the problem and their probable sources: for example, nutrients from septic systems adjoining the lake and runoff from a nearby horse pasture.

3. Determine how each pollutant is delivered to the water (e.g., soluble nutrients from septic tank drain fields rise to the surface when the systems are overloaded and are carried to the lake by overland flow during rainstorms and snowmelt).

4. Set a reasonable water-quality goal for the water body and determine the level of treatment needed to meet that goal.

5. Evaluate feasible BMPs for water-quality effectiveness, effect on ground water, economic feasibility, and suitability of the practice to the site.

When selecting BMPs to prevent a problem, a more technology-based approach can be followed. Such an approach divides agricultural land into use categories, such as irrigated cropland and rangeland, and specifies a minimum level of treatment necessary to protect the resource base. This type of planning is used by the U.S. Department of Agriculture in developing conservation plans for soil conservation. Recent experiences in the Nordic and Baltic areas of Europe have demonstrated that implementation of agricultural BMPs such as reduced fertilizer application and decreased agricultural intensity are very slow to take effect in reducing the level of pollution in receiving waters, especially in medium-sized and large catchments (Granlund et al., 2005). The reason for this slow response has been attributed to the large nutrient pools in soils and the considerable potential for nutrient losses due to mineralization and erosion.

Agricultural BMPs can be divided into source controls, hydrologic modifications, reduction of delivery, and storage and treatment. Several of the more common BMPs are described below.

Cropping Practices Cover crops, crop rotation, and conservation tillage are cropping practices that are used as BMPs. These practices stress vegetative cover during critical times, and their primary objective is to reduce erosion and associated soil loss.

Cover crops are close-growing grasses, legumes, or small grain crops that cover the soil during the critical erosion period for the area. An example of a cover crop is shown in Figure 9.38(*a*), where close-growing grasses are covering the preemergent crop between the trees. Cover crops have been found to be 40 to 60% effective in reducing sediment and 30 to 50% effective in removing total phosphorus.

Crop rotation is a system of periodically changing the crops grown on a particular field. When legumes are used as a cover crop or as part of a crop rotation, they provide a nitrogen

(*a*)   (*b*)

FIGURE 9.38 Cropping practices: (*a*) cover cropping; (*b*) conservation tillage. (From NRCS, 2005e.)

source for the subsequent crop, thus minimizing the addition of commercial fertilizers. The most effective crop rotations for water-quality protection involve at least 2 years of grass or legumes in a 4-year rotation. Crop rotations reduce erosion and associated adsorbed pollutant loads by improving soil structure. Some rotations have been estimated to reduce nitrogen demand by 50% and phosphorus demand by 30% on an annual basis.

Conservation tillage is any tillage method that leaves at least 30% of the soil surface covered with crop residue after planting; in these cases the soil is tilled only to the extent needed to prepare the seedbed. An example of conservation tillage is shown in Figure 9.38(*b*). Conservation tillage has been found to be highly effective in sediment reduction but of little effect in controlling soluble nutrients and pesticides.

Integrated Pest Management   Integrated pest management (IPM) is a combination of practices to control crop pests (insects, weeds, diseases) while minimizing pollution. IPM uses practices such as the selection of resistant crop varieties and crop rotations and modified planting dates, along with sophisticated pesticide application management. IPM works primarily by decreasing the amount of pesticide or crop-protection chemical available and by selecting the least toxic, least mobile, and/or least persistent material. An example of IPM is shown in Figure 9.39, where a wiper attachment applies contact herbicide only to those weeds that grow above the soybean canopy. Complete studies on the effectiveness of IPM show mixed results.

Nutrient Management   Nutrient management is a series of practices designed to decrease the availability of excess nutrients through improvements in timing, application rates, and location selection for fertilizer application. Current nutrient management is based on the limiting-nutrient concept that the fertilizer application rate should be based on the nutrient most needed by the plant for optimum growth, usually nitrogen. This practice is especially effective in controlling the soluble phases of nutrients. An example of nutrient management is shown in Figure 9.40, where a nitrogen fertilizer is being applied to growing corn in a contoured, no-tilled field. Applying smaller amounts of nitrogen

FIGURE 9.39   Integrated pest management. (From NRCS, 2005e.)

FIGURE 9.40   Nutrient management. (From NRCS, 2005e.)

several times over the growing season rather than all at once at or before planting helps the plants use the nitrogen rather than have it enter water supplies.

Terraces and Diversions   A *terrace* is an earthen embankment, channel, or a combination ridge and channel constructed across the slope to intercept runoff. Terraces reduce the slope effect on erosion by dividing fields into segments with less steep or nearly horizontal slopes. Soil particles and adsorbed pollutants are thus not transported from the field, with terraces retaining all water applied to the fields, hence reducing the loss of dissolved pollutants. Hillside vineyard terraces in Sonoma County, California, are shown in Figure 9.41. Level terraces can remove up to 95% of the sediment, up to 90% of its associated adsorbed nutrients, and between 30 and 70% of dissolved nutrients. Diversions can reduce sediment movement by 30 to 60% and adsorbed nutrients by 20 to 45%.

Critical-Area Treatment   Grade-stabilization structures and critical-area planting and pollutant delivery control are both considered critical-area treatment. A grade-stabilization structure is used to control the grade in natural or artificial channels. These structures reduce water velocity, thus preventing additional sediment detachment and decreasing the transport capacity of water. Structures can be constructed of metal, wood, rocks, concrete, or earth, and the area surrounding the structure must be stabilized. A drop structure is a type of grade-stabilization structure that is sometimes built at the outlet of a grassed waterway to stabilize the waterway and allow runoff to leave the waterway without causing gully erosion. A typical drop structure is shown in Figure 9.42. Critical-area planting involves planting suitable vegetation (such as grass, trees, or shrubs) on highly unstable sites. Intensive methods such as the use of erosion blankets, tie-downs, gabions, mulches, hydromulching, increased seeding rate, and hand planting are often needed.

Sediment Basins and Detention–Retention Ponds   Basins and ponds generally consist of earthen embankments designed to trap and store sediment and other pollutants.

FIGURE 9.41   Terraces. (From NRCS, 2005e.)

Tailwater detention ponds are used to collect runoff from agricultural fields for storage and pollution control. A tailwater detention pond to promote sedimentation of irrigation runoff is shown in Figure 9.43, where the banks of the pond are stabilized with rye grass for erosion control. Such detention ponds are sometimes called *tailwater recovery ponds*. Sediment basins can remove 40 to 90% of the incoming sediment, up to 30% of the adsorbed nitrogen, and 40% of the adsorbed phosphorus. The effectiveness of detention–retention ponds, especially those with aquatic vegetation, is generally higher. Sedimentation basins may not be effective for removal of toxicity or dissolved toxic compounds (Novotny, 2003).

FIGURE 9.42   Drop structure. (From NRCS, 2005e.)

FIGURE 9.43   Tailwater recovery pond. (From NRCS, 2005e.)

Animal Waste Storage and Treatment   Practices for management of animal wastes include storage and land disposal of treated waste (Miner et al., 2000). Lined manure storage bins and liquid-waste storage ponds hold waste until it can be land-applied without causing a water-pollution problem. Waste-treatment processes include lagoons that treat liquid waste biologically to reduce the nutrient and BOD content, anaerobic treatment in digesters, aerobic treatment by trickling filters and suspended-growth processes such as activated sludge, and composting. A state-of-the-art lagoon waste management system for a 900-head hog farm is shown in Figure 9.44, where this facility is completely automated and temperature controlled.

Animal wastes are frequently classified as point sources and require a permit in many states. However, pollution is generated by runoff from the contaminated premises and typically requires runoff control BMPs in addition to treatment and/or safe disposal. The basic principles of runoff control are that "clean" runoff originating outside the feedlot or storage area should be diverted so that it does not come in contact with the contaminated soil, and runoff originating inside the feedlot should be disposed of in a way that minimizes its pollution potential. Animal waste systems (all practices combined) have been estimated to reduce pollutant loadings by the following average amounts (Novotny, 2003): bacteria 74%, sediment 64%, total nitrogen 62%, and total phosphorus 21%. Most BMP recommendations suggest applying manure to satisfy crop nitrogen needs; however, this often results in excess phosphorus being applied.

Livestock Exclusion Fences   Keeping animals out of a stream by installing fences and crossings will reduce deposition of fecal material (a nutrient and bacterial source), eliminate turbidity from instream trampling, and eliminate detachment of sediment from streambanks. An example of the unsanitary practice of allowing livestock to move freely in a stream is shown in Figure 9.45(*a*). Reductions of 50 to 90% of suspended solids and total phosphorus have been reported as a result of fencing (Novotny, 2003), as shown in Figure 9.45(*b*).

FIGURE 9.44   Animal-waste treatment lagoon. (From NRCS, 2005e.)

Filter Strips and Field Borders   Filter strips and field borders utilize strips of closely growing vegetation such as sod or bunch grasses or small-grain crops with the primary purpose of water-quality protection. Filter strips are generally placed between agricultural land being used and a water body being protected. They are designed to remove sediment and other pollutants from sheet runoff by slowing the water velocity and allowing the suspended material and any adsorbed pollutants to drop out. A grass filter strip in Jasper County, Iowa, is shown in Figure 9.46.

(a)   (b)

FIGURE 9.45   Livestock control: (*a*) cattle wading in a stream; (*b*) exclusion fence. (From NRCS, 2005e.)

FIGURE 9.46   Filter strip. (From NRCS, 2005e.)

Field borders normally consist of perennial vegetation planted at the edge of fields to control erosion regardless of their proximity to water. Field borders are effective in preventing detachment of soil particles on the areas covered by the borders but have little effect on controlling erosion or pollutant detachment or transport from the fields they surround. Filter strips are very effective in removing sediment and sediment-bound nitrogen (35 to 90%) but less effective in removing phosphorus, fine sediment, and soluble nutrients such as nitrate (14%) or orthophosphate (5 to 50%) (Novotny, 2003).

Wetland Rehabilitation   Rehabilitation and development of wetlands involves restoring, rehabilitating, or enhancing existing wetlands to function as self-sustaining ecosystems that process, remove, transfer, and store pollutants. A restored wetland in northern California is shown in Figure 9.47. Wetlands are 80 to 90% effective in sediment removal, 40 to 80% effective in nitrogen removal, and 10 to 70% effective in phosphorus removal.

Riparian Buffer Zones   A riparian buffer zone is a vegetated area along the perimeter of a water body. The grasses and low vegetation in the buffer zone filter both surface and subsurface flow, while roots of taller vegetation take up and transform pollutants and nutrients from shallow ground water. A buffer along Bear Creek in Story County, Iowa, is shown in Figure 9.48. This buffer was established for about four years at the time of photo in 1997. Bear Creek is a national demonstration area for conservation buffers. Sediment-removal efficiencies of 80 to 90%, total phosphorus removal efficiencies of 50 to 75%, and total nitrogen removal efficiencies of 80 to 90% have been measured (Novotny, 2003).

FIGURE 9.47   Restored wetland. (From NRCS, 2005e.)

The wider the buffer, the greater the opportunity for the buffer to perform its functions. The desirable width of the buffer can be related to its function as follows:

- *Stream bank stabilization:* 3 to 10 m
- *Water-temperature moderation:* 3 to 20 m
- *Water quality:* 15 to 30 m
- *Nitrogen removal:* 8 to 40 m
- *Sediment removal:* 10 to 50 m
- *Wildlife habitat:* 10 to 90 m

Irrigation Water Management   Irrigation water management (IWM) consists of a combination of practices that control irrigation water to prevent pollution and reduce water loss. IWM includes proper scheduling, efficient application, efficient transport systems, utilization and reuse of tailwater and runoff, and management of drainage water. IWM is particularly effective in reducing nitrogen and pesticide loadings to ground water and salt loading to surface waters. Tailwater pits (which are pits to catch water that runs out of the furrows at the end of a field) are about 50% effective in sediment removal and can be highly effective in nutrient and pesticide pollution control as long as the water collected is reused and not discharged. An example of a tailwater pit under construction is shown in Figure 9.49.

Stream Bank Stabilization   Stream bank stabilization consists of structural or vegetative methods to protect the stream bank from erosion. Riprap, concrete, wood, or rock gabions can be used, but vegetative stabilization is the most effective for pollution control. Effective stabilization can reduce the sediment loading by 90% but is highly dependent on the type of vegetation used and the stability of the reclaimed area. Stabilization of the banks of a stream using riprap and willows is illustrated in Figure 9.50.

Range and Pasture Management   Range and pasture management consists of systems of practices to protect the vegetative cover on pastureland and rangeland. It includes

FIGURE 9.48   Filter strip adjacent to Bear Creek. (From NRCS, 2005e.)

such practices as seeding and reseeding, brush management, proper stocking rates, proper grazing use, and deferred-rotation systems. Rangeland is generally in native grass and managed as a range, whereas pastures are normally seeded to an improved grass variety and managed with agronomic practices. Keeping land permanently covered in high-quality closely spaced vegetation decreases soil loss to negligible amounts, and sediment and adsorbed pollutants are not lost from the land surface. Since rangeland is not fertilized, it can represent a low-input system if managed properly. Typical rangeland is illustrated in Figure 9.51(*a*), and a grass drill being used to interseed native grasses into a pasture is illustrated in Figure 9.51(*b*).

9.6 AIRSHEDS

An airshed is commonly defined as the geographic area responsible for emitting 75% of the air pollution reaching a body of water. Since different pollutants behave differently in the atmosphere, the airshed of a given body of water may vary depending on the pollutant of interest. Whereas watersheds are actual physical features of the landscape, airsheds are determined using mathematical models of atmospheric deposition. Airsheds are very useful in explaining the transportation of pollutants and can help manage a body of water much more effectively. Airborne pollution can fall to the ground in raindrops, in dust, or simply due to gravity. As the pollution falls, it may end up in streams, lakes, or estuaries and can affect the water quality. As an example of the significance of airsheds, studies show that 21% of the nitrogen pollution entering Chesapeake Bay comes from the air (USEPA, 2000d).

FIGURE 9.49 Tailwater pit. (From NRCS, 2005e.)

The deposition of an air pollutant on land or water can take several forms. *Wet deposition* occurs when air pollutants fall with rain, snow, or fog. *Dry deposition* is the deposition of pollutants as dry particles or gases. The pollutants can reach bodies of water as *direct deposition* falling directly into the water, or as *indirect deposition*, falling onto land and washing into the body of water as runoff.

There are five categories of atmospheric pollutants with the greatest potential to affect water quality: nitrogen compounds, mercury, other metals, pesticides, and combustion emissions. These categories are based on both method of emission and other characteristics

FIGURE 9.50 Stabilized channel banks. (From NRCS, 2005f.)

(a)   (b)

FIGURE 9.51   Range and pasture management: (*a*) rangeland; (*b*) grass drill in a pasture. (From NRCS, 2005e.)

of the pollutants. Mercury is in its own category since it behaves so much differently in the environment than do other metals. Combustion of fossil fuels is a major source of nitrogen oxides to the atmosphere. However, nitrogen is in its own category since its effects on ecosystems are so different from those of other combustion emissions. Pesticides and combustion emissions are exclusively human-made, whereas mercury, other metals, and nitrogen compounds arise from both natural and human-made sources.

9.6.1   Nitrogen Compounds

The largest single source of NO_x to the atmosphere is the combustion of fossil fuels (such as coal, oil, and gas) by automobiles and electric power plants (Schlesinger, 1997). In some water bodies, nitrogen deposited from the atmosphere is a large percentage of the total nitrogen load. For instance, Albemarle and Pamlico Sounds in North Carolina receive 38% of their nitrogen from the atmosphere (USEPA, 2000d). Table 9.25 shows several selected U.S. estuaries and the estimated percentage of total nitrogen entering the watershed as atmospheric deposition.

As human sources of nitrogen compounds to the atmosphere increase, the importance of atmospheric deposition of nitrogen to bodies of water will increase as well. The largest sources of NH_3 emissions are fertilizers and domesticated animals (such as hogs, chickens, and cows). Atmospheric deposition of nitrogen compounds can lead to accelerated eutrophication or harmful increases in the growth of algae. In some cases, nitrogen pollution can contribute to acidification of water bodies.

9.6.2   Mercury

Mercury is able to travel great distances in the atmosphere, and mercury pollution can arise through atmospheric deposition. Its unique chemical characteristics greatly influence its behavior in the environment and distinguish it from other metals. For example, biological processes can transform mercury into a very toxic compound known as methyl mercury.

Although mercury is found naturally in the environment, human activities have greatly increased its atmospheric concentration. It is estimated that human-made emissions have

TABLE 9.25 Atmospheric Loading of Nitrogen in Estuaries

Estuary	Atmospheric Loading of Nitrogen (%)
Albemarle and Pamlico Sounds	38
Chesapeake Bay	21
Delaware Bay	15
Long Island Sound	20
Narragansett Bay	12
New York Bight	38
Waquoit Bay, MA	29
Delaware inland bays	21
Flanders Bay, NY	7
Guadalupe River estuary, TX	2–8
Massachusetts bays	5–27
Narragansett Bay	4
Newport River coastal waters, NC	>35
Potomac River, MD	5
Sarasota Bay, FL	26
Tampa Bay, FL	28

Source: USEPA (2000d).

tripled mercury concentrations in the air and in the ocean surface since 1900 (Mason et al., 1994). Human activities presently account for about 75% of worldwide mercury emissions, and human-made sources include incinerators, coal-burning facilities, certain industrial processes, and household items. Burning coal for electric power generation and municipal waste incineration are the largest combustion sources of mercury emission.

Atmospheric deposition plays a major role in delivering mercury to ecosystems. Up to 83% of the mercury load to the Great Lakes comes from atmospheric deposition (Shannon and Voldner, 1995). Approximately half of the mercury in Chesapeake Bay is deposited from the atmosphere directly to the surface of the bay (Mason et al., 1997). The National Atmospheric Deposition Program (NADP) estimated that mercury was deposited at the rate of 4 to 20 μg/m^2 in the United States in 1998.

The unique chemistry of mercury affects its toxicity and how it travels in the atmosphere. Elemental mercury (Hg^0) is able to travel great distances but is not absorbed readily into biological tissues and is not very toxic. Other forms of mercury that may be emitted are divalent mercury (Hg^{2+}) and mercury that is bound to particles. These forms of mercury do not travel far in the atmosphere and tend to deposit very close to the source of emission. The forms of mercury emitted by different sources are not well characterized. In most cases, a combination of Hg^0, Hg^{2+}, and particle-bound mercury are emitted from most sources. Hg^{2+} dissolves quickly in water and is often the form of mercury found in bodies of water. When mercury becomes deposited within a water body, microorganisms can transform it into a very toxic substance known as methyl mercury. Methyl mercury (CH_3Hg) tends to remain dissolved in water and does not travel very far in the atmosphere. However, it can be converted back into elemental mercury and emitted again to the atmosphere. Mercury emitted from a single source can be deposited and reemitted many times. The cycles of mercury allow it to travel great distances and make it very difficult to track in the environment, including the atmospheric deposition of mercury into water bodies. This process, called the *leapfrog effect*, emphasizes the tendency of mercury to become deposited and reemitted many times. There is some evidence that the process continues until the mercury comes to rest at high latitudes or high altitudes.

9.6.3 Other Metals

Industrial processes have led to an increase in the environmental concentration of a number of metals. Lead contamination results from incinerating material that contains the metal (such as solder and paint) and from gasoline additives. Lead contamination peaked in the United States around 1970 and has been declining steadily since then. In Lake Michigan, for example, atmospheric deposition of lead decreased from over 500,000 kg in 1988 to under 90,000 kg in 1994 (USEPA, 1997f). Cadmium pollution also arises from incinerating cadmium-containing waste. Cadmium is used in batteries, in electroplating, and in many types of solder. Cadmium is also a significant by-product of zinc purification. Sources of cadmium include incinerators, smelters, and coal-burning facilities. Atmospheric deposition adds about 1000 kg of cadmium to the Chesapeake Bay each year (USEPA, 1997f).

9.6.4 Pesticides

Atmospheric deposition of pesticides is recognized as a source of toxic substances to water bodies. The likelihood that a pesticide will become an atmospheric deposition problem depends on its use, its chemical characteristics, how much pesticide already exists in a receiving water body, and how it reaches the water body (direct deposition versus indirect deposition through agricultural runoff). Although the concentrations of pesticides that are deposited in rainwater are usually low, there is a marked seasonality peaking in the summer months. On several occasions, the concentration of the pesticides 2,4-dichlorophenoxyacetic acid and atrazine in rainwater has exceeded levels considered safe by USEPA maximum contaminant levels (USGS, 2005a).

Pesticides linked to water-quality problems which are potentially transported through the atmosphere are chlordane, DDT/DDE, aldrin/dieldrin, hexachlorobenzene, α-hexachlorocyclohexane, and toxaphene. Besides these pesticides, chlorpyrifos, atrazine, and methoxychlor are currently being studied for their adverse effects in the environment. Chlorpyrifos is one of the most commonly used insecticides in the world today. It is the major ingredient in the pesticides Dursban and Lorsban.

9.6.5 Combustion Emissions

Pollutants that are released by incineration of waste are known as *combustion emissions*. Although many compounds are released during the incineration of solid waste, several classes of compounds pose a significant threat to water quality and human health. Dioxins, furans, polycyclic aromatic hydrocarbons (PAHs), and polychlorinated biphenyls (PCBs) are pollutants that degrade very slowly in the environment, can build up in the tissues of humans and wildlife, and have adverse effects on human and ecosystem health. Both dioxins and furans are families of chemicals that are formed completely as by-products of industrial processes and have no use as products. Both families of chemicals enter the atmosphere predominantly during incineration and may become a deposition problem for some bodies of water. Dioxins are very stable and may travel great distances in the atmosphere. It has been estimated that 60% of all dioxins deposited in the Great Lakes arise from only 10 sources (Commoner et al., 1996).

Polycyclic aromatic hydrocarbons (PAHs) are a family of over 100 compounds that arise from the incomplete combustion of fuel, garbage, coal, and other materials. They usually occur in complex mixtures such as soot. Typically, concentrations of 2 to 6 mg/L of individual PAHs are found in rainwater throughout the Great Lakes and the Chesapeake Bay regions (USEPA, 1997f).

Polychlorinated biphenyls (PCBs) are a class of chemicals that were manufactured for such industrial purposes as coolants and lubricants for electrical equipment until they were phased out in 1977. These highly persistent chemicals can still be found in older electrical equipment and in industrial waste sites. The primary means of introduction to the environment today, however, is through incineration of material that contains PCBs.

9.7 COMPUTER MODELS

Several computer codes are available for simulating the water quality of surface runoff. In engineering practice, the use of computer codes to apply the fundamental principles covered in this chapter is sometimes essential. In choosing a code for a particular application, there is usually a variety of codes to choose from. However, in doing work that is to be reviewed by regulatory agencies or where professional liability is a concern, codes developed and maintained by the U.S. government have the greatest credibility and, perhaps more important, are almost universally acceptable in developing permit applications and defending design protocols on liability issues.

Urban models are based primarily on lumped-parameter formulations, whereas agricultural models are commonly based on both lumped-parameter and distributed-parameter formulations. Operational models are defined as computer models that have an available user's manual and documentation, have been used by persons other than the model developer, and for which continuous support is available. Commonly used operational water-quality models are listed in Table 9.26 and described below.

The Agricultural Nonpoint Source Pollution Model (AGNPS) was developed and is maintained by the U.S. Department of Agriculture (Young et al., 1987, 1994). The primary emphasis of this model is on nutrients and sediment and on comparison of the effects of various best management practices on pollutant loadings. AGNPS can simulate sediment

TABLE 9.26 Operational Computer Models for Urban and Agricultural Applications

Type of Area	Model	Sponsoring Agency	Simulation Type[a]
General	BASINS	USEPA	C,SE
Urban	HSPF	USEPA	C,SE
	SWMM	USEPA, Proprietary	C,SE
Agricultural	AGNPS	USDA-ARS	C,SE
	ARM	USEPA	C
	CREAMS–GLEAMS	USDA-ARS	C,SE
	HSPF	USEPA	C
	SWRRB	USDA-ARS	C
	WEPP	USDA-ARS	C,SE

[a]C, continuous simulation; SE, single-event simulation.

and nutrient loads from agricultural watersheds for a single storm event or for a continuous simulation. The model does not simulate pesticides. The watershed must be divided into uniform square cells, with cells grouped according to division of the basin into sub-watersheds. AGNPS is capable of accepting and handling point inputs such as feedlots, wastewater discharges, and streambank and gully erosion. The modified universal soil loss equation is used for predicting soil loss in five different particle sizes (clay, silt, sand, small aggregates, and large aggregates). The pollutant transport portion of the model is subdivided into one part handling soluble pollutants and another part handling sediment adsorbed pollutants. The input data requirements are extensive, but most of the data can be retrieved from topographic and soil maps, local meteorological information, field observations, and various publications, tables, and graphs provided in the manual or references. Both mainframe and PC versions of the model are available.

Better Assessment Science Integrating Point and Nonpoint Sources (BASINS) is a multipurpose environmental analysis system that integrates into one convenient package a geographical information system (GIS), national watershed data, and state-of-the-art environmental assessment and modeling tools. This system makes it possible quickly to assess large amounts of point-source and nonpoint-source data in a format that is easy to use and understand. This tool integrates environmental data, analytical tools, and modeling programs to support development of cost-effective approaches to watershed management and environmental protection. BASINS has three objectives: (1) to facilitate examination of environmental information, (2) to provide an integrated watershed modeling framework, and (3) to support analysis of point- and nonpoint-source management alternatives. BASINS supports the TMDLs that integrate point and nonpoint sources, and its modeling components have been used by consultants as a principal tool in TMDL preparation (Novotny, 2003). BASINS has also been useful in identifying impaired surface waters from point and nonpoint pollution, wet-weather combined sewer overflows, stormwater management issues, drinking-water source protection, and the development of best management practices. A comprehensive assessment of BASINS can be found in Whittemore and Beebe (2000). BASINS was developed and is maintained by the USEPA (www.epa.gov/ost/basins).

HSPF (Hydrological Simulation Program–Fortran) is the flagship model in the USEPA watershed-modeling system BASINS. The HSPF package comprises a series of computer codes designed to simulate watershed hydrology, land-surface runoff, and fate and transport of pollutants in receiving waters. The hydrologic portion of the model includes a hydrology model, a nonpoint-source estimation model, and a stream-routing component based on the kinematic-wave approximation. The water-quality portion of the model includes sediment-transport simulation (settling–deposition–scouring) for three classes of particle sizes. Adsorption–desorption processes are calculated separately for each particle class in both the water column and the sediments. Degradation-transformation processes include hydrolysis, photolysis, oxidation, volatization, and biodegradation. Kinetic reactions are simulated as second-order processes. Chemical reaction simulation includes up to two secondary chemicals for each primary component. The model is limited in application to one-dimensional systems in which the pollutants are well mixed over the entire section. Zero-dimensional representation (lakes and other impoundments) implies that pollutants are completely mixed throughout the water body and that the water body is not stratified. The model inputs are numerous and depend on the application. They may include one or more of the following groups:

- Time-series inputs of air temperature, rainfall, evapotranspiration, channel inflows, and wind

- Parameters such as channel geometry, soil moisture and vegetative conditions, infiltration/interflow data, overland sediment mobility, and washoff potential
- Water-quality constituents in interflow and ground water
- Agrichemical quality constituents data
- Reach/reservoir water-quality characteristics, coefficients, and kinetic rates

Typical outputs are pollutant concentrations. Yearly summaries include statistical analysis of time-varying contaminant concentrations that can be used for risk assessment.

The Agricultural Runoff Management Model (ARM) is a version of the HSPF model which can be run independently or included in HSPF. The model simulates runoff (including snow accumulation and melt), sediment, pesticides, and nutrient loadings from surface and subsurface sources (Donigian and Davis, 1978). The ARM model (in the form of the more complex HSPF model, which incorporates ARM) requires extensive calibration. The PC version is available from the EPA (www.epa.gov) only as a part of HSPF and BASINS.

The Chemicals, Runoff, and Erosion from Agricultural Management Systems Model (CREAMS) was developed by the U.S. Department of Agriculture (Knisel, 1980, 1985; Leonard and Knisel, 1986). CREAMS is a field-scale model that uses separate hydrology, erosion, and chemistry submodels connected by pass files. The hydrology component has two options, depending on the availability of rainfall data. Option 1 estimates storm runoff when only daily rainfall data are available. This is accomplished using the NRCS runoff curve number model. When hourly rainfall data are available, option 2 estimates runoff using the Green–Ampt equation to calculate the rainfall abstraction. The erosion component of the model considers the basic processes of soil detachment, transport, and deposition. Detachment of soil particles is modeled by the modified universal soil loss equation for a single storm event. The basic concepts for nutrient modeling treat their transport as proceeding separately in adsorbed (with sediment) and dissolved (with runoff) phases. Soil nitrate is lost with both surface and subsurface flows. Soil nitrogen is modified by nitrification–denitrification processes and by plant uptake. The pesticide component estimates concentrations of pesticides in runoff (water and sediment phases) and total mass carried from the field for each storm during the period of interest. Pesticides in runoff are partitioned between the solution and sediment phases using a simplified isotherm model. The model has the capability of simulating up to 20 quality components at one time. User-defined management practices can be simulated by CREAMS. These activities include aerial spraying (foliar or soil directed) or soil incorporation of pesticides, animal waste management, and agricultural best management practices.

The Groundwater Loading Effects of Agricultural Management Systems Model (GLEAMS) was developed by the U.S. Department of Agriculture to utilize the management-oriented CREAMS model. GLEAMS is essentially a vadose-zone component for CREAMS (Leonard et al., 1987). The soil column is divided into 3 to 12 layers of variable thickness in which pesticide and nutrient mass balance and routing is executed. The input data requirement for CREAMS–GLEAMS simulations are extensive and detailed. The maximum size of the watershed is limited to a field plot consisting of a maximum of three segments. The CREAMS–GLEAMS model is available for personal computers (PCs) and there is a very active users group, encompassing several hundred users throughout the world.

The Simulator for Water Resources in Rural Basins Model (SWRRB), developed from the CREAMS model is used to evaluate basin-scale quality in rural watersheds (Williams et al., 1985). SWRRB operates on a daily time step and simulates meteorology, hydrology, crop growth, sedimentation, floodplain degradation and aggradation, and nitrogen, phosphorus, and pesticide movement. The model was developed by modifying the CREAMS daily rainfall hydrology model for applications to large complex rural basins.

The Pesticide Runoff Simulator Model (PRS), developed from the CREAMS model, simulates pesticide runoff and adsorption onto the soil in a small agricultural watershed.

The Water Erosion Prediction Project Model (WEPP) is a new-generation water erosion model developed by the U.S. Department of Agriculture Soil Erosion Research Laboratory (Foster and Lane, 1987; Lane and Nearing, 1989; Laflen et al., 1991). It is a continuous-simulation model, although it can be run on a single-storm basis. By continuous simulation the model "mimics" the processes important to erosion prediction as a function of time and as affected by management decisions and the climatic environment. The output of the continuous simulation are time-integrated estimates of erosion. The model calculates both detachment and deposition; hence, the delivery process is considered. The output includes both on- and off-site erosion effects. The on-site effects of erosion include the time-integrated (average annual) soil loss over the area on the hill slope of the net soil loss, which is analogous (but not identical) to the universal soil loss equation estimates. The output also includes deposition over the areas on the hill slope with net deposition. The output describing off-site effects includes sediment loads and particle-size information. The output options also include the potential for obtaining monthly or daily (storm by storm) estimates of on- and off-site effects of erosion.

The Stormwater Management Model (SWMM) is regarded as the standard modeling tool for urban drainage-system evaluation. The basic SWMM code was developed and is maintained by the U.S. Environmental Protection Agency (USEPA). SWMM is used primarily to simulate the runoff from a continuous rainfall record and consists principally of a runoff module, a transport module, and a storage/treatment module. The catchment is separated into subareas that are connected to small gutter/pipe elements leading to an outfall location. The runoff module simulates runoff by modeling each subarea as a nonlinear reservoir with precipitation as input and infiltration, evaporation, and surface runoff as outflows. Infiltration is modeled by either the Green–Ampt or Horton equation; and infiltrated water is routed through upper and lower subsurface zones and may contribute to runoff. The transport module receives subarea runoff hydrographs and uses either kinematic or dynamic routing to calculate sewer flows and the hydraulic head at key junctions. The storage/treatment module simulates the effects of detention basins and combined-sewer overflows. Quality constituents modeled by SWMM include suspended solids, settleable solids, biochemical oxygen demand, nitrogen, phosphorus, and grease. A variety of options are provided for determining surface-runoff loads, including constant concentration, regression relationships of load versus flow, and buildup/wash-off. Many commercial versions of SWMM have added attractive output presentations, including elaborate graphics, with the basic USEPA SWMM as the main computing engine.

SUMMARY

Contaminant sources contributing to pollution of a water body are usually located in the watershed surrounding the water body. In cases where the water body does not meet its water-quality standards, the relationship between pollutant concentrations in the water body and the distribution, magnitude, and timing of point and nonpoint sources of pollution within the watershed must be quantified such that appropriate reductions in contaminant loadings can be made to bring the water body into compliance with water-quality standards. The allowable contaminant loadings, called total maximum daily loads

(TMDLs), must be allocated separately to point sources and nonpoint sources. The types of land uses that have the greatest impact on water quality are urban areas and agricultural areas. In urban areas, contaminant loads in surface runoff are commonly estimated using event-mean concentrations, and buildup and wash-off models are commonly used to quantify the accumulation and subsequent transport of pollutants in the urban area. Best management practices for controlling the quality of urban runoff include source-control measures such as street sweeping, hydrologic modifications such as decreasing directly connected impervious area, attenuation of pollutants such as by using filter strips, collection-system pollution control such as by using grassed waterways, and using detention–retention facilities. In agricultural areas, increased sediment load, nutrients, pesticides, pathogens, and salinity are all by-products of agricultural operations that have negative consequences on water quality. Many of the contaminants of concern in agricultural areas are adsorbed onto sediments, so contaminant transport models first estimate the sediment load and then contaminant transport is estimated from the sediment load based on the sorption characteristics of the contaminants. Sediment loads are commonly estimated using the universal soil loss equation, and contaminant sorption characteristics are described using potency factors. Retention of contaminants in soils is described by isotherm models. A wide variety of best management practices are used to control the fate and transport of contaminants on agricultural lands, including cropping practices, detention–retention ponds, animal waste storage and treatment, and riparian buffer zones. Several sophisticated computer models are available to simulate contaminant-transport processes on urban and agricultural watersheds.

PROBLEMS

9.1. Analysis of a water-quality impaired lake indicates that the maximum contaminant load consistent with meeting the applicable water-quality standard is 2000 kg/yr. Background loading of the contaminant is estimated as 200 kg/yr, and current NPDES permits for discharges into the water body allow an annual loading of 1400 kg/yr.

  (a) Determine the TMDL of the lake and the acceptable loading from nonpoint sources in the watershed.

  (b) If the current diffuse loading from nonpoint sources is estimated to be 700 kg/yr, recommend an equitable reduction in the load allocations. Assume a 15% factor of safety in the estimated loading capacity of the water body.

9.2. An urban (residential) area is estimated to be 45% impervious and to have an average annual rainfall of 700 mm, and it is estimated that 60% of the annual rainfall does not produce runoff. Assuming that the EMC distribution of phosphorus is normal, estimate the median and 90th percentile load of phosphorus contained in the runoff.

9.3. A 150-m segment of a 10-m-wide urban roadway drains into a nearby river. Field reconnaissance indicates that the roadway segment has a solids deposition rate of 15 g/m · day, a dry-deposition rate of 5 g/m^2 · day, a traffic emission rate of 0.1 g/axle · m, a traffic density of 100 axles/day, and a road condition factor of 0.5. The height of the curb is 15 cm, the average traffic speed is 80 km/h, and the average wind speed is 10 km/h. It is estimated that the urban washoff coefficient is 0.15 mm^{-1}. Five days before a recent 3-h storm produced 100 mm of runoff, the solids on the roadway segment was estimated to be 1 kg/m.

(**a**) Estimate the mass of solids entering the river from the storm runoff.

(**b**) What is the equilibrium level of solids on the roadway?

9.4. A 150-ha wheat farm in central Iowa consists of predominantly loamy sand soil with approximately 1% organic matter, and the ground surface has approximately a 1% slope. During the off-season, plowing is up and down the land slope. The annual runoff volume is 10 cm and the maximum runoff rate is 3 cm/h.

(**a**) If the typical overland flow distance from the beginning of overland flow to a drainage stream is 1.1 km, estimate the average annual soil loss.

(**b**) If the delivery ratio is 5%, what is the sediment load on the receiving water body?

9.5. The soil in a watershed has an average mercury concentration of 1.2 μg/kg, annual sediment load in the runoff of 7.2×10^5 kg, and a mean enrichment ratio of 3.5. Estimate the annual mercury loading from sediment-laden runoff into the receiving water body.

9.6. A 0.5-m^3 sample of soil is found to contain 32 g of carbofuran (an insecticide) and to have a water content of 0.2 and a bulk (dry) density of 1600 kg/m^3. If the soil has an estimated distribution coefficient of 0.3 mL/g, estimate the concentration of carbofuran in the pore water. The solubility of carbofuran is 700 mg/L.

9.7. Repeat Problem 9.6, accounting for vaporization of carbofuran by taking $K_H = 7.9 \times 10^6$ and a soil porosity of 0.25.

9.8. The half-life of carbofuran is estimated to be 35 days. If the pore-water concentration of carbofuran in a soil is found to be 5 mg/L, estimate how long it will take for the concentration of carbofuran to decrease to the drinking-water standard of 0.040 mg/L.

9.9. Fertilizer is applied to a field at an annual rate of 70 kg/ha and is plowed in and distributed uniformly to a depth of 0.2 m. The soil has a porosity of 0.25, a bulk density of 1500 kg/m^3, and consists of silt loam with 20% clay, 60% silt, and 20% sand. The pH of the pore water is 6.2, and the organic carbon content is 2%.

(**a**) Assuming that the pores are saturated with water, estimate the concentration of phosphorus in the pore water.

(**b**) What is the retention capacity of the soil for phosphorus?

(**c**) If the annual plant uptake of phosphorus is 60 kg/ha, estimate the time for the soil to be saturated with phosphorus.

9.10. A silt soil contains 2.5% organic matter and 0.15% total nitrogen measured as total Kjeldahl nitrogen, of which 75% is organic nitrogen.

(**a**) If the moisture content of the soil is 0.2, the porosity of the soil is 0.35, the wilting point of the soil is 10%, and the bulk density of the soil is 1500 kg/m^3, determine the amount of mobile and immobile nitrogen.

(**b**) If a 5-cm storm results in 2 cm of runoff and 2 tonnes/ha of soil loss, and the enrichment ratio for organic nitrogen is 2.0, estimate the nitrogen loss in the runoff. Assume that the runoff mixes with the soil moisture over a topsoil depth of 0.5 cm.

9.11. A broccoli farmer spreads 5000 tonnes/ha of alfalfa meal over his land each year. If alfalfa meal contains 6% nitrogen, is 70% organic, 10% of the nitrogen volatilizes, and 80% of the applied nitrogen is taken up by the broccoli crop, estimate the nitrogen loading on the ground water.

APPENDIX A

UNITS AND CONVERSION FACTORS

A.1 UNITS

The Système International d'Unités (International System of Units or SI system) was adopted by the 11th General Conference on Weights and Measures (CGPM) in 1960 and is now used by almost the entire world. In the SI system, all quantities are expressed in terms of seven base (fundamental) units. These base units and their standard abbreviations are as follows (Dean, 1985):

- *Meter* (m): distance light travels in a vacuum during 1/299,792,458 of a second.[1]
- *Kilogram* (kg): mass of a cylinder of platinum–iridium alloy kept in Paris.
- *Second* (s): duration of 9,192,631,770 cycles of the radiation corresponding to the transition between two hyperfine levels of the ground state of the cesium-133 atom.
- *Ampere* (A): magnitude of the current that when flowing through each of two long, parallel wires of negligible cross section separated by 1 meter in a vacuum results in a force between the two wires of 2×10^{-7} newton for each meter of length.
- *Kelvin* (K): defined in the thermodynamic scale by assigning 273.16 K to the triple point of water (freezing point, 273.16 K = 0°C).
- *Candela* (cd): luminous intensity of 1/600,000 of a square meter of a radiating cavity at the temperature of freezing platinum (2042 K).
- *Mole* (mol): amount of substance that contains as many specified entities (molecules, atoms, ions, electrons, photons, etc.) as there are atoms in exactly 0.012 kg of carbon 12.

---

[1]"Meter" is the accepted spelling in the United States; the rest of the world uses the spelling "metre."

Water-Quality Engineering in Natural Systems, by David A. Chin
Copyright © 2006 John Wiley & Sons, Inc.

TABLE A.1   SI Derived Units

Unit	Quantity	Symbol	In Terms of Base Units
becquerel	Activity of a radionuclide	Bq	s^{-1}
coulomb	Quantity of electricity, electric charge	C	$A \cdot s$
degree Celsius	Celsius temperature	°C	K
farad	Capacitance	F	C/V
gray	Absorbed dose of ionizing radiation	Gy	J/kg
henry	Inductance	H	Wb/A
hertz	Frequency	Hz	s^{-1}
joule	Energy, work, quantity of heat	J	$N \cdot m$
lumen	Luminous flux	lm	$cd \cdot sr$
lux	Illuminance	lx	lm/m^2
newton	Force	N	$kg \cdot m/s^2$
ohm	Electric resistance	Ω	V/A
pascal	Pressure, stress	Pa	N/m^2
siemens[a]	Conductance	S	A/V
sievert	Dose equivalent of ionizing radiation	Sv	J/kg
telsa	Magnetic flux density	T	Wb/m^2
volt	Electric potential, potential difference	V	W/A
watt	Power, radiant flux	W	J/s
weber	Magnetic flux	Wb	$V \cdot s$

Source: Wandmacher and Johnson (1995) and Szirtes (1998).

[a]The siemens was previously called the mho.

In addition to the seven base units of the SI system, there are two supplementary SI units: the radian and the steradian. The radian (rad) is defined as the angle at the center of a circle subtended by an arc equal in length to the radius, and the steradian (sr) is defined as the solid angle with its vertex at the center of a sphere that is subtended by an area of the spherical surface equal to the radius squared. The SI units should not be confused with the now obsolete *metric units*, which were developed in Napoleonic France over 200 years ago. The primary difference between metric and SI units is that the former uses centimeters and grams to measure length and mass, whereas these quantities are measured in meters and kilograms in SI units. The United States is gradually moving toward the use of SI units; however, there is still widespread use of the "English" system of units, also referred to as "U.S. Customary" or "British Gravitational" units.

In addition to the seven SI base units, there are several derived units that are given special names. These derived units are used for convenience rather than necessity and are listed in Table A.1.

A.2   CONVERSION FACTORS

In most cases, application of unit conversion factors results in converted numbers that have more significant digits than the original numbers. In these cases, the converted number should be rounded off such that rounding error is consistent with the rounding

TABLE A.2 Multiplicative Factors for Unit Conversion

Quantity	Convert From:	Convert To:	Multiply By:
Area	acre	ha	0.404687
	mi^2	km^2	2.59000
Energy	Btu	J	1054.350264
	cal	J	4.184^a
Energy/area	ly^b	kJ/m^2	41.84^a
Flow rate	cfs	m^3/s	0.02831685
	gpm^b	L/s	0.06309
	mgd^b	m^3/s	0.04381
		m^3/day	3785.412
Force	lbf	N	4.4482216152605^a
Length	ft	m	0.3048^a
	in.	m	0.0254^a
	mi (U.S. statute)	km	1.609344^a
	mi (U.S. nautical)	km	1.852000^a
	yd	m	0.9144^a
Mass	g	kg	0.001^a
	lbm	kg	0.45359237^a
Permeability	darcy	m^2	0.987×10^{-12}
Power	hp	W	745.69987
Pressure	atm	kPa	101.325^a
	bar	kPa	100.000^a
	mmHg (at 0°C)	kPa	0.133322
	psi	kPa	6.894757
	torr	kPa	0.133322
Speed	knot	m/s	0.514444444
Viscosity (dynamic)	cP	Pa·s	0.001^a
Viscosity (kinematic)	cS	m^2/s	10^{-6a}
Volume	gal	L	3.785411784^a

[a]Exact conversion.
[b]ly, langley; gpm, gallons per minute; mgd, million gallons per day.

error of the converted number (Taylor, 1995). Common conversion factors are listed in Table A.2.

Example A.1 The height of a water-control structure is reported as 19.3 feet. Convert this dimension to meters.

SOLUTION The conversion factor is given in Table A.2 as 1 foot = 0.3048 meter; hence,

$$19.3\,ft = 19.3(0.3048) = 5.88264\,m$$

Since 19.3 ft could have resulted from rounding any number between 19.25 and 19.35 ft, the maximum possible rounding error is $\pm 0.05/19.3 = \pm 0.26\%$. Similarly, rounding 5.88264 m to 5.88 m gives a maximum rounding error of $\pm 0.005/5.88 = \pm 0.085\%$, and rounding to 5.9 m gives an error of $\pm 0.05/5.9 = \pm 0.85\%$. Hence, accuracy is lost by taking 19.3 ft as 5.9 m, while 5.88 m is more accurate than is indicated by 19.3 ft. It is usually prudent not to discard accuracy, so take 19.3 ft = 5.88 m.

A good rule of thumb is that the converted number should have the same number of significant digits as the original number, assuming that the conversion factor is always more accurate than the original number.

APPENDIX B

FLUID PROPERTIES

B.1 WATER

TABLE B.1 Properties of Pure Water

Temperature (°C)	Density (kg/m³)	Dynamic Viscosity (mPa · s)	Heat of Vaporization (MJ/kg)	Saturation Vapor Pressure (kPa)	Surface Tension (mN/m)	Bulk Modulus (10^6 kPa)
0	999.8	1.781	2.499	0.611	75.6	2.02
5	1000.0	1.518	2.487	0.872	74.9	2.06
10	999.7	1.307	2.476	1.227	74.2	2.10
15	999.1	1.139	2.464	1.704	73.5	2.14
20	998.2	1.002	2.452	2.337	72.8	2.18
25	997.0	0.890	2.440	3.167	72.0	2.22
30	995.7	0.798	2.428	4.243	71.2	2.25
40	992.2	0.653	2.405	7.378	69.6	2.28
50	988.0	0.547	2.381	12.340	67.9	2.29
60	983.2	0.466	2.356	19.926	66.2	2.28
70	977.8	0.404	2.332	31.169	64.4	2.25
80	971.8	0.354	2.307	47.367	62.6	2.20
90	965.3	0.315	2.282	70.113	60.8	2.14
100	958.4	0.282	2.256	101.325	58.9	2.07

Source: Density, viscosity, surface tension, and bulk modulus data from Finnemore and Franzini (2002); heat of vaporization and saturation vapor pressure data from Viessman and Lewis (1996).

The properties given in Table B.1 are for pure water. Pure water seldom exists in nature, where the density of water can be influenced significantly by salinity, temperature, and

TABLE B.2 Properties of Organic Compounds Found in Contaminated Water

Compound	Formula	Molecular Weight	Boiling Point (°C)	Density at 20°C (kg/m³)	Dynamic Viscosity at 20°C (mPa·s)	Solubility in Water at 20°C (mg/L)	Sorption Coefficient, log K_{oc} (log mL/g)	Saturation Vapor Pressure at 20°C (kPa)	Henry's Law Constant at 20°C (Pa·m³/mol)
Acetone (dimethyl ketone)	CH_3COCH_3	58.08	56.1[a]–56.2[b]	790[b]	—	440,600[c]	−0.59[b]–0.34[d]	3.07[c]–35.97[d]	3.3[c]–10[l]
Benzene	C_6H_6	78.11	80.1[b]	870[l]–880[b]	—	1710[b]–1796[b]	1.39[b]–2.95[b]	1.20[a]–12.67[d]	458[b]–557[g]
Bis(2-ethylhexyl)phthalate (DEHP)	$C_{24}H_{38}O_4$	390.57	230[h]–387[h]	985[b]	—	0.041[b]–0.285[d]	3.77[d]–5.15[b]	1.09×10^{-9b}–26.7×10^{-9b}	0.0011[c]
Chlorobenzene	C_6H_5Cl	112.56	132[i]	1106[b]	0.8[i]	466[d]–500[i]	1.92[b]–2.63[k]	1.17[i]–1.2[b]	263[g]–288[b]
Chloroethane	CH_3CH_2Cl	64.5	—	897[b]–920[l]	—	5710[b]–5740[k]	0.51[b]–1.57[k]	133.3[i]–133.7[d]	1030[m]
Chloroform (trichloromethane)	$CHCl_3$	119.4	62[i]	1480[i]–1490[b]	0.56[i]	8000[b]–8200[b]	1.46[b]–1.94[b]	20.2[d]–25.9[i]	278[i]–486[g]
1,1-Dichloroethane	$C_2H_4Cl_2$	99.0	57.3[i]	1174[i]–1180[n]	0.5[i]	5100[i]–5500[b]	1.48[b]–1.80[k]	23.9[i]–29.5[i]	435[i]–550[i]
1,2-Dichloroethane	$C_2H_4Cl_2$	99.0	83.5[i]	1235[b]–1253[b]	0.84[i]	8000[k]–8800[b]	1.15[d]–1.57[k]	8.13[i]–10.9[i]	92[q]–152[i]
Ethylbenzene	C_8H_{10}	106.2	—	867[b]–870[n]	—	150[i]–152[n]	1.98[b]–3.13[k]	0.911[d]–1.28[e]	559[m]–882[q]
Gasoline[o]		—	—	680[p]–750[q]	0.29[p]–0.31[g]	—	—	55.2[p]	—
Methyl tert-butyl ether (MTBE)	$(CH_3)_3COCH_3$	88.2	—	740[r]	—	48,000[d]	0.55[i]–1.05[d]	32.7[d]	64.3[s]
Methylene chloride (dichloromethane)	CH_2Cl_2	84.9	—	1327[b]–1336[b]	—	13,000[n]–20,000[d]	0.94[b]	48.2[d]	229[b]
Naphthalene	$C_{10}H_8$	128.2	—	1030[n]–1162[b]	—	31.7[d]–38[b]	2.62[b]–5.00[b]	0.0113[c]–0.0307[d]	56.0[s]
Phenol	C_6H_6O	94.1	—	1058[b]–1070[b]	—	93,000[k]	1.15[d]–3.49[b]	0.027[q]–0.045[d]	0.030
Tetrachloroethene (PCE)	CCl_2CCl_2	165.8	121.4[i]	1623[i]–1630[i]	0.9[i]	149[b]–200[k]	2.25[b]–2.99[b]	1.87[i]–2.53[c]	1327[i]–1763[i]
Toluene (methylbenzene)	$C_6H_5CH_3$	92.1	—	867[b]–870[n]	—	500[i]–535[b]	1.57[b]–3.05[b]	2.93[a]–3.75[d]	529[m]–578[a]
1,1,1-Trichloroethane	CCl_3CH_3	133.4	—	1339[b]–1350[i]	0.84[i]	480[b]–4400[k]	2.02[b]–3.40[b]	13.3[i]–16.6[i]	1337[i]–1824[q]
Trichloroethene (TCE)	C_2HCl_3	131.5	86.7[a]	1460[n]–1460[i]	0.57[i]	1000[n]–1100[i]	1.66[b]–2.83[b]	7.70[a]–10.0[i]	722[i]–1013[q]
Vinyl chloride (chloroethene)	CH_2CHCl	62.5	—	910[a]–912[b]	—	90[i]–267[d]	0.39[b]–1.76[d]	355[d]–400[e]	2200[m]
o-Xylene (1,2-dimethylbenzene)	$C_6H_4(CH_3)_2$	106.2	—	880[b]	—	152[b]–175[n]	1.92[b]–2.92[d]	0.667[a]–0.912[d]	409[m]–537[q]
p-Xylene (1,4-dimethylbenzene)	$C_6H_4(CH_3)_2$	106.2	—	861[b]–881[b]	—	160[b]	2.31[b]–2.72[b]	—	555[m]

[a]From Sage and Sage (2000).
[b]From Montgomery (2000).
[c]From Montgomery (2000) at 25°C.
[d]From Charbeneau (2000).
[e]From USEPA (1996a).
[f]From Davis and Masten (2004).
[g]From Jackson et al. (1985).
[h]From Montgomery (2000) at 5 mmHg.
[i]From Pankow and Cherry (1996).
[j]From Pankow and Cherry (1996) at 25°C.
[k]From Schnoor (1996).
[l]From Fetter (1999).
[m]From Rathbun (1998).
[n]From Hemond and Fechner (1993).
[o]From Typical properties at 15.6°C. Properties of petroleum products vary.
[p]From Finnemore and Franzini (2002) at 20°C.
[q]From Munson et al. (1990).
[r]From USEPA (1998).
[s]From Rathbun (1998) at 25°C.
From Nyer et al. (1991).

possibly other properties through an equation of state. The general dependence of water density on temperature has been found to be approximately parabolic, with a maximum at 4°C. However, the temperature corresponding to the maximum density of water changes with increasing salinity, decreasing to about 0°C for highly saline systems. To a first-order approximation, density is linearly dependent on salinity over much of the normal range of interest.

B.2 ORGANIC COMPOUNDS FOUND IN WATER

The properties of several organic compounds commonly found in contaminated waters are given in Table B.2. Substances with boiling points at or below 100°C are classified as *volatile organic compounds* (VOCs). The chemical properties given in Table B.2 are those used most frequently in quantitative analysis of the fate of organic contaminants in the environment.

B.3 AIR AT STANDARD ATMOSPHERIC PRESSURE

See Table B.3.

TABLE B.3   Properties of Air at Standard Atmospheric Pressure

Temperature (°C)	Density (kg/m^3)	Dynamic Viscosity (mPa · s)	Specific Heat Ratio	Speed of Sound (m/s)
−40	1.514	0.0157	1.401	306.2
−20	1.395	0.0163	1.401	319.1
0	1.292	0.0171	1.401	331.4
5	1.269	0.0173	1.401	334.4
10	1.247	0.0176	1.401	337.4
15	1.225	0.0180	1.401	340.4
20	1.204	0.0182	1.401	343.3
25	1.184	0.0185	1.401	346.3
30	1.165	0.0186	1.400	349.1
40	1.127	0.0187	1.400	354.7
50	1.109	0.0195	1.400	360.3
60	1.060	0.0197	1.399	365.7
70	1.029	0.0203	1.399	371.2
80	0.9996	0.0207	1.399	376.6
90	0.9721	0.0214	1.398	381.7
100	0.9461	0.0217	1.397	386.9
200	0.7461	0.0253	1.390	434.5
300	0.6159	0.0298	1.379	476.3
400	0.5243	0.0332	1.368	514.1
500	0.4565	0.0364	1.357	548.8
1000	0.2772	0.0504	1.321	694.8

Source: Derived from Blevins (1984).

APPENDIX C

U.S. WATER-QUALITY STANDARDS

C.1 WATER-QUALITY CRITERIA FOR SURFACE WATERS

Water-quality criteria recommended by the U.S. Environmental Protection Agency can be found in USEPA (2002a). See the summaries in Tables C.1 to C.3.

TABLE C.1 Water-Quality Criteria for Priority Toxic Pollutants[a]

	Freshwater		Saltwater		Human Health for the Consumption of:	
Priority Pollutant	CMC (μg/L)	CCC (μg/L)	CMC (μg/L)	CCC (μg/L)	Water + Organism (μg/L)	Organism Only (μg/L)
Antimony	—	—	—	—	5.6	640
Arsenic	340	150	69	36	0.018	0.14
Beryllium	—	—	—	—	—	—
Cadmium	2.0	0.25	40	8.8	—	—
Chromium(III)	570	74	—	—	—	—
Chromium(IV)	16	11	1100	50	—	—
Copper	65	2.5	210	8.1	—	—
Mercury	1.4	0.77	1.8	0.94	—	—
Methyl mercury	—	—	—	—	—	0.3 mg/kg
Nickel	470	52	74	8.2	610	4,600
Selenium	—	5.0	290	71	170	4,200
Silver	3.2	—	—	1.9	—	—
Thallium	—	—	—	—	1.7	6.3

(continued)

TABLE C.1 (*continued*)

Priority Pollutant	Freshwater		Saltwater		Human Health for the Consumption of:	
	CMC (μg/L)	CCC (μg/L)	CMC (μg/L)	CCC (μg/L)	Water + Organism (μg/L)	Organism Only (μg/L)
Zinc	120	120	90	81	7400	26,000
Cyanide	22	5.2	1	1	700	220,000
Asbestos	—	—	—	—	7×10^6 fibers/L	—
2,3,7,8-TCDD (dioxin)	—	—	—	—	5.0×10^{-9}	5.1×10^{-9}
Acrolein	—	—	—	—	190	290
Acrylonirile	—	—	—	—	0.051	0.25
Benzene	—	—	—	—	2.2	51
Bromoform	—	—	—	—	4.3	140
Carbon tetrachloride	—	—	—	—	0.23	1.6
Chlorobenzene	—	—	—	—	680	21,000
Chlorodibromomethane	—	—	—	—	0.40	13
Chloroethane	—	—	—	—	—	—
2-Chloroethylvinyl ether	—	—	—	—	—	—
Chloroform	—	—	—	—	5.7	470
Dichlorobromomethane	—	—	—	—	0.55	17
1,1-Dichloroethane	—	—	—	—	—	—
1,2-Dichloroethane	—	—	—	—	0.38	37
1,1-Dichloroethylene	—	—	—	—	0.057	3.2
1,2-Dichloropropane	—	—	—	—	0.50	15
1,3-Dichloropropene	—	—	—	—	10	1,700
Ethylbenzene	—	—	—	—	3,100	29,000
Methyl bromide	—	—	—	—	47	1,500
Methyl chloride	—	—	—	—	—	—
Methylene chloride	—	—	—	—	4.6	590
1,1,2,2-Tetrachloroethane	—	—	—	—	0.17	4.0
Tetrachloroethylene	—	—	—	—	0.69	3.3
Toluene	—	—	—	—	6,800	200,000
1,2-*trans*-Dichloroethylene	—	—	—	—	700	140,000
1,1,1-Trichloroethane	—	—	—	—	—	—
1,1,2-Trichloroethane	—	—	—	—	0.59	16
Trichloroethylene	—	—	—	—	2.5	30
Vinyl chloride	—	—	—	—	2.0	530
2-Chlorophenol	—	—	—	—	81	150
2,4-Dichlorophenol	—	—	—	—	77	290
2,4-Dimethylphenol	—	—	—	—	380	850
2-Methyl-4,6-dinitrophenol	—	—	—	—	13	280
2,4-Dinitrophenol	—	—	—	—	69	5,300
2-Nitrophenol	—	—	—	—	—	—
4-Nitrophenol	—	—	—	—	—	—

(*continued*)

TABLE C.1 (*continued*)

3-Methyl-4-chlorophenol	—	—	—	—	—	—
Pentachlorophenol	19	15	13	7.9	0.27	3.0
Phenol	—	—	—	—	21,000	1,700,000
2,4,6-Trichlorophenol	—	—	—	—	1.4	2.4
Acenaphthene	—	—	—	—	670	990
Acenaphthylene	—	—	—	—	—	—
Anthracene	—	—	—	—	8,300	40,000
Benzidine	—	—	—	—	0.000086	0.00020
Benzo[*a*]anthracene	—	—	—	—	0.0038	0.018
Benzo[*a*]pyrene	—	—	—	—	0.0038	0.018
Benzo[*b*]fluoranthene	—	—	—	—	0.0038	0.018
Benzo[*ghi*]perylene	—	—	—	—	—	—
Benzo[*k*]fluoranthene	—	—	—	—	0.0038	0.018
Bis(2-chloroethoxy) methane	—	—	—	—	—	—
Bis(2-chloroethyl)ether	—	—	—	—	0.030	0.53
Bis(2-chloroisopropyl) ether	—	—	—	—	1,400	65,000
Bis(2-ethlyhexyl) phthalate	—	—	—	—	1.2	2.2
4-Bromophenyl phenyl ether	—	—	—	—	—	—
Butylbenzyl phthalate	—	—	—	—	1,500	1,900
2-Chloronapthalene	—	—	—	—	1,000	1,600
4-Chlorophenyl phenyl ether	—	—	—	—	—	—
Chrysene	—	—	—	—	0.0038	0.018
Dibenzo[*a,h*]anthracene	—	—	—	—	0.0038	0.018
1,2-Dichlorobenzene	—	—	—	—	2,700	17,000
1,3-Dichlorobenzene	—	—	—	—	320	960
1,4-Dichlorobenzene	—	—	—	—	400	2,600
3,3-Dichlorobenzidine	—	—	—	—	0.021	0.028
Diethyl phthalate	—	—	—	—	17,000	44,000
Dimethyl phthalate	—	—	—	—	270,000	1,100,000
Di-*n*-butyl phthalate	—	—	—	—	2,000	4,500
2,4-Dinitrotoluene	—	—	—	—	0.11	3.4
2,6-Dinitrotoluene	—	—	—	—	—	—
Di-*n*-octyl phthalate	—	—	—	—	—	—
1,2-Diphenylhydrazine	—	—	—	—	0.036	0.20
Fluoranthene	—	—	—	—	130	140
Fluorene	—	—	—	—	1,100	5,300
Hexachlorobenzene	—	—	—	—	0.00028	0.00029
Hexachlorobutadiene	—	—	—	—	0.44	18
Hexachlorocyclo-pentadiene	—	—	—	—	240	17,000
Hexachloroethane	—	—	—	—	1.4	3.3
Ideno[1,2,3-*cd*]pyrene	—	—	—	—	0.0038	0.018
Isophorone	—	—	—	—	35	960

(*continued*)

TABLE C.1 (*continued*)

Priority Pollutant	Freshwater		Saltwater		Human Health for the Consumption of:	
	CMC (μg/L)	CCC (μg/L)	CMC (μg/L)	CCC (μg/L)	Water + Organism (μg/L)	Organism Only (μg/L)
Naphthalene	—	—	—	—	—	—
Nitrobenzene	—	—	—	—	17	690
N-Nitrosodiethylamine	—	—	—	—	0.00069	3.0
N-Nitrosodi-n-propylamine	—	—	—	—	0.0050	0.51
N-Nitrosodiphenylamine	—	—	—	—	3.3	6.0
Phenanthrene	—	—	—	—	—	—
Pyrene	—	—	—	—	830	4,000
1,2,4-trichlorobenzene	—	—	—	—	260	940
Aldrin	3.0	—	1.3	—	0.000049	0.000050
α-BHC	—	—	—	—	0.0026	0.0049
β-BHC	—	—	—	—	0.0091	0.017
γ-BHC (lindane)	0.95	—	0.16	—	0.019	0.063
δ-BHC	—	—	—	—	—	—
Chlordane	2.4	0.0043	0.09	0.004	0.00080	0.00081
4,4'-DDT	1.1	0.001	0.13	0.001	0.00022	0.00022
4,4'-DDE	—	—	—	—	0.00022	0.00022
4,4'-DDD	—	—	—	—	0.00031	0.00031
Dieldrin	0.24	0.056	0.71	0.0019	0.000052	0.000054
α-Endosulfan	0.22	0.056	0.034	0.0087	62	89
β-Endosulfan	0.22	0.056	0.034	0.0087	62	89
Endosulfan sulfate	—	—	—	—	62	89
Endrin	0.086	0.036	0.037	0.0023	0.76	0.81
Endrin aldehyde	—	—	—	—	0.29	0.30
Heptachlor	0.52	0.0038	0.053	0.0036	0.000079	0.000079
Heptachlor epoxide	0.52	0.0038	0.053	0.0036	0.000039	0.000039
Polychlorinated biphenyls (PCBs)	—	0.014	—	0.03	0.000064	0.000064
Toxaphene	0.73	0.0002	0.21	0.0002	0.00028	0.00028

[a]CMC, criteria for maximum concentration; CCC, criteria for continuous concentration.

TABLE C.2   Water-Quality Criteria for Nonpriority Pollutants[a]

Non priority Pollutant	Freshwater		Saltwater		Human Health for the Consumption of:	
	CMC (μg/L)	CCC (μg/L)	CMC (μg/L)	CCC (μg/L)	Water + Organism (μg/L)	Organism Only (μg/L)
Alkalinity	—	20,000	—	—	—	—
Aluminum, pH 6.5–9.0	750	87	—	—	—	—
Ammonia	Freshwater criteria are pH, temperature, and life-stage dependent; see EPA-822-R-99-014. Saltwater criteria are pH and temperature dependent; see EPA-440-5-88-004.					
Aesthetic qualities	Narrative statement; see Gold Book.					
Bacteria	For primary recreation and shellfish uses, see Gold Book.					
Barium	—	—	—	—	1,000	—
Boron	Narrative statement; see Gold Book.					
Chloride	860000	230,000	—	—	—	—
Chlorine	19	11	13	7.5	—	—
Chlorophenoxy herbicide (2,4,5-TP)	—	—	—	—	10	—
Chlorophenoxy herbicide (2,4-D)	—	—	—	—	100	—
Chloropyrifos	0.083	0.041	0.011	0.0056	—	—
Color	Narrative statement; see Gold Book.					
Demeton	—	0.1	—	0.1	—	—
Ether, bis (chloromethyl)	—	—	—	—	0.00010	0.00029
Gases, total dissolved	Narrative statement; see Gold Book.					
Guthion	—	0.01	—	0.01	—	—
Hardness	Narrative statement; see Gold Book.					
Hexachloro-cyclohexane, technical	—	—	—	—	0.0123	0.0414
Iron	—	1,000	—	—	300	—
Malathion	—	0.1	—	0.1	—	—
Manganese	—	—	—	—	50	100
Methoxychlor	—	0.03	—	0.03	100	—
Mirex	—	0.001	—	0.001	—	—
Nitrates	—	—	—	—	10,000	—
Dinitrophenols	—	—	—	—	0.0008	1.24
Nitrosamines	—	—	—	—	69	5300

(*continued*)

TABLE C.2   (*continued*)

Non priority Pollutant	Freshwater		Saltwater		Human Health for the Consumption of:	
	CMC (μg/L)	CCC (μg/L)	CMC (μg/L)	CCC (μg/L)	Water + Organism (μg/L)	Organism Only (μg/L)
N-Nitrosodibutyl-amine	—	—	—	—	0.0063	0.22
N-Nitrosopyr-rolidine	—	—	—	—	0.016	34
Oil and grease Narrative statement; see Gold Book.						
Oxygen, dissolved Warmwater and coldwater matrix; see Gold Book. (freshwater)						
Oxygen, dissolved Saltwater, see Gold Book. (saltwater)						
Parathion	0.065	0.013	—	—	—	—
Pentachloro-benzene	—	—	—	—	1.4	1.5
pH	—	6.5–9	—	6.5–8.5	5–9	—
Phosphorus elemental	—	—	—	0.1	—	—
Nutrients See USEPA's ecoregional criteria for total phosphorus, total nitrogen, chlorophyll *a*, and water clarity (Secchi depth for lakes; turbidity for rivers and streams).						
Solids dissolved and salinity	—	—	—	—	250,000	—
Solids suspended Narrative statement; see Gold Book. and turbidity						
Sulfide–hydrogen sulfide	—	2.0	—	2.0	—	—
Tainting substances Narrative statement; see Gold Book.						
Temperature Species-dependent criteria; see Gold Book.						
Tetrachloro-benzene	—	—	—	—	0.97	1.1
Tributyltin (TBT)	0.46	0.063	0.37	0.010	—	—
2,4,5-Trichloro-phenol	—	—	—	—	1,800	3,600

[a]CMC, criteria for maximum concentration; CCC, criteria for continuous concentration.

TABLE C.3   Water-Quality Criteria for Organoleptic Effects

Pollutant	Organoleptic Effect Criteria (μg/L)
Acenaphthene	20
Monochlorobenzene	20
3-Chlorophenol	0.1
4-Chlorophenol	0.1
2,3-Dichlorophenol	0.04
2,5-Dichlorophenol	0.5
2,6-Dichlorophenol	0.2
3,4-Dichlorophenol	0.3
2,4,5-Trichlorophenol	1
2,4,6-Trichlorophenol	2
2,3,4,6-Tetrachlorophenol	1
2-Methyl-4-chlorophenol	1800
3-Methyl-4-chlorophenol	3000
3-Methyl-6-chlorophenol	20
2-Chlorophenol	0.1
Copper	1000
2,4-Dichlorophenol	0.3
2,4-Dimethylphenol	400
Hexachlorocyclopentadiene	1
Nitrobenzene	30
Pentachlorophenol	30
Phenol	300
Zinc	5000

C.2   WATER-QUALITY CRITERIA FOR DRINKING WATER

Drinking-water quality in the United States is regulated by the U.S. Environmental Protection Agency (USEPA) under the Safe Drinking Water Act (SDWA). Primary drinking-water standards are currently set for 92 contaminants. These include turbidity, 8 microbial or indicator organisms, 4 radionuclides, 20 inorganic contaminants, and 60 organic contaminants. Maximum contaminant levels (MCLs) have been set for 83 contaminants, and nine have treatment-technique requirements. MCLs and treatment-technique requirements are enforceable by USEPA, whereas MCL goals (MCLGs) are health-based goals that are not enforceable. Secondary standards are recommended for 15 contaminants to ensure the aesthetic quality of drinking water, although a few states have adopted them as enforceable standards. The primary drinking-water standards given in Table C.4 are current as of April 2006. Secondary standards are given in Table C.5.

TABLE C.4 U.S. Primary Drinking-Water Standards

Contaminant	MCLG[a] (mg/L)	MCL[b] (mg/L)	Treatment[c]
Inorganic Substances			
Antimony	0.006	0.006	C–F, RO
Arsenic	0	0.01	AA, C–F, EDR, IX, LS, RO
Asbestos	7 MFL[d]	7 MFL	CC, C–F, DEF, DF, IX, RO
Barium	2	2	IX, LS, RO
Beryllium	0.004	0.004	AA, IX, RO, LS
Bromate	0	0.010	DC
Cadmium	0.005	0.005	C–F, IX, LS, RO
Chlorite	0.8	1.0	DC
Chromium (total)	0.1	0.1	C–F, IX, LS, RO
Copper	1.3	TT[e]	CC, SWT
Cyanide (as free cyanide)	0.2	0.2	CL, IX, RO
Fluoride	4.0	4.0	AA, RO
Lead	0	TT	CC, LSLR, PE, SWT
Mercury (inorganic)	0.002	0.002	C–F, GAC, LS, RO
Nickel	0.1	0.1	IX, LS, RO
Nitrate (as N)	10	10	ED, IX, RO
Nitrite (as N)	1	1	IX, RO
Nitrate + nitrite (as N)	10	10	IX, RO
Selenium	0.05	0.05	AA, C–F, ED
Thallium	0.0005	0.002	AA, IX
Organic Substances			
Acrylamide	0	TT	PAP
Alachlor	0	0.002	GAC
Benzene	0	0.005	GAC, PTA
Benzo[a]pyrene	0	0.0002	GAC
Carbofuran	0.04	0.04	GAC
Carbon tetrachloride	0	0.005	GAC, PTA
Chlordane	0	0.002	GAC
2,4-D	0.07	0.07	GAC
Dalapon	0.2	0.2	GAC
Di(2-ethylhexyl)adipate	0.4	0.4	GAC, PTA
Di(2-ethylhexyl)phthalate	0	0.006	GAC
1,2-Dibromo-3-chloropropane (DBCP)	0	0.0002	GAC, PTA
o-Dichlorobenzene	0.6	0.6	GAC, PTA
p-Dichlorobenzene	0.075	0.075	GAC, PTA
1,2-Dichloroethane	0	0.005	GAC, PTA
1,1-Dichloroethylene	0.007	0.007	GAC, PTA
cis-1,2-Dichloroethylene	0.07	0.07	GAC, PTA
trans-1,2-Dichloroethylene	0.1	0.1	GAC, PTA
Dichloromethane	0	0.005	PTA
1,2-Dichloropropane	0	0.005	GAC, PTA
Dinoseb	0.007	0.007	GAC
Diquat	0.02	0.02	GAC

(*continued*)

TABLE C.4   *(continued)*

Endothall	0.1	0.1	GAC
Endrin	0.002	0.002	GAC
Epichlorohydrin	0	TT	PAP
Ethylbenzene	0.7	0.7	GAC, PTA
Ethlylene dibromide	0	0.00005	GAC, PTA
Glyphosate	0.7	0.7	OX
HAA5[f]	—	0.060	EC
Heptachlor	0	0.0004	GAC
Heptachlor epoxide	0	0.0002	GAC
Hexachlorobenzene	0	0.001	GAC
Hexachlorocyclopentadiene	0.05	0.05	GAC, PTA
Lindane	0.0002	0.0002	GAC
Methoxychlor	0.04	0.04	GAC
Monochlorobenzene	0.1	0.1	GAC, PTA
Oxamyl (Vydate)	0.2	0.2	GAC
Pentachlorophenol	0	0.001	GAC
Picloram	0.5	0.5	GAC
Polychlorinated biphenlys (PCBs)	0	0.0005	GAC
Simazine	0.004	0.004	GAC
Styrene	0.1	0.1	GAC, PTA
2,3,7,8-TCDD (dioxin)	0	5×10^{-8}	GAC
Tetrachloroethylene	0	0.005	GAC, PTA
Toluene	1	1	GAC, PTA
Toxaphene	0	0.005	GAC
2,4,5-TP (silvex)	0.05	0.05	GAC
1,2,4-Trichlorobenzene	0.07	0.07	GAC, PTA
1,1,1-Trichloroethane	0.2	0.2	GAC, PTA
1,1,2-Trichloroethane	0.003	0.005	GAC, PTA
Trichloroethylene	0	0.005	GAC, PTA
Vinyl chloride	0	0.002	PTA
Xylenes (total)	10	10	GAC, PTA
Total trihalomethanes (TTHMs)	—	0.08	AD, PR, SPC, EC

Radionuclides

Beta particles and photon emitters	0	4 mrem/yr	IX, RO
Gross alpha-particle activity	0	15 pCi/L	RO
Radium 226 and 228	0	5 pCi/L	LS, IX, RO
Uranium	0	30 µg/L	LS, RO, EC, IX

Microorganisms

Cryptosporidium	0	TT	C–F, SSF, DEF, DF, D
Escherechi coli	0	TT	D
Fecal coliforms	0	TT	D
Giardia lambia	0	TT	C–F, SSF, DEF, DF, D

(continued)

TABLE C.4 (*continued*)

Contaminant	MCLG[a] (mg/L)	MCL[b] (mg/L)	Treatment[c]
Heterotrophic bacteria	N/A	TT	C–F, SSF, DEF, DF, D
Legionella	0	TT	C–F, SSF, DEF, DF, D
Total coliforms	0	[g]	D
Turbidity	N/A	TT	C–F, SSF, DEF, DF, D
Viruses	0	TT	C–F, SSF, DEF, DF, D

[a]Maximum contaminant level goal.

[b]Maximum contaminant level.

[c]AA, activated alumina; AD, alternative disinfectants; CC, corrosion control; C–F, coagulation and filtration; CL, chlorination; D, disinfection; DC, disinfection system control; DEF, diatomaceous earth filtration; DF, direct filtration; EC, enhanced coagulation; ED, electrodialysis; EDR, electrodialysis reversal; GAC, granular activated carbon; IX, ion exchange; LS, lime softening; LSLS, lead service line replacement; OX, oxidation; PAP, polymer addition practices; PE, public education; PR, precursor removal; PTA, packed tower aeration; RO, reverse osmosis; SPC, stop prechlorination; SSF, slow sand filtration; SWT, source-water treatment.

[d]Million fibers per liter longer than $10\,\mu m$.

[e]Treatment technique requirement. Action level for copper is 1.3 mg/L, lead is 0.015 mg/L, acrylamide is 0.05% dosed at 1 mg/L, epichlorohydrin is 0.01% dosed at 20 mg/L *Giardia lamblia* is 99.9% killed/inactivated, viruses is 99.99% killed/inactivated, *Legionella* is *Giardia* and virus inactivation, turbidity is 1 NTU, and heterotrophic plate count is 500 bacterial colonies per milliliter.

[f]Sum of five haloacetic acids—sum of concentrations of mono-, di-, and trichloroacetic acids and mono- and dibromoacetic acids.

[g]No more than 5% of the samples per month may be positive. For systems collecting fewer than 40 samples per month, no more than 1 sample per month may be positive.

TABLE C.5   U.S. Secondary Drinking-Water Standards

Contaminant	Secondary Standard	Effect
Aluminum	0.05–0.2 mg/L	Colored water
Chloride	250 mg/L	Salty taste
Color	15 color units	Visible tint
Copper	1.0 mg/L	Metallic taste, blue-green stain
Corrosivity	Noncorrosive	Metallic taste, corrosion, fixture staining
Fluoride	2.0 mg/L	Tooth decoloration
Foaming agents	0.5 mg/L	Frothy, cloudy, bitter taste, odor
Iron	0.3 mg/L	Rusty color, sediment, metallic taste, reddish or orange staining
Manganese	0.05 mg/L	Black to brown color, black staining, bitter metallic taste
Odor	3 threshold odor number	"Rotten egg," musty, or chemical smell
pH	6.5–8.5	Low pH: bitter metallic taste, corrosion; high pH: slippery feel, soda taste, deposits
Silver	0.10 mg/L	Skin decoloration, graying of the white of the eye
Sulfate	250 mg/L	Salty taste
Total dissolved solids	500 mg/L	Hardness, deposits, colored water, staining, salty taste
Zinc	5 mg/L	Metallic taste

C.3   PRIORITY POLLUTANTS

The U.S. Environmental Protection Agency's list of priority pollutants, subdivided according to method of analysis is as follows:

Base-Neutral Extractables

Acenaphthene
Acenaphthylene
Anthracene
Benzidine
Benzo[*a*]anthracene
Benzo[*b*]fluoranthene
Benzo[*k*]fluoranthene
Benzo[*ghi*]perylene
Benzo[*a*]pyrene
Bis(2-chloro-
 ethoxy)methane
Bis(2-chloroethyl)ether
Bis(2-chloroisopropyl)ether
Bis(2-ethylhexyl)phthalate
4-Bromophenyl phenyl
 ether
Butyl benzyl phthalate

2-Chloronaphthalene
4-Chlorophenyl phenyl
 ether
Chrysene
Dibenzo[*a,h*]anthracene
Di-*n*-butyl phthalate
1,2-Dichlorobenzene
1,3-Dichlorobenzene
1,4-Dichlorobenzene
3,3'-Cichlorobenzidine
Diethyl phthalate
Dimethyl phthalate
2,4-Dinitrotoluene
2,6-Dinitrotoluene
Di-*n*-octyl phthalate
1,2-Diphenylhydrazine
Fluoranthene

Fluorene
Hexachlorobenzene
Hexachlorobutadiene
Hexachlorocyclopentadiene
Hexachloroethane
Indeno[1,2,3-*cd*]pyrene
Isophorone
Naphthalene
Nitrobenzene
N-Nitrosodimethylamine
N-Nitrosodiphenylamine
N-Nitrosodi-*n*-propylamine
Phenanthrene
Pyrene
2,3,7,8-Tetrachlorodibenzo-
 p-dioxin
1,2,4-Trichlorobenzene

Acid Extractables

P-Chloro-*m*-cresol
2-Chlorophenol
2,4-Dichlorophenol
2,4-Dimethylphenol

4,6-Dinitro-*o*-cresol
2,4-Dinitrophenol
2-Nitrophenol
4-Nitrophenol

Pentachlorophenol
Phenol
2,4,6-Trichlorophenol
Total phenols

Volatiles

Acrolein
Acrylonitrile
Benzene
Bis(chloromethyl)ether
Bromodichloromethane
Bromoform
Bromomethane
Carbon tetrachloride
Chlorobenzene
Chloroethane
2-Chloroethyl vinyl ether

Chloroform
Chloromethane
Dibromochloromethane
Dichlorodifluoromethane
1,1-Dichloroethane
1,2-Dichloroethane
1,1-Dichloroethylene
trans-1,2-Dichloroethylene
1,2-Dichloropropane
cis-1,3-Dichloropropene
trans-1,3-Dichloropropene

Ethylbenzene
Methylene chloride
1,1,2,2-Tetrachloroethane
Tetrachloroethene
Toluene
1,1,1-Trichloroethane
1,1,2-Trichloroethane
Trichloroethylene
Trichlorofluoromethane
Vinyl chloride

Pesticides

Aldrin	Dieldrin	PCB-1221
α-BHC	α-Endosulfan	PCB-1232
β-BHC	β-Endosulfan	PCB-1242
γ-BHC	Endosulfan sulfate	PCB-1248
δ-BHC	Endrin	PCB-1254
Chlordane	Endrin aldehyde	PCB-1260
4,4'-DDD	Heptachlor	Toxaphene
4,4'-DDE	Heptachlor epoxide	
4,4'-DDT	PCB-1016	

Note: PCB compounds are not pesticides.

Inorganics

Antimony	Chromium	Nickel
Arsenic	Copper	Selenium
Asbestos	Cyanide	Silver
Beryllium	Lead	Thallium
Cadmium	Mercury	Zinc

APPENDIX D

STATISTICAL TABLES

D.1 AREAS UNDER THE STANDARD NORMAL CURVE

See Table D.1

TABLE D.1 Areas Under the Standard Normal Curve

z	0.00	0.01	0.02	0.03	0.04	0.05	0.06	0.07	0.08	0.09
−4.0	.0000	.0000	.0000	.0000	.0000	.0000	.0000	.0000	.0000	.0000
−3.9	.0000	.0000	.0000	.0000	.0000	.0000	.0000	.0000	.0000	.0000
−3.8	.0001	.0001	.0001	.0001	.0001	.0001	.0001	.0001	.0001	.0001
−3.7	.0001	.0001	.0001	.0001	.0001	.0001	.0001	.0001	.0001	.0001
−3.6	.0002	.0002	.0001	.0001	.0001	.0001	.0001	.0001	.0001	.0001
−3.5	.0002	.0002	.0002	.0002	.0002	.0002	.0002	.0002	.0002	.0002
−3.4	.0003	.0003	.0003	.0003	.0003	.0003	.0003	.0003	.0003	.0002
−3.3	.0005	.0005	.0005	.0004	.0004	.0004	.0004	.0004	.0004	.0003
−3.2	.0007	.0007	.0006	.0006	.0006	.0006	.0006	.0005	.0005	.0005
−3.1	.0010	.0009	.0009	.0009	.0008	.0008	.0008	.0008	.0007	.0007
−3.0	.0013	.0013	.0013	.0012	.0012	.0011	.0011	.0011	.0010	.0010
−2.9	.0019	.0018	.0018	.0017	.0016	.0016	.0015	.0015	.0014	.0014
−2.8	.0026	.0025	.0024	.0023	.0023	.0022	.0021	.0021	.0020	.0019
−2.7	.0035	.0034	.0033	.0032	.0031	.0030	.0029	.0028	.0027	.0026
−2.6	.0047	.0045	.0044	.0043	.0041	.0040	.0039	.0038	.0037	.0036
−2.5	.0062	.0060	.0059	.0057	.0055	.0054	.0052	.0051	.0049	.0048
−2.4	.0082	.0080	.0078	.0075	.0073	.0071	.0069	.0068	.0066	.0064

(continued)

Water-Quality Engineering in Natural Systems, by David A. Chin
Copyright © 2006 John Wiley & Sons, Inc.

TABLE D.1 (*continued*)

z	0.00	0.01	0.02	0.03	0.04	0.05	0.06	0.07	0.08	0.09
−2.3	.0107	.0104	.0102	.0099	.0096	.0094	.0091	.0089	.0087	.0084
−2.2	.0139	.0136	.0132	.0129	.0125	.0122	.0119	.0116	.0113	.0110
−2.1	.0179	.0174	.0170	.0166	.0162	.0158	.0154	.0150	.0146	.0143
−2.0	.0228	.0222	.0217	.0212	.0207	.0202	.0197	.0192	.0188	.0183
−1.9	.0287	.0281	.0274	.0268	.0262	.0256	.0250	.0244	.0239	.0233
−1.8	.0359	.0351	.0344	.0336	.0329	.0322	.0314	.0307	.0301	.0294
−1.7	.0446	.0436	.0427	.0418	.0409	.0401	.0392	.0384	.0375	.0367
−1.6	.0548	.0537	.0526	.0516	.0505	.0495	.0485	.0475	.0465	.0455
−1.5	.0668	.0655	.0643	.0630	.0618	.0606	.0594	.0582	.0571	.0559
−1.4	.0808	.0793	.0778	.0764	.0749	.0735	.0721	.0708	.0694	.0681
−1.3	.0968	.0951	.0934	.0918	.0901	.0885	.0869	.0853	.0838	.0823
−1.2	.1151	.1131	.1112	.1093	.1075	.1056	.1038	.1020	.1003	.0985
−1.1	.1357	.1335	.1314	.1292	.1271	.1251	.1230	.1210	.1190	.1170
−1.0	.1587	.1562	.1539	.1515	.1492	.1469	.1446	.1423	.1401	.1379
−0.9	.1841	.1814	.1788	.1762	.1736	.1711	.1685	.1660	.1635	.1611
−0.8	.2119	.2090	.2061	.2033	.2005	.1977	.1949	.1922	.1894	.1867
−0.7	.2420	.2389	.2358	.2327	.2297	.2266	.2236	.2206	.2177	.2148
−0.6	.2743	.2709	.2676	.2643	.2611	.2578	.2546	.2514	.2483	.2451
−0.5	.3085	.3050	.3015	.2981	.2946	.2912	.2877	.2843	.2810	.2776
−0.4	.3446	.3409	.3372	.3336	.3300	.3264	.3228	.3192	.3156	.3121
−0.3	.3821	.3783	.3745	.3707	.3669	.3632	.3594	.3557	.3520	.3483
−0.2	.4207	.4168	.4129	.4090	.4052	.4013	.3974	.3936	.3897	.3859
−0.1	.4602	.4562	.4522	.4483	.4443	.4404	.4364	.4325	.4286	.4247
0.0	.5000	.5040	.5080	.5120	.5160	.5199	.5239	.5279	.5319	.5359
0.1	.5398	.5438	.5478	.5517	.5557	.5596	.5636	.5675	.5714	.5753
0.2	.5793	.5832	.5871	.5910	.5948	.5987	.6026	.6064	.6103	.6141
0.3	.6179	.6217	.6255	.6293	.6331	.6368	.6406	.6443	.6480	.6517
0.4	.6554	.6591	.6628	.6664	.6700	.6736	.6772	.6808	.6844	.6879
0.5	.6915	.6950	.6985	.7019	.7054	.7088	.7123	.7157	.7190	.7224
0.6	.7257	.7291	.7324	.7357	.7389	.7422	.7454	.7486	.7517	.7549
0.7	.7580	.7611	.7642	.7673	.7704	.7734	.7764	.7794	.7823	.7852
0.8	.7881	.7910	.7939	.7967	.7995	.8023	.8051	.8078	.8106	.8133
0.9	.8159	.8186	.8212	.8238	.8264	.8289	.8315	.8340	.8365	.8389
1.0	.8413	.8438	.8461	.8485	.8508	.8531	.8554	.8577	.8599	.8621
1.1	.8643	.8665	.8686	.8708	.8729	.8749	.8770	.8790	.8810	.8830
1.2	.8849	.8869	.8888	.8907	.8925	.8944	.8962	.8980	.8997	.9015
1.3	.9032	.9049	.9066	.9082	.9099	.9115	.9131	.9147	.9162	.9177
1.4	.9192	.9207	.9222	.9236	.9251	.9265	.9279	.9292	.9306	.9319
1.5	.9332	.9345	.9357	.9370	.9382	.9394	.9406	.9418	.9429	.9441
1.6	.9452	.9463	.9474	.9484	.9495	.9505	.9515	.9525	.9535	.9545
1.7	.9554	.9564	.9573	.9582	.9591	.9599	.9608	.9616	.9625	.9633
1.8	.9641	.9649	.9656	.9664	.9671	.9678	.9686	.9693	.9699	.9706
1.9	.9713	.9719	.9726	.9732	.9738	.9744	.9750	.9756	.9761	.9767
2.0	.9772	.9778	.9783	.9788	.9793	.9798	.9803	.9808	.9812	.9817
2.1	.9821	.9826	.9830	.9834	.9838	.9842	.9846	.9850	.9854	.9857
2.2	.9861	.9864	.9868	.9871	.9875	.9878	.9881	.9884	.9887	.9890
2.3	.9893	.9896	.9898	.9901	.9904	.9906	.9909	.9911	.9913	.9916

(*continued*)

TABLE D.1 (*continued*)

2.4	.9918	.9920	.9922	.9925	.9927	.9929	.9931	.9932	.9934	.9936
2.5	.9938	.9940	.9941	.9943	.9945	.9946	.9948	.9949	.9951	.9952
2.6	.9953	.9955	.9956	.9957	.9959	.9960	.9961	.9962	.9963	.9964
2.7	.9965	.9966	.9967	.9968	.9969	.9970	.9971	.9972	.9973	.9974
2.8	.9974	.9975	.9976	.9977	.9977	.9978	.9979	.9979	.9980	.9981
2.9	.9981	.9982	.9982	.9983	.9984	.9984	.9985	.9985	.9986	.9986
3.0	.9987	.9987	.9987	.9988	.9988	.9989	.9989	.9989	.9990	.9990
3.1	.9990	.9991	.9991	.9991	.9992	.9992	.9992	.9992	.9993	.9993
3.2	.9993	.9993	.9994	.9994	.9994	.9994	.9994	.9995	.9995	.9995
3.3	.9995	.9995	.9995	.9996	.9996	.9996	.9996	.9996	.9996	.9997
3.4	.9997	.9997	.9997	.9997	.9997	.9997	.9997	.9997	.9997	.9998
3.5	.9998	.9998	.9998	.9998	.9998	.9998	.9998	.9998	.9998	.9998
3.6	.9998	.9998	.9999	.9999	.9999	.9999	.9999	.9999	.9999	.9999
3.7	.9999	.9999	.9999	.9999	.9999	.9999	.9999	.9999	.9999	.9999
3.8	.9999	.9999	.9999	.9999	.9999	.9999	.9999	.9999	.9999	.9999
3.9	1.0000	1.0000	1.0000	1.0000	1.0000	1.0000	1.0000	1.0000	1.0000	1.0000
4.0	1.0000	1.0000	1.0000	1.0000	1.0000	1.0000	1.0000	1.0000	1.0000	1.0000

APPENDIX E

SPECIAL FUNCTIONS

E.1 ERROR FUNCTION

The error function, erf(z), is defined by the relation

$$\text{erf}(z) = \frac{2}{\sqrt{\pi}} \int_0^z e^{-x^2}\, dx \qquad (E.1)$$

The error function is closely related to the cumulative distribution function of a normal probability distribution and is defined for $-\infty \leq z \leq \infty$. The error function is antisymmetric, such that

$$\text{erf}(-z) = -\text{erf}(z) \qquad (E.2)$$

The constant before the integral sign in Equation E.1 is simply a normalizing constant such that erf(z) approaches 1 as z approaches infinity. For small values of z, it is convenient to use the series expansion for e^{-x^2} to obtain (Carslaw and Jaeger, 1959)

$$\text{erf}(z) = \frac{2}{\sqrt{\pi}} \int_0^z \left[\sum_{n=0}^{\infty} \frac{(-1)^n x^{2n}}{n!} \right] dx \qquad (E.3)$$

Since the series is uniformly convergent, it can be integrated term by term to yield

$$\text{erf}(z) = \frac{2}{\sqrt{\pi}} \left[\sum_{n=0}^{\infty} \frac{(-1)^n z^{2n+1}}{(2n+1)n!} \right] \qquad (E.4)$$

Water-Quality Engineering in Natural Systems, by David A. Chin
Copyright © 2006 John Wiley & Sons, Inc.

TABLE E.1 Error Function

z	erf(z)	z	erf(z)	z	erf(z)	z	erf(z)
0.0	0.00000	0.8	0.74210	1.6	0.97635	2.4	0.99931
0.1	0.11246	0.9	0.79691	1.7	0.98379	2.5	0.99959
0.2	0.22270	1.0	0.84270	1.8	0.98909	2.6	0.99976
0.3	0.32863	1.1	0.88021	1.9	0.99279	2.7	0.99987
0.4	0.42839	1.2	0.91031	2.0	0.99532	2.8	0.99992
0.5	0.52050	1.3	0.93401	2.1	0.99702	2.9	0.99996
0.6	0.60386	1.4	0.95229	2.2	0.99814	3.0	0.99998
0.7	0.67780	1.5	0.96611	2.3	0.99886	∞	1.00000

which can also be written in the form (Hermance, 1999)

$$\text{erf}(z) = \frac{2}{\sqrt{\pi}}\left(z - \frac{z^3}{3} + \frac{z^5}{2!5} - \frac{z^7}{3!7} + \cdots\right) \tag{E.5}$$

This relationship is particularly useful in estimating erf(z) for small values of z. A closely related function to the error function is the *complementary error function*, erfc(z), which is defined by

$$\text{erfc}(z) = 1 - \text{erf}(z) \tag{E.6}$$

Values of the error function, erf(z), are listed in Table E.1.

E.2 BESSEL FUNCTIONS

E.2.1 Definition

A second-order linear homogeneous differential equation of the form

$$x^2\frac{d^2y}{dx^2} + x\frac{dy}{dx} + (x^2 - n^2)y = 0, \qquad n \geq 0 \tag{E.7}$$

is called *Bessel's equation*. The general solutions of Bessel's equation are

$$y = AJ_n(x) + BJ_{-n}(x), \qquad n \neq 1, 2, \ldots \tag{E.8}$$

$$y = AJ_n(x) + BY_n(x), \qquad \text{all } n \tag{E.9}$$

where $J_n(x)$ is called the *Bessel function of the first kind of order n* and $Y_n(x)$ is called the *Bessel function of the second kind of order n*.

If Bessel's equation (Equation E.7) is slightly modified and written in the form

$$x^2\frac{d^2y}{dx^2} + x\frac{dy}{dx} - (x^2 + n^2)y = 0, \qquad n \geq 0 \tag{E.10}$$

this equation is called the *modified Bessel's equation*. The general solutions of the modified Bessels equation are

$$y = AI_n(x) + BI_{-n}(x), \qquad n \neq 1, 2, \ldots \tag{E.11}$$

$$y = AI_n(x) + BK_n(x), \qquad \text{all } n \tag{E.12}$$

where $I_n(x)$ is called the *modified Bessel function of the first kind of order n*, and $K_n(x)$ is called the *modified Bessel function of the second kind of order n*.

E.2.2 Evaluation of Bessel Functions

The solution of Bessel's equations can be found in most calculus texts [e.g., Hildebrand (1976)]. The Bessel functions cannot generally be expressed in closed form and are usually presented as infinite series.

***Bessel Function of the First Kind of Order* n**   This function is given by

$$J_n(x) = \frac{x^n}{2^n \Gamma(n+1)} \left[1 - \frac{x^2}{2(2n+2)} + \frac{x^4}{2 \cdot 4 (2n+2)(2n+4)} - \cdots \right]$$

$$= \sum_{k=0}^{\infty} \frac{(-1)^k (x/2)^{n+2k}}{k! \, \Gamma(n+k+1)} \tag{E.13}$$

The Bessel function $J_{-n}(x)$ can be derived from Equation E.13 simply by replacing n by $-n$ in the formula. A convenient relationship to note is

$$J_{-n}(x) = (-1)^n J_n(x), \qquad n = 0, 1, 2, \ldots \tag{E.14}$$

***Bessel Function of the Second Kind of Order* n**   This function is given by

$$Y_n(x) = \begin{cases} \dfrac{J_n(x) \cos n\pi - J_{-n}(x)}{\sin n\pi}, & n \neq 0, 1, 2, \ldots \\[3mm] \lim_{p \to n} \dfrac{J_p(x) \cos p\pi - J_{-p}(x)}{\sin p\pi}, & n = 0, 1, 2, \ldots \end{cases} \tag{E.15}$$

***Modified Bessel Function of the First Kind of Order* n**   This function is given by

$$I_n(x) = \frac{x^n}{2^n \Gamma(n+1)} \left[1 + \frac{x^2}{2(2n+2)} + \frac{x^4}{2 \cdot 4 (2n+2)(2n+4)} + \cdots \right]$$

$$= \sum_{k=0}^{\infty} \frac{(x/2)^{n+2k}}{k! \, \Gamma(n+k+1)} \tag{E.16}$$

The Bessel function $I_{-n}(x)$ can be derived from Equation E.16 simply by replacing n by $-n$ in the formula. A convenient relationship to note is

$$I_{-n}(x) = I_n(x), \qquad n = 0, 1, 2, \ldots \tag{E.17}$$

Modified Bessel Function of the Second Kind of Order **n** This function is given by

$$K_n(x) = \begin{cases} \dfrac{\pi}{2\sin n\pi}[I_{-n}(x) - I_n(x)], & n \neq 0, 1, 2, \ldots \\[2ex] \lim_{p \to n} \dfrac{\pi}{2\sin p\pi}[I_{-p}(x) - I_p(x)], & n = 0, 1, 2, \ldots \end{cases} \tag{E.18}$$

Tabulated Values of Useful Bessel Functions See Table E.2.

TABLE E.2 Useful Bessel Functions

x	$I_0(x)$	$K_0(x)$	$I_1(x)$	$K_1(x)$
0.001	1.0000	7.0237	0.0005	999.9962
0.002	1.0000	6.3305	0.0010	499.9932
0.003	1.0000	5.9251	0.0015	333.3237
0.004	1.0000	5.6374	0.0020	249.9877
0.005	1.0000	5.4143	0.0025	199.9852
0.006	1.0000	5.2320	0.0030	166.6495
0.007	1.0000	5.0779	0.0035	142.8376
0.008	1.0000	4.9443	0.0040	124.9782
0.009	1.0000	4.8266	0.0045	111.0871
0.010	1.0000	4.7212	0.0050	99.9739
0.020	1.0001	4.0285	0.0100	49.9547
0.030	1.0002	3.6235	0.0150	33.2715
0.040	1.0004	3.3365	0.0200	24.9233
0.050	1.0006	3.1142	0.0250	19.9097
0.060	1.0009	2.9329	0.0300	16.5637
0.070	1.0012	2.7798	0.0350	14.1710
0.080	1.0016	2.6475	0.0400	12.3742
0.090	1.0020	2.5310	0.0450	10.9749
0.100	1.0025	2.4271	0.0501	9.8538
0.110	1.0030	2.3333	0.0551	8.9353
0.120	1.0036	2.2479	0.0601	8.1688
0.130	1.0042	2.1695	0.0651	7.5192
0.140	1.0049	2.0972	0.0702	6.9615
0.150	1.0056	2.0300	0.0752	6.4775
0.160	1.0064	1.9674	0.0803	6.0533
0.170	1.0072	1.9088	0.0853	5.6784
0.180	1.0081	1.8537	0.0904	5.3447
0.190	1.0090	1.8018	0.0954	5.0456
0.200	1.0100	1.7527	0.1005	4.7760
0.210	1.0111	1.7062	0.1056	4.5317
0.220	1.0121	1.6620	0.1107	4.3092
0.230	1.0133	1.6199	0.1158	4.1058
0.240	1.0145	1.5798	0.1209	3.9191

(*continued*)

TABLE E.2 (*continued*)

0.250	1.0157	1.5415	0.1260	3.7470
0.260	1.0170	1.5048	0.1311	3.5880
0.270	1.0183	1.4697	0.1362	3.4405
0.280	1.0197	1.4360	0.1414	3.3033
0.290	1.0211	1.4036	0.1465	3.1755
0.300	1.0226	1.3725	0.1517	3.0560
0.310	1.0242	1.3425	0.1569	2.9441
0.320	1.0258	1.3136	0.1621	2.8390
0.330	1.0274	1.2857	0.1673	2.7402
0.340	1.0291	1.2587	0.1725	2.6470
0.350	1.0309	1.2327	0.1777	2.5591
0.360	1.0327	1.2075	0.1829	2.4760
0.370	1.0345	1.1832	0.1882	2.3973
0.380	1.0364	1.1596	0.1935	2.3227
0.390	1.0384	1.1367	0.1987	2.2518
0.400	1.0404	1.1145	0.2040	2.1844
0.410	1.0425	1.0930	0.2093	2.1202
0.420	1.0446	1.0721	0.2147	2.0590
0.430	1.0468	1.0518	0.2200	2.0006
0.440	1.0490	1.0321	0.2254	1.9449
0.450	1.0513	1.0129	0.2307	1.8915
0.460	1.0536	0.9943	0.2361	1.8405
0.470	1.0560	0.9761	0.2415	1.7916
0.480	1.0584	0.9584	0.2470	1.7447
0.490	1.0609	0.9412	0.2524	1.6997
0.500	1.0635	0.9244	0.2579	1.6564
0.510	1.0661	0.9081	0.2634	1.6149
0.520	1.0688	0.8921	0.2689	1.5749
0.530	1.0715	0.8766	0.2744	1.5364
0.540	1.0742	0.8614	0.2800	1.4994
0.550	1.0771	0.8466	0.2855	1.4637
0.560	1.0800	0.8321	0.2911	1.4292
0.570	1.0829	0.8180	0.2967	1.3960
0.580	1.0859	0.8042	0.3024	1.3638
0.590	1.0889	0.7907	0.3080	1.3328
0.600	1.0920	0.7775	0.3137	1.3028
0.610	1.0952	0.7646	0.3194	1.2738
0.620	1.0984	0.7520	0.3251	1.2458
0.630	1.1017	0.7397	0.3309	1.2186
0.640	1.1051	0.7277	0.3367	1.1923
0.650	1.1084	0.7159	0.3425	1.1668
0.660	1.1119	0.7043	0.3483	1.1420
0.670	1.1154	0.6930	0.3542	1.1181
0.680	1.1190	0.6820	0.3600	1.0948
0.690	1.1226	0.6711	0.3659	1.0722
0.700	1.1263	0.6605	0.3719	1.0503
0.710	1.1301	0.6501	0.3778	1.0290
0.720	1.1339	0.6399	0.3838	1.0083
0.730	1.1377	0.6300	0.3899	0.9882

(*continued*)

TABLE E.2 (*continued*)

x	$I_0(x)$	$K_0(x)$	$I_1(x)$	$K_1(x)$
0.740	1.1417	0.6202	0.3959	0.9686
0.750	1.1456	0.6106	0.4020	0.9496
0.760	1.1497	0.6012	0.4081	0.9311
0.770	1.1538	0.5920	0.4142	0.9130
0.780	1.1580	0.5829	0.4204	0.8955
0.790	1.1622	0.5740	0.4266	0.8784
0.800	1.1665	0.5653	0.4329	0.8618
0.810	1.1709	0.5568	0.4391	0.8456
0.820	1.1753	0.5484	0.4454	0.8298
0.830	1.1798	0.5402	0.4518	0.8144
0.840	1.1843	0.5321	0.4581	0.7993
0.850	1.1889	0.5242	0.4646	0.7847
0.860	1.1936	0.5165	0.4710	0.7704
0.870	1.1984	0.5088	0.4775	0.7564
0.880	1.2032	0.5013	0.4840	0.7428
0.890	1.2080	0.4940	0.4905	0.7295
0.900	1.2130	0.4867	0.4971	0.7165
0.910	1.2180	0.4796	0.5038	0.7039
0.920	1.2231	0.4727	0.5104	0.6915
0.930	1.2282	0.4658	0.5171	0.6794
0.940	1.2334	0.4591	0.5239	0.6675
0.950	1.2387	0.4524	0.5306	0.6560
0.960	1.2440	0.4459	0.5375	0.6447
0.970	1.2494	0.4396	0.5443	0.6336
0.980	1.2549	0.4333	0.5512	0.6228
0.990	1.2604	0.4271	0.5582	0.6122
1.000	1.2661	0.4210	0.5652	0.6019
1.100	1.3262	0.3656	0.6375	0.5098
1.200	1.3937	0.3185	0.7147	0.4346
1.300	1.4693	0.2782	0.7973	0.3725
1.400	1.5534	0.2437	0.8861	0.3208
1.500	1.6467	0.2138	0.9817	0.2774
1.600	1.7500	0.1880	1.0848	0.2406
1.700	1.8640	0.1655	1.1963	0.2094
1.800	1.9896	0.1459	1.3172	0.1826
1.900	2.1277	0.1288	1.4482	0.1597
2.000	2.2796	0.1139	1.5906	0.1399
2.100	2.4463	0.1008	1.7455	0.1227
2.200	2.6291	0.0893	1.9141	0.1079
2.300	2.8296	0.0791	2.0978	0.0950
2.400	3.0493	0.0702	2.2981	0.0837
2.500	3.2898	0.0623	2.5167	0.0739
2.600	3.5533	0.0554	2.7554	0.0653
2.700	3.8417	0.0493	3.0161	0.0577
2.800	4.1573	0.0438	3.3011	0.0511
2.900	4.5027	0.0390	3.6126	0.0453
3.000	4.8808	0.0347	3.9534	0.0402

(*continued*)

TABLE E.2   (*continued*)

3.100	5.2945	0.0310	4.3262	0.0356
3.200	5.7472	0.0276	4.7343	0.0316
3.300	6.2426	0.0246	5.1810	0.0281
3.400	6.7848	0.0220	5.6701	0.0250
3.500	7.3782	0.0196	6.2058	0.0222
3.600	8.0277	0.0175	6.7927	0.0198
3.700	8.7386	0.0156	7.4357	0.0176
3.800	9.5169	0.0140	8.1404	0.0157
3.900	10.3690	0.0125	8.9128	0.0140
4.000	11.3019	0.0112	9.7595	0.0125
5.000	27.2399	0.0037	24.3356	0.0040
6.000	67.2344	0.0012	61.3419	0.0013
7.000	168.5939	0.0004	156.0391	0.0005
8.000	427.5641	0.0001	399.8731	0.0002
9.000	1093.5880	0.0001	1030.9150	0.0001
10.000	2815.7170	0.0000	2670.9880	0.0000

E.3 Gamma Function

See Table E.3.

TABLE E.3   Gamma Function

z	$\Gamma(z)$	z	$\Gamma(z)$	z	$\Gamma(z)$
1.00	1.00000	1.34	0.89222	1.68	0.90500
1.01	0.99433	1.35	0.89115	1.69	0.90678
1.02	0.98884	1.36	0.89018	1.70	0.90864
1.03	0.98355	1.37	0.88931	1.71	0.91057
1.04	0.97844	1.38	0.88854	1.72	0.91258
1.05	0.97350	1.39	0.88785	1.73	0.91467
1.06	0.96874	1.40	0.88726	1.74	0.91683
1.07	0.96415	1.41	0.88676	1.75	0.91906
1.08	0.95973	1.42	0.88636	1.76	0.92137
1.09	0.95546	1.43	0.88604	1.77	0.92376
1.10	0.95135	1.44	0.88581	1.78	0.92623
1.11	0.94740	1.45	0.88566	1.79	0.92877
1.12	0.94359	1.46	0.88560	1.80	0.93138
1.13	0.93993	1.47	0.88563	1.81	0.93408
1.14	0.93642	1.48	0.88575	1.82	0.93685
1.15	0.93304	1.49	0.88595	1.83	0.93969
1.16	0.92980	1.50	0.88623	1.84	0.94261
1.17	0.92670	1.51	0.88659	1.85	0.94561
1.18	0.92373	1.52	0.88704	1.86	0.94869
1.19	0.92089	1.53	0.88757	1.87	0.95184

(*continued*)

TABLE E.3 (*continued*)

z	$\Gamma(z)$	z	$\Gamma(z)$	z	$\Gamma(z)$
1.20	0.91817	1.54	0.88818	1.88	0.95507
1.21	0.91558	1.55	0.88887	1.89	0.95838
1.22	0.91311	1.56	0.88964	1.90	0.96177
1.23	0.91075	1.57	0.89049	1.91	0.96523
1.24	0.90852	1.58	0.89142	1.92	0.96877
1.25	0.90640	1.59	0.89243	1.93	0.97240
1.26	0.90440	1.60	0.89352	1.94	0.97610
1.27	0.90250	1.61	0.89468	1.95	0.97988
1.28	0.90072	1.62	0.89592	1.96	0.98374
1.29	0.89904	1.63	0.89724	1.97	0.98768
1.30	0.89747	1.64	0.89864	1.98	0.99171
1.31	0.89600	1.65	0.90012	1.99	0.99581
1.32	0.89464	1.66	0.90167	2.00	1.00000
1.33	0.89338	1.67	0.90330		

APPENDIX F

PIPE SPECIFICATIONS

F.1 PVC PIPE

Pipe dimensions of interest to engineers are usually the diameter and wall thickness. The pipe diameter is typically specified by the *nominal pipe size*, and the wall thickness is usually specified by the *schedule*. Nominal pipe sizes are typically given in either inches or millimeters and represent rounded approximations to the inside diameter of the pipe. The schedule of a pipe is a number that approximates the value of the expression $1000P/S$, where P is the service pressure and S is the allowable stress. Higher schedule numbers correspond to thicker pipes, and schedule numbers in common use are 5, 5S, 10, 10S, 20, 20S, 30, 40, 40S, 60, 80, 80S, 100, 120, 140, and 160. The schedule numbers followed by the letter "S" are intended primarily for use with stainless steel pipe (ASME 1985). PVC-pipe dimensions are given in Table F.1, and physical properties of PVC and other pipe materials are given in Table F.2.

F.2 DUCTILE IRON PIPE

Ductile iron pipe is manufactured in diameters from 100 to 1200 mm (4 to 48 in.), and for diameters from 100 to 500 mm (4 to 20 in.) standard commercial sizes are available in increments of 50 mm (2 in.), while for diameters from 600 to 1200 mm (24 to 48 in.) the size increments are 150 mm (6 in.). The standard lengths of ductile iron pipe are 5.5 m (18 ft) and 6.1 m (20 ft).

TABLE F.1 PVC Pipe Dimensions

Nominal Pipe Size (mm)	Outside Diameter (mm)	Schedule 5		Schedule 10		Schedule 40		Schedule 80	
		Wall Thickness (mm)	Inside Diameter (mm)	Wall Thickness (mm)	Inside Diameter (mm)	Wall Thickness (mm)	Inside Diameter (mm)	Wall Thickness (mm)	Inside Diameter (mm)
50	60	1.7	57	2.8	55	3.9	53	5.5	49
80	90	2.1	85	3.0	83	5.5	78	7.6	74
100	114	2.1	110	3.0	108	6.0	102	8.6	97
125	141	2.8	136	3.4	134	6.6	128	9.5	122
150	168	2.8	163	3.4	161	7.1	154	11.0	146
2	2.375	0.065	2.245	0.109	2.157	0.154	2.067	0.218	1.939
3	3.500	0.083	3.334	0.120	3.260	0.216	3.068	0.300	2.900
4	4.500	0.083	4.334	0.120	4.260	0.237	4.026	0.337	3.826
5	5.563	0.109	5.345	0.134	5.295	0.258	5.047	0.375	4.813
6	6.625	0.109	6.407	0.134	6.357	0.280	6.065	0.432	5.761

Source: Fetter (1999).

F.3 CONCRETE PIPE

Common sizes of concrete pipe are listed in Table F.3.

TABLE F.2 Physical Properties of Common Pipe Materials

Material	Young's Modulus, E (GPa)	Poisson's Ratio
Concrete	14–30	0.10–0.15
Concrete (reinforced)	30–60	—
Ductile iron	165–172	0.28–0.30
PVC	2.4–3.5	0.45–0.46
Steel	200–207	0.30

TABLE F.3 Commercially Available Sizes of Concrete Pipe

Nonreinforced Pipe		Reinforced Pipe	
Diameter (mm)	Diameter (in.)	Diameter (mm)	Diameter (in.)
100	4	—	—
150	6	—	—
205	8	—	—
255	10	—	—
305	12	305	12
380	15	380	15
455	18	455	18
535	21	535	21
610	24	610	24
685	27	685	27
760	30	760	30
840	33	840	33
915	36	915	36
—	—	1065	42
—	—	1220	48
—	—	1370	54
—	—	1525	60
—	—	1675	66
—	—	1830	72
—	—	1980	78
—	—	2135	84
—	—	2285	90
—	—	2440	96
—	—	2590	102
—	—	2745	108

REFERENCES

Aber, J.S., Autumn colors of Estonian peat bogs, http://www.geospectra.net/kite/estonia/color/color.htm, 2001.

———, Algal mat on Devils Lake, http://www.emporia.edu/kas/trans100/sympos96/aber/algae.htm, 2002.

Abernathy, A.R., *Oxygen-Consuming Organics in Nonpoint Source Runoff*, Technical Report EPA-600-S3-81-033, U.S. Environmental Protection Agency, Washington, DC, 1981.

Abriola, L.M., and S.A. Bradford, Experimental investigations of the entrapment and persistence of organic liquid contaminants in the subsurface environment, *Environmental Health Perspectives Supplements*, 106(S4):1083–1095, August 1998.

ACF Environmental, Inc., Stormwater solutions, http://www.acfenvironmental.com/bmp_siltfence.htm, 2005.

Ackers, P., Sediment aspects of drainage and outfall design, in J.H.W. Lee and Y.K. Cheung, editors, *Environmental Hydraulics*, Volume 1, pages 34–42, A.A. Balkema, Rotterdam, The Netherlands, 1991.

Adeel, Z., J.W. Mercer, and C.R. Faust, Models for describing multiphase flow and transport of contaminants, in J.J. Kaluarachchi, editor, *Groundwater Contamination by Organic Pollutants: Analysis and Remediation*, ASCE Manuals and Reports on Engineering Practice 100, pages 1–39, American Society of Civil Engineers, Reston, Virginia, 2000.

Alberta Geological Survey, Deep injection of acid gas (H_2S) in western Canada, http://www.ags.gov.ab.ca/activities/CO2/acidgas.shtml, 2005.

Alberts, E.E., G.E. Schuman, and R.E. Burwell, Seasonal runoff losses of nitrogen and phosphorus from Missouri Valley loess watersheds, *Journal of Environmental Quality*, 7:203–208, 1978.

Aller, L., T.W. Bennett, G. Hackett, R.J. Petty, J.H. Lehr, H. Sedoris, D.M. Nielsen, and J.E. Denne, *Handbook of Suggested Practices for the Design and Installation of Ground-Water Monitoring Wells*, National Water Well Association, Dublin, Ohio, 1989.

Allied Biological, Inc., Nutrient inactivation: representative project, http://www.alliedbiological.com/Nutrient%20Inactivation.html, 2005.

Al-Suwaiyan, M.S., Nondeterministic evaluation of field-scale dispersivity relation, *Journal of Hydrologic Engineering*, 3(3):215–217, 1998.

Ambrose, R.B., T.A. Wool, J.L. Martin, J.P. Connolly, and R.W. Schanz, *WASP5.x, A Hydrodynamic and Water Quality Model: Model Theory, User's Manual, and Programmer's Guide*, Technical Report, Environmental Research Laboratory, U.S. Environmental Protection Agency, Athens, Georgia, 1992.

American Academy of Environmental Engineers, Excellence in environmental engineering 2005, *Environmental Engineer*, 41(2):21, 2005.

American Petroleum Institute, *A Guide to the Assessment and Remediation of Underground Petroleum Releases*, Publication 1628, API Marketing Department, Washington, DC, 1989.

Anderson, M.P., Using models to simulate the movement of contaminants through groundwater flow systems, *Critical Reviews in Environmental Controls*, 9(2):97–156, 1979.

APHA (American Public Health Association), *Standard Methods for the Examination of Water and Wastewater*, 18th ed., APHA, Washington, DC, 1992.

ASCE (American Society of Civil Engineers), *Quality of Ground Water: Guidelines for Selection and Application of Frequently Used Models*, Manual of Practice 85, ASCE, New York, 1996.

Ashworth, R.A., G.B. Howe, M.E. Mullins, and T.N. Rogers, Air–water partitioning coefficients of organics in dilute aqueous solutions, *Journal of Hazardous Materials*, 18:25–36, 1988.

ASME (American Society of Mechanical Engineers), *Stainless Steel Pipe*, ASME B36.19M, ASME, New York, 1985.

Assabet River Stream Watch, Riffle along Elizabeth Brook, Stow, http://www.assabetriver.org/streamwatch/howindex_b.html, 2005a.

———, Pool in Elizabeth Brook, http://www.assabetriver.org/streamwatch/howindex_b.html, 2005b.

ASTM (American Society for Testing and Materials), *Standard Guide for Risk-Based Corrective Action Applied at Petroleum Release Sites*, Technical Report E 1739-95, ASTM, Philadelphia, 1995.

———, *Remediation of Ground Water by Natural Attenuation at Petroleum Release Sites*, Technical Report E 1943-98, ASTM, Philadelphia, 1998a.

———, *Standard Provisional Guide for Risk-Based Corrective Action*, Technical Report PS 104-98, ASTM, Philadelphia, 1998b.

Atlas, R.M., *Microbiology: Fundamentals and Applications*, 2nd ed., Macmillan, New York, 1988.

Australia Waters and Rivers Commission, River restoration case studies: south west case study, http://www.wrc.wa.gov.au/protect/waterways/casestudies_sw.html, 2005.

AWWA (American Water Works Association), *Water Quality and Treatment*, 4th ed., McGraw-Hill, New York, 1990.

———, *Water Quality: Principles and Practices of Water Supply Operations*, 3rd ed., AWWA, Denver, Colorado, 2003.

Baehr, A.L., and M.Y. Corapcioglu, A compositional multiphase model for groundwater contamination by petroleum products, 2: Numerical solution, *Water Resources Research*, 23(1):201–214, 1987.

Baily, R.G., *Description of the Ecoregions of the United States*, Miscellaneous Publication 1391, U.S. Department of Agriculture, Washington, DC, March 1995.

Bakr, A.A., Stochastic analysis of the effects of spatial variations of hydraulic conductivity on groundwater flow, Ph.D. dissertation, New Mexico Institute of Mining and Technology, Socorro, New Mexico, 1976.

Bales, J.D., C.J. Oblinger, and A.H. Sallenger, Jr., *Two Months of Flooding in Eastern North Carolina, September–October 1999: Hydrologic Water-Quality, and Geologic Effects of Hurricanes Dennis, Floyd, and Irene*, Water-Resources Investigations Report 00-4093, U.S. Geological Survey, Raleigh, North Carolina, 2000.

Banerjee, P., M.D. Piwoni, and K. Ebeid, Sorption of organic contaminants to a low carbon subsurface core, *Chemosphere*, 8:1057–1067, 1985.

Bannerman, R., K. Baun, P. Hughes, and D. Graczyk, *Nationwide Urban Runoff Program, Milwaukee, Wisconsin: Evaluation of Urban Nonpoint Source Pollution Management in Milwaukee County, Wisconsin*, Volume 1, *Urban Stormwater Characteristics, Sources and Pollutant Management by Street Sweeping*; Volume 2, *Feasibility and Application of Urban Nonpoint Source Water Pollution Abatement Measures*, Technical Report Contract EPA-P-005432-01-5, Wisconsin Department of Natural Resources, Madison, Wisconsin, 1983.

Bannerman, R.T., D.W. Owens, and R.B. Dobbs, Source of pollution in Wisconsin stormwater, *Water Science and Technology*, 28(3/5):241–259, 1993.

Bartsch, A.F., and J.H. Gakstatter, Management desisions for lake systems on a survey of trophic status, limiting nutrients, and nutrient loadings, in *American–Soviet Symposium on Use of Mathematical Models to Optimize Water Quality Management, 1975*, Technical Report EPA-600-9-78-024, Office of Research and Development, Environmental Research Laboratory, U.S. Environmental Protection Agency, Gulf Breeze, Florida, 1978.

Bass, D., and R. Brown, Air sparging case study data base update, presented at the First International Symposium on In Situ Air Sparging for Site Remediation, Las Vegas, Nevada, October 26–27, 1996.

Bathurst, J.C., Flow resistance estimation in mountain rivers, *Journal of Hydraulic Engineering*, 111(4):625–643, 1985.

Baumgartner, D.J., Surface water pollution, in I.L. Pepper, C.P. Gerba, and M.L. Brusseau, editors, *Pollution Science*, pages 189–209, Academic Press, New York, 1996.

Baumgartner, D.J., W.E. Frick, and P.J.W. Roberts, *Dilution Models for Effluent Discharges*, 3rd ed., Technical Report EPA-600-R-94-086, ERL-N, U.S. Environmental Protection Agency, Newport, Oregon, 1994.

Bear, J., *Dynamics of Fluids in Porous Media*, Dover Publications, New York, 1972.

———, *Hydraulics of Groundwater*, McGraw-Hill, New York, 1979.

Bedient, P.B., H.S. Rifai, and C.J. Newell, *Ground Water Contamination*. Prentice Hall, Upper Saddle River, New Jersey, 1994.

———, *Ground Water Contamination*, 2nd ed., Prentice Hall, Upper Saddle River, New Jersey, 1999.

Beller, H.R., E.A. Edwards, D. Grbic-Galic, H.R. Hutchins, and M. Reinhard, *Microbial Degradation of Alkylbenzenes Under Sulfate-Reducing and Methanogenic Conditions*, Technical Report EPA-600-S2-91-1027, ERL-N, U.S. Environmental Protection Agency, Washington, DC, 1991.

Bennett, J.P., and R.E. Rathbun, *Reaeration in Open Channel Flow*, Professional Paper 737, U.S. Geological Survey, Washington, DC, 1972.

Bernhardt, H., Control of water quality in lakes and reservoirs, in M. Hino, editor, *Water Quality and Its Control*, pages 77–109, A.A. Balkema, Brookfield, Vermont, 1994.

Blatchley, E.R., III, and J.E. Thompson, Groundwater contaminants, in J.W. Delleur, editor, *The Handbook of Groundwater Engineering*, pages 13.1–13.30, CRC Press, Boca Raton, Florida, 1998.

Blevins, R.D., *Applied Fluid Dynamics Handbook*, Van Nostrand Reinhold, New York, 1984.

Blowes, D.W., R.W. Gillham, C.J. Ptacek, R.W. Puls, T.A. Bennett, S.F. O'Hannesin, C.J. Hanton-Fong, and J.G. Bain, *An In-Situ Permeable Reactive Barrier for the Treatment of Hexavalent Chromium and Trichloroethylene in Ground Water*, Volume 1, *Design and Installation*, Technical

Report EPA-600-R-99-095a, Office of Research and Development, U.S. Environmental Protection Agency, Washington, DC, September 1999.

Boano, F., R. Revelli, and L. Ridolfi, Source identification in river pollution problems: a geostatistical approach, *Water Resources Research*, 41, W07023, doi:10.1029/2004WR003754, 2005.

Boggs, J.M., S.C. Young, D.J. Benton, and Y.C. Chung, *Hydrogeologic Characterization of the MADE Site*, Interim Report EPRI EN-6915, Project 2485-5, Electric Power Research Institute, Palo Alto, California, 1990.

Borden, R.C., Natural bioattenuation of aromatic hydrocarbons and chlorinated solvents in groundwater, in J.J. Kaluarachchi, editor, *Groundwater Contamination by Organic Pollutants: Analysis and Remediation*, ASCE Manuals and Reports on Engineering Practice 100, pages 121–151, American Society of Civil Engineers, Reston, Virginia, 2000.

Bouwer, E.J., Bioremediation of chlorinated solvents using alternative electron acceptors, in R.D. Norris, R.E. Hinchee, R. Brown, P.L. McCarty, L. Semprini, J.T. Wilson, D.H. Kampbell, M. Reinhard, E.J. Bouwer, R.C. Borden, T.M. Vogel, J.M. Thomas, and C.H. Ward, editors, *Handbook of Bioremediation*, pages 149–176, Lewis Publishers, Boca Raton, Florida, 1994.

Bow River Basin Council, Flow control, http://www.urbanswm.ab.ca/bmp-3.3.1.asp, 2005.

Boxall, J.B., and I. Guymer, Analysis and prediction of transverse mixing coefficients in natural channels, *Journal of Hydraulic Engineering*, 129(2):129–139, February 2003.

Brach, J., *Agriculture and Water Quality: Best Management Practices for Minnesota*, Division of Water, Minnesota Pollution Control Agency, St. Paul, Minnesota, 1990.

Brazier, R.E., A.L. Heathwaite, and S. Liu, Scaling issues relating to phosphorus transfer from land to water in agricultural catchments, *Journal of Hydrology*, 304:330–342, 2005.

Brewster, J., J.A. Oleszkiewicz, K.M. Coombs, and A. Nartey, Enteric virus indicators: reovirus versus poliovirus, *Journal of Environmental Engineering*, 131(7):1010–1013, July 2005.

Briggs, G.G., A simple relationship between soil adsorption of organic chemicals and their octanol/water partitioning coefficients, in *Proceedings of the 7th British Insecticide and Fungicide Conference*, Volume 1, Boots Company, Ltd., Nottingham, England, 1973.

Britton, J.P., G.M. Filz, and J.C. Little, The effect of variability in hydraulic conductivity on contaminant transport through soil–bentonite cutoff walls, *Journal of Geotechnical and Geoenvironmental Engineering*, 131(8):951–957, August 2005.

Brocksen, R., M. Marcus, and H. Olem, *Practical Guide to Managing Acidic Surface Waters and Their Fisheries*, Lewis Publishers, Chelsea, Michigan, 1992.

Brooks, N.H., Diffusion of sewage effluent in an ocean current, in E.A. Pearson, editor, *Proceedings of the First International Conference on Waste Disposal in the Marine Environment*, pages 246–267, Pergamon Press, New York, 1960.

Brown, M., Design, construction and operation of sea outfalls in southeast England, in *Proceedings of the International Conference on Marine Disposal of Wastewater*, Wellington, New Zealand, 1988.

Brown, L.C., and T.O. Barnwell, Jr., *The Enhanced Stream Water Quality Models QUAL2E and QUAL2E-UNCAS: Documentation and User Manual*, Technical Report EPA-600-3-87-007, U.S. Environmental Protection Agency, Athens, Georgia, 1987.

Brown, D.S., and E.W. Flagg, Empirical prediction of organic pollutant adsorption in natural sediments, *Journal of Environmental Quality*, 10:382–386, 1981.

Brown, K.W., H.W. Wolf, K.C. Donnelly, and J.F. Slowey, The movement of fecal coliforms and coliphages below septic lines, *Journal of Environmental Quality*, 5:121–125, 1979.

Browns Drilling, The drilling process, http://www.brownsdrilling.com/photo_gallery.htm, 2005.

Bufe, M., The Everglades are forever, *Water Environment and Technology*, 17(4):40–43, April 2005.

Burns, N.M., and F. Rosa, In situ measurement of settling velocity of organic carbon particles and 10 species of phytoplankton, *Limnology and Oceanography*, 25:855–864, 1980.

Butts, T.A., *Development of Design Criteria for Sidestream Elevated Pool Aeration Stations*, Technical Report, Illinois State Water Survey, Champaign, Illinois, 1988.

Butts, T.A., D.B. Shackleforf, and T.R. Bergerhouse, *Evaluation of Reaeration Efficiencies of Sidestream Elevated Pool Aeration (SEPA) Stations*, Technical Report, Illinois State Water Survey, Champaign, Illinois, 1999.

Byers, E., and D.B. Stephens, Statistical and stochastic analyses of hydraulic conductivity and particle size in a fluvial sand, *Soil Science Society of America Journal*, 47:1072–1081, 1983.

Cabelli, V.J., *Health Effects Criteria for Marine Recreational Waters*, Technical Report EPA-600-1-80-031, U.S. Environmental Protection Agency, Research Triangle Park, North Carolina, 1983.

Canale, R.P., M.T. Auer, E.T. Owens, T.M. Heidtke, and S.W. Effler, Modeling fecal coliform bacteria, ii: Model development and application, *Water Research*, 27:703–714, 1993.

Canter, L.W., and R.C. Knox, Groundwater pollution from septic tank systems, in *Septic Tank System Effects on Groundwater Quality*, Lewis Publishers, Chelsea, Michigan, 1985.

Canter, L.W., R.C. Knox, and D.M. Fairchild, *Ground Water Quality Protection*, Lewis Publishers, Chelsea, Michigan, 1987.

Carlson, R.E., A trophic state index for lakes, *Limnology and Oceanography*, 22:361–369, 1977.

Carslaw, H.W., and J.C. Jaeger, *Conduction of Heat in Solids*, 2nd ed., Oxford University Press, New York, 1959.

Carson Dunlop Consulting Engineers, Septic systems overview, part one of three, traditional systems, http://www.usinspect.com/car/1203Septic1.asp, 2005.

Caruso, B.S., Simulation of metals total maximum daily loads and remediation in a mining-impacted stream, *Journal of Environmental Engineering*, 131:777–789, May 2005.

Carvalho, J.L.B., P.J.W. Roberts, and J. Roldao, Field observations of Ipanema Beach outfall, *Journal of Hydraulic Engineering*, 128(2):151–160, February 2002.

Challenger Oceanic, Deployment of two samplers, http://www.challengeroceanic.com/samplingsystems/filtration/radionuclidesmplrs.htm, 2005.

Chang, M., *Forest Hydrology: An Introduction to Water and Forests*, CRC Press, Boca Raton, Florida, 2002.

Chao, H.-C., H. Rajaram, and T. Illangasekare, Intermediate-scale experiments and numerical simulations of transport under radial flow in a two-dimensional heterogeneous porous medium, *Water Resources Research*, 36(10):2869–2884, 2000.

Chapra, S.C., *Surface Water-Quality Modeling*, McGraw-Hill, New York, 1997.

Chapra, S.C., and D.M. Di Toro, Delta method for estimating primary production, respiration, and reaeration in streams, *Journal of Environmental Engineering*, 117(5):640–655, 1991.

Chapra, S.C., and K.H. Reckhow, *Engineering Approaches for Lake Management*, Volume 2, Butterworth, Boston, 1983.

Charbeneau, R.J., *Groundwater Hydraulics and Pollutant Transport*, Prentice Hall, Upper Saddle River, New Jersey, 2000.

Chen, C.I., Flow resistance in broad shallow grassed channels, *Journal of the Hydraulics Division, ASCE*, 102(3):307–322, 1976.

Chen, X., and Y.P. Sheng, Three-dimensional modeling of sediment and phosphorus dynamics in Lake Okeechobee, Florida: spring 1989 simulation, *Journal of Environmental Engineering*, 131(3): 359–374, March 2005.

Chescheir, G.M., R.W. Skaggs, J.W. Gilliam, and R.G. Broadhead, The hydrology of wetland buffer areas for pumped agricultural drainage water, in D.D. Hook, editor, *The Ecology and Management of Wetlands*, pages 260–274, Croom Helm, Beckenham, Kent, England, 1987.

Chicago Public Library, Digital collections, http://www.chipublib.org/digital/sewers/history5.html, 2005.

Chin, D.A., Outfall dilution: the role of a far-field model, *Journal of Environmental Engineering*, 111(4):473–486, 1985.

———, Influence of surface waves on outfall dilution, *Journal of Hydraulic Engineering*, 113(8): 1005–1017, 1987.

———, Model of buoyant jet–surface wave interaction, *Journal of Waterway, Port, Coastal and Ocean Engineering*, 114(3):331–345, 1988.

———, An assessment of first-order stochastic dispersion theories in porous media, *Journal of Hydrology*, 199:53–73, 1997.

———, *Water-Resources Engineering*, Prentice Hall, Upper Saddle River, New Jersey, 2000.

———, *Water-Resources Engineering*, 2nd ed., Prentice Hall, Upper Saddle River, New Jersey, 2006.

Chin, D.A., and P.J.W. Roberts, Time series modeling of coastal currents, *Journal of Waterway, Port, Coastal and Ocean Engineering*, 111(6):954–972, November 1985.

Chin, D.A., and T. Wang, An investigation of the validity of first-order stochastic dispersion theories in isotropic porous media, *Water Resources Research*, 28(6):1531–1542, June 1992.

Chin, D.A., L. Ding, and H. Huang, Ocean-outfall mixing zone delineation using Doppler radar, *Journal of Environmental Engineering*, 123(12):1217–1226, December 1997.

Chou, C.T., and P.C. Jurs, Computer assisted computation of partition coefficients from molecular structures using fragment constants. *Journal of Chemical Information and Computer Sciences*, 19:172–178, 1979.

Christchurch City Council, Ocean outfall construction, http://www.ccc.govt.nz/HaveYourSay/OceanOutfall/construction/, 2005.

Churchill, M.A., H.L. Elmore, and R.A. Buckingham, Prediction of stream reaeration rates, *Journal of the Sanitary Engineering Division, ASCE*, SA4:1, 1962.

Citizens for Safe Water, Whats wrong with drinking treated Willamette River water? http://www.hevanet.com/safewater/willamette.htm, 2005.

City of Austin, Texas, Watershed signs, http://www.ci.austin.tx.us/watershed/watershed_signs.htm, 2005.

City of Greensboro, North Carolina, National pollutant discharge elimination system (NPDES), http://www.greensboro-nc.gov/stormwater/Quality/npdes.htm, 2005.

City of Vernon, British Columbia, Canada, Street sweeping and flushing, http://www.vernon.ca/services/public_works/sweeping.html, 2005.

Clark, R.B., *Marine Pollution*, 4th ed., Oxford University Press, New York, 1997.

Cohen, J., A stubborn amoeba takes center stage, *Science*, 267:822–824, 1995.

Cohen, R.M., and J.W. Mercer, *DNAPL Site Investigation* (in Russian), CRC Press, Boca Raton, Florida, 1993.

Cole, G.A., *Textbook of Limnology*, C.V. Mosby, St. Louis, Missouri, 1979.

Cole, T.M., and E.M. Buchak, *Ce-qual-w2: A Two-Dimensional, Laterally Averaged, Hydrodynamic and Water Quality Model, Version 2.0: User Manual*, Technical Report EL-95-1, U.S. Army Engineers Waterways Experiment Station, Vicksburg, Mississippi, 1995.

Commission for European Communities, *Environmental Programme for the Danube River Basin, Danube Integrated Environmental Study, Phase I*, Technical Report, CEC, Brussels, Belgium, 1994.

Commoner, B., M. Cohen, P.W. Bartlett, A. Dickar, H. Eisland, C. Hill, and J. Rosenthal, Dioxin fallout in the Great Lakes, http://qcpages.qc.cuny.edu/CBNS/dxnsum.html, 1996.

Connolly, J.P., and R.V. Thomann, Toxicity in contaminated sediments: models for criteria, exposure, and remediation, *Continuing Education Seminar Notes*, 1991.

Cornell University, Pollution prevention, http://www.sustainablecampus.cornell.edu/pollution-prevent-stormwater-bmps.htm, 2005.

Corwin, D.L., K. Loague, and T.R. Ellsworth, Introduction: assessing non-point source pollution in the vadose zone with advanced information technologies, in D.L. Corwin, K. Loague, and T.R. Ellsworth, editors, *Assessment of Non-point Source Pollution in the Vadose Zone*, pages 1–20, American Geophysical Union, Washington, DC, 1999.

Council on Environmental Quality, *Environmental Quality,* Twentieth Annual Report of the Council on Environmental Quality, U.S. Government Printing Office, Washington, DC, 1978.

Cowley, W.P., A global numerical ocean model, part 1, *Journal of Computational Physics*, 3:111–147, 1968.

Coyle, J.J., *Grassed Waterways and Outlets: Engineering Field Manual*, U.S. Soil Conservation Service, Washington, DC, April 1975.

Crites, R.W., and G. Tchobanoglous, *Small and Decentralized Wastewater Management Systems*, McGraw-Hill, New York, 1998.

Crump, K.S., An improved procedure for low-dose carcinogenic risk assessment from animal data, *Journal of Environmental Pathology and Toxicology*, 5:339–348, 1984.

CSIRO Marine Research, Marine pests, http://www.marine.csiro.au/media/archive/99releases/ seastar4jun99/dock_sign.html, 2005.

Cullum, M.G., *Evaluation of the Water Management System at a Single Family Residential Site: Water Quality Analysis for Selected Storm Events at Timbercreek Subdivision in Boca Raton, Florida*, Technical Publication 84-11, Volume II, South Florida Water Management District, West Palm Beach, Florida, November 1984.

Cunningham, P.A., *Nonpoint Source Impacts on Aquatic Life: Literature Review*, technical Report prepared for Monitoring and Data Support Division, Office of Water Regulations and Standards, U.S. Environmental Protection Agency, Research Triangle Institute, Research Triangle Park, North Carolina, 1988.

Dagan, G., *Flow and Transport in Porous Formations*, Springer-Verlag, New York, 1989.

Davis, S.N., and R.J.M. DeWiest, *Hydrogeology*, Wiley, New York, 1966.

Davis, M.L., and S.J. Masten, *Principles of Environmental Engineering and Science*, McGraw-Hill, New York, 2004.

Dean, C.M., J.J. Sansalone, F.K. Cartledge, and J.H. Pardue, Influence of hydrology on rainfall-runoff metal element speciation, *Journal of Environmental Engineering*, 131(4):632–642, April 2005.

Dean, J.A., *Lange's Handbook of Chemistry*, 13th ed., McGraw-Hill, New York, 1985.

Dean Bennett Supply, Baroid's bentonite tablets, http://www.deanbennett.com/bentonite-compressed-tablets.htm, 2005.

DeBarry, P.A., *Watersheds: Processes, Assessment, and Management*, Wiley, Hoboken, New Jersey, 2004.

Delleur, J.W., Elementary groundwater flow and transport processes, in J.W. Delleur, editor, *The Handbook of Groundwater Engineering*, pages 2.1–2.40, CRC Press, Boca Raton, Florida, 1998.

Delta Contracting Company of New Jersey, Case 9040 B Long Reach Excavators, with 60 feet booms removing silt and sediment from Strawbridge Lake, Moorestown, NJ, http://www.deltacontracting.us/lake1.html, 2005.

Deng, Z., V.P. Singh, and L. Bengtsson, Longitudinal dispersion coefficient in straight rivers, *Journal of Hydraulic Engineering*, 127(11):919–927, November 2001.

Dillon, P.J., and F.H. Rigler, The phosphorous–chlorophyll relationship in lakes, *Limnology and Oceanography*, 19(4):767–773, 1974.

Dirnberger, J.M., Lake restoration and applied limnology, http://science.kennesaw.edu/jdirnber/limno/LecApplied/LecApplied.html, 2005.

Di Toro, D.M., Algae and dissolved oxygen, *Summer Institute in Water Pollution Notes*, 1975.

Domenico, P.A., An analytical model for multidimensional transport of a decaying contaminant species, *Journal of Hydrology*, 91:49–58, 1987.

Domenico, P.A., and V.V. Palciauskas, Alternative boundaries in solid waste management, *Ground Water*, 20:303–311, 1982.

Domenico, P.A., and G.A. Robbins, A new method of contaminant plume analysis, *Ground Water*, 23(4):476–485, 1985.

Domenico, P.A., and F.W. Schwartz, *Physical and Chemical Hydrogeology*, Wiley, New York, 1990.

———, *Physical and Chemical Hydrogeology*, 2nd ed., Wiley, New York, 1998.

Donigan, A.S., and H.H. Davis, *User's Manual for Agricultural Runoff Management (ARM) Model*, Technical Report EPA-600-3-78-080, U.S. Environmental Protection Agency, Athens, Georgia, 1978.

Donigan, A., Jr., W. Huber, and T. Barnwell, Modeling of nonpoint source water quality in urban and nonurban areas, in V. Novotny, editor, *Nonpoint Pollution and Urban Stormwater Management*, pages 293–345, Technomic Publishing Company, Lancaster, Pennsylvania, 1995.

Donohue, I., D. Styles, C. Coxon, and K. Irvine, Importance of spatial and temporal patterns for assessment of risk of diffuse nutrient emissions to surface waters, *Journal of Hydrology*, 304:183–192, 2005.

Douglas, M.S., *The Everglades: River of Grass*, Ballantine, New York, 1947.

Driscoll, F.G., *Ground Water and Wells*, 2nd ed., Johnson Division, St. Paul, Minnesota, 1986.

Droste, R.L., *Theory and Practice of Water and Wastewater Treatment*, Wiley, New York, 1997.

Dufour, A.P., *Health Effects Criteria for Fresh Recreational Waters*, Technical Report EPA-600-1-84-004, U.S. Environmental Protection Agency, Research Triangle Park, North Carolina, 1984.

Durham Geo Slope Indicator, 3.25-inch hand auger, http://www.durhamgeo.com/testing/soils/field-testing-handaug.html, 2005.

Earth Action Partnership, Inc., Reifsnyder storm water wetland, http://www.earthactionpartnership.org/reifsnyder.htm, 2005.

Economic Commission for Europe, *Protection of Inland Water Against Eutrophication*, Technical Report ECE/ENVWA/26, United Nations, Geneva, Switzerland, 1992.

Elder, J.W., The dispersion of marked fluid in turbulent shear flow, *Journal of Fluid Mechanics*, 5:544–560, 1959.

Ellis, J.B., and P. Chatfield, Oil and hydrocarbons, in B.J. D'Arcy, J.B. Ellis, R.C. Ferrier, A. Jenkins, and R. Dills, editors, *Diffuse Pollution Impacts: The Environmental and Economic Impacts of Diffuse Pollution in the U.K.*, Terence Dalton Publishers, Lavenham, Suffolk, England, 2000.

———, Diffuse urban oil pollution in the UK, in *Proceedings of the 5th International Conference on Diffuse Pollution and Watershed Management*, International Water Association, Milwaukee, Wisconsin, June 10–15, 2001.

Emmons and Oliver Resources, Inc., Making a difference through integrated resource management, http://www.eorinc.com/Services/Watershed_Eng.htm, 2005.

Energy Information Administration, Landfill gas, http://www.eia.doe.gov/cneaf/solar.renewables/page/landfillgas/landfillgas.html, 2005.

EnviroTools, A citizen's guide to pump and treat, http://www.envirotools.org/factsheets/Remediation/pump_and_treat.shtml, 2005.

Esry, D.H., and D.J. Cairns, Overview of the Lake Jackson Restoration Project with artificially created wetlands for treatment of urban runoff, in D.W. Fisk, editor, *Proceedings of the American Water Resources Association Conference: Wetlands Concerns and Successes*, pages 247–257, American Water Resources Association, Tampa, Florida, 1989.

Falta, R.W., Steam flooding for environmental remediation, in J.J. Kaluarachchi, editor, *Groundwater Contamination by Organic Pollutants: Analysis and Remediation*, ASCE Manuals and Reports on Engineering Practice 100, pages 153–192, American Society of Civil Engineers, Reston, Virginia, 2000.

Fan, L.N., Design and modeling of lake wastewater disposal systems, http://www.cormix.info/picgal/lakes.php, 2005.

Fan, L.N., and N.H. Brooks, *Numerical Solutions of Turbulent Buoyant Jet Problems*, Report KH-R-18, W.M. Keck Laboratory of Hydraulics and Water Resources, California Institute of Technology, January 1969, 94 pp.

Federal Remediation Technologies Roundtable, Figure, http://www.frtr.gov/matrix2/section4/D01-4-37b.html, 2005a.

Federal Remediation Technologies Roundtable, Figure, http://www.frtr.gov/matrix2/section4/D01-4-34.html, 2005b.

Feenstra, S., D.M. Mackay, and J.A. Cherry, Presence of residual NAPL based on organic chemical concentrations in soil samples, *Ground Water Monitoring Review*, 11(2):128–136, 1991.

Fergen, R.E., P. Vinci, and F. Bloetscher, Water plant membrane reject water in an ocean outfall, *Florida Water Resources Journal*, pages 24–26, June 1999.

Fetter, C.W., *Contaminant Hydrogeology*, Macmillan, New York, 1992.

———, *Applied Hydrogeology*, 3rd ed., Macmillan, New York, 1993.

———, *Applied Hydrology*, 3rd ed., Prentice Hall, Upper Saddle River, New Jersey, 1994.

———, *Contaminant Hydrogeology*, 2nd ed., Prentice Hall, Upper Saddle River, New Jersey, 1999.

———, *Applied Hydrogeology*, 4th ed., Prentice Hall, Upper Saddle River, New Jersey, 2001.

FHWA (Federal Highway Administration, U.S. Department of Transportation), *Urban Drainage Design Manual*, HEC-22, Technical Report FHWA-SA-96-078, FHWA, Washington, DC, 1996.

Fick, A., On liquid diffusion. *Philosophical Magazine*, 4(10):30–39, 1855.

Field, R., J.P. Heaney, and R. Pitt, *Innovative Urban Wet-Weather Flow Management Systems*, Technomic Publishing Company, Lancaster, Pennsylvania, 2000.

Finnemore, E.J., and J.B. Franzini, *Fluid Mechanics with Engineering Applications*, 10th ed., McGraw-Hill, New York, 2002.

Fischer, H.B., E.J. List, R.C.Y. Koh, J. Imberger, and N.H. Brooks, *Mixing in Inland and Coastal Waters*, Academic Press, New York, 1979.

Flores, H.E., Equations for estimating the reaeration-rate coefficient in natural streams developed from the USGS national data base, Master's thesis, University of Illinois, Urbana, Illinois, 1998.

Florida DEP (Department of Environmental Protection), Permits, *Florida Administrative Code*, Chapter 62-4, State of Florida, Tallahassee, Florida, 1995.

———, Surface water quality standards, *Florida Administrative Code*, Chapter 62-302, State of Florida, Tallahassee, Florida, 1996.

Fondriest Environmental, Inc., YSI 600 OMS rhodamine dye tracer system, http://www.fondriest.com/hydrology/hydrology_products.htm, 2005.

Foster, G.R., and L.J. Lane, *User Requirements: USDA Water Erosion Prediction Project* (*WEPP*), NSERL Report 1, USDA-ARS, West Lafayette, Indiana, 1987.

Foster, G.R., L.D. Meyer, and C.A. Onstad, An erosion equation devised from basic erosion principles, *Transactions of the American Society of Agricultural Engineers*, 20:678–682, 1977.

Foth, H.D., *Fundamentals of Soil Science*, 8th ed., Wiley, New York, 1990.

Freeze, R.A., A stochastic-conceptual analysis of one-dimensional groundwater flow in nonuniform homogeneous media, *Water Resources Research*, 11:725–741, 1975.

French, R.H., S.C. McCutcheon, and J.L. Martin, Environmental hydraulics, in L.W. Mays, editor, *Hydraulic Design Handbook*, pages 5.1–5.33. McGraw-Hill, New York, 1999.

Fried, J.J., *Groundwater Pollution*, Elsevier Scientific Publishing Company, Amsterdam, 1975.

Frind, E.O., Simulation of long-term transient density-dependent transport in groundwater, *Advances in Water Resources*, 5:73–88, 1982.

Gallant, A.L., T.R. Whittier, D.P. Larsen, J.M. Omernik, and R.M. Hughes, *Regionalization as a Tool for Managing Environmental Resources*, Technical Report EPA-600-3-89-060, U.S. Environmental Protection Agency, Corvallis, Oregon, 1989.

Galli, J., *Preliminary Analysis of the Performance and Longevity of Urban BMPs Installed in Prince George County, Maryland*, Department of Environmental Resources, Prince George County, Maryland, 1992.

Geldreich, E.E., Water-borne pathogens, in R. Mitchel, editor, *Water Pollution Microbiology*, pages 207–241, Wiley-Interscience, New York, 1972.

Gelhar, L.W., *Stochastic Subsurface Hydrology*, Prentice Hall, Englewood Cliffs, New Jersey, 1993.

Gelhar, L.W., and C.L. Axness, Three-dimensional stochastic analysis of dispersion in aquifers, *Water Resources Research*, 19(1):161–180, January 1983.

Gelhar, L.W., C. Welty, and K.R. Rehfeldt, A critical review of data on field-scale dispersion in aquifers, *Water Resources Research*, 28(7):1955–1974, 1992.

Genereux, D.P., Fields studies of streamflow generation using natural and injected tracers on Bicford and Walker branch watersheds, Ph.D. dissertation, Massachusetts Institute of Technology, Boston, 1991.

Geophysical and Environmental Research Corp., History of Power Plants on the Hudson: Hudson River Settlement Agreement, http://www.riverkeeper.org/campaign.php/fishkills/thefacts/565, 2005.

Gillham, R.W., and S.F. O'Hannesin, Metal-catalysed abiotic degradation of halogenated organic compounds, in *Modern Trends in Hydrogeology*, International Association of Hydrogeologists, Hamilton, Ontario, May 10–13, 1992.

Gin, K.Y.-H., and S.Y. Neo, Microbial populations in tropical reservoirs using flow cytometry, *Journal of Environmental Engineering*, 131(8):1187–1193, August 2005.

Glass, R.L., J.R. Gentsch, and B. Ivanoff, New lessons for rotavirus vaccines, *Science*, 272:46–48, 1996.

Gleick, P.H., An introduction to global fresh water issues, in P.H. Gleick, editor, *Water in Crisis*, pages 3–12, Oxford University Press, New York, 1993.

Global Security, List of figures and tables, http://www.globalsecurity.org/military/library/policy/army/fm/5-484/Listfig.htm, 2005.

Goggin, D.J., M.A. Chandler, G. Kacurek, and L.W. Lake, Patterns of permeability in Eolian deposits: Page sandstone (Jurassic), northeastern Arizona, *SPE Formation Evaluation*, pages 297–306, June 1988.

Goode, D.J., and L.F. Konikow, *Modification of a Method-of-Characteristics Solute-Transport Model to Incorporate Decay and Equilibrium Controlled Sorption or Ion Exchange*, Water-Resources Investigation 89-4030, U.S. Geological Survey, Washington, DC, 1989.

Gordon, D.G., and N. Hansen, *Managing Nonpoint Source Pollution: An Action Plan Handbook for Puget Sound Watersheds*, Puget Sound Water Quality Authority, Seattle, Washington, 1989.

Gosselink, J.G., *The Ecology of Delta Marshes of Coastal Louisiana: A Community Profile*, Technical Report FWS/OBS-84/09, Biological Services Program, U.S. Fish and Wildlife Service, Washington, DC, 1984.

Graciela Tiburcio Pintos, Estero de San Jose del Cabo, Baja California Sur, http://members.cox.net/bajaturtles/Estuary/Estuary.htm, 2005.

Granlund, K., A. Räike, P. Ekholm, K. Rankinen, and S. Rekolainen, Assessment of water protection targets for agricultural nutrient loading in Finland, *Journal of Hydrology*, 304:251–260, 2005.

Hach, Inc., BOD test, http://www.arachem.com.my/hachr_bodtesting.htm, 2005.

Hammer, D.E., and R.H. Kadlec, A model for wetland surface water dynamics, *Water Resources Research*, 22(13):1951–1958, 1986.

Hampton, D.R., and P.D.G. Miller, Laboratory investigation of the relationship between actual and apparent product thickness in sands, in *Petroleum Hydrocarbons and Organic Chemicals in Ground Water*, pages 521–546, National Water Well Association, Dublin, OH, 1989.

Hansch, C., and A. Leo, *Substitute Constants for Correlation Analysis in Chemistry and Biology*, Wiley, New York, 1979.

Harbaugh, A.W., and M.G. McDonald, *User's Documentation for MODFLOW-96: An Update to the U.S. Geological Survey Modular Finite-Difference Ground-water Flow Model*, Open File Report 96-485, U.S. Geological Survey, Washington, DC, 1996.

Harper, H.H., Y.A. Yousef, and M.P. Wanielista, Effectiveness of detention/retention basins for removal of heavy metals in highway runoff, in T.O. Barnwell, Jr. and W.C. Huber, editors, *Proceedings of the Stormwater and Water Quality Model Users Group Meeting*, EPA-600-9-86-023, U.S. Environmental Protection Agency, Washington, DC, 1986.

Harrington, B.W., Design and construction of infiltration trenches, in L.A. Roesner, B. Urbonas, and M.B. Sonnen, editors, *Design of Urban Runoff Quality Controls*, pages 290–304, American Society of Civil Engineers, New York, 1989.

Hartigan, J.P., Basis for design of wet detention basin BMP's, in L.A. Roesner, B. Urbonas, and M.B. Sonnen, editors, *Design of Urban Runoff Quality Controls*, pages 122–143, American Society of Civil Engineers, New York, 1989.

Harvey, J.W., J.E. Saiers, and J.T. Newlin, Solute transport and storage mechanisms in wetlands of the Everglades, south Florida, *Water Resources Research*, 41, W05009, doi:10.1029/2004WR003507, 2005.

Hassett, J.J., J.C. Means, W.L. Banwart, and S.G. Wood, *Sorption Properties of Sediments and Energy-Related Pollutants*, Technical Report EPA-600-3-80-041, U.S. Environmental Protection Agency, Washington, DC, 1980.

Health and Development Initiative, HIV structure, http://www.healthinitiative.org/html/hiv/firstcontact/hivbig.htm, 2005.

Healthgate Resources Pty. Ltd. The Beverly uranium mine, http://www.heathgateresources.com.au/public/content/gallery_popup.asp?xid=28, 2005.

Hemond, H.F., and E.J. Fechner, *Chemical Fate and Transport in the Environment*, Academic Press, New York, 1993.

Hermance, J.F., *A Mathematical Primer on Groundwater Flow*, Prentice Hall, Upper Saddle River, New Jersey, 1999.

Herrmann, R., and F.G. Kari, Grain size dependent transport of nonpolar trace pollutants (PAH, PCB) by suspended solids during urban storm runoff, in Y. Iwasa and T. Sueishi, editors, *Proceedings of the 5th International Conference on Urban Storm Drainage*, pages 499–503, University of Osaka, Suita-Osaka, Japan, 1990.

Herson, L., Riparian restoration: a guide to reforestation planning, in P. Kumble and T. Schueler, editors, *Watershed Restoration Sourcebook: Collected Papers Presented at the Conference on Restoring Our Home River: Water Quality and Habitat in the Anacostia*, Publication 92701, pages 217–231, Metropolitan Washington Council of Governments, Washington, DC, November 6–7, 1992.

Hess, K.M., Use of a borehole flowmeter to determine spatial homogeneity of hydraulic conductivity and macrodispersion in a sand and gravel aquifer, Cape Cod, Massachusetts, in F.J. Moltz, J.G. Melville, and O. Guven, editors, *Proceedings of the Conference on New Field Techniques for Quantifying the Physical and Chemical Properties of Heterogeneous Aquifers*, pages 497–508, National Water Well Association, Dublin, Ohio, 1989.

Hildebrand, F., *Advanced Calculus for Applications*, 2nd ed., Prentice Hall, Englewood Cliffs, New Jersey, 1976.

Hill, D.R., *Giardia lamblia*, in G.L. Mandell, R.G. Douglas, Jr., and J.E. Bennett, editors, *Principles and Practice of Infectious Diseases*, 3rd ed., pages 2110–2115. Churchill Livingstone, New York, 1990.

Hino, M., and T. Matsuo, Water purification in rivers, in M. Hino, editor, *Water Quality and Its Control*, pages 129–146, A.A. Balkema, Brookfield, Vermont, 1994.

Horton, R.E., Erosional development of streams and their drainage basins: hydrophysical approach to quantitative morphology, *Geological Society of America Bulletin*, 56:275–370, 1945.

Howard, P.H., R.S. Boethling, W.F. Jarvis, W.M. Meylan, and E.M. Michalenko, *Handbook of Environmental Degradation Rates*, Lewis Publishers, Chelsea, Michigan, 1991.

Hsu, P.H., and M.L. Jackson, Inorganic phosphorus transformation by chemical weathering in soils as influenced by soil pH, *Soil Science*, 90:15–24, 1960.

Huang, H., R.E. Fergen, J.R. Proni, and J.J. Tsai, Probabilistic analysis of ocean outfall mixing zones, *Journal of Environmental Engineering*, 122(5):359–367, 1996.

————, Initial dilution equations for buoyancy-dominated jets in current, *Journal of Hydraulic Engineering*, 124(1):105–108, January 1998.

Hubbert, M.K., The theory of ground water motion, *Journal of Geology*, 48:785–944, 1940.

Hufschmied, P., Estimation of three-dimensional, statistically anisotropic hydraulic conductivity field by means of single well pumping tests combined with flow meter measurements, *Hydrogeologie*, 2:163–174, 1986.

Hunt, B., Dispersion model for mountain streams, *Journal of Hydraulic Engineering*, 125(2):99–105, February 1999.

Hydro Geo Chem, Inc., Efficiency of air sparging trenches, http://www.hgcinc.com/summer97.htm, 1997.

Illangasekare, T., and D.D. Reible, Pump-and-treat for remediation and plume containment: applications, limitations, and relevant processes, in J.J. Kaluarachchi, editor, *Groundwater Contamination by Organic Pollutants: Analysis and Remediation*, ASCE Manuals and Reports on Engineering Practice 100, pages 79–119, American Society of Civil Engineers, Reston, Virginia, 2000.

Illinois Leader, EPA wastewater fees stink, downstate senator says, http://www.illinoisleader.com/news/newsview.asp?c=14614, May 4 2004.

India Together, Rain barrels catalyse water harvesting, http://www.indiatogether.org/2005/mar/env-barrel.htm, 2005.

Iwasa, Y., and S. Aya, Predicting longitudinal dispersion coefficient in open-channel flows, in *Proceedings of the International Symposium on Environmental Hydraulics*, pages 505–510, Hong Kong, 1991.

Jackson, R.E., R.J. Patterson, B.W. Graham, J. Bahr, D. Belanger, J. Lockwood, and M. Priddle, *Contaminant Hydrogeology of Toxic Organic Chemicals at a Disposal Site, Gloucester, Ontario*, Volume 1, *Chemical Concepts and Site Assessment*, Inland Waters Directorate Scientific Series 141, Environment Canada, Ottawa, Ontario, Canada, 1985.

Jackson Bottom Wetlands Preserve, Photo gallery: birds at the preserve, http://www.jacksonbottom.org/photo_birds.htm, 2005.

James, A., Modelling water quality in lakes and reservoirs, in A. James, editor, *An Introduction to Water Quality Modelling*, 2nd ed., pages 233–260, Wiley, New York, 1993.

Jassby, A., and T. Powell, Vertical patterns of eddy diffusion during stratification in Castle Lake, *Limnology and Oceanography*, 20:530–543, 1975.

Ji, W., A. Dahmani, D.P. Ahlfeld, J.D. Lin, and E. Hill III, Laboratory study of air sparging: air flow visualization, *Ground Water Monitoring and Remediation*, Fall:115–126, 1993.

Jirka, G.H., and J.H.-W. Lee, Waste disposal in the ocean, in M. Hino, editor, *Water Quality and Its Control*, pages 193–242, A.A. Balkema, Brookfield, Vermont, 1994.

Jobson, H.E., and W.W. Sayre, Vertical transfer in open channel flow, *Journal of the Hydraulics Division, ASCE*, 96:703, 1970.

Johansen, O.J., and D.A. Carlson, Characterization of landfill leachate, *Water Research*, 10:1129–1134, 1976.

Johnson, P.C., Hydraulic design for groundwater contamination, in L.W. Mays, editor, *Hydraulic Design Handbook*, pages 23.1–23.71, McGraw-Hill, New York, 1999.

Johnson, P.C., M.W. Kemblowski, and J.D. Colthart, Practical screening models for soil venting applications, in *Petroleum Hydrocarbons and Organic Chemicals in Ground Water*, pages 521–546, National Water Well Association, Dublin, Ohio, 1989.

Johnson, P.C., C.C. Stanley, M.W. Kemblowski, D.L. Byers, and J.D. Colthart, A practical approach to the design, operation, and monitoring of in-situ soil-venting systems, *Ground Water Monitoring Review*, 10(2):159–178, 1990.

Johnson, R.L., P.C. Johnson, D.B. McWhorter, R.E. Hinchee, and I. Goodman, An overview of in situ air sparging, *Ground Water Monitoring Review*, Fall:127–135, 1993.

Jørgensen, S.E., H. Löffler, W. Rast, and M. Straškraba, *Land and Reservoir Management*, Elsevier, Amsterdam, The Netherlands, 2005.

Kadlec, R.H., Denitrification in wetland treatment systems, presented at the 61st Conference of the Water Pollution Control Federation, Session 26, Dallas, Texas, 1988.

————, Overland flow in wetlands: vegetation resistance, *Journal of Hydraulic Engineering*, 116:691–705, 1990.

Kadlec, R.H., and R.L. Knight, *Treatment Wetlands*, CRC Press/Lewis Publishers, Boca Raton, Florida, 1996.

Kadlec, R.H., D.E. Hammer, I.S. Nam, and J.O. Wilkes, The hydrology of overland flow in wetlands, *Chemical Engineering Communications*, 9:331–334, 1981.

Kantrud, H.A., J.B. Millar, and A.G. van der Valk, Vegetation of wetlands of the prairie pothole region, in A.G. van der Valk, editor, *Northern Prairie Wetlands*, pages 132–187, Iowa State University Press, Ames, Iowa, 1989.

Karickhoff, S.W., Estimation of sorption of hydrophobic semi-empirical pollutants on natural sediments and soils, *Chemosphere*, 10(8):833–846, 1981.

Karickhoff, S.W., D.S. Brown, and T.A. Scott, Sorption of hydrophobic pollutants on natural sediments, *Water Resources*, 13:241–248, 1979.

Karr, J.R., and D.R. Dudley, Ecological perspectives on water quality goals, *Environmental Management*, 5(6):55–68, 1981.

Karr, J.R., and I.J. Schlosser, *Impact of Nearstream Vegetation and Stream Morphology on Water Quality and Stream Biota*, Technical Report EPA-600-3-77-097, U.S. Environmental Protection Agency, Athens, Georgia, 1977.

Keeney, D.R., The nitrogen cycle in sediment–water systems, *Journal of Environmental Quality*, 2(1):15–19, 1973.

Keeney, D.R., S. Schmidt, and C. Wilkinson, *Concurrent Nitrification–Denitrification at the Sediment–Water Interface as a Mechanism for Nitrogen Losses from Lakes*, Technical Report 75-07, Water Resources Center, University of Wisconsin, Madison, Wisconsin, 1975.

Kelley, S.P., CMA contamination images, http://geotech.ecs.umass.edu/spkelley/researchpics.htm, 2005.

Kemblowski, M.W., and C.Y. Chiang, Hydrocarbon thickness fluctuations in monitoring wells, *Ground Water*, 28(2):244–252, 1990.

Kemblowski, M.W., and G.E. Urroz, Subsurface flow and transport, in L.W. Mays, editor, *Hydraulic Design Handbook*, pages 4.1–4.26, McGraw-Hill, New York, 1999.

Kenaga, E.E., and C.A.I. Goring, Relationship between water solubility, soil sorption, octanol–water partitioning, and bioconcentration of chemicals in biota, in *Third Aquatic Toxicology Symposium, Proceedings of the American Society of Testing and Materials*, number 707 in STP, pages 78–115, 1980.

Kiely, G., *Environmental Engineering*, McGraw-Hill, New York, 1997.

Klaine, S.J., D. Winkelmann, K. Sauser, J. Martin, and L.W. Moore, Characterization of agricultural nonpoint pollution: pesticide migration in a west Tennessee watershed, *Environmental Toxicology and Chemistry*, 7:609–614, 1988.

Klamath Resource Information System, Fish and aquatic life: fish populations, http://www.krisweb.com/aqualife/fishpop.htm, 2005.

Klotz, D., K.P. Seiler, H. Moser, and F. Neumaier, Dispersivity and velocity relationship from laboratory and field relationships, *Journal of Hydrology*, 45(3):169–184, 1980.

Klusman, R.W., and K.W. Edwards, Toxic metals in groundwater of Front Range, Colorado, *Ground Water*, 15:160–169, 1977.

Knight, R.L., Wetland systems, in *Natural Systems for Wastewater Collection*, Manual of Practice FD-16, pages 211–260. Water Pollution Control Federation, Alexandria, Virginia, 1990.

Knisel, W.G., *CREAMS: A Field Scale Model for Chemicals, Runoff, and Erosion from Agricultural Management Systems*, Conservation Research Report 26, U.S. Department of Agriculture, Washington, DC, 1980.

————, Use of computer models in managing nonpoint pollution from agriculture, in *Proceedings of the Non-point Pollution Abatement Symposium*, Marquette University, Milwaukee, Wisconsin, April 23–25, 1985.

Koh, R.C.Y., Wastewater field thickness and initial dilution, *Journal of Hydraulic Engineering*, 109:1232–1240, September 1983.

Konikow, L.F., and J.D. Bredehoeft, Computer model of two-dimensional solute transport and dispersion in ground water, in *Techniques of Water-Resources Investigations*, Book 7, Chapter C2, U.S. Geological Survey, Washington, DC, 1978.

Konikow, L.F., and T.E. Reilly, Groundwater modeling, in J.W. Delleur, editor, *The Handbook of Groundwater Engineering*, pages 20.1–20.40, CRC Press, Boca Raton, Florida, 1998.

Konikow, L.F., G.E. Granato, and G.Z. Hornberger, *User's Guide to Revised Method-of-Characteristics Solute-Transport Model (MOC, Version 3.1)*, Water-Resources Investigation 94-4115, U.S. Geological Survey, Washington, DC, 1994.

Konikow, L.F., D.J. Goode, and G.Z. Hornberger, *A Three-Dimensional Method-of-Characteristics Solute-Transport Model (MOC3D)*, Water-Resources Investigation 96-4267, U.S. Geological Survey, Washington, DC, 1996.

Koussis, A., and J. Rodríguez-Mirasol, Hydraulic estimation of dispersion coefficient for streams, *Journal of Hydraulic Engineering*, 124(3):317–320, March 1998.

Kouwen, N., R.M. Li, and D.B. Simons, *Velocity Measurements in a Channel Lined with Flexible Plastic Roughness Elements*, Technical Report CER79-80-RML-DBS-11, Department of Civil Engineering, Colorado State University, Fort Collins, Colorado, 1980.

Kreissl, J.F., and D.S. Brown, "Blue collar" workers, *Water Environment and Technology*, 12(10): 37–40, 2000.

Krenkel, P.A., and V. Novotny, *Water Quality Management*, Academic Press, New York, 1980.

Kuhnt, G., Long term fate of pesticides in soil, in W. Salomons and B. Stigliani, editors, *Biogeodynamics of Pollutants in Soils and Sediments*, Springer-Verlag, New York, 1995.

Kuo, J.-T., and M.-D. Yang, Water quality modeling in reservoirs, in H.H. Shen, A.H.D. Cheng, K.-H. Wang, M.H. Teng, and C.C.K. Liu, editors, *Environmental Fluid Mechanics: Theories and Applications*, pages 377–420, American Society of Civil Engineers, Reston, Virginia, 2002.

KVMR ( ), Algae bloom fed by nutrient-rich runoff from the coast of Norway, http://www.kvmr.org/programs/soundings/, 2005.

LaBaugh, J.W., T.C. Winter, and D.O. Rosenberry, Hydrological functions of prairie wetlands, *Great Plains Research*, 8:17–37, 1998.

Laflen, J.M., L.J. Lane, and G.R. Foster, WEPP: a new generation of erosion prediction technology, *Journal of Soil and Water Conservation*, 46(1):34–38, 1991.

Lake superior streams, Lake Superior Streams: Community Partnerships For Understanding Water Quality and Stormwater Impacts at the Head of the Great Lakes (http://lakesuperiorstreams.org). University of Minnesota-Duluth, Duluth, MN 55812. Authors: Axler, R., M. Lonsdale, C. Hagley, G. Host, V. Reed, V. Schomberg, E. Ruzycki, N. Will, B. Munson, and C. Richards, 2005.

Lamb, J.C., *Water Quality and Its Control*, Wiley, New York, 1985.

Lane, L.J., and M.A. Nearing, *USDA Water Erosion Prediction Project: Hillslope Profile Model Documentation*, NSERL Report 2, National Soil Erosion Research Laboratory, USDA-ARS, West Lafayette, Indiana, 1989.

Lau, Y.L., and B.G. Krishnappan, Transverse dispersion in rectangular channels, *Journal of the Hydraulics Division, ASCE*, 103(HY10):1173–1189, 1977.

Laws, E.A., *Aquatic Pollution: An Introductory Text*, 3rd ed., Wiley, New York, 2000.

Lee, J.H.-W., and G. Jirka, Vertical round buoyant jet in shallow water, *Journal of the Hydraulics Division, ASCE*, 107(HY12):1651–1675, 1981.

Lee, J.H.-W., and P. Neville-Jones, Initial dilution of horizontal jet in crossflow, *Journal of Hydraulic Engineering*, 113(5):615–629, May 1987.

Lee, L.S., P.S.C. Rao, and L. Okuda, Equilibrium partitioning of polycyclic aromatic hydrocarbons from coal tar into water, *Environmental Science and Technology*, 26:2110–2115, 1992.

Lee, J.K., L.C. Roig, H.L. Jenter, and H.M. Visser, Vertically averaged flow resistance in free surface flow through emergent vegetation at low Reynolds numbers, *Ecological Engineering*, 22:237–248, 2004.

Leonard, R.A., and W.G. Knisel, Selection and application of models for nonpoint source pollution and resource conservation, in A. Giorgini and F. Zingales, editors, *Agricultural Nonpoint Source Pollution Model Selection and Application*, Elsevier, Amsterdam, 1986.

Leonard, R.A., W.G. Knisel, and D.A. Still, GLEAMS: groundwater loading effects of agricultural management systems, *Transactions of the American Society of Agricultural Engineers*, 30(5): 1403–1418, 1987.

Lewis, W.K., and W.G. Whitman, Principles of gas absorption, *Industrial and Engineering Chemistry*, 16(12):1215–1220, 1924.

Liban, C.B., Air sparging technology: theory and modeling of remediation design systems, in J.J. Kaluarachchi, editor, *Groundwater Contamination by Organic Pollutants: Analysis and Remediation*, ASCE Manuals and Reports on Engineering Practice 100, pages 193–231, American Society of Civil Engineers, Reston, Virginia, 2000.

Liu, H., Prediction dispersion coefficient of stream, *Journal of the Environmental Engineering Division, ASCE*, 103(1):59–69, 1977.

Lohman, S.W., Ground-Water Hydraulics, professional paper 70B, U.S. Geological Survey, Washington, DC, 1972.

Long, R.R., Velocity concentrations in stratified fluids, *Journal of the Hydraulics Division, ASCE*, 88:15, 1962.

Lu, G., C. Zheng, and A. Wolfsberg, Effect of uncertain hydraulic conductivity on the simulated fate and transport of BTEX compounds at a field site, *Journal of Environmental Engineering*, 131(5):767–776, May 2005.

Lung, W.-S., *Water Quality Modeling for Watershed Allocations and TMDLs*, Wiley, New York, 2001.

Lung, W.-S., and R.G. Sobeck, Jr., Renewed use of BOD/DO models in water quality management, *Journal of Water Resources Planning and Management*, 125(4):222–227, July/August 1999.

Mace, R.E., R.S. Fischer, D.M. Welch, and S.P. Parra, *Extent, Mass, and Duration of Hydrocarbon Plumes from Leaking Underground Storage Tank Sites in Texas*, Geologic Circular 97-1, Bureau of Economic Geology, University of Texas at Austin, Austin, Texas, 1997.

MacKay, D., *Multimedia Environmental Models: The Fugacity Approach*, Lewis Publishers, Chelsea, Michigan, 1991.

MacKay, D., W.Y. Shiu, A. Maijanen, and S. Feenstra, Dissolution of non-aqueous phase liquids in groundwater, *Journal of Contaminant Hydrology*, 8:23–42, 1991.

Mahin, T., and O. Pancorbo, Waterborne pathogens, *Water Environment and Technology*, 11(4):51–55, April 1999.

Majeti, V.A., and C.S. Clark, *Potential Health Effects from Viable Emissions and Toxins Associated with Wastewater Treatment Plants and Land Application Sites*, Technical Report EPA-600-S1-81-006, U.S. Environmental Protection Agency, Cincinnati, Ohio, 1981.

Malaysia University, College of Engineering and Technology, Bringing life back to a recreation park, http://cee.kuktem.edu.my/article.cfm?id=23, 2003.

Martin, J.L., and S.C. McCutcheon, *Hydrodynamics and Transport for Water Quality Modeling*, Lewis Publishers, Boca Raton, Florida, 1998.

Mason, C.F., *Biology of Freshwater Pollution*, 2nd ed., Wiley, New York, 1991.

Mason, R.P., W.F. Fitzgerald, and F.M.M. Morel, The biogeochemical cycling of elemental mercury: anthropogenic influences, *Geochimica et. Cosmochimica Acta*, 58:3191–3198, 1994.

Mason, R.P., N.M. Lawson, and K.A. Sullivan, The concentration, speciation and sources of mercury in Chesapeake Bay precipitation, *Atmospheric Environment*, 31(21):3541–3550, 1997.

Massmann, J., Applying groundwater flow models in vapor extraction system design, *Journal of Environmental Engineering*, 115(1):129–149, 1989.

Massmann, J., S. Shock, and L. Johannesen, Uncertainties in cleanup times for soil vapor extraction, *Water Resources Research*, 36(3):679–692, March 2000.

Massoudieh, A., X. Huang, T.M. Young, and M.A. Marino, Modeling fate and transport of roadside-applied herbicides, *Journal of Environmental Engineering*, 131(7):1057–1067, July 2005.

McBride, G.B., and S.C. Chapra, Rapid calculation of oxygen in streams: approximate delta method, *Journal of Environmental Engineering*, 131(3):336–342, March 2005.

McCall, P.J., R.L. Swann, and D.A. Laskowski, Partition models for equilibrium distribution of chemicals in environmental compartments, in R.L. Swann and A. Eschenroder, editors, *Fate of Chemicals in the Environment*, pages 105–123. American Chemical Society, Washington, DC, 1983.

McCarty, P.L., Biotic and abiotic transformations of chlorinated solvents in ground water, in *Symposium on the Natural Attenuation of Chlorinated Organics in Ground Water*, Technical Report EPA-540-R-96-509, U.S. Environmental Protection Agency, Washington, DC, 1996.

McCutcheon, S.C., *Water Quality and Streamflow Data for the West Fork Trinity River in Fort Worth, TX*, Water-Resources Investigations Report 84-4330, U.S. Geological Survey, Washington, DC, 1985.

McDowell, L.L., G.H. Wills, and C.E. Murphree, Nitrogen and phosphorus yields in run-off from silty soil in the Mississippi Delta, *Agriculture, Ecosystems and Environment*, 25:119, 1989.

McGlynn, B., Stillwater River headwaters–Montana, http://landresources.montana.edu/watershed/Lab_photos.htm, 2005.

McPherson, T.N., S.J. Burian, M.K. Stenstrom, H.J. Turin, M.J. Brown, and I.H. Suffet, Trace metal pollutant load in urban runoff from a southern California watershed, *Journal of Environmental Engineering*, 131(7):1073–1080, July 2005.

Meier, W.K., and P. Reichert, Mountain streams: modeling hydraulics and substance transport, *Journal of Environmental Engineering*, 131(2):252–261, February 2005.

Melching, C.S., and H.E. Flores, Reaeration equations derived from U.S. Geological Survey database, *Journal of Environmental Engineering*, 125(5):407–414, May 1999.

Melching, C.S., and C.G. Yoon, Key sources of uncertainty in QUAL2E model of Passaic River, *Journal of Water Resources Planning and Management*, 112(2):105–113, 1997.

Méndez-Díaz, M.M., and G.H. Jirka, Buoyant plumes from multiport diffuser discharge in deep coflowing water, *Journal of Hydraulic Engineering*, 122(8):428–435, August 1996.

Mercer, J.W., and R.M. Cohen, A review of immiscible fluids in the subsurface: properties, models, characterization, and remediation, *Journal of Contaminant Hydrology*, 6:107–163, 1990.

Metcalf & Eddy, Inc., *Wastewater Engineering: Treatment, Disposal, Reuse*, 2nd ed., McGraw-Hill, New York, 1989.

MetExperts, X-ray fluorescence analyzer MetExpert–ECO, http://www.metexperts.com/Devices/ ECO.htm, 2005.

Mihelcic, J.R., *Fundamentals of Environmental Engineering*, Wiley, New York, 1999.

Mills, W.B., J.D. Dean, D.B. Porcella, S.A. Gherini, R.J.M. Hudson, W.E. Frick, G.L. Rupp, and G.L. Bowie, *Water Quality Assessment: A Screening Procedure for Toxic and Conventional Pollutants*, Technical Report EPA-600-6-82-004a and b, Volumes I and II, U.S. Environmental Protection Agency, Athens, Georgia, 1982.

Miner, J.R., F.J. Humenik, and M.R. Overcash, *Managing Livestock Waste to Preserve Environmental Quality*, Iowa State University Press, Ames, Iowa, 2000.

Minnesota Pollution Control Agency, *Protecting Water Quality in Urban Areas*, Division of Water Quality, Minnesota Pollution Control Agency, St. Paul, Minnesota, 1989.

Mitsch, W.J., and J.G. Gosselink, *Wetlands*, 3rd ed., Wiley, New York, 2000.

Mitsch, W.J., and S.E. Jørgensen, *Ecological Engineering: An Introduction to Ecotechnology*, Wiley, New York, 1989.

Mojid, M.A., and H. Vereecken, On the physical meaning or retardation factor and velocity of a non-linearly sorbing solute, *Journal of Hydrology*, 302:127–136, 2005.

Montgomery, J.H., *Groundwater Chemicals Desk Reference*, 3rd ed., Lewis Publishers, Boca Raton, Florida, 2000.

Morrison, M.P., D.B. Rewittand J.B. Ellis, G. Svensson, and P. Balmér, The physicochemical speciation of zinc, cadmium, lead and copper in urban stormwater, in *Proceedings of the 3rd International Conference on Urban Storm Drainage*, Volume 3, pages 989–1000, Chalmers University of Technology, Göteborg, Sweden, 1984.

Mortellaro, S., S. Krupa, L. Fink, and J. VanArman, *Literature Review of the Effects of Groundwater Drawdowns on Isolated Wetlands*, Technical Publication 96-01, South Florida Water Management District, West Palm Beach, Florida, 1995.

Morton, B.R., G.I. Taylor, and J.S. Turner, Turbulent gravitational convection from maintained and instantaneous sources, *Proceedings of the Royal Society*, A234:1–23, 1956.

Munson, B.R., D.F. Young, T.H. Okiishi, *Fundamentals of Fluid Mechanics*, Wiley, Hoboken, New Jersey, 1990.

Musèo Nazionale Dell'Antartide, Zooplancton and fitoplancton, http://www.mna.it/italiano/ Scopri_Antartide/CatenaTrofica_main.htm, 2005.

NASA (National Aeronautics and Space Administration), Protecting and restoring the South Florida Everglades, http://www.nemw.org/everglades.htm, 2005a.

———, Radar image, part of California coast, http://www2.cerritos.edu/earth-science/tutor/Coastal/ Estuaries%20page%203a.htm, 2005b.

———, Phytoplankton bloom near Norway, http://earthobservatory.nasa.gov/NaturalHazards/ shownh.php3?img_id=12287, 2005c.

———, Algal mat on Devils Lake, http://www.emporia.edu/kas/trans100/sympos96/aber/algae.htm, 2002.

National Academy of Sciences, *Use of Reclaimed Water and Sludge in Food Crop Production*, Technical Report, National Research Council, Washington, DC, 1996.

National Geographic Adventure, The Everglades moment, http://www.nationalgeographic.com/ adventure/0201/everglades.html, 2002.

National Oceanography Centre, Southhampton, Classification by salinity and stratification structures, http://www.soes.soton.ac.uk/research/groups/soton_water/salinity.html, 2005.

National Research Council, *Natural Attenuation for Ground Water Remediation*, National Academy Press, Washington, DC, 2000.

National Sanitation Foundation, *National Sanitation Foundation Standard 14*, Ann Arbor, Michigan, 1988.

National Water Research Institute, *Briefings*, Fall 2004.

Nazaroff, W.W., and L. Alvarez-Cohen, *Environmental Engineering Science*, Wiley, New York, 2001.

Neal, C., and A.L. Heathwaite, Nutrient mobility within river basins: a European perspective, *Journal of Hydrology*, 304:477–490, 2005.

Neuman, S.P., Universal scaling of hydraulic conductivities and dispersivities in geologic media, *Water Resources Research*, 26(8):1749–1758, 1990.

NEWater, Overseas experiences: Orange County Wastewater District, http://www.pub.gov.sg/NEWater_files/overseas_experiences/mapus_orange.html, 2005.

Newbould, D., Sunlight on the Mawddach Estuary, http://www.davenewbould.co.uk/posters-gallery/pages/Sunlight%20on%20the%20Mawddach%20Estuary.htm, 2005.

Ng, H.Y.F., Rainwater contribution to the dissolved chemistry of storm runoff, in W. Gujer and V. Krejci, editors, *Urban Storm Water Quality, Planning and Management*, pages 21–26, IAHR-IAWPRC, École Polytechnique Fédérale, Lausanne, Switzerland, 1987.

Nieswand, G.H., R.M. Hordon, T.B. Shelton, B.B. Chavooshian, and S. Blarr, Buffer strips to protect water supply and reservoirs: a model and recommendations, *Water Resources Bulletin*, 26(6):959–966, 1990.

NOAA (National Oceanic and Atmospheric Administration), Turbulent dispersion, http://pafc.arh.noaa.gov/puff/intro/intro03.html, 2005a.

———, Biological resources, http://www.csc.noaa.gov/acebasin/plants.htm, 2005b.

North American Lake Management Society, *The Lake and Reservoir Restoration Guidance Manual*, Technical Report EPA-440-4-90-006, U.S. Environmental Protection Agency, Washington, DC, 1990.

North Carolina State University, SW centennial campus GIS and environmental modeling with GRASS, http://skagit.meas.ncsu.edu/helena/wrriwork/cenntenial/centhome.html, 2005.

Nova Scotia Agricultural College, Home page of Steven Tattrie, http://www.nsac.ns.ca/eng/gradstudents/sta/, 2005.

Novotny, V., *Nonpoint Pollution and Urban Stormwater Management*, Technomic Publishing Company, Lancaster, Pennsylvania, 1995.

———, Integrated water quality management, *Water Science and Technology*, 33(4/5):1–7, 1996.

———, Integrating diffuse/nonpoint pollution control and water body restoration into watershed management, *Journal of the American Water Resources Association*, 35(4):717–727, 1999.

———, *Water Quality*, 2nd ed., Wiley, New York, 2003.

Novotny, V., and H. Olem, *Water Quality: Prevention, Identification and Management of Diffuse Pollution*, Van Nostrand Reinhold, New York, 1994.

Novotny, V., H.M. Sung, R. Bannerman, and K. Baum, Estimating nonpoint pollution from urban watersheds, *Journal of the Water Pollution Control Federation*, 57(4):339–348, 1985.

Novotny, V., G.V. Simsima, and G. Chesters, Delivery of pollutants from nonpoint sources, in *Proceedings of the Symposium on Drainage Basin Sediment Delivery*, Publication 159, International Association of Hydrological Sciences, Wallingford, Berkshire, England, 1986.

Novotny, V., K.R. Imhoff, M. Olthof, and P.A. Krenkel, *Karl Imhoff's Handbook of Urban Drainage and Wastewater Disposal*, Wiley, New York, 1989.

Novotny, V., D.W. Smith, D.A. Kuemmel, J. Mastriano, and A. Bartošová, *Urban and Highway Snowmelt: Minimizing the Impact on Receiving Water*, Number 94-IRM-2, Water Environment Research Foundation, Alexandria, Virginia, 1999.

NRCS (Natural Resources Conservation Service), NRCSIA99415, http://photogallery.nrcs.usda.gov/Detail.asp, 2005a.

———, NRCSKS02100, http://photogallery.nrcs.usda.gov/Detail.asp, 2005b.

———, Riparian buffers, http://www.oh.nrcs.usda.gov/programs/Lake_Erie_Buffer/riparian.html, 2005c.

———, Urban and recreation photo gallery, http://www.al.nrcs.usda.gov/technical/photo/urb_r.html, 2005d.

———, NRCS photo gallery, http://photogallery.nrcs.usda.gov/, 2005e.

———, Emergency Watershed Protection Program, http://www.ga.nrcs.usda.gov/programs/emergencyprog.html, 2005f.

NRCS (Natural Resources Conservation Service), *Vegetated Buffers Indicator Information*, http://www.csc.noaa.gov/alternatives/buffer.html, 2005g.

Nriagu, J.O., and J.M. Pacyna, Quantitative assessment of worldwide contamination of air, water, and soils by trace metals, *Nature*, 333:134–139, 1988.

Nürnberg, G.K., Trophic state of clear colored, soft- and hardwater lakes with special consideration of nutrients, anoxia, phytoplankton and fish, *Journal of Lake and Reservoir Management*, 12(4): 432–447, 1996.

Nyer, E., G. Boettcher, and B. Morello, Using the properties of organic compounds to help design a treatment system, *Ground Water Monitoring Review*, 11(4):115–120, 1991.

O'Connor, D.J., The bacterial distribution in a lake in the vicinity of a sewage discharge, in *Proceedings of the 2nd Purdue Industrial Waste Conference*, West Lafayette, Indiana, 1962.

O'Connor, D.J., and W.E. Dobbins, Mechanism of reaeration in natural streams, *Transactions of the American Society of Civil Engineers*, 123:655, 1958.

Okubo, A., Ocean diffusion diagrams, *Deep-Sea Research*, 18:789–802, 1971.

Olem, H., *Liming Acidic Surface Waters*, Lewis Publishers, Chelsea, Michigan, 1991.

Olem, H., and G. Flock, *The Lake and Reservoir Restoration Guidance Manual*, 2nd ed., Technical Report EPA-440-4-90-006, prepared by the North American Lake Management Society for the U.S. Environmental Protection Agency, Washington, DC, 1990.

Olem, H., R.K. Schreiber, R.W. Brocksten, and D.B. Porcella, *International Lake and Watershed Liming Practices*, Terrene Institute, Washington, DC, 1991.

Omernik, J.M., Ecoregions of the coterminous United States, *Annals of the Association of American Geographers*, 77(1):118–125, 1987.

———, Ecoregions: a framework for environmental management, in W. Davis and T. Simon, editors, *Biological Assessment and Criteria: Tools for Water Resources Planning and Decision Making*, Lewis Publishers, Chelsea, Michigan, 1995.

Ongley, E.D., *Control of Water Pollution for Agriculture*, Technical Report, Food and Agriculture Organization of the United Nations, Rome, Italy, 1996.

Ontario Ministry of the Environment, *Stormwater Management Planning and Design Manual, 2003*, http://www.ene.gov.on.ca/envision/gp/4329e_4.htm, 2003.

Osgood, R.A., Using differences among Carlson's trophic state index values in regional water quality assessment, *Water Resources Bulletin*, 18(1):67–74, 1982.

Owens, M., R. Edwards, and J. Gibbs, Some reaeration studies in streams, *International Journal of Air and Water Pollution*, 8:469–486, 1964.

Pacific Northwest National Laboratory, Battelle chlorinated solvent bioremediation design service, http://bioprocess.pnl.gov/isbio.htm, 2005.

Pankow, J.F., and J.A. Cherry, *Dense Chlorinated Solvents and Other DNAPLs in Groundwater*, Waterloo Press, Portland, Oregon, 1996.

Pankow, J.F., R.L. Johnson, and J.A. Cherry, Air sparging in gate wells and cutoff walls and trenches for control of plumes of volatile organic compounds, *Ground Water*, 31(4):654–663, 1993.

Papanicolaou, P.N., Mass and momentum transport in a turbulent buoyant vertical axisymmetric jet, Ph.D. dissertation, California Institute of Technology, Pasadena, California, 1984.

Parker, L.V., A.D. Hewitt, and T.F. Jenkins, Influence of casing materials on trace-level chemicals in ground water, *Ground Water Monitoring Review*, 10(2):146–156, 1990.

Parsons, F., P.R. Wood, and J. DeMarco, Transformation of tetrachloroethylene and trichloroethylene in microcosms and groundwater, *Journal of the American Water Works Association*, 26(2):56f, 1984.

Patterson, J., Rational environmental protection, *Water Environment Research*, 72(4):387, July/August 2000.

Payne, F.C., A.R. Blaske, and G.A. van Houten, Contamination removal rates in pulsed and steady flow aquifer sparging, in R.E. Hinchee, R.N. Miller, and P.C. Johnson, editors, *In Situ Aeration: Air Sparging, Bioventing, and Related Remediation Processes*, pages 177–183, Battelle Press, Columbus, Ohio, 1995.

Penobscot Bay Watch, GAC Corporation (General Aluminum and Chemical Corp.), http://www.penbay.org/gacalum.html, 2005.

Perkins, T.K., and O.C. Johnson, A review of diffusion and dispersion in porous media, *Society of Petroleum Engineers Journal*, 3:70–84, 1963.

Permeable Reactive Barrier Network, Photos, http://www.prbnet.qub.ac.uk/eerg/dissemination/wpm/Results/images/photos.htm, 2005.

Phelps, E.B., *Stream Sanitation*, Wiley, New York, 1944.

Pitt, R., and P. Baron, *Assessment of Urban and Industrial Stormwater Runoff Toxicity and the Evaluation/Development of Treatment for Runoff Toxicity Abatement: Phase I*, Technical Report, Storm and Combined Sewer Pollution Program, U.S. Environmental Protection Agency, Edison, New Jersey, 1989.

Plumb, R.H., Jr., and A.M. Pitchford, Volatile organic scans: implication for ground-water monitoring, in *Proceedings of Petroleum Hydrocarbons and Organic Chemicals in Ground Water*, pages 207–222, National Water Well Association, Houston, Texas, November 13–15, 1985.

Pontolillo, J. and R.P. Eganhouse, The Search for Reliable Aqueous Solubility (S_w) and Octanol-Water Partition Coefficient (K_{ow}) Data for Hydrophobic Organic Compounds: DDT and DDE as a Case Study, Water-Resources Investigations Report 01-4201, U.S. Geological Survey, Reston, Virginia, 2001.

Poulsen, T., J.W. Massmann, and P. Moldrup, Effects of vapor extraction on contaminant fluxes to atmosphere and groundwater, *Journal of Environmental Engineering*, 122(8):700–706, 1996.

Powell, G.M., *Land Treatment of Municipal Wastewater Effluents: Design Factors II*, Technical Report, U.S. Environmental Protection Agency, Washington, DC, 1976.

Prakash, A., *Water Resources Engineering*, ASCE Press, New York, 2004.

Prepas, E.E., and J.M. Burke, Effects of hypolimnetic oxygenation on water quality in Amisk Lake, Alberta, a deep eutrophic lake with high internal phosphorus loading rates, *Canadian Journal of Fisheries and Aquatic Science*, 54(9):2111–2120, 1997.

Preul, H.C., and G.J. Schoepfer, Travel of nitrogen in soils, *Journal of the Water Pollution Control Federation*, 40:30–48, 1968.

Prince William Conservation Alliance, Photos, http://www.pwconserve.org/photo/, 2005.

Pringle, C., What is hydrologic connectivity and why is it ecologically important? *Hydrologic Processes*, 17:2685–2689, 2003.

Purdue Research Foundation, Grass-lined channel case study, http://pasture.ecn.purdue.edu/sedspec/sedspec/html/case/grass/grass_case.shtml, 2005.

Queens University, Faculty of Applied Science, Construction-site services, http://appsci.queensu.ca/ilc/archives/siteservices/mechservices/mechservices_03.php, 2005.

Ramaswami, A., and R.G. Luthy, Mass transfer and bioavailability of PAH compounds in coal tar NAPL–slurry systems, 1: Model development, *Environmental Science and Technology*, 31: 2260–2267, 1997.

Ramaswami, A., J.B. Milford, and M.J. Small, *Integrated Environmental Modeling*, Wiley, New York, 2005.

Ranney, T.A., and L.V. Parker, Comparison of fiberglass and other polymeric well casings, I: Susceptibility to degradation by chemicals, *Ground Water Monitoring and Remediation*, 17(1): 97–103, 1997.

Rao, P.S.C., and J.M. Davidson, Estimation of pesticide retention and transformation parameters required in nonpoint source pollution models, in M.R. Overcash and J.M. Davidson, editors, *Environmental Impact of Nonpoint Source Pollution*, pages 23–67, Ann Arbor Science, Ann Arbor, Michigan, 1980.

Rast, W., and G.F. Lee, *Summary Analysis of the North American (U.S. Portion) OECD Eutrophication Project: Nutrient Loading-Lake Response Relationships and Trophic State Indices*, Technical Report EPA-600-3-78-008, U.S. Environmental Protection Agency, Corvallis, Oregon, 1978.

Rathbun, R.E., *Transport, Behavior, and Fate of Volatile Organic Compounds in Streams*, Professional Paper 1589, U.S. Geological Survey, Washington, DC, 1998.

Rathbun, R.E., and D.Y. Tai, Gas-film coefficients for streams, *Journal of Environmental Engineering*, 109(5):1111–1127, 1983.

Rathfelder, K., W.W.-G. Yeh, and D. Mackay, Mathematical simulation of soil vapor extraction systems: model development and numerical examples, *Journal of Contaminant Hydrology*, 8:263–297, 1991.

Reed, S.C., R.W. Crites, and E.J. Middlebrooks, *Natural Systems for Waste Management and Treatment*, 2nd ed., McGraw-Hill, New York, 1995.

Rehfeldt, K.R., and L.W. Gelhar, Stochastic analysis of dispersion in unsteady flow in heterogeneous aquifers, *Water Resources Research*, 28(8):2085–2099, 1992.

Rehfeldt, K.R., L.W. Gelhar, J.B. Southard, and A.M. Dasinger, *Estimates of Macrodispersivity Based on Analyses of Hydraulic Conductivity Variability at the MADE Site*, Technical Report EPRI EN-6405, Project 2485-5, Electric Power Research Institute, Palo Alto, California, 1989.

Reneau, R.B., and D.E. Petry, Phosphorus distribution from septic tank effluents in coastal plain soils, *Journal of Environmental Quality*, 5:34–39, 1976.

Restrepo, J.I., A.M. Montoya, and J. Obeysekera. A wetland simulation module for MODFLOW ground water model, *Ground Water*, 36(5):764–770, 1998.

Ricassi, A.L., and R.W. Schaffranek, *Flow Velocity, Water Temperature, and Conductivity in Shark River Slough, Everglades National Park, Florida: August 2001–June 2002*, Open File Report 04-348, U.S. Geological Survey, Reston, Virginia, 2003.

Rice, D.W., R.D. Grose, J.C. Michaelsen, B.P. Dooher, D.H. MacQueen, S.J. Cullen, W.E. Kastenberg, L.G. Everett, and M.A. Marino, *California Leaking Underground Fuel Tank Historical Case Analyses*, Technical Report UCRL-AR-122207, Environmental Protection Department, Environmental Restoration Division, Lawrence Berkeley Laboratories, University of California, Berkeley, California, 1995.

Richardson, L.F., Atmospheric diffusion shown on a distance-neighbour graph, *Proceedings of the Royal Society*, A110:709, 1926.

Robbins, G.A., S. Wang, and J.D. Stuart, Using the static headspace method to determine Henry's law constants, *Analytical Chemistry*, 65(21):3113–3118, 1993.

Roberts, P.J.W., Line plume and ocean outfall dispersion, *Journal of the Hydraulics Division, ASCE*, 105(HY4):313–331, April 1977.

——, Modeling Mamala Bay outfall plumes, II: Far field, *Journal of Hydraulic Engineering*, 125(6):574–583, June 1999.

Roberts, P.J.W., and D.R. Webster, Turbulent diffusion, in H.H. Shen, A.H.D. Cheng, K.-H. Wang, M.H. Teng, and C.C.K. Liu, editors, *Environmental Fluid Mechanics: Theories and Applications*, pages 7–45, American Society of Civil Engineers, Reston, Virginia, 2002.

Roberts, P.J.W., W.H. Snyder, and D.J. Baumgartner, Ocean outfalls, I: Submerged wastefield formation, *Journal of Hydraulic Engineering*, 115(1):1–25, 1989.

Robin, M.J.L., E.A. Sudicky, R.W. Gillham, and R.G. Kachanaski, Spatial variability of strontium distribution coefficients and their correlation with hydraulic conductivity in the Canadian force's base Borden aquifer, *Water Resources Research*, 27(10):2619–2632, 1991.

Robins, J.W.D., Best management practices for animal production, in *Proceedings, Non-point Pollution Abatement Symposium*, pages P–III to C–1–11, Marquette University, Milwaukee, Wisconsin, April 23–25, 1985.

Rogue Valley Council of Governments, Impacts of stormwater, http://www.rvcog.org/MN.asp?pg=WR_Stormwater, 2000.

Rouse, H., C.S. Yih, and H.W. Humphreys, Gravitational convection from a boundary source, *Tellus*, 4:201–210, 1952.

Ruane, R.J., and P.A. Krenkel, Nitrification and other factors affecting nitrogen in the Holston River, in *Proceedings of the IAWPR Conference on Nitrogen as a Water Pollutant*, Copenhagen, Denmark, 1975.

Rubin, E.S., *Introduction to Engineering and the Environment*, McGraw-Hill, New York, 2001.

Rubin, H., and J. Atkinson, *Environmental Fluid Mechanics*, Marcel Dekker, New York, 2001.

Rutherford, J.C., *River Mixing*, Wiley, New York, 1994.

Ryther, J.H., and W.M. Dunstan, Nitrogen, phosphorus, and eutrophication in the coastal marine environment, *Science*, 171:1008–1013, 1971.

Sage, M.L., and G.W. Sage, Vapor pressure, in R.S. Boethling and D. Mackay, editors, *Handbook of Property Estimation Methods for Chemicals*, pages 53–65. Lewis Publishers, Boca Raton, Florida, 2000.

Salomons, W., and B. Stol, Soil pollution and its mitigation: impact of land use changes on soil storage of pollutants, in V. Novotny, editor, *Nonpoint Pollution and Urban Stormwater Management*, Technomic Publishing Company, Lancaster, Pennsylvania, 1995.

San Marcos Growers, Plant images, http://www.smgrowers.com/info/images.asp?strLetter=V, 2005.

Sano, D., K. Fukushi, K. Yano, Y. Yoshida, and T. Omura, Enhanced virus recovery from municipal sewage sludge with a combination of enzyme and cation exchange resin, *Water Science and Technology*, 43(2):75–82, 2001.

Sartor, J.D., and G.B. Boyd, *Water Pollution Aspects of Street Surface Contamination*, Technical Report EPA-R-272-081, U.S. Environmental Protection Agency, Washington, DC, November 1972.

Sartor, J.D., G.B. Boyd, and F.J. Agardy, Water pollution aspects of street surface contamination, *Journal of the Water Pollution Control Federation*, 46:458, 1974.

Satkin, R.L., and P.B. Bedient, Effectiveness of various aquifer restoration schemes under variable hydrogeologic conditions, *Ground Water*, 26(4):488–498, 1988.

Sawyer, C.H., Fertilization of lakes by agricultural and urban drainage, *New England Water Works Association*, 61:109–127, 1947.

Scalf, M.R., J.F. McNabb, W.J. Dunlap, R.L. Cosby, and J. Fryberger, *Manual of Ground-Water Quality Sampling Procedures*, National Water Well Association, Worthington, Ohio, 1981.

Schelske, C.L., Assessment of nutrient effects and nutrient limitation in Lake Okeechobee, *Water Resources Bulletin*, 25(6):1119–1130, 1989.

Schindler, D.W., The coupling of chemical cycles by organisms: evidence from whole lake chemical perturbations, in W. Stumm, editor, *Chemical Processes in Lakes*, pages 225–250. Wiley, New York, 1985.

Schlesinger, W.H., *Biogeochemistry: An Analysis of Global Change*, 2nd ed., Academic Press, San Diego, California, 1997.

Schneider, R.L., T.L. Negley, and C. Wafer, Factors influencing groundwater seepage in a large, mesotrophic lake in New York, *Journal of Hydrology*, 310:1–16, 2005.

Schnoor, J.L., *Environmental Modeling: Fate and Transport of Pollutants in Water, Air, and Soil*, Wiley, New York, 1996.

Schnoor, J.L., C. Sato, D. McKetchnie, and D. Sahoo, *Processes, Coefficients, and Models for Simulating Toxic Organics and Heavy Metals in Surface Waters*, Technical Report EPA-600-3-87-015, U.S. Environmental Protection Agency, Athens, Georgia, 1987.

Schroepfer, G.J., M.L. Robins, and R.H. Susag, Research program on the Mississippi River in the vicinity of Minneapolis and St. Paul, *Advances in Water Pollution Research*, 1(1):145, 1964.

Schueler, T.R., *Controlling Urban Runoff: A Practical Manual for Planning and Designing Urban BMPs*, Metropolitan Washington Council of Governments, Washington, DC, July 1987.

———, *Site Planning for Urban Stream Protection*, Metropolitan Washington Council of Governments, Washington, DC, 1995.

Schueler, T.R., P.A. Kumble, and M.A. Heraty, *Current Assessment of Urban Best Management Practices: Techniques for Reducing Non-point Source Pollution in the Coastal Zones*, Metropolitan Washington Council of Governments, and Office of Wetlands, Oceans, and Watersheds, U.S. Environmental Protection Agency, Washington, DC, 1991.

Schwarzenbach, R.P., and J. Westall, Transport of non-polar organic compounds from surface water to groundwater: laboratory sorption studies, *Environmental Science and Technology*, 15(11): 1360–1367, 1981.

Schwarzenbach, R.P., P.M. Gschwend, and D.M. Imboden, *Environmental Organic Chemistry*, Wiley, New York, 1993.

Schwille, F., *Dense Chlorinated Solvents in Porous and Fractured Media: Model Experiments* (English Translation), Lewis Publishers, Boca Raton, Florida, 1988.

Scientific Software Group, MODFLOWT, http://www.scisoftware.com/environmental_software/index.php?cPath=21_27, 2005.

Seattle Daily Journal of Commerce, State considers unique MUDS disposal facility, http://www.djc.com/special/environment2000/muds.html, 2005.

Seo, I.W., and T.S. Cheong, Predicting longitudinal dispersion coefficient in natural streams, *Journal of Hydraulic Engineering*, 124(1):25–32, January 1998.

Serrano, S.E., *Hydrology for Engineers, Geologists, and Environmental Professionals*, Hydro-Science, Inc., Lexington, Kentucky, 1997.

Shanahan, P., Groundwater remediation and modeling, in V. Novotny and L. Somlyódy, editors, *Remediation and Management of Degraded River Basins*, NATO ASI Series, Springer-Verlag, Berlin, Germany, 1995.

Shannon, J.D., and E.C. Voldner, Modeling atmospheric concentrations of mercury and deposition to the Great Lakes, *Atmospheric Environment*, 29(14):1649–1661, 1995.

Shaver Lake Power Center, Snow removal service, http://www.cmstoys.com/snow.htm, 2005.

Shell Research and Technology Center, BIOVENT: bioremediation of contaminated soils, http://www.clues.abdn.ac.uk:8080/tltp/biovent2.html, 2005.

Shifrin, N.S., Pollution management in the twentieth century, *Journal of Environmental Engineering*, 131(5):676–691, May 2005.

Shutes, R.B.E., Artificial wetlands and water quality improvement, *Environment International*, 26:441–447, 2001.

Sikora, L.J., M.G. Bent, R.B. Corey, and D.R. Keeney, Septic nitrogen and phosphorus removal test system, *Ground Water*, 14:304–314, 1976.

Sloey, W.E., and F.L. Spangler, Trophic status of the Winnebago pool lakes, in *Proceedings of the 2nd Annual Conference, Wisconsin Section, AWRA*, Water Resources Center, University of Wisconsin, Madison, Wisconsin, 1978.

Smith, L., A stochastic analysis of steady state groundwater flow in a bounded domain, Ph.D. dissertation, University of British Columbia, Vancouver, British Columbia, Canada, 1978.

———, Spatial variability of flow parameters in a stratified sand, *Mathematical Geology*, 13(1): 1–21, 1981.

Smith, V.H., and J. Shapiro, *A Retrospective Look at the Effects of Phosphorous Removal in Lakes, in Restoration of Lakes and Inland Waters*, Technical Report EPA-440-5-81-010, U.S. Environmental Protection Agency, Office of Water Regulations and Standards, Washington, DC, 1981.

Smolen, M.D., G.D. Jennings, and R.L. Huffman, Impact of nonpoint sources of pollution on aquatic systems: agricultural land uses, in *Nonpoint Source Impact Assessment*, Report 90-5 by CH_2M-Hill, Water Environment Research Federation, Alexandria, Virginia, 1990.

Sokolofsky, S., and G.H. Jirka, Environmental Fluid Mechanics 1: Mass Transfer and Diffusion, http://www.ifh.unikarlsruhe.de/lehre/envfluI/Relatedresources/photos.htm, 2005.

Solley, W.B., C.F. Merk, and R.R. Pierce, *Estimated Use of Water in the United States in 1985*, Circular 1004, U.S. Geological Survey, Reston, Virginia, 1988.

South Dakota DNR (Department of Natural Resources), Wellhead Protection Program, http://www.state.sd.us/denr/DES/Ground/Wellhead/Wellhead.htm, 2005.

South West Water, Ltd. Treating the land, http://www.swwater.co.uk/index.cfm?articleid=239, 2005.

St. Petersburg Times, Fuel spill streaks the river, http://www.sptimes.com/2002/06/25/TampaBay/Fuel_spill_streaks_th.shtml, June 25, 2002.

Stanford, G., M.H. Free, and D.H. Swaininger, Temperature coefficient of soil nitrogen mineralization, *Soil Science*, 115(4):321–328, 1973.

State of Arkansas, What are the main sources of NPS pollution? http://www.aswcc.arkansas.gov/NPS_Webpage/NPS%20Sources.htm, 2005.

State of California, Cantara Loop/Dunsmuir chemical spill, http://www.dfg.ca.gov/ospr/organizational/scientific/nrda/NRDAcantara.htm., 2005a.

———, Riparian habitat along San Joaquin River, Central Valley, California, http://virtual.yosemite.cc.ca.us/randerson/Great%20Valley%20Museum/habitats/10San%20Joaquin, 2005b.

State of Iowa, How do you determine water quality? http://www.iowadnr.com/water/tmdlwqa/wqa/, 2005.

State of Michigan, Michigan environmental education curriculum, http://techalive.mtu.edu/meec/module03/WhattoRemove.htm, 2005.

State of North Carolina, Silt fences, http://www.dfr.state.nc.us/forestry_glossary.htm, 2005a.

———, Turbidity, http://www.dfr.state.nc.us/forestry_glossary.htm, 2005b.

State of Oregon, Coliform tests, http://dl.clackamas.cc.or.us/wqt111/unit-8-coliformtest.htm, 2005.

State of Texas, Ecoregions of Texas, http://www.tpwd.state.tx.us/edu/trunks.phtml, 2005.

State of Vermont, Stormwater Section, http://www.anr.state.vt.us/dec/waterq/stormwater.htm, 2005.

State of Victoria, Australia, 2005, Department of Sustainability and Environment, Waterbird monitoring in Barmah Forest, http://www.dse.vic.gov.au/dse/nrenfor.nsf/LinkView/6428B4E89058E1CDCA256E75002115641A7728FC84, 2005.

State of Virginia, Laurel Bed lake liming project, http://csm.jmu.edu/laurelbedlake/Photo_Album/Laurel_Bed_Lake_-_Photo_Album/laurel_bed_lake_-_photo_album_26.html, 2005.

State of Washington, aquatic plant management bottom screening, http://www.ecy.wa.gov/programs/wq/plants/management/aqua023.html, 2005.

State of Wisconsin, Mixing and stratification: understanding lake data, http://www.dnr.state.wi.us/org/water/fhp/lakes/under/mixing.htm, 2005.

Stefan, H.G., Lake and reservoir eutrophication: prediction and protection, in M. Hino, editor, *Water Quality and Its Control*, pages 45–76, A.A. Balkema, Brookfield, Vermont, 1994.

Steiner, G.R., and R.J. Freeman, Configuration and substrate design considerations for constructed wetlands for wastewater treatment, in D.A. Hammer, editor, *Constructed Wetlands for Wastewater Treatment*, pages 363–378, Lewis Publishers, Chelsea, Michigan, 1989.

Stewart, R.E., Unusual plume behavior from an ocean outfall off the east coast of Florida, *Journal of Physical Oceanography*, 3(2):241–243, doi: 10.1175/1520-0485(1973)003, 1973.

Stewart, B.A., D.A. Woolhiser, W.H. Wischmeier, J.H. Caro, and M.H. Frere, *Control of Pollution from Cropland*, Technical Report USEPA 600/2-75-026 or USDA ARS-H-5-1, U.S. Environmental Protection Agency, Washington, DC, 1975.

Stoltenberg, N.I., and J.L. White, Selective loss of plant nutrients by erosion, *Soil Science Society of America Proceedings*, 27:406–410, 1953.

Stone, W.J., *Hydrogeology in Practice*, Prentice Hall, Upper Saddle River, New Jersey, 1999.

Strahler, A.N., Quantitative analysis of watershed geomorphology, *Transactions of the American Geophysical Union*, 38:913–920, 1957.

Strecker, E., J. Kersnar, E. Driscoll, and R. Horner, *The Use of Wetlands for Controlling Stormwater Pollution*, Terrene Institute, Washington, DC, 1992.

Streeter, H.W., and E.B. Phelps, *A Study of the Pollution and Natural Purification of the Ohio River, iii: Factors Concerned in the Phenomena of Oxidation and Reaeration*, Bulletin 146, U.S. Public Health Service, Washington, DC, 1925.

Stroud Water Research Center, Riparian buffers, http://www.stroudcenter.org/portrait/11.htm, 2005.

Stumm, W., and J.J. Morgan, Stream pollution by algal nutrients, in *Transactions of the 12th Annual Conference on Sanitary Engineering*, University of Kansas, Lawrence, Kansas, 1962.

Stylianou, C., and B.A. DeVantier, Relative air permeability as a function of saturation in soil venting, *Journal of Environmental Engineering*, 121(4):337–347, 1995.

Sudicky, E.A., A natural gradient experiment on solute transport in a sand aquifer: spatial variability of hydraulic conductivity and its role in the dispersion process, *Water Resources Research*, 22(13):2069–2082, 1986.

Suffolk County Soil and Water Conservation District, Long Island Regional Envirothon: soils, http://www.co.suffolk.ny.us/webtemp3.cfm?dept=18&id=822, 2005.

Sundaram, T.R., C.C. Easterbrook, K.R. Piech, and G. Rudinger, *An Investigation of the Physical Effects of Thermal Discharges into Cuyuga Lake*, Technical Report VT-2626-0-2, Cornell Aeronautical Laboratory, Ithaca, NY, 1969.

Szirtes, T., *Applied Dimensional Analysis and Modeling*, McGraw-Hill, New York, 1998.

Taylor, B.N., *Guide for the use of the international system of units (SI)*, NIST Special Publication 811, National Institute of Standards and Technology, Reston, Virginia, 1995.

Taylor, S.W., P.C.D. Milly, and P.R. Jaffe, Biofilm growth and the related changes in the physical properties of a porous medium, *Water Resources Research*, 26:2161–2169, 1990.

Tchobanoglous, G., and E.D. Schroeder, *Water Quality*, Addison-Wesley, Reading, Massachusetts, 1985.

Tebbutt, T.H.Y., *Principles of Water Quality Control*, 5th ed., Butterworth-Heinemann, Oxford, 1998.

Thibault, D.H., M.I. Sheppard, and P.A. Smith, *A Critical Compilation and Review of Default Soil Solid/Liquid Partition Coefficients, k_d, for Use in Environment Assessments*, Technical Report

AECL-10125, ROE 1L0, Atomic Energy of Canada Limited Research Company, Pinawa, Manitoba, Canada, March 1990.

Thom, R.M., A.B. Borde, K.O. Richter, and L.F. Hibler, Influence of urbanization on ecological processes in wetlands, in M.S. Wigmosta and S.J. Burges, editors, *Land Use and Watersheds*, pages 5–16, American Geophysical Union, Washington, DC, 2001.

Thomann, R.V., *Systems Analysis and Water Quality Management*, Technical Report, Environmental Research Applications, New York, 1972.

————, *A Statistical Model of Environmental Contaminants Using Variance Spectrum Analysis*, Technical Report NTIS PB 88-235130/A09, report to National Science Foundation, Washington, DC, 1987.

————, Bioaccumulation model of organic chemical distribution in aquatic food chains, *Environmental Science and Technology*, 23:699–707, 1989.

Thomann, R.V., and J.A. Mueller, *Principles of Surface Water Quality Modeling and Control*, Harper & Row, New York, 1987.

Tiedeman, C.R., and P.A. Hsieh, Evaluation of longitudinal dispersivity estimates from simulated forced- and natural-gradient tracer tests in heterogeneous aquifers, *Water Resources Research*, 40, W01512, doi:10.1029/2003WR002401, 2004.

Tindall, J.A., and J.R. Kunkel, *Unsaturated Zone Hydrology for Scientists and Engineers*, Prentice Hall, Upper Saddle River, New Jersey, 1999.

Town of Chapel Hill, Erosion and sedimentation control, http://townhall.townofchapelhill.org/stormwater/regs_ords.html, 2005.

Trust for Public Land, *Source Protection Handbook*, Technical Report, Trust for Public Land and American Water Works Association, West Palm Beach, Florida, 2005.

Tsivoglou, E.C., and S.R. Wallace, Characterization of stream reaeration capacity, Technical Report EPA-R3-72-012, U.S. Environmental Protection Agency, Washington, DC, 1972.

Tung, W.-S., *Water Quality Modeling for Wasteload Allocations and TMDLs*, Wiley, New York, 2001.

Uniformed Services University, Scanning electron micrograph of a pair of *Schistosoma mansoni*, http://www.usuhs.mil/mic/davies.html, 2005.

United Nations Environment Programme, The watershed: water from the mountains into the sea, http://www.unep.or.jp/ietc/publications/short_series/lakereservoirs-2/3.asp, 2005.

University of Arizona, Water Resources Research Center, Roles of citizens and government in water policy, http://ag.arizona.edu/AZWATER/publications/sustainability/report_html/chap7_03.html, 2005.

University of British Columbia, Institute for Resources Environment and Sustainability, Constructed wetlands: design options, http://www.ires.ubc.ca/projects/ponds/WetDes.htm, 2005.

University of Nevada, Cooperative Extension, Home page, http://www.unce.unr.edu/western/, 2005.

Urase, T., Final examination for urban environmental engineering, http://www.cv.titech.ac.jp/ turase/e/lecture/uese3.html, 2005.

U.S. Army, Multi-service procedures for well-drilling operations, Technical Report Field Manual No. 5–404, USDOA, Washington, DC, 1994.

USACE (U.S. Army Corps of Engineers), *Urban Stormwater Runoff: STORM*, Technical Report, Hydrologic Engineering Center, Davis, California, 1975.

————, *Corps of Engineers Wetlands Delineation Manual*, Technical Report Y-87-1, U.S. Army Corps of Engineers Waterways Experiment Station, Vicksburg, Mississippi, 1987.

————, *Scour and Deposition in Rivers and Reservoirs*, HEC-6 Users Manual, Technical Report, Hydrologic Engineering Center, Davis, California, 1991.

————, Regulatory permits, http://www.swl.usace.army.mil/regulatory/Missouri/missouriindex.html, 2005a.

————, Floodwall and Line Creek, http://www.nwk.usace.army.mil/projects/1385/photos_floodwall.htm, 2005b.

————, Restoration Advisory Board for the former Walker AFB, http://www.spa.usace.army.mil/ec/walker-rab/geoproberigphoto.html, 2005c.

————, Environmental Remediation Branch, http://www.nwo.usace.army.mil/html/pm-h/homefeb.htm, 2005d.

————, Portland District, Aerial view: fish hatchery, http://www.bpa.gov/Power/pl/columbia/gallery/hatchery.jpg, 2005e.

U.S. Air Force, Report to stakeholders, http://www.edwards.af.mil/penvmng/Documents/RTS/2003/NOV03/Nov03pg3.htm, 2005.

USAWES (U.S. Army Waterways Experiment Station), *A Numerical One-Dimensional Model of Reservoir Water Quality: CE-QUAL-R1*, Technical Report E-82-1, USAWES, Vicksburg, Mississippi, 1986.

USDA (U.S. Department of Agriculture), *Field Office Technical Guide: Section III*, USDA Soil Conservation Service, Washington, DC, 1980.

————, *National Handbook of Conservation Practices*, USDA Soil Conservation Service, Washington, DC, 1988.

————, Rill erosion, http://topsoil.nserl.purdue.edu/nserlweb/weppmain/overview/rill.html, 2005.

————, News and events, http://www.ars.usda.gov/is/graphics/photos/sep02/k9986-1.htm, 2005b.

USEPA (U.S. Environmental Protection Agency), *The Relationships of Phosphorus and Nitrogen to the Trophic State of Northeast and North-Central Lakes and Reservoirs*, National Eutrophication Survey, Working Paper 23, USEPA, Washington, DC, 1974.

————, *A Compendium of Lake and Reservoir Data Collected by the National Eutrophication Survey in the Northeast and North-Central United States*, Working Paper 474, USEPA, Corvalis, Oregon, 1975.

————, *Report to Congress on Waste Disposal Practices and Their Effects on Groundwater*, Technical Report, USEPA, Washington, DC, 1977.

————, *Report to Congress on Control of Combined Sewer Overflow in the United States*, Technical Report EPA-430/9-78/006, USEPA, Washington, DC, 1978.

————, *Ambient Water Quality Criteria for Aldrin/Dieldrin*, Technical Report EPA-440-5-80-019, USEPA, Washington, DC, 1980.

————, *Results of the Nationwide Urban Runoff Program*, Volume I, *Final Report*, Technical Report, Water Planning Division, USEPA, Washington, DC, 1983a.

————, *Technical Support Document for Waterbody Surveys and Assessments for Conducting Use Attainability Analysis*, Technical Report, Office of Water Regulations and Standards, USEPA, Washington, DC, 1983b.

————, *Technical Support Manual: Waterbody Surveys and Assessment for Conducting Use Attainability Analyses*, Volume II, *Estuarine Systems*, Technical Report, Office of Water Regulations and Standards, USEPA, Washington, DC, 1984a.

————, *Ground-Water Protection Strategy*, Technical Report, Office of Ground-Water Protection, USEPA, Washington, DC, 1984b.

————, *Water Quality Assessment: A Screening Procedure for Toxic and Conventional Pollutants in Surface and Ground Water*, Technical Report Part I, EPA-600-6-85-002a, and Part II, EPA-600-6-85-002b, Environmental Research Laboratory, USEPA, Athens, Georgia, 1985a.

————, *Methods for Measuring the Acute Toxicity of Effluents to Freshwater and Marine Organisms*, Technical Report EPA-600-4-85-013, USEPA, Washington, DC, 1985b.

————, *Quality criteria for water*. Technical Report EPA 440-5-86-001, Office of Water Regulations and Standards, USEPA, Washington, DC, 1986.

————, *WASP4, A Hydrodynamic and Water Quality Model: Model Theory, User's Manual, and Programmer's Guide*, Technical Report EPA-600-3-86-034, USEPA, Athens, Georgia, 1988.

————, *National Water Quality Inventory: 1988 Report to Congress*, Technical Report EPA-440-4-90-003, USEPA, Washington, DC, 1990a.

————, *Laboratory Investigation of Residual Liquid Organics from Spills, Leaks, and Disposal of Hazardous Wastes in Groundwater*, Technical Report EPA-600-6-90-004, USEPA, Washington, DC, 1990b.

————, *Urban Targeting and BMP Selection: An Information and Guidance Manual for State NPS Program Staff Engineers and Managers*, Technical Report Contract 68-C8-0034, USEPA, Washington, DC, 1990d.

————, *Technical Support Document for Water-Quality Based Toxics Control*, Technical Report EPA-505-2-90-001, Office of Water, USEPA, Washington, DC, 1991a.

————, *Guidance for Water Quality-Based Decisions: The TMDL Process*, Technical Report, Assessment and Watershed Protection Division, USEPA, Washington, DC, 1991b.

————, *Estimating Potential for Occurrence of DNAPL at Superfund Sites*, Technical Report, Publication 9355.4-07FS, USEPA, Washington, DC, 1992a.

————, *Technical Guidance Manual for Performing Waste Load Allocation*, Book III. *Estuaries*, Part 4, Critical Review of Coastal Embayment and Estuarine Waste Load Allocation Modeling, Technical Report EPA-823-R-92-005, Office of Water, USEPA, Washington, DC, 1992b.

————, *Remediation Technologies Screening Matrix and Reference Guide, Version I*, Technical Report EPA-542-B-93-005, Solid Waste and Emergency Response, USEPA, Washington, DC, 1993.

————, *Water Quality Standards Handbook*, 2nd ed., Technical Report EPA-823-B-94-005a, Office of Water, USEPA, Washington, DC, August 1994.

————, *How to Effectively Recover Free-Products for Leaking Underground Storage Tank Sites: A Guide for State Regulators*, Draft Technical Report, Office of Underground Storage Tanks, USEPA, Washington, DC, August 1995.

————, *ICR Microbial Laboratory Manual*, Technical Report EPA-600-R-95-178, Office of Research and Development, USEPA, Washington, DC, 1996a.

————, *Superfund Chemical Data Matrix (SCDM)*, downloaded from USEPA Web site www.epa.gov, cited in Johnson (1999), 1996b.

USEPA (U.S. Environmental Protection Agency), *National Water Quality Inventory—1996 Report to Congress*, http://www.epa.gov/305b/, 1996c.

————, *State Methods for Delineating Source Water Protection Areas for Surface Water Supplied Sources of Drinking Water*, Technical Report EPA-816-R-97-008, Office of Water, USEPA, Washington, DC, August 1997a.

————, *Delineation of Source Water Protection Areas: A Discussion for Managers*, Part 1, A Conjunctive Approach for Ground Water and Surface Water, Technical Report EPA-816-R-97-012, Office of Water, USEPA, Washington, DC, October 1997b.

————, *State Source Water Assessment and Protection Programs Guidance*, Technical Report EPA-816-R-97-009, Office of Water, USEPA, Washington, DC, 1997c.

————, *The Quality of Our Nation's Water: 1996*, Technical Report EPA-841-S-97-001, USEPA, Washington, DC, 1997d.

————, *Compendium of Tools for Watershed Assessment and TMDL Development*, Technical Report EPA-841-B-97-006, Office of Water, USEPA, Washington, DC, 1997e.

————, *Deposition of Air Pollutants to the Great Waters: Second Report to Congress*, Technical Report EPA-453-R-97-011, Office of Air Quality, USEPA, Research Triangle Park, North Carolina, June 1997f.

————, *Oxygenates in Water: Critical Information and Research Needs*, Technical Report EPA-600-R-98-048, Office of Research and Development, USEPA, Washington, DC, 1998.

————, National pollution elimination system regulation for revision of the water pollution control program addressing storm water discharges; final rule, report to Congress on the Phase II Storm Water Regulations, 40 CFR Parts 9, 122, 123 and 124, *Federal Register*, 64(235):68722–68851, 1999a.

————, Proposed revisions to the Water Quality Planning and Management Regulations, proposed Rule 40 CFR Part 130, *Federal Register*, 64(162), 1999b.

————, *Guiding Principles for Constructed Treatment Wetlands*, Technical Report EPA-843-B-00-003, Office of Wetlands, Oceans and Watersheds, USEPA, Washington, DC, October 2000a.

————, *National Water Quality Inventory: 1998 Report to Congress*, Technical Report EPA 841-R-00-001, Office of Water, USEPA, Washington, DC, 2000b.

————, *Ambient Aquatic Life Water Quality Criteria for Dissolved Oxygen (Saltwater): Cape Cod to Cape Hatteras*, Technical Report EPA-882-R-00-012, Office of Water, USEPA, Washington, DC, 2000c.

————, *Deposition of Air Pollutants to the Great Waters: Third Report to Congress*, Technical Report EPA-453-R-00-005, Office of Air Quality, USEPA, Research Triangle Park, North Carolina, 2000d.

————, A citizen's guide to pump and treat, Technical Report EPA-542-F-01-025, USEPA, Office of Solid Waste and Emergency Response, Washington, DC, December, 2001.

————, *National Recommended Water Quality Criteria: 2002*, Technical Report EPA-822-R-02-047, Office of Water, Office of Science and Technology, USEPA, Washington, DC, November 2002a.

————, *National Management Measures to Control Nonpoint Source Pollution from Urban Areas: Draft*, Technical Report EPA-842-B-02-003, Office of Wetlands, Oceans, and Watersheds, USEPA, Washington, DC, July 2002b.

————, *AQUATOX (Release 2): Modeling Environmental Fate and Ecological Effects in Aquatic Ecosystems*, Volume 1, *User's Manual*, Technical Report EPA-823-R-04-001, Office of Water, USEPA, Washington, DC, January 2004a.

————, *AQUATOX (Release 2): Modeling Environmental Fate and Ecological Effects in Aquatic Ecosystems*, Volume 2, *Technical Documentation*, Technical Report EPA-823-R-04-002, Office of Water, USEPA, Washington, DC, January 2004b.

————, What is a watershed? http://www.epa.gov/owow/watershed/whatis.html, 2005a.

————, Combined sewer overflows demographics, http://cfpub.epa.gov/npdes/cso/demo.cfm?program_id=5, 2005b.

————, Biocriteria basics, http://www.epa.gov/waterscience/biocriteria/basics/, 2005c.

————, GWERD research on stream and riparian restoration to benefit water quality, http://www.epa.gov/ada/topics/riparian.html, 2005d.

————, Leaking underground storage tanks (LUST), http://www.epa.gov/reg5rcra/wptdiv/r5lust/, 2005e.

————, Assessed rivers, lakes, and estuaries meeting all designated uses, http://www.epa.gov/iwi/1999april/iii1_usmap.html, 2005f.

————, Bacteria and viruses, http://omp.gso.uri.edu/doee/policy/orgalb.htm, 2005g.

————, Kingman Lake, http://www.epa.gov/owow/nps/Section319III/DC.htm, 2005h.

————, Pollution reports, http://yosemite1.epa.gov/r6/polreps.nsf/0/53e9b246c16f701786256cf40062751b?OpenDocument, 2005i.

————, U.S. titanium, http://epa.gov/reg3hwmd/super/sites/VAD980705404/, 2005j.

U.S. Fish and Wildlife Service, Fish passage projects funded and/or completed in 2004, http://www.fws.gov/r5crc/Habitat/fish_passage.htm, 2005a.

————, Chase Lake prairie project, http://arrowwood.fws.gov/ChaseLake_PP/predator_fences. html, 2005b.

U.S. Forest Service, George Washington and Jefferson National Forests, http://www.fs.fed.us/r8/gwj/ recreation/fishing/projects_issues.shtml, 2005.

USGS (U.S. Geological Survey), *Estimated Use of Water in the United States in 1995*, Circular 1200, USGS, Denver, Colorado, 1998.

————, *The Quality of Our Nation's Waters: Nutrients and Pesticides*, Circular 1225, USGS, Washington, DC, 1999.

————, Pesticides in the atmosphere, Fact Sheet FS-152-95, http://ca.water.usgs.gov/pnsp/atmos/, 2005a.

————, Metal-rich drainage from the old Howardsville mill site, near Silverton, Colorado, http:// amli.usgs.gov/pictures/aimage5.html, 2005b.

————, Controlling the noxious sea lamprey on the Great Lakes, http://www.usgs.gov/125/articles/ sea_lamprey.html, 2005c.

————, Georgia drought watch: June 2000, http://ga.water.usgs.gov/news/drought99/poster/june/ 02347500.html, 2005d.

————, Water quality sampling, http://www.umrba.org/wq.htm, 2005e.

————, Dissolved oxygen measurement, http://www.umrba.org/wq.htm, 2005f.

————, Coastal zone, http://sofia.usgs.gov/publications/reports/rali/coastdredge.html, 2005g.

————, Determination of water depth at the deep edge of Seagrass Meadows in Tampa Bay using GPS carrier phase processing, http://gulfsci.usgs.gov/tampabay/conf2002/se_joha2/results2.html, 2005h.

————, Landsat™ scene of Tampa Bay, Florida, http://gulfsci.usgs.gov/maps/sat/satmap.html, 2005i.

————, Environmental quality, http://sofia.usgs.gov/publications/reports/rali/eqpollution.html, 2005j.

————, Key science issues, http://sofia.usgs.gov/publications/ofr/01-180/issue2.html, 2005k.

————, Natural attenuation strategy for groundwater cleanup focuses on demonstrating cause and effect, http://toxics.usgs.gov/pubs/eos-v82-n5-2001-natural/, 2005l.

U.S. National Park Service, Katmai: building in an ashen land—historic resource study, http://www. nps.gov/katm/hrs/hrst.htm, 2005a.

————, Typical lake view: Whiskeytown, http://www.nps.gov/whis/exp/photos/gal1/, 2005b.

U.S. Navy, Soil Vapor Extraction (in-situ), http://enviro.nfesc.navy.mil/scripts/WebObjects.exe/erbweb. woa/2/wa/DisplayIndex?pageShortName=Soil+Vapor+Extraction & Page ID=97&wosid= 7L3UofzuiVvnJt7C5YV5W0, 2005.

U.S. Minerals Management Service, http://www.mms.gov/taroilspills/, 2005.

Viessman, W., and G.L. Lewis, *Introduction to Hydrology*, 4th ed., Harper Collins, New York, 1996.

Vertex Water Features, Inc., Pond aerators and lake aeration systems, http://www.vertexwaterfeatures. com/pond_aerator_gallery.php, 2005.

Vesilind, P.A., and S.M. Morgan, *Introduction to Environmental Engineering*, Brooks/Cole–Thomson Learning, Belmont, California, 2004.

Vollenweider, R.A., *The Scientific Basis of Lake and Stream Eutrophication with Particular Reference to Phosphorus and Nitrogen as Eutrophication Factors*, Technical Report DAS/CSI/68.27, Organisation for Economic Co-operation and Development, Paris, 1968.

————, Input–output models with special reference to the phosphorus loading concept in limnology, *Journal of Hydrology*, 37:53–84, 1975.

————, Advances in defining critical loading levels for phosphorus in lake eutrophication, *Memorie dell'Istituto Italiano di Idrobiologia*, 33:53–83, 1976.

Wandmacher, C., and A.I. Johnson, *Metric Units in Engineering: Going SI*, ASCE Press, New York, 1995.

Wanner, O., T. Egli, T. Fleischmann, K. Lanz, P. Reichert, and R.P. Schwarzenbach, Behavior of the insecticides Disulfoton and Thiometon in the Rhine River: a chemodynamic study, *Environmental Science and Technology*, 23(10):1232–1242, 1989.

Water Environment and Technology, 12, Number (3): 54, May 2000.

Water Environment Federation, *Natural Systems for Wastewater Treatment*, WEF Manual of Practice FD-16, WEF, Alexandria, Virginia, 2001.

Water Pollution Control Federation, *Design and Construction of Urban Stormwater Management Systems*, WPCF, Washington, DC, 1992.

Weibel, S.R., Urban drainage as a factor in eutrophication, in *Eutrophication: Causes, Consequences, Correctives*, pages 383–403, National Academy of Sciences, Washington, DC, 1969.

Wetlands Connection Center, Local wetland loss and concerns, http://www.sci.tamucc.edu/ccs/Refugio/ wetland%20loss.html, 2005.

Wetzel, R.G., *Limnology*, Saunders College Publishing, Philadelphia, 1975.

Wezernak, C.T., and J.J. Gannon, Evaluation of nitrification in streams, *Journal of Sanitation Engineering Division, ASCE*, 94(SA5):883–895, 1968.

Whalen, P.J., and M.G. Cullum, *An Assessment of Urban Land Use/Stormwater Runoff Quality Relationships and Treatment Efficiencies of Selected Stormwater Management Systems*, Technical Publication 88–9, South Florida Water Management District, West Palm Beach, Florida, July 1988.

Wharton, C.H., W.M. Kitchens, E.C. Pendleton, and T.W. Sipe, *The Ecology of Bottomwood Hardwood Swamps of the Southeast: A Community Profile*, Technical Report FWS/OBS-81/37, Biological Services Program, U.S. Fish and Wildlife Service, Washington, DC, 1982.

Whittemore, R.C., and J. Beebe, EPA's BASINS model: good science or serendipitous modeling, *Journal of the American Water Resources Association*, 363:493–499, 2000.

Wiedemeier, T.H., J.T. Wilson, D.H. Kampbell, R.N. Miller, and J.E. Hansen, *Technical Protocol for Implementing Intrinsic Remediation with Long-Term Monitoring for Natural Attenuation of Fuel Contamination Dissolved in Groundwater*, Volume 1, Technical Report, Air Force Center for Environmental Excellence, Brooks Air Force Base, San Antonio, Texas, 1995.

Wiedemeier, T.H., M.A. Swanson, D.E. Moutoux, E.K. Gordon, J.T. Wilson, B.H. Wilson, and D.H. Kampbell, *Technical Protocol for Evaluating Natural Attenuation of Chlorinated Solvents in Groundwater*, Volume 1, Technical Report, Air Force Center for Environmental Excellence, Brooks Air Force Base, San Antonio, Texas, 1996.

Wiedemeier, T.H., H.S. Rifai, C.J. Newell, and J.W. Wilson, *Natural Attenuation of Fuels and Chlorinated Solvents*, Wiley, New York, 1999.

Wilkinson, D.L., The internal hydraulics of tunnelled outfalls: lessons from the model studies of the Sydney outfalls, in *Proceedings of the International Conference on Physical Modelling of Transport and Dispersion*, pages 12A1–6, MIT, Boston, August 7–10 1990.

Williams, J.R., A.D. Nicks, and J.G. Arnold, Simulator for water resources in rural basins, *Journal of the Hydraulics Division, ASCE*, 111(6):970–986, 1985.

Wilson, L.G., Sediment removal from flood water, *Transactions of the American Society of Agricultural Engineers*, 10(1):35–37, 1967.

Wischmeier, W.H., Estimating the soil loss equation's cover and management factor for undisturbed areas, in *Proceedings of the Sediment Yield Workshop*, Oxford, Mississipi, 1972.

———, Use and misuse of the universal soil loss equation, *Journal of Soil and Water Conservation*, 31(1):5–9, 1976.

Wischmeier, W.H., and D.D. Smith, *Predicting Rainfall-Erosion Losses from Cropland East of the Rocky Mountains*, Agricultural Handbook 282, U.S. Department of Agriculture, Washington, DC, 1965.

Wood, I., Buoyant spreading at the surface of a wastewater discharge into the ocean, http://www. cormix.info/picgal/density.php, 2005a.

————, A coastal multiport diffuser dye test reveals an obstructed port, http://www.cormix.info/picgal/oceans.php, 2005b.

————, Near-field flow processes, http://www.cormix.info/picgal/nearfield.php, 2005c.

Wood, I.R., R.G. Bell, and D.L. Wilkinson, *Ocean Disposal of Wastewater*, World Scientific, Singapore, 1993.

Woodbury, A.D., and E.A. Sudicky, The geostatistical characteristics of the Borden aquifer, *Water Resources Research*, 27(4):533–546, 1991.

Worrall, F., and T. Besien, The vulnerability of groundwater to pesticide contamination estimated directly from observations of presence or absence in wells, *Journal of Hydrology*, 303:92–107, 2005.

Wright, S.J., Mean behavior of buoyant jets in a crossflow, *Journal of the Hydraulics Division, ASCE*, 103(HY5):499–513, 1977. Subsequent closure 1978(HY9):1359–1360.

Wright, S.J., *Effects of Ambient Crossflows and Density Stratification on the Characteristic Behavior of Round Turbulent Buoyant Jets*, Report KH-R-36, W.M. Keck Laboratory of Hydraulics and Water Resources, California Institute of Technology, May 1977, 254 pp.

Wright, S.J., P.J.W. Roberts, Y. Zhongmin, and N.E. Bradley, Surface dilution of round submerged buoyant jets, *Journal of Hydraulic Research*, 29(1):67–89, 1991.

Yang, C.T., *Sediment Transport: Theory and Practice*, McGraw-Hill, New York, 1996.

Yotsukura, N., and W.W. Sayre, Transverse mixing in natural channels, *Water Resources Research*, 12(4):695–704, 1976.

Young, R.A., C.A. Onstad, D.D. Bosch, and W.P. Anderson, *Agricultural Nonpoint Source Pollution Model: A Watershed Analysis Tool*, Technical Conservation Research Report 35, Agricultural Research Service, U.S. Department of Agriculture, Morris, Minnesota, 1987.

————, *Agricultural Non-point Source Pollution Model, Version 4.03: AGNPS User's Guide*, U.S. Department of Agriculture, Agricultural Research Service, Morris, Minnesota, 1994.

YSI, Inc., YSI 55 dissolved oxygen instrument, http://www.ysi.com/environmental-monitoring/dissolved-oxygen-sensors-ysi.htm, 2005.

Zedler, P.H., *The Ecology of Southern California Vernal Pools: A Community Profile*, Technical Biological Report 85(7.11), U.S. Fish and Wildlife Service, Washington, DC, 1987.

Zheng, C., and G.D. Bennett, *Applied Contaminant Transport Modeling*, Wiley, New York, 1995.

Zimmer, D.W., and R.W. Bachman, *A Study of the Effects of Stream Channelization and Bank Stabilization on Warmwater Sport-Fish in Iowa, Subproject 4: The Effect of Long-Reach Channelization on Habitat and Invertebrate Drift in Some Iowa Streams*, Technical Report FWS/OBS-76/14, U.S. Fish and Wildlife Service, Washington, DC, 1976.

————, Channelization and invertebrate drift in Iowa streams, *Water Resources Bulletin*, 14:868, 1978.

Zimmerman, J.T.F., The tidal whirlpool: a review of horizontal dispersion by tidal and residual currents, *Netherlands Journal of Sea Research*, 20:133–156, 1986.

Zumwalt, G.S., A.P. Krishna, and J.D. Nelson, Air distribution within a sparging cone of influence, in *In Situ and On-Site Bioremediation*, 4th International In Situ and On-Site Bioremediation Symposium, New Orleans, Louisiana, April 28–May 1, Volume 1, pages 141–146, Batelle Press, Columbus, Ohio, 1997.

INDEX